Forestry and Wood Production

Forestry and Wood Production

Edited by Malcolm Fisher

SYRAWOOD
PUBLISHING HOUSE

New York

Published by Syrawood Publishing House,
750 Third Avenue, 9th Floor,
New York, NY 10017, USA
www.syrawoodpublishinghouse.com

Forestry and Wood Production
Edited by Malcolm Fisher

International Standard Book Number: 978-1-68286-541-5 (Hardback)

Cataloging-in-Publication Data

Forestry and wood production / edited by Malcolm Fisher.
     p. cm.
Includes bibliographical references and index.
ISBN 978-1-68286-541-5
1. Forests and forestry. 2. Wood. I. Fisher, Malcolm.
SD373 .F67 2018
634.9--dc23

TABLE OF CONTENTS

PREFACE

The process of managing and conserving forests with the goal of environmental stability and proper resource allocation is known as forestry. Resources such as timber, wood, paper or lumber continue to be used for subsistence as well as for commercial purposes. The topics covered herein deal with the core subjects of forestry and wood production. In this book, using case studies and examples, constant effort has been made to make the understanding of the difficult concepts of forestry and wood production as easy and informative as possible, for the readers.

This book is the end result of constructive efforts and intensive research done by experts in this field. The aim of this book is to enlighten the readers with recent information in this area of research. The information provided in this profound book would serve as a valuable reference to students and researchers in this field.

At the end, I would like to thank all the authors for devoting their precious time and providing their valuable contribution to this book. I would also like to express my gratitude to my fellow colleagues who encouraged me throughout the process.

Editor

Conservation Strategies for Orangutans: Reintroduction versus Habitat Preservation and the Benefits of Sustainably Logged Forest

Howard B. Wilson[1]*, Erik Meijaard[1,2], Oscar Venter[1,3], Marc Ancrenaz[4], Hugh P. Possingham[1]

1 Australian Research Council Centre of Excellence for Environmental Decisions, School of Biological Sciences, The University of Queensland, Brisbane, Australia, 2 People and Nature Consulting International, Jakarta, Indonesia, 3 Centre for Tropical Environmental and Sustainability Science and the School of Marine and Tropical Biology, James Cook University, Cairns, Australia, 4 Kinabatangan Orangutan Conservation Project, Sandakan, Malaysia

Abstract

The Sumatran orangutan is currently listed by the IUCN as critically endangered and the Bornean species as endangered. Unless effective conservation measures are enacted quickly, most orangutan populations without adequate protection face a dire future. Two main strategies are being pursued to conserve orangutans: (i) rehabilitation and reintroduction of ex-captive or displaced individuals; and (ii) protection of their forest habitat to abate threats like deforestation and hunting. These strategies are often mirrored in similar programs to save other valued and endangered mega-fauna. Through GIS analysis, collating data from across the literature, and combining this information within a modelling and decision analysis framework, we analysed which strategy or combination of strategies is the most cost-effective at maintaining wild orangutan populations, and under what conditions. We discovered that neither strategy was optimal under all circumstances but was dependent on the relative cost per orangutan, the timescale of management concern, and the rate of deforestation. Reintroduction, which costs twelve times as much per animal as compared to protection of forest, was only a cost-effective strategy at very short timescales. For time scales longer than 10–20 years, forest protection is the more cost-efficient strategy for maintaining wild orangutan populations. Our analyses showed that a third, rarely utilised strategy is intermediate: introducing sustainable logging practices and protection from hunting in timber production forest. Maximum long-term cost-efficiency is achieved by working in conservation forest. However, habitat protection involves addressing complex conservation issues and conflicting needs at the landscape level. We find a potential resolution in that well-managed production forests could achieve intermediate conservation outcomes. This has broad implications for sustaining biodiversity more generally within an economically productive landscape. Insights from this analysis should provide a better framework to prioritize financial investments, and facilitate improved integration between the organizations that implement these strategies.

Editor: Sadie Jane Ryan, SUNY College of Environmental Science and Forestry, United States of America

Funding: This work was supported by the Centre for Applied Environmental Decision Analysis, a Commonwealth Environmental Research Facility Hub funded by the Australian Government Department of Environment, Water, Heritage and the Arts. The funders had no role in study design, data collection and analysis, decision to publish, or preparation of the manuscript.

Competing Interests: Co-author Erik Meijaard is employed by People and Nature Consulting International, Jakarta, Indonesia. There are no patents, products in development or marketed products to declare.

* Email: h.wilson1@uq.edu.au

Introduction

Orangutans *Pongo* spp. are severely threatened by habitat loss and hunting [1–3] and populations without adequate protection face a dire future [2]. Even for most orangutan populations in areas with legally recognized conservation status, habitat management and law enforcement need to be improved to prevent further population declines [4–5].

Two strategies are currently being pursued to conserve wild orangutans: (i) rehabilitating and returning orangutans back into the wild and (ii) preserving current orangutan-populated forest. Rehabilitation centres were initially set up for welfare reasons and as a tool for dealing with confiscated animals held illegally in captivity [6]. In South East Asia, these centres mostly take in animals that have been displaced by deforestation activities [1,7].

Following rehabilitation, animals are then reintroduced back into their historical range.

As opposed to reintroduction, the management of wild populations is focused on habitat loss and other threats to wild populations, such as hunting. The key to this strategy is to ensure that the quantity and quality of habitat remains sufficient for long-term population viability, without necessarily requiring that an area is legally set aside for conservation. For example, well managed logging concessions provide sufficient resources for orangutans to survive [8], with the revenues from sustainable timber extraction offsetting some of the opportunity costs (i.e. loss of potential revenue) that would occur if the area was fully protected [9].

Reintroduction of orangutans is a widespread strategy that attracts large financial support [10–11], but is also being questioned in terms of its contribution to conservation goals

[1,12]. Protecting wild populations also receives substantial investment. Considering that funding for reintroduction or habitat protection is at least partly fungible, planning for optimal conservation outcomes requires that limited funds are allocated wisely. The aim of this study is to investigate how the costs and benefits of a reintroduction strategy compare to those of preserving wild orangutan populations in their natural habitat. We combined GIS analysis and data collated from across the literature within a modelling and decision analysis framework to find the circumstances under which either strategy is optimal.

Methods

The geographic scope of this study is the two islands of Borneo and Sumatra, South East Asia, currently the only areas with wild orangutan populations. We defined a model in which the forest can be in a number of different states. First, forests are either with or without breeding orangutan populations [3]. Second, forests can be in one of three land uses, i) legal conservation status, ii) production of natural timber (but not industrial tree plantations), or iii) available for conversion to agri- or silviculture (oil palm or industrial tree plantations), but not yet cleared (Table 1). We consider that although industrial tree plantations and oil palm plantations are sometimes used by orangutans [13], these intensive land uses cannot sustain orangutan populations in the long-term and are not considered any further. Third, forests are either with or without "extra protection", a layer of management specifically to protect the forest for orangutans. These extra protection measures include prevention of illegal logging, fires, and agricultural encroachment, as well as implementing anti-poaching patrols, and human-orangutan conflict management. We considered this protection as an option in all of the forest land uses above including conservation area forest, as the legal land use status of forest is not necessarily related to the quality of forest management and law enforcement regarding orangutans [14,15].

Our study looked at two different ways of maintaining the number of wild orangutans. The first was to provide extra protection for wild orangutans in their forest habitat (strategy P). This can be done in: (i) conservation area forests (provide extra protection only); (ii) timber production forest (introduce reduced impact logging practices, for example as prescribed by the Tropical Forest Foundation and under the Forest Stewardship Council certification schemes [16], and also provide extra protection); or (iii) forest available for conversion to agriculture (purchase forest and provide extra protection). We show in the results section that due to the high opportunity costs, working in conversion forest was never the most efficient strategy, so we don't present further details here. The second strategy was to rehabilitate orangutans that have been rescued from captivity or forest under conversion, and reintroduce them into orangutan-free forest and then provide extra protection (strategy R). Extra protection is needed for the release site otherwise the same

processes of hunting and illegal logging will wipe out reintroduced orangutans [17] just as it does for current wild populations [18]. A third strategy, keeping orangutans in permanent captive conditions, makes no direct contribution to the survival of wild orangutans so is not discussed further.

For both strategies, the conservation objective was to maximise the number of wild orangutans alive at a specific management time horizon, t_H, for a given budget. All effects later than the time horizon were not considered. A particular time horizon may be chosen as conditions may significantly change after this point in time (for example, Indonesia has made a legally binding decree to empty, through reintroductions, all rehabilitation centers by 2015, and to protect all wild populations by 2017 [19], although it is difficult to see how these goals will be achieved given current conditions). We assess the effectiveness of previous conservation policies at this time horizon.

The spatial variation of orangutan abundance depends on forest types and stages [20], and the degree of hunting. In our models, we used an average orangutan density calculated across this variation in density (see *Parameter estimates* below). The number of orangutans then equalled the number of hectares of orangutan populated forest multiplied by the average density of orangutans per hectare. Our goal of maximizing the number of wild orangutans is equivalent here to maximizing the total number of hectares of any forest type with orangutans present (without regard to whether they have extra protection or not). Later we include a higher orangutan population growth in forests with the extra layer of management, which resulted in a different density of orangutans in different forest types. We also return to this issue of spatial variation in density in the discussion.

Conservation forest only

Our simplest model includes only forest that is already legally, but not effectively, conserved. We assumed that: 1) there was always enough conservation forest free of orangutans for reintroduction (see File S1 for justification, although relaxing this assumption does not qualitatively change our results); 2) there were always enough orangutans in rehabilitation centres for reintroduction (see File S1); and 3) there was enough orangutan-populated forest for protection (we relax this in the next section). We modelled the amount of conservation forest with orangutans but without extra protection, CF, and the amount of conservation forest with both orangutans and extra protection, CF_p:

$$\frac{dCF}{dt} = -d_1 CF - \gamma\alpha_1$$
$$\frac{dCF_P}{dt} = -(1-e_1)d_1 CF_P + \gamma\alpha_1 + (1-\gamma)\beta \tag{1}$$

CF and CF_p are functions of time and CF_p at time 0 equals zero, d_1 is the rate of conservation forest loss, e_1 ($0<e_1<1$) is the efficiency

Table 1. Definitions of land management categories considered in this study.

Conservation area	Areas legally gazetted for the conservation of nature and environmental services (National Park, Nature Reserve, Wildlife Reserve, watershed protection, etc.).
Timber Production forest	Any natural forest area legally gazetted for selective timber harvest (no mono-culture timber species, clear cutting or conversion to agriculture).
Conversion forest	Forest areas not yet cleared but ultimately slated for conversion for agricultural uses, such as oil palm, or silvicultural use such as softwood plantations.

of extra protection in reducing the rate of forest loss ($e_I = 0$ is when extra protection provides no benefit in reducing the rate of forest loss, and $e_I = 1$ is when there is no loss of forest with extra protection), α_I is the total potential amount of conservation area that can be protected per year if all the budget was used for this purpose, β is the total potential amount of orangutan-free conservation forest that can be converted to protected forest per year by reintroducing orangutans (both α_I and β have units of hectares per year), and γ is a control parameter that changes the proportion of the total budget spent on the two strategies (a proportion γ is spent on protection, P, and $[1-\gamma]$ is spent on rehabilitation and reintroduction, R). The parameter α_I equals B/C_P, where B is the total budget per year and C_P is the cost of extra protection per hectare of forest, and $\beta = B/C_R$, where C_R is the reintroduction cost per hectare (i.e. rehabilitation cost per orangutan plus the cost of forest protection after release per orangutan all multiplied by the average orangutan density per hectare).

Parameter estimates

We determined the area of Indonesian and Malaysian forest housing orangutans in 2010 by overlaying, through geographic information system analyses, a map of 2010 forest cover [21] with a map of the distribution of the Sumatran and Bornean orangutans [22]. We determined the area of this forest that was currently under some form of conservation management by overlaying a map of IUCN category I–VI protected areas [23]. We find that, in 2010, there was a total 12,177,153 ha of forest within the mapped range of orangutans, of this, 3,398,392 ha are in some form of conservation management.

Regardless of the legal status of land in Indonesia and Malaysia, forests are being lost due to unsustainable logging, anthropogenic fires and agricultural conversion. We calculated the rate at which forests are being lost by determining the proportion of forest in these countries that was cleared, or otherwise transitioned to non-forest, between 2000 and 2010 [21] using high resolution land cover maps for this period [21]. These maps are not able to distinguish between regenerating natural forests and industrial timber plantations. Our estimate of forest loss therefore represents the annual percent conversion of all forest types (primary, regenerating and plantation) to non-forest land covers. As we had data from two time periods only, we were unable to establish whether habitat loss is a linear process, or follows an alternate trajectory. This data source is currently the highest quality land cover data available for Indonesian Borneo and Sumatra, with validated accuracy for 2000 for forest/non-forest of 91.7%, and for 2010 forest/non-forest 93.6%. We found that the rate of forest loss was lower in conservation forest (16,487 ha yr^{-1}, or 0.485% yr^{-1}), than in non-conservation forest (164,949 ha yr^{-1}, or 1.879% yr^{-1}).

To estimate the ongoing management costs of effectively protecting forest, we used data from McQuistan et al. (2006) [24]. We extracted the optimal budget of effectively managing a strict no-take protected forest area. The cost of managing forest that is currently under legal conservation status was used to be the optimal per hectare budget for a national park [24] in Indonesia, and the cost of managing forest that is available for conversion was taken as the optimal budget for a forest park [24]. We stress that the cost figure is not what is currently being spent, rather it is an estimate of the optimal budget required to make sure these parks are effectively managed to fulfill the park's objectives. It's the best case scenario cost for effective forest protection, i.e. for $e_i = 1$. Total forest protection can, and has been achieved in practice: in Kalimantan the Wehea protection forest has had zero forest loss

between 2004 and 2014; as has the Sungai Wain protection forest between 2000 and 2014; in Malaysian Borneo the Danum Valley, Tabin Wildlife Reserve, and Sepilok forest have had no forest loss. Hence achieving $e_i = 1$ is possible, although it is probable that protection would be less than 100% effective in many, or most, cases. We have estimated the loss rate of legally protected forests to be one quarter the loss rate of legally unprotected forests (0.486% vs. 1.88%, above). We have used this as a guideline for the efficiency of our extra layer of forest management, and have used $e_i = 0.75$ (i.e. a reduction in forest loss of 75%).

Management costs of effectively protecting forest were calculated as the amount of money needed today to fund all the future costs of management up to the time horizon. This uses a discount rate for future costs. Costs were estimated by an initial setup cost ($52.4/ha) and an ongoing management cost ($3.87/ha) [24], and a discount rate of 10% (a value typically used by Indonesian companies, [25]). All costs have been converted to 2010 US$. One should note that C_P and C_R are dependent on t_H (the time horizon), although as t_H increases, then this dependence is very small ($t_H > 20$).

The costs of protecting forests available for conversion to agriculture were taken from [26]. These figures were based on a literature review of the revenues derived from intensive logging, and from the financial reports of oil palm companies. We assumed that logging and clearing of the land would take place over five years, and that oil palms take five years to reach maturity (as in [26]). The opportunity costs of purchasing these forests were estimated as the net present value of the annual profits discounted at a rate of 10% per annum.

Rehabilitation and reintroduction costs (β) have been estimated from the operating costs of the Borneo Orangutan Survival Foundation (BOS), the largest primate rescue and rehabilitation organization in the world. The operating budget for BOS in 2007 was $4,322,026 [27], or $5,403 per orangutan. When calculating the total cost of rehabilitating an orangutan, we estimated that the cost of releasing an orangutan into the wild is $5,000, and 10% of captive orangutans would be released after one year, a further 20% after two years, 30% after three, 20% after four and 10% after five. We assumed that the remaining 10% of orangutans would never be fully rehabilitated and would remain in captivity for the duration of their life. After release, we assumed that rehabilitated orangutans would suffer a 50% mortality rate [7]. From these parameters, the average cost per successfully released orangutan is $44,121. In addition, there was the cost of extra forest protection after release per orangutan (the same as for the P strategy).

The average orangutan density across the two islands was estimated to be 42 ha/animal, which is a population size weighted average for the densities of Sumatra (25 ha/animal) and Borneo (44 ha/animal). The average density from each island was based on the mid-point of the densities found on those islands: 1–7 animals/km^2 for Sumatra; and 0.5–4 animals/km^2 for Borneo [20].

Conservation and timber production forest

It is always more cost efficient to reintroduce orangutans into conservation forest rather than timber production forest (see costs in Table 2). For P, the cost of providing extra protection for timber production forest was higher than for conservation areas, and there was an extra cost for introducing reduced impact logging practices that maintain forest structure. However, there was also an advantage of protecting timber production forest, as a higher rate of forest destruction could be prevented. The dynamics of the four kinds of forest (conservation forest, no extra protection;

production forest, no extra protection; conservation forest, extra protection; and production forest, extra protection) are described by the equations:

$$\frac{dCF}{dt} = -d_1\,CF - \gamma(1-\theta)\alpha_1$$

$$\frac{dTP}{dt} = -d_2\,TP - \gamma\theta\alpha_2$$

$$\frac{dCF_P}{dt} = -(1-e_1)d_1\,CF_P + \gamma(1-\theta)\alpha_1 + (1-\gamma)\beta \qquad (2)$$

$$\frac{dTP_P}{dt} = -(1-e_2)d_2\,TP_P + \gamma\theta\alpha_2$$

where TP is timber production forest populated with orangutans, TP_P is timber production forest populated with orangutans with extra protection, d_2 is the rate of timber production forest loss, α_2 is the total potential amount of timber production forest that can be protected in any one year if all the budget was used for this purpose, e_2 is the efficiency of extra protection in reducing the rate of timber production forest loss, and θ is a control parameter that changes the proportion of the P budget spent on conservation or timber production forest. There is a cost in timber production forest for introducing sustainable logging practices ($20.9 ha^{-1}, [28]), and the extra protection costs per hectare are also higher ($16.86 ha^{-1} yr^{-1} versus $3.87 ha^{-1} yr^{-1} for conservation forest). All other parameters and states are as in model 1. The costs of implementing reduced impact logging practices are not the full opportunity costs of conventional logging, as these practices usually produce a similar timber yield to conventional logging practices [29]. Instead, the cost presented in Table 2 is the additional cost of pre-harvest planning, vine cutting, felling, skidding, supervising and training associated with reduced impact logging. A key point is that these logging practices are not introduced into primary forest areas, rather they are introduced

into areas that are already being logged and will continue to be logged.

Hunting

Hunting of orangutans, especially in Kalimantan, is a major threat [1,18] and we assume that nearly all orangutan populations are below carrying capacity because of past hunting [30]. We modelled hunting by including extra loss terms for the dynamics of non-protected, orangutan-populated forest ($-h_1CF$ and $-h_2TP$ in the equations for dCF/dt and dTP/dt). Recent studies have indicated that the rate of loss of orangutans due to hunting is of a similar order of magnitude as to forest destruction [18]. Here we have used a conservative estimate of hunting as being of a similar order to the loss of conservation forest (0.485% p.a).

Orangutan population growth

Some natural repopulation of the forest by orangutan population growth in well managed (i.e. with extra protection) forest would be expected as current population levels are probably lower than carrying capacity [30]. We modeled this by including an orangutan growth rate, r, in forest with extra protection only. This growth rate thus represented an increase in the density of orangutans within these areas, as opposed to a colonization rate of new areas. We kept track of when each area of forest received extra protection, and the density of orangutans in these areas was multiplied by a factor r^t, where t was how long it has been protected. Orangutan populations under no external threats can grow at a maximum of 2% annually, although very few wild populations probably achieve this maximum theoretical rate [31]. We used a more biologically reasonable growth rate of 0.75% p.a.

Leverage

A possible extra benefit of the rehabilitation and reintroduction strategy is that it attracts media attention, thereby raising awareness about the orangutan's plight, putting pressure on

Table 2. The parameter estimates, probable range of values, and the critical value at which the optimal strategy changes.

Parameter	Estimated value – conservation forest (range)	Critical value -with hunting and orangutan popn growth (no hunting or popn growth)	Estimated value – timber production forest (range)	Critical value- with hunting and orangutan popn growth (no hunting or popn growth)
Time horizon, (yrs)	50	12	50	25 (52)
	(5–100)	(49)	(5–100)	(52)
Rate of forest loss (yr^{-1}), d_i	0.00485	never (0.0044)	0.0188	never (0.0196)
	(0.0046–0.0052)		(0.0173–0.0203)	
Protection management cost	94.6	459.4 (103.4)	257.2	420.0 (250.0)
(US$ ha^{-1})	(81.5–126.6)		(202.4–396.5)	
Opportunity cost (US$ ha^{-1})	0	N/A	20.9	176 (13.1)
			(9.6–32.3)	
Efficiency of protection, e_i	0.75	0.22[a] (0.69)	0.75	0.52 (0.76)
	(0.5–1.0)		(0.5–1.0)	
Rehab cost, ($ orangutan^{-1})	44,121	9,124 (38,705)	44,121	26,900 (45,500)
	(33,091–55,151)		(33,091–55,151)	

The optimal strategy using the estimated values was protection. The critical value at which reintroduction resulted in more wild orangutans than protection was calculated by keeping all the other parameters constant at the estimated values, and then varying one parameter to find when the optimal strategy changed. Values were calculated by simulation (although the formula $t_H \approx 2C_P/d\,e\,C_R$ can also be used as an approximation to the critical point). The protection cost has three underlying parameters that were varied; the initial setup cost, the cost per hectare, and the discount rate. For clarity, we have summarized these into variation in the overall protection cost. Hunting was assumed to result in a loss of 0.485% p.a., population growth was 0.75% p.a., and a budget of $5M p.a. was used.
[a] when the efficiency is <1, we assumed that the orangutan growth rate was e_1 * 0.75%.

conservation authorities, and providing fund raising opportunities. In this case, instead of a set budget amount being split between the two strategies, the more reintroduction is used as a strategy then the bigger the budget. We modelled this by assuming the total budget increases by a factor l, in proportion to the fraction of the reintroduction strategy being adopted; $B_T = \gamma\,B + (1-\gamma)\,l\,B$, where γ is the proportion of the budget spent on P.

Parameter error estimates

It's possible that some institutions might have short-time horizons dictated by funding or voting cycles, whilst others might consider perpetuity as the only appropriate horizon for biodiversity conservation. To reflect this uncertainty we chose a range for the time horizon of 5–100 years.

For the range of the rate of forest loss, we used the maximum errors in the land cover maps (8%, [21]). We don't have any a priori reason to think that errors will be biased on one direction or another, so a pixel that is thought to be forest is as likely to be non-forest as a pixel that is thought to be non-forest is to be forest. In general a land cover classification algorithm will be established such that it equalizes the rates of false negatives and positives, and thereby minimizes the overall error rate. If there is no bias in the error, then the land cover errors will tend to cancel out.

The range for the protection cost was estimated by varying the initial setup cost, the cost per hectare, and the discount rate separately by ±25% each to find the maximum possible range. This fits our practical experience of the costs of protective management in different parts of Indonesian and Malaysian Borneo. Also, upfront costs tend to be higher than long-term maintenance costs, and the 25% variation captures this adequately. The budget for extra protection is the estimate for what is needed for full and effective protection, so the efficiency should be close to 1. However, we took a precautionary approach and analyzed a range between 0.5–1.0. The cost for implementing reduced impact logging was varied from \$9.6–\$32.3 (±54%, [28]). Finally, for the costs of rehabilitation and reintroduction we did not have any data that would allow us to estimate variability. Consequently, we varied the reintroduction cost by ±25%, the same as for the protection costs.

Sensitivity analysis

All the parameters used in the above models have been systematically varied. We primarily looked at how variation in the parameters affected our analytical results, but we also checked our analytical results against simulations. We performed five different sensitivity analyses on the parameters. First, we varied a single parameter at a time. Second, we randomly selected a value for each parameter from their range and then calculated whether protection or reintroduction was the best strategy. This was done 50,000 times to find the probability of each strategy being optimal, and to sample the whole parameter space. Hence there was no need for a complex sampling technique. Third, one particular parameter was selected and its value fixed, whilst selecting random values for the other parameters. This was done 50,000 times to find the mean strategy probabilities for a fixed value of the chosen parameter. We then systematically selected different values for the chosen parameter across its range to find how the optimal strategy changes. This approach demonstrated how sensitive the optimal strategy was to variation in each parameter, averaged across the sensitivity in the response to all the other parameters. Fourth, during each simulation, we randomly selected a new value for each parameter from their range every year. This was done 50,000 times to find the probability of each strategy being optimal when the parameters are randomly changing with time. Fifth, an

alternative to a fixed time horizon is to allow uncertainty in the end point, as certain knowledge of when conditions might change is rare. We investigated this by using a fixed probability that in any one year the model ends. If this probability was 0.05, this would equate to an average time horizon of 20 years. We then ran 1,000 simulations of the model (each will have a different end time) and determined how well each strategy performed in terms of the number of wild orangutans alive at the end averaged over all the simulations.

We looked at pursuing either a single strategy (P or R), a mixed strategy (a fixed proportion of each) for the whole time, or switching between the two strategies over the time horizon (in which case the control parameter, γ, above is a function of time).

Results

Conservation forest only

The total number of hectares of forest populated by orangutans at the time horizon, $CF_T(t_H)$, was found by integrating model (1) with respect to time and summing CF and CF_p. The strategy that maximised $CF_T(t_H)$ is also the one that maximised the number of wild orangutans, as the two are proportional. This maximum can be determined by differentiating $CF_T(t_H)$ with respect to γ (see File S1). As this is linear with respect to γ, then the optimal solutions occurred at the limiting values for γ and there was no optimal mixed strategy which would have allocated a proportion of the budget to both strategies. More complex solutions might occur if there were greater complexities in the cost functions, C_P and C_R (e.g. the cost of extra protection per hectare decreased as the amount of forest protected increased). Using our estimates of the parameters (Table 2), the optimum was $\gamma = 1$ (P). However, different parameter values resulted in a different optimal strategy. The critical point at which the optimal strategy changed was when $t_H \approx 2C_P/d\ e\ C_R$ (found by solving $d\ CF(t_H)/d\gamma = 0$, see File S1). Rearranging this equation for any particular parameter will give the critical value for that parameter (Table 2). The critical time horizon was approximately 49 years. We have verified this analytical result using numerical simulations (see Fig. 1a). For time horizons shorter than this, the optimal strategy was to allocate the entire budget to R. For longer time horizons, the optimal strategy was to spend the entire budget on P. We emphasise that the critical time at which the optimal strategy changes does not represent a time at which to switch strategies, rather if a timescale longer that 49 years is considered, P represents the best strategy adopted for the whole time; for a shorter timescale, R is the best strategy.

Protecting conservation or timber production forest

Analytical work (see File S1) shows that the optimal protection strategy is to concentrate 100% of the resource allocation into protecting the one forest type that delivers the most benefit (in terms of the number of wild orangutans) for a fixed cost; i.e. there is no optimal mixed protection strategy. The optimal strategy is the one which has the highest value of the quantity $\varphi_i\,d_i\,e_i/C_i$, where φ_i is the density of orangutans in forest type i relative to pristine forest. The best forests to protect are ones where there is a high density of orangutans, where there is a high rate of destruction in the absence of protection, high efficiency of protection, and low cost. Frequently flooded lowland swamp forest with low agricultural potential would be an example.

Protecting timber production forest was more cost-effective than providing protection for conservation forest ($\varphi_i d_i\,e_i/C_i = 4.8e\text{-}5$ and $3.8e\text{-}5$ respectively). However, our analytical results used an approximation, $O(d_i^2 t^3) \approx 0$ (see File S1), which was less accurate at long time horizons or when there were high rates of forest

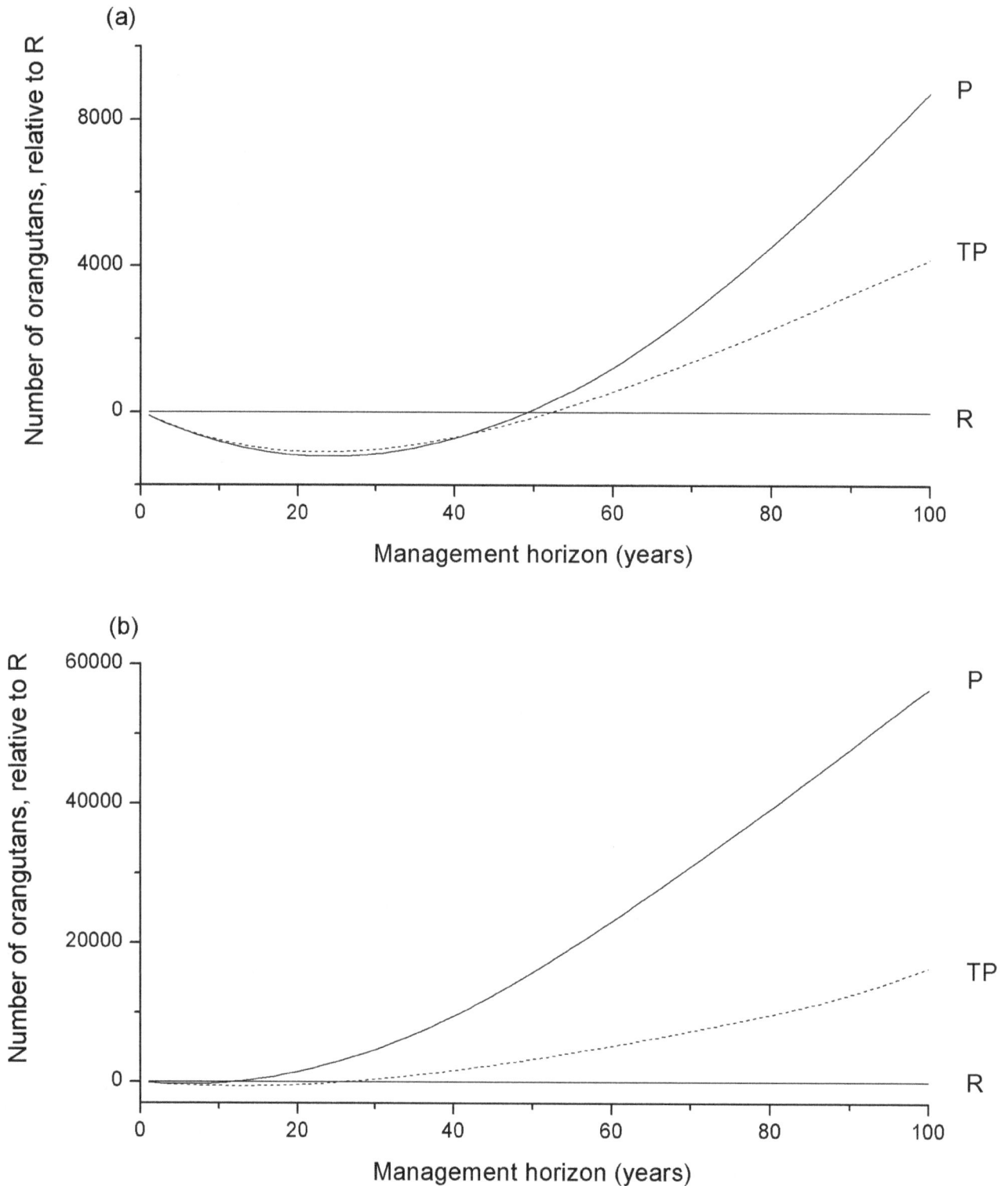

Figure 1. The relative performance of different strategies. The y-axis shows the difference between the number of orangutans for each strategy relative to the R strategy (with reintroduction into conservation areas). R compared to itself is a straight line at zero, above zero a strategy performed better than R, below zero a strategy performed worse than R. The strategy P protects conservation areas first. The strategy TP introduces sustainable logging and protection into timber production forest first. The budget was $5M per year, other parameters are in Table 2. (a) Without hunting or orangutan population growth. The critical time horizon is 49 years for P, and 52 years for TP. (b) Hunting and orangutan population growth included. The critical time horizon is 12 years for P, and 25 years for TP.

destruction. Simulations confirmed that timber production forest was more cost-effective only for shorter timescales (less than 40 years, see Fig. 1a, and less than 55 years if $e_i = 1$), whereas for longer timescales extra protection for conservation forest is more cost-effective.

Purchasing conversion forest that would otherwise be converted to silviculture and agriculture results in a high opportunity cost of removing potential timber revenues (estimated to be $2,268 ha^{-1}, [9]) and agricultural rents (estimated to be $6,766 ha^{-1} for oil palm, [9]). This means the product $\varphi_i\, d_i\, e_i/C_i$ ($= 5.3e\text{-}6$ using the opportunity cost for timber revenues only) was an order of magnitude lower as compared to protecting other forest types and the purchase and protection of conversion forest was never a cost-effective solution as compared to protection of the other two forest types.

Hunting and wild orangutan growth rate

Hunting has structurally the same effect in the models as an increase in the value of d_i, the rate of forest loss. Hence, including hunting made the protection strategy more likely to be optimal. If the loss of orangutans due to hunting is of a similar order of magnitude as for forest destruction, as recent studies have indicated [18], then the critical values of other parameters were significantly changed (Table 2). Inclusion of the orangutan growth rate had a similar effect; the protection strategy was more likely to be optimal and the impact on critical values was significant (Table 2). Inclusion of both effects resulted in the critical time horizon dropping to only 10 years in conservation forest only, and to 18 years in timber production forest. Simulations showed that protection of conservation forest is always more cost-effective than protection of timber production forest when an orangutan population growth rate was included, with or without hunting (Fig. 1b).

Leverage

Leverage favoured the reintroduction strategy, approximately in proportion to the increase in the budget for reintroduction. A 50% increase in the budget ($l = 1.5$) resulted in a 50% increase in the critical time horizon. For long time horizons approximately greater than 50 years (which favoured protection), and with a budget approximately twice as large for reintroduction ($l>2$), a mixed strategy of both protection and reintroduction was optimal. The proportion of the budget spent on each depended on the time horizon (as it increased then the proportion of the budget spent on protection increased), and the degree of leverage (as leverage increased then the proportion of the budget spent on reintroduction was higher).

Sensitivity analysis

We performed five different sensitivity analyses. (i) Varying one parameter at a time. Analysis of the parameters showed reintroduction was more likely to be the optimal strategy if: (a) the time horizon was small; (b) the rate of forest destruction d was small; (c) the cost of protection was high; (d) efficiency of protection was low; or (e) the cost of reintroduction was low (see Table 2 for the critical values). However, inspection of the range of possible parameter values showed that only a short time horizon can result in reintroduction being the optimal strategy when hunting and population growth are included. All other critical values were outside the estimated range, typically by a considerable amount. (ii) When all the parameters were allowed to vary randomly, the probability of protection being the optimal strategy was 0.93 (with hunting or population growth) and 0.53 (no hunting and population growth). (iii) When one parameter was fixed and the

others chosen randomly, only the time horizon significantly influenced whether protection or reintroduction was the optimal strategy (Figure 2). (iv) Randomly changing the parameters in every year produced very similar results to fixing the parameter at the estimated value. (v) When there was variability in the end point, the qualitative results were the same as before; although protection being the optimal strategy occurred at shorter time-frames than previously. As for fixed time horizons, there was no optimal mixed strategy.

Discussion

Protection of forest is a long term strategy for conserving orangutans. Reintroduction seeks to increase the wild population of orangutans via re-establishing viable populations in areas where they have vanished. However, it is approximately twelve times more expensive than protection (per orangutan), which means less forest can be protected for the same budget, and so a short-term gain occurs at the expense of more forest and orangutan loss in the future. In effect, for long time horizons prevention is better than cure. These results show that the timescale over which conservation goals are being assessed is critical to understanding what type of management approach is cost-effective [32–33]. When all the effects we studied were included (hunting, orangutan population growth, leverage), protection is a more cost-effective strategy when the timescale is greater than 10 (no leverage for reintroduction) to 20 years (the budget for reintroduction is twice that of protection). This is a timescale short enough to be realistic and relevant for organisations working in the field.

For both strategies, (reintroduction and habitat protection), maximum long-term cost-efficiency is achieved by working in conservation forest. This means that a proper network of protected areas remains the ultimate goal for long-term orangutan protection. However, introducing reduced impact logging practices coupled with additional protection for orangutans in timber production forest is a strategy intermediate in performance between reintroduction and protecting conservation forest, and in some cases can outperform protecting conservation forest (at intermediate time scales when there is a high efficiency of protection). Timber production forest is more expensive to effectively protect per hectare, but there is a benefit as a relatively higher rate of destruction of orangutan habitat can be prevented. This is similar to work in conservation planning where prioritising areas is a combination of conservation value (how much we stand to lose) and threat (how likely we are to lose it without intervention) [34–35].

Although conservationists and the public are generally keener to protect what is perceived as vast and genuine patches of wilderness, our results reveal that there may well be a role for well managed production forests for orangutan conservation in some instances. We recognize that opening up forests for timber exploitation or other types of industry brings people, roads and infrastructure into orangutan habitat and results in increased poaching [36]. However, we are not suggesting that the reduced impact logging practices be introduced into primary forest areas, rather we've considered their introduction into areas that are already being logged. Orangutans are hunted for food throughout their range, especially in the areas where the commercial timber industry operates [18,37]. Orangutans would be hunted in these areas irrespective of whether conventional logging or reduced impact logging would be implemented. Reduced impact logging would in fact have a better chance of reducing hunting as compared to conventional logging because of the often related requirements to close up skid trails and logging roads. There

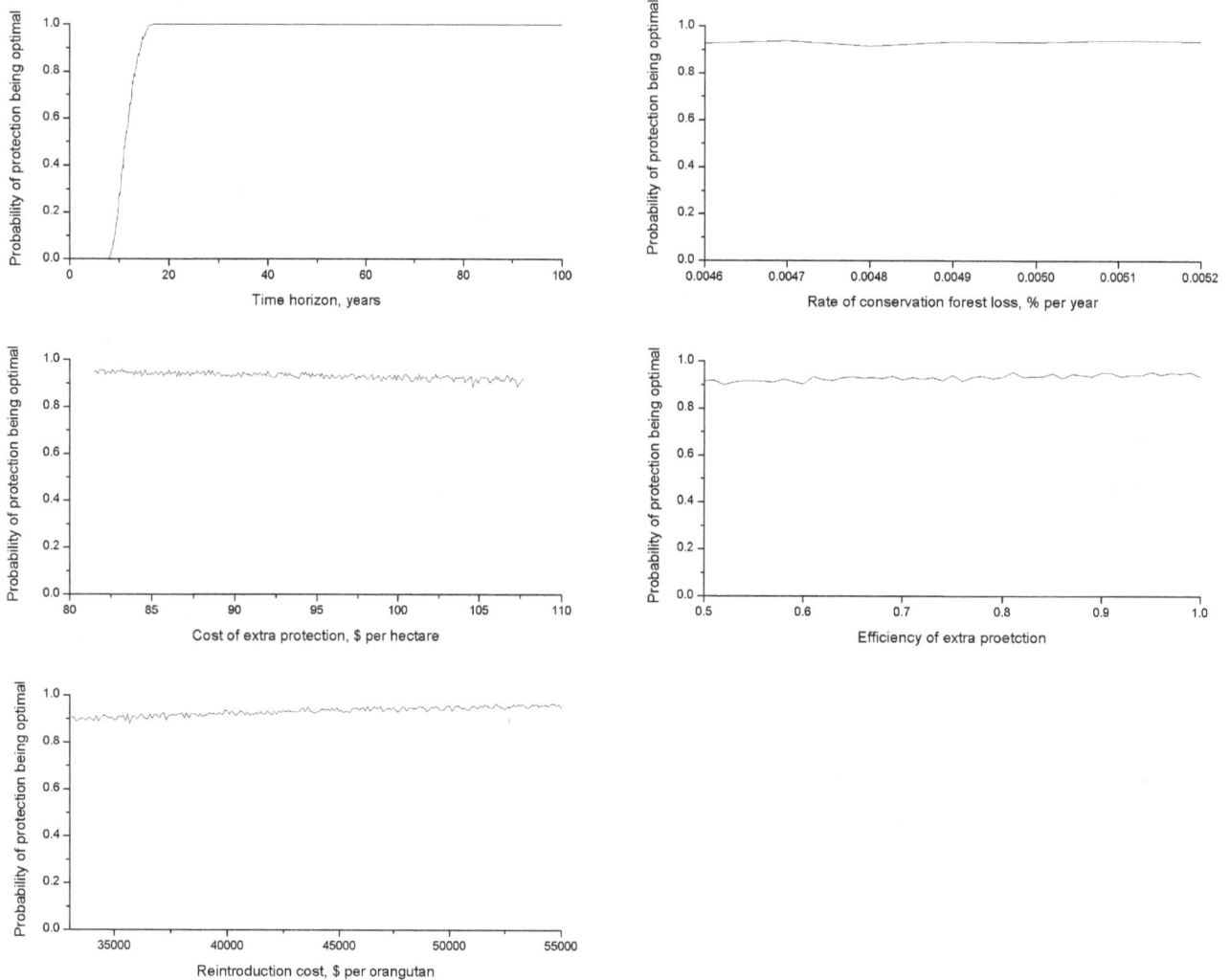

Figure 2. Sensitivity analysis. Each figure shows the probability of protection being the best strategy, when holding one parameter fixed whilst varying all the others. The x-axis gives the fixed value of the parameter in question, all other parameters were randomly chosen from their range. The y-axis is the probability of protection being the optimal strategy, averaged over 50,000 random selections. When every parameter was allowed to vary randomly, the probability of protection was 0.93. The parameters are for conservation forest (see Table 2), with hunting and population growth included.

would also be additional advantages if sustainable forest management certification was sought, through organizations such as the Tropical Forest Foundation (TFF) or the Forest Stewardship Council (FSC). For the FSC certification in particular, timber extraction needs to comply with a series of non-harvesting related practices that are likely to reduce hunting pressure on orangutans, such as control of illegal hunting in the concession area, wildlife monitoring and community development programs [16]. Furthermore, certification would allow a premium to be fetched for the timber produced. Overall, sustainable logging practices in concessions that enforce a zero-killing policy are compatible with the maintenance of viable orangutan populations [8].

We have addressed uncertainty in our parameter estimates, parameters changing through time, interactions between parameters, and stochastic endpoints. This systematic look at parameter variation showed that the time horizon was the most significant parameter influencing which strategy was optimal. An explanation for this lies in the analytical result for the critical values. How the optimal strategy changes is linear with respect to all the parameters, so that the parameter elasticities (the ratio of the percentage change of the result with respect to the percentage change in the parameter) are all equal to one (so that a 10% change in a parameter will change the result by 10%). Hence the parameter which varied the most (the time horizon), was the most important in influencing the result. Interaction effects between the other parameters did not significantly influence the optimal strategy, as when all the parameters were randomly varied, the probability of protection being the optimal strategy was 0.93 (with hunting and population growth) or 0.53 (without hunting and population growth). This value is dominated almost exclusively by whether the chosen time horizon was above or below the critical value of 12 and 49 years (randomly choosing from the range 5–100 years resulted in $t_H > 12$ years 93% of the time and $t_H > 49$ years 53%), despite potential interactions between other parameters. Finally, if hunting and the orangutan population growth were included, changes to our estimates of the parameters (apart from the time horizon) would have had to be very large (greater than 75%) to influence the optimal strategy (Table 2).

One uncertainty, environmental variability or catastrophes, was not analysed as population viability modeling done for orangutans have assumed that severe climatic events, such as very dry El Niño years, could kill up to 3.5% of all orangutans [31]. This is similar to the annual combined losses due to hunting and forest destruction already included within our model (Table 2), yet the environmental catastrophes are rare. Hence they are unlikely to be significant relative to other threats.

Two issues we have not addressed in this paper are leakage and spatial variation of orangutan abundance. In our model, leakage (where protection of one area of forest can lead to increased destruction elsewhere, e.g. [38]) would mean that the extra layer of management specifically to protect the forest for orangutans would be less efficient. Both strategies would therefore be affected, although reintroduction would be affected less as it protects less forest area. There is not much quantitative work in Borneo on leakage, so it is hard to estimate the level of this effect at present. The current spatial variation of orangutan abundance depends on forest types and stages [20] and the degree of hunting. A future refinement of the model could take spatial variation in density into consideration, but we did not address this in the present study because there were insufficient data to analyze the extent to which density variation is caused by ecological conditions (e.g., food availability) or low threats (e.g., limited hunting pressure). Recent analyses [37] suggest that hunting pressure for food is generally high in areas with low orangutan densities, and low in areas with high densities. Other reasons for killing (e.g., crop conflict) are, however, often concentrated in areas of high threat and high densities. At the moment, we do not sufficiently understand the interplay between ecology, killing rates, and densities, making it difficult to incorporate this into our present model. However, the reintroduction strategy puts orangutans into unpopulated forest, so spatial variation in orangutan abundance is irrelevant for this strategy. A strategy that accounted for spatial variation in abundance would therefore only increase the benefits to the protection strategy.

Both Indonesia and Malaysia call for sustainable management of natural resources for the benefit of present and future generations of people. Conservation therefore needs to deal with time frames of 100 years or more. Our models suggest that protection of wild populations and their habitat is a better strategy over such time frames. Why then is reintroduction presently employed as a strategy for conserving orangutans?

Although conservation planning might be thought of as always long-term, the reality is that decisions are typically made to time frames that are often relatively short. Different groups value rewards over different time horizons and conservation funding often tends to be short-term in nature. For short time scales, our results showed a cost-efficient allocation of funds would be to fund rehabilitation and reintroduction of orangutans. There are also other benefits of orangutan rehabilitation centres: the survival and welfare of the reintroduced individuals; improved enforcement of wildlife protection laws and a solution for law enforcement when animals that are kept illegally are confiscated by relevant authorities; and increased public awareness about conservation [7]. Increased ability to raise conservation dollars could also be a significant benefit. Rehabilitation centres provide visible evidence of the impacts of poaching and habitat destruction, and a strongly emotional call for constructive conservation solutions. We recognise that funding available for orangutan rehabilitation and reintroduction can originate from different sources than the funding available for forest protection. Because some of the money available for orangutan conservation has a different origin and is made available for a different motivation, part of the funding allocated to rehabilitation will never be available for forest protection.

There are no reliable data on how much conservation funding is presently spent on orangutan reintroduction versus habitat protection, but we think overall funding (including government and non-government sources) is presently about equally divided between the two strategies. From an animal welfare perspective, rehabilitation centres are valuable. From a purely conservation point of view, however, funding should be allocated primarily towards habitat protection and management. If funding is fungible across the two strategies, the emotive aspects of rehabilitation programs can be counter-productive for the long-term conservation of wild orangutans.

The fact that the majority of wild orangutans currently live outside forests with conservation status [3] implies that we must develop innovative strategies to manage orangutan habitats in landscapes potentially threatened by silvi- and agricultural production. This requires long-term solutions for managing remaining habitats more sustainably [39] and addressing complex conservation issues at the landscape level. Well managed production forests using reduced impact logging techniques are one possible solution.

Insights from the present analysis provide conservation authorities and non-governmental organizations a more rational framework for prioritizing their investments to different strategies. Aside from providing important feedback to donors financing these activities, this information should help develop an improved integration between the organizations that implement these strategies and hopefully lead to optimal outcomes for orangutans and other users of forest services.

Author Contributions

Conceived and designed the experiments: EM OV MA. Performed the experiments: HW OV HPP. Analyzed the data: HW EM OV. Contributed reagents/materials/analysis tools: EM OV MA. Wrote the paper: HW EM OV MA HPP.

References

1. Rijksen HD, Meijaard E (1999) Our vanishing relative. The status of wild orangutans at the close of the twentieth century. Kluwer Academic Publishers, Dordrecht, The Netherlands.

2. Singleton I, Wich SA, Husson S, Atmoko SU, Leighton M, et al. (2004) Orangutan Population and Habitat Viability Assessment: Final Report. IUCN/ SSC Conservation Breeding Specialist Group, Apple Valley, MN, USA.

3. Wich SA, Meijaard E, Marshall AJ, Husson S, Ancrenaz M, et al. (2008) Distribution and conservation status of the orang-utan (Pongo spp.) on Borneo and Sumatra: how many remain? Oryx 42: 329–339.

4. Corlett RT (2007) The impact of hunting on the mammalian fauna of tropical Asian forests. Biotropica 39: 292–303.

5. Linkie M, Smith RJ, Zhu Y, Martyr DJ, Suedmeyer B, et al. (2008) Evaluating biodiversity conservation around a large Sumatran protected area. Cons Biol 22: 683–690.

6. Tutin CEG, Paredes G, Vacher-Vallas M, Vidal C, Goossens B, et al. (2001) Release of wild-born orphaned chimpanzees into the Conkouati reserve, Congo: a conservation biological assessment. Cons Biol 15: 1247–1257.

7. Russon A (2009) Orangutan rehabilitation and reintroduction: successes, failure and role in conservation. Geographic variation in behavioral ecology and conservation (ed. S. . Wich, et al.), pp 327–350. Oxford University Press, Oxford, UK.

8. Ancrenaz M, Ambu L, Sunjoto I, Ahmad K, Manokaran K, et al. (2010) Recent surveys in the forests of Ulu Segama Malua, Sabah, Malaysia, show that orangutans (P. p. morio) can be maintained in slightly logged forests. PLOS One 5:e11510.

9. Ruslandi O, Venter O, Putz FE (2011) Overestimating conservation costs in Southeast Asia. Front Ecol Env 9:542–544.

10. Rijksen HD (1982) How to save the mysterious man of the forest? The Orangutan: its biology and conservation (ed. L.E.M. . de Boer,), pp 317–341. Dr W Junk Publishers, The Hague, The Netherlands.

11. Lardoux-Gilloux I (1995) Rehabilitation Centers: their struggle, their future. The Neglected Ape (eds R.D. . Nadler, et al.), pp 61–68. Plenum Press, New York.

12. Aveling RJ, Mitchell A (1982) Is rehabilitating orangutans worthwhile? Oryx, 16, 263–271.

13. Meijaard E, Albar G, Rayadin Y, Nardiyono, Ancrenaz M, et al. (2010a) Unexpected ecological resilience in Bornean Orangutans and implications for pulp and paper plantation management. PLOS ONE 5: e12813.

14. Jepson P, Momberg F, van Noord H (2002) A review of efficacy of the protected area system of East Kalimantan Province. Nat Areas J 22: 28–42.

15. Gaveau DLA, Curran LM, Paoli GD, Carlson KM, Wells P, et al. (2012) Examining protected area effectiveness in Sumatra: importance of regulations governing unprotected lands. Cons Lett 5:142–148.

16. Lagan P, Mannan S, Matsubayashi H (2007) Sustainable use of tropical forests by reduced-impact logging in Deramakot Forest reserve, Sabah, Malaysia. Ecol Res 22: 414–421.

17. Grundmann E (2006) Back to the wild: will reintroduction and rehabilitation help the long-term conservation of orang-utans in Indonesia? Soc Science Inform 45: 265–284.

18. Meijaard E, Buchori D, Hadiprakoso Y, Utami-Atmoko SS, Tjiu A, et al. (2011) Quantifying killing of orangutans and human-orangutan conflict in Kalimantan, Indonesia PLOS ONE 6: e27491.

19. Soehartono T, Susilo HD, Andayani N, Atmoko SSU, Sihite J, et al. (2007) Strategi dan rencana aksi konservasi orangutan Indonesia 2007–2017. Ministry of Forestry of the Republic of Indonesia, Jakarta, Indonesia.

20. Husson SJ, Wich SA, Marshall AJ, Dennis RA, Ancrenaz M, et al. (2009) Orangutan distribution, density, abundance and impacts of disturbance. Orangutans: geographic variation in behavioral ecology and conservation (eds. S.A. . Wich, et al.), pp.77–96. Oxford University Press, Oxford.

21. Miettinen J, Chenghua S, Liew SC (2011) Deforestation rates in insular Southeast Asia between 2000 and 2010. Glob Change Biol 17: 2261–2270.

22. Caldecott J, Miles L (2005) World Atlas of Great Apes and their Conservation. University of California Press, Berkeley.

23. WDPA (2009) World Database on Protected Areas. UNEP-WCMC, Cambridge, UK. Available: http://www.wdpa.org.

24. McQuistan CI, Fahmi Z, Leisher C, Halim A, Adi SW (2006) Protected area funding in Indonesia: A study implemented under the Programmes of Work on Protected Areas of the Seventh Meeting of the Conference of Parties on the Convention on Biological Diversity. Ministry of Environment republic of Indonesia, PHPA, Departemen Kelautan dan Perikanan Jakarta, Indonesia.

25. Venter O, Possingham HP, Hovani L, Dewi S, Griscom B, et al. (2013) Using systematic conservation planning to minimize REDD+ conflict with agriculture and logging in the tropics. Cons Lett 6: 116–124.

26. Venter O, Meijaard E, Possingham HP, Dennis R, Sheil D, et al. (2009) Carbon payments as a safeguard for threatened tropical mammals. Cons Lett 2: 123–129.

27. BOS (2008) Independent auditor's report on Financial Statements of Borneo Orangutan Survival Foundation. Available from Erik Meijaard, Jakarta, Indonesia.

28. Applegate GB (2002) Financial costs of reduced impact timber harvesting in Indonesia: case study comparisons. Applying reduced impact logging to advance sustainable forest management (eds. Enters, T., et al.) (downloaded from http://www.fao.org/DOCREP/005/AC805E/ac805e0j.htm) FAO, Rome, Italy.

29. Putz FE, Sist P, Fredericksen T, Dykstra D (2008) Reduced-impact logging: Challenges and opportunities. For Ecol Manag 256: 1427–1433.

30. Meijaard E, Welsh A, Ancrenaz M, Wich S, Nijman V, et al. (2010b) Declining orangutan encounter rates from Wallace to the present suggest the species was once more abundant. PLOS ONE 5: e12042.

31. Marshall AJ, Lacy R, Ancrenaz M, Byers U, Husson S, et al. (2009) Orangutan population biology, life history, and conservation. Perspectives from population viability analysis models. Orangutans: geographic variation in behavioral ecology and conservation. (eds S. . Wich, et al.), pp 311–326. Oxford University Press, Oxford, UK.

32. Hartig F, Drechsler M (2008) The time horizon and its role in multiple species conservation planning. Biol Cons 141: 2625–2631.

33. Wilson HB Joseph LN, Moore AL, Possingham HP (2011) When should we save the most critically endangered species? Ecol Lett 14: 886–890.

34. Pressey RL, Watts ME, Barrett TW (2004) Is maximizing protection the same as minimizing loss? Efficiency and retention as alternative measures of the effectiveness of proposed reserves. Ecol Lett 7: 1035–1046.

35. Game ET, McDonald-Madden E, Puotinen ML, Possingham HP (2008) Should we protect the strong or the weak? Risk, resilience and the selection of marine protected areas. Cons Biol 22: 1619–1629.

36. Marshall AJ, Nardiyono, Engström LM, Pamungkas B, Palapa J, et al. (2006) The blowgun is mightier than the chainsaw in determining population density of Bornean orangutans (Pongo pygmaeus morio) in the forests of East Kalimantan. Biol Cons 129: 566–578.

37. Davis JT, Mengersen K, Abram N, Ancrenaz M, Wells J, et al. (2013) It's not just conflict that motivates killing of orangutans. PLOS ONE 8: e75373.

38. Gaveau DLA, Epting J, Lyne O, Linkie M, Kumara I, et al. (2009) Evaluating whether protected areas reduce tropical deforestation in Sumatra. J Biogeog 36: 2165–2175.

39. Meijaard E, Wich S, Ancrenaz M, Marshall AJ (2012) Not by science alone: Why orangutan conservationists must think outside the box. Ann NY Acad 1249:29–44.

2

Economic Impacts of Non-Native Forest Insects in the Continental United States

Juliann E. Aukema[1]*, **Brian Leung**[2,3], **Kent Kovacs**[4], **Corey Chivers**[2], **Kerry O. Britton**[5], **Jeffrey Englin**[6], **Susan J. Frankel**[7], **Robert G. Haight**[8], **Thomas P. Holmes**[9], **Andrew M. Liebhold**[10], **Deborah G. McCullough**[11], **Betsy Von Holle**[12]

1 The National Center for Ecological Analysis and Synthesis, Santa Barbara, California, United States of America, 2 Department of Biology, McGill University, Montreal, Quebec, Canada, 3 School of Environment, McGill University, Montreal, Quebec, Canada, 4 Department of Applied Economics and Institute on the Environment, University of Minnesota, St. Paul, Minnesota, United States of America, 5 U.S. Forest Service, Research and Development, Arlington, Virginia, United States of America, 6 Morrison School of Agribusiness and Resource Management, Arizona State University, Mesa, Arizona, United States of America, 7 U.S. Forest Service, Pacific Southwest Research Station, Albany, California, United States of America, 8 U.S. Forest Service, Northern Research Station, St. Paul, Minnesota, United States of America, 9 U.S. Forest Service, Southern Research Station, Research Triangle Park, North Carolina, United States of America, 10 U.S. Forest Service, Northern Research Station, Morgantown, West Virginia, United States of America, 11 Department of Entomology and Department of Forestry, Michigan State University, East Lansing, Michigan, United States of America, 12 Department of Biology, University of Central Florida, Orlando, Florida, United States of America

Abstract

Reliable estimates of the impacts and costs of biological invasions are critical to developing credible management, trade and regulatory policies. Worldwide, forests and urban trees provide important ecosystem services as well as economic and social benefits, but are threatened by non-native insects. More than 450 non-native forest insects are established in the United States but estimates of broad-scale economic impacts associated with these species are largely unavailable. We developed a novel modeling approach that maximizes the use of available data, accounts for multiple sources of uncertainty, and provides cost estimates for three major feeding guilds of non-native forest insects. For each guild, we calculated the economic damages for five cost categories and we estimated the probability of future introductions of damaging pests. We found that costs are largely borne by homeowners and municipal governments. Wood- and phloem-boring insects are anticipated to cause the largest economic impacts by annually inducing nearly $1.7 billion in local government expenditures and approximately $830 million in lost residential property values. Given observations of new species, there is a 32% chance that another highly destructive borer species will invade the U.S. in the next 10 years. Our damage estimates provide a crucial but previously missing component of cost-benefit analyses to evaluate policies and management options intended to reduce species introductions. The modeling approach we developed is highly flexible and could be similarly employed to estimate damages in other countries or natural resource sectors.

Editor: Brian Gratwicke, Smithsonian's National Zoological Park, United States of America

Funding: This work is the product of a National Center for Ecological Analysis and Synthesis (NCEAS) Working Group supported by The Nature Conservancy and NCEAS, which is funded by the National Science Foundation (Grant #DEB-0553768), the University of California Santa Barbara, and the State of California. The funders had no role in study design, data collection and analysis, decision to publish, or preparation of the manuscript.

Competing Interests: The authors have declared that no competing interests exist.

* E-mail: aukema@nceas.ucsb.edu

Introduction

Invasive species are widely recognized as among the greatest threats to biodiversity and ecosystem stability worldwide, and they impose serious economic and social costs [1,2,3]. Global trade yields enormous economic benefits, but a side effect can be the inadvertent transport of organisms from one region to another [4,5]. Impacts of invasive species have not been adequately accounted for in trade policy, in part because the economic impacts of invaders have not been reliably quantified. Strategies for internalizing the costs of invaders, including pricing, quarantines and tariffs may be the most effective means of avoiding impacts of invasive species if implemented vigorously [6]. An economic rationale for such efforts requires consideration of projected benefits (economic damages avoided) compared to implementation costs. Thus, quantifying the economic damages caused by biological invasions is critical to informing these strategies.

The few studies that have calculated aggregate costs of invasive species have been useful for drawing attention to the economic significance of biological invasions [7,8], but they have been plagued with difficulties such as double counting certain costs and failing to account for uncertainty and the ability to substitute one resource for another [9,10]. The difficulties of conducting rigorous economic analysis are compounded by the scarcity of economic data, which are only available for perhaps 1–2% of invaders [11]. Although most non-native species cause low or intermediate impacts [12], in combination these costs can accumulate. To avoid a downward bias, it is critical to model the entire range of impacts rather than assuming that no damages are caused by species for which economic impacts are unknown.

Despite conceptual challenges, economic assessments of the impacts of non-native species are needed to provide credible information to policy makers and to justify costs associated with management efforts [13]. Decisions must often be made in the absence of complete data but it is important to explicitly identify and address the uncertainty inherent in the data [14]. Risk analyses in general, and Bayesian approaches in particular, offer a coherent means of incorporating uncertainty into decision-making. Specifically, it is possible to integrate across an uncertainty distribution, rather than assuming point estimates are correct or being incapacitated in the face of large uncertainties [e.g.15].

We estimated total direct annual costs of non-native forest insects established in the United States. Forests and urban trees provide important economic and social benefits, as well as ecosystem services [13,16]. Non-native forest insects often encounter evolutionarily naive, vulnerable host trees and few natural enemies when they arrive in a new habitat. These invaders may kill their host trees or affect tree health, growth or appearance. Our analysis is based on an exhaustive database of non-native forest insects in the continental U.S., which enabled us to standardize the area of analysis and to take advantage of available data.

Our objective is to provide improved cost estimates that policy makers can use to inform decision–making in a framework that can be updated and improved as new data become available. In constructing our approach, we advance previous work in three ways. First, we stratify analyses by insect feeding guild. Pests in the same feeding guild generally cause similar types of damage and often share some biological traits. Moreover, guilds are associated with probable pathways of introduction, and therefore are relevant units for trade policy considerations. Second, we separate analyses by economic cost categories to avoid double counting (such as those federal expenditures which subsidize local expenditures) and to highlight the relevance of invasive forest pests to different sectors of society. Finally, we quantify uncertainty in our estimates to reflect the limits of data used in our models.

Methods

Established non-native forest insects

We used a database of 455 non-native phytophagous forest insect species known to be established in the continental United States, compiled using published sources and expert input [17]. While the majority of the 455 species have not caused detectable damage, we identified a subset of 62 species that have been reported to cause noticeable impacts (above background levels) to live forest trees [17, part II of Appendix S1]. We assigned each of the 455 species to a feeding guild based on their dominant or most damaging feeding mode – phloem and wood borers (hereafter borers) (71 species), sap feeders (192 species), foliage feeders (155 species), or other (37 species) [17]. For each of the three main feeding guilds, we identified one high impact "poster pest" that was the most damaging species of its guild to date: emerald ash borer (*Agrilus planipennis* Fairmaire: borer), hemlock woolly adelgid (*Adelges tsugae* Annand: sap feeder), and gypsy moth (*Lymantria dispar* L.: foliage feeder) [Appendix S1].

Economic assessments of "poster pests"

We selected five cost categories for analysis for each poster pest, based on data availability. Cost categories included: (1) federal government expenditures (survey, research, regulation, management, and outreach), (2) local government expenditures (tree removal, replacement, and treatment), (3) household expenditures (tree removal, replacement, and treatment), (4) residential property value losses and (5) timber value losses to forest landowners.

Dead and dying trees reduce the value of homes due to lost aesthetic value, create hazards that must be removed by governments and homeowners, and have lower timber value than healthy trees. Although there are political considerations in the allocation of government funding for surveys, research, and outreach activities related to invasive species, we counted these as costs because they expend resources that could have been used for other public services if those invasive species had not arrived. We restrict our analyses to these five cost categories because they cover a significant fraction of the direct costs of forest pests and because data were available. We recognize that there are other indirect costs, secondary effects, and non-market ecosystem services (e.g., changes in water quality, altered species composition) that can be important. Data for assessing these impacts are scarce, however, and methods for scaling local studies up to the national level have not been developed, which would have potentially compounded the uncertainty of our estimates [Appendix S1]. Thus, our analysis should be viewed as providing a lower bound cost estimate. A management action or policy implementation that is worthwhile based on these available direct costs would certainly be deemed valuable if the full range of possible impacts were known.

We estimated short-run (ten year) economic impacts for each cost category using a partial equilibrium framework in which interactions between costs were not considered and which is appropriate when the short-term linkages between economic categories are weak [18]. All economic impacts reflect changes from a baseline scenario reflecting the absence of economic impacts from the poster pests (see part I of Appendix S1, Tables S1,S2,S3,S4,S5, and Figure S1 for a detailed description of the methods and data sources used to estimate economic impacts). Changes in local government and household expenditures were estimated using a dynamic optimization model that captures the economic trade-offs between protecting tree health and the costs of tree removal and replacement. Changes in property values due to changes in tree health were based on economic welfare estimates obtained from published non-market valuation studies. Changes in timber harvesting levels were based on estimates of timber mortality from non-native forest insects, and mortality induced harvest reductions were small enough to have no impact on timber prices. Changes in federal expenditures were based on historical data, as it was deemed to be infeasible to model the budget decision process.

We chose a ten-year horizon to represent the short-run because: (1) this time span encompasses periodic or cyclical behavior typical in forest pest dynamics, (2) uncertainty is constrained by not extrapolating too far into the future, and (3) shorter time horizons could be greatly influenced by stochastic factors, such as weather, or a particular phase of a pest outbreak. Because each pest is at a different stage of invasion, for each poster pest, we selected a ten-year period that would closely reflect average pest-related damages, management options and costs, and for which data were available or could be projected (Table S1). For each poster pest, we converted estimated impacts to constant 2009 US dollars using a 2% real discount rate. We obtained annual costs by calculating an annuity for our discounted damages over a ten-year time horizon.

For all cost categories except federal government expenditures, we estimated economic impacts using spatial data and dynamic models of infestation extent. We did not sum economic impacts across categories to avoid double counting. For example, double counting could occur between federal and local government expenditures due to transfers between the government bodies; homeowner expenditures and residential property value losses could overlap, because property values capitalize the potential real estate losses, including expenditures on tree removal and treatment. This approach facilitates comparison across guilds, within cost categories, but we caution against adding across cost categories. If data related to the extent of overlap between categories become available in the future, adjusted cost categories could be summed.

Bayesian modeling of total impacts

For each insect guild and cost category, we estimated the total annual costs (expenditures or losses) across the entire guild and quantified uncertainty given the available data. We began by asserting that there is a frequency distribution of annual costs (the cost curve) (Fig 1A). We assumed that introduced phytophagous species do not have net positive effects on our economic cost categories (cost >0), and that species causing little damage are more common than species that cause intermediate or high impacts, while only a few species cause severe damage [12]. Given these constraints, the possible functional forms that describe the

cost curves are limited. Because the exact forms of the cost curves are unknown, we examined several alternative models. We considered 39 parametric families of curves [19], and reduced these to four non-redundant families with appropriate theoretic properties: the gamma, Weibull, power function and log-normal distributions. Although we did not have cost estimates for each species, we used the frequency of species in our database, partitioned into low, intermediate and high damage classes to fit the curve (Fig. 1A). We used expert opinion to define the thresholds between pests that cause low and intermediate costs (Table S6), and our detailed economic analysis of the poster pest for each guild to define the thresholds between intermediate and high costs. By calculating the expected value of each cost curve, and multiplying by the posterior probability, we could then estimate the expected cost of a single species, as well as the total

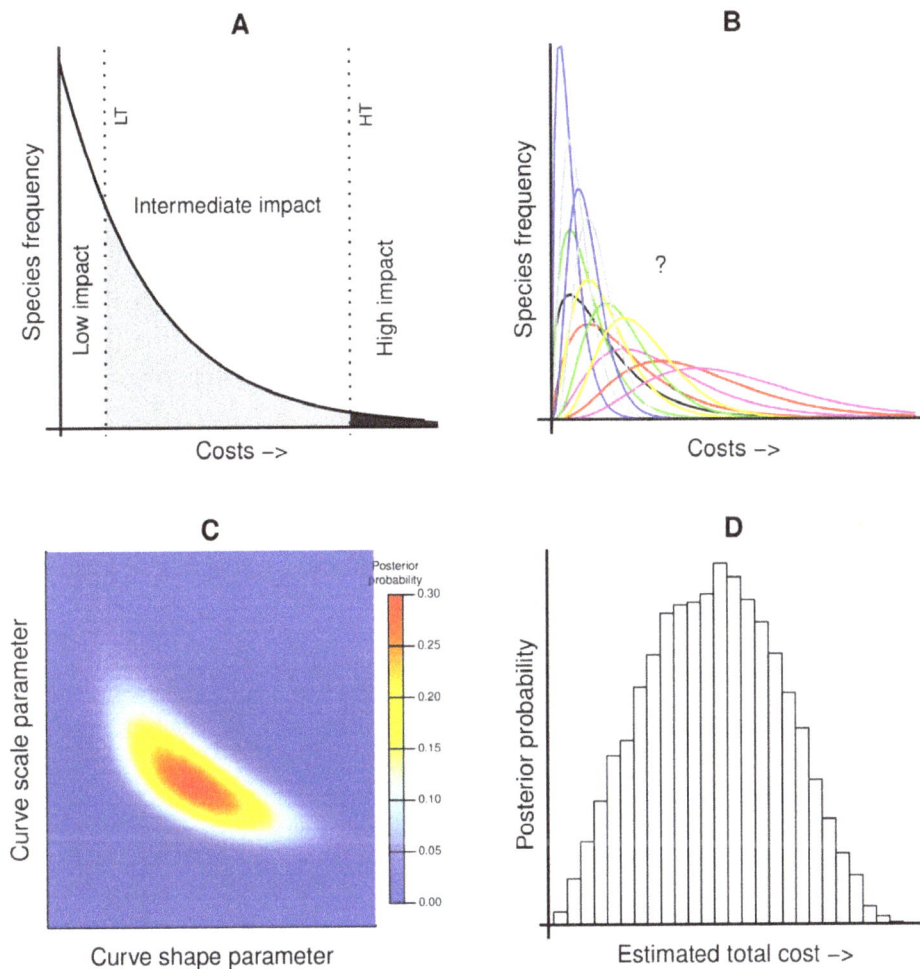

Figure 1. Framework for estimating economic costs of invasive species. A) The hypothetical cost curve is the frequency distribution of annual economic cost caused by invasive species belonging to a feeding guild. The counts of low and intermediate impact species, as well as the low impact threshold (LT) and the level of damages caused by the poster pest (HT) are known; however, the exact shape of the curve is unknown. B) Alternative cost curves. The data are fit using different parameter values, for four alternative models: gamma (illustrated), log-normal, power and Weibull distributions. C) Illustrative Bayesian posterior probability distribution of cost curve parameters. The posterior probability is the relative probability for each cost curve (defined by parameter values). Some cost curves are more likely than others, given the observed data. The posterior probability allows us to consider and incorporate the relative evidence for each cost curve, thereby accounting for parameter uncertainty. This process is repeated for all four models (Weibull, log-normal, gamma (shown), and power function), and then integrated using Bayesian model averaging, which accounts for model uncertainty. The relative probabilities are shown as a heat graph. D) Probability distribution of total annual cost across species in the guild. We converted the cost curves from the Bayesian analysis into a more meaningful metric - total costs from invaders (Appendix S1). Each cost curve and its corresponding total cost has a relative probability of being true given the observed data. The entire process is repeated for each guild and cost category.

annual cost of all known pests in each guild. Once the shape of the cost curve (and associated uncertainty) has been characterized, any number of derived values of interest can be extracted in a similar way to the expected and total costs. Here, we also present estimates of the probability that a new invader will be more costly than the poster pest (i.e., the area under the curve to the right of the poster pest).

We accounted for uncertainty by quantifying variability among species (the cost curve; Fig. 1B), parameter uncertainty (Bayesian analysis; Fig 1C), and model uncertainty (Bayesian model averaging across the four families of curves; Fig 1D, Fig 2 A,B). Further, because they were classified by expert opinion, we performed sensitivity analysis on the lower threshold, spanning the threshold value by two orders of magnitude (Fig. 2C,D, S2, S3). As new data become available in the future, damage estimates can be readily updated to re-evaluate total cost estimates. We report the Bayesian expectations in the text (i.e., the mean of the Bayesian posterior distribution), as well as the 90% credible intervals in Table 1. For further details of the framework and complete model specification, see part III of Appendix S1.

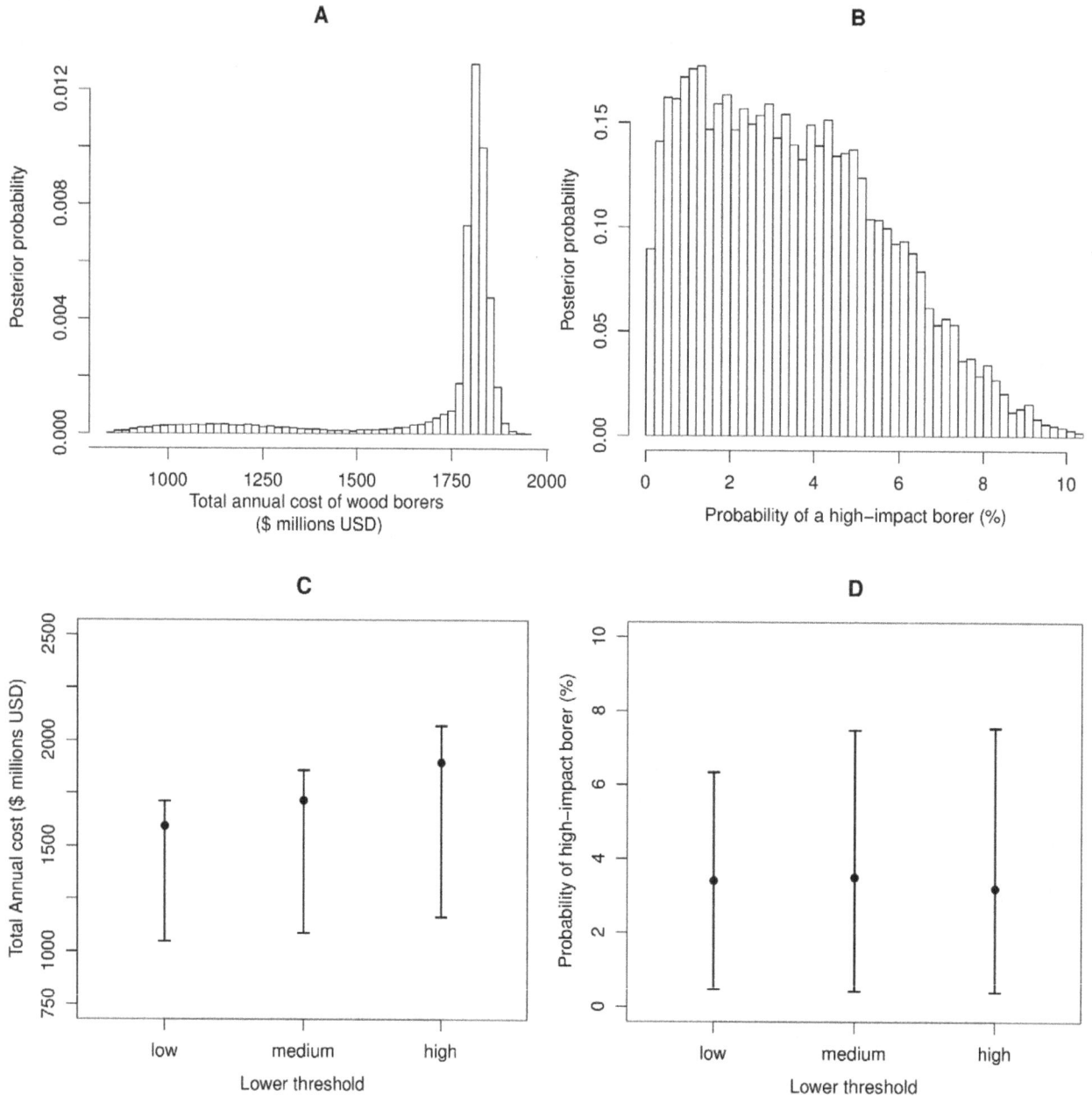

Figure 2. Results for borer feeding guild and local government cost category showing the posterior probability distributions of a) estimated total annual cost of all known borer species, b) probability that a newly introduced borer will cost local governments as much or more than the poster pest (emerald ash borer). Panels c) and d) show the low sensitivity of the posterior predictions to alternative specifications of the low impact threshold (LT) which was based on expert opinion (Appendix S1). Shown are alternative specifications for total annual guild costs (c) and probability of a high impact borer (d) across two orders of magnitude, where low, medium and high costs are defined as 150, 1,500 and 15,000 annual USD, respectively (Table S6).

Table 1. Annualized damage in U.S. $1,000,000 associated with each guild and cost category.

Guild	Federal Government Expenditures	Local Government Expenditures	Household Expenditures	Residential Property Value Loss	Forest Landowner Timber Loss
BORERS (N = 71, N$_i$ = 14)					
Poster: emerald ash borer damages ($10^6)	38	850	350	380	60
Total damage ($10^6)	92 [62–97]	1700 [1100–1900]	760 [460–820]	830 [510–900]	130 [81–150]
P>poster (%)	3.5 [0.47–8.1]	3.4 [0.43–7.2]	3.4 [0.39–7.6]	3.3 [0.41–7.8]	3.3 [0.42–7.5]
SAP FEEDERS (N = 192, N$_i$ = 19)					
Poster: hemlock woolly adelgid damages ($10^6)	4.3	66	44	100	1.1
Total damage ($10^6)	14[6.6–15]	170 [85–190]	130 [62–140]	260 [130–290]	4.2 [2.1–4.6]
P>poster (%)	1.1 [0.14–2.8]	1.1 [0.12– .8]	1.1 [0.13– .1]	1.0 [0.14–2.5]	1.1 [0.15–2.9]
FOLIAGE FEEDERS (N = 155, N$_i$ = 25)					
Poster: gypsy moth damages ($10^6)	33	50	46	120	4.6
Total damage ($10^6)	110 [52–120]	170 [75–180]	160 [72–180]	410 [190–450]	18 [8.2–20]
P>poster (%)	1.3 [0.21–3.4]	1.4 [0.20–3.6]	1.2 [0.21–3.1]	1.6 [0.22–4.7]	1.3 [0.23–3.4]

The poster pest damage was calculated for each cost category (See Appendix S1 for detailed methods). We report the posterior mean of total damage (see Figure 1) and the probability that a newly introduced pest will be as damaging or more damaging than the poster pest for that cost category (P>poster). N is the total number of species in the guild and N$_i$ shows the number of intermediate pests in each guild. The 90% Bayesian credible intervals are in brackets.

Results

Pests from the borer guild, which often arrive on wood packaging materials, generally exacted the highest total costs across categories (Table 1). At an estimated $1.7 billion in local government expenditures and approximately $830 million in lost residential property values each year, borers' economic impacts were several times greater than impacts from other guilds. Of the three guilds, borers were represented by the fewest species, but a high proportion of them (20%) are damaging [17]. Furthermore, integrated across the uncertainty distribution, the probability that the next species to invade will cause damages at least as great as the poster pest was substantially higher (3.4%) for borers than for the other guilds, although Bayesian credible intervals overlap for the average probability (Table 1).

Sap feeders accounted for the largest proportion of the insects in our database, but relatively few cause tree mortality or substantial damage (Table 1). The high frequency of sap feeder invasions may be attributed in part to the historical trade in live plants, a pathway for introduction of these insects [20]. Of the three guilds, sap feeders caused the least timber value loss; the timber value loss caused by sap feeders was less than 5% of that caused by borers. Efforts to control or manage sap feeders received the fewest federal dollars ($14 million annually), although they caused substantial losses in real estate values - approximately $260 million per year.

Foliage feeders, also frequently introduced with live plants, were almost as abundant in the dataset as sap feeders. Costs associated with foliage feeders were substantially lower than costs associated with borers for all categories except annual federal expenditures, which were slightly greater ($110 million for foliage feeders and $92 million for borers) (Table 1). Foliage feeders were estimated to cause approximately $410 million per year in lost property value. Foliage feeders, such as gypsy moth, typically cause mortality only if consecutive years of severe defoliation occur or under exacerbating circumstances such as drought, which is reflected in the lower costs of this guild.

Discussion

Government officials, resource managers and property owners routinely make decisions about trade or regulatory policies and about whether or how to manage an invasive forest pest. These decisions should ideally consider the costs of specific actions as well as the benefits to be gained by them. This process is challenging, particularly given the limited information about the current or potential costs of pest impacts. We identified costs likely to be incurred by different societal sectors that can be used in such cost-benefit analyses and the damages associated with major guilds of insects, while incorporating uncertainty to the extent possible.

Our analysis indicates that the cost of non-native forest insects is largely borne by homeowners and municipal governments, large constituencies that may not be adequately considered in most analyses [10]. For all guilds, local government expenditures and residential property value losses were the two highest cost categories. Household expenditures were also high, which was partially reflected in property value loss. The costs of tree removal, replacement and treatment outweighed the costs of federal government containment programs by at least one order of magnitude.

In contrast, we found that timber value losses are relatively modest, often an order of magnitude lower than local government expenditures. This reflects timber values of the tree species attacked by the poster pests. Timber mortality induced by poster pests constituted a small proportion of overall timber harvest volumes across tree species, so we assumed timber supply curves were unaltered. However, we recognize that future biological invasions could have more severe impacts on timber species. In the case where a biological invasion causes catastrophic mortality of valuable timber species, timber buyers and forest owners with non-impacted stands face changes in economic welfare due to market price impacts [21,22]. A previous estimate of nation-wide economic impacts of non-native forest insects based on timber losses was $2.1 billion annually. However, this estimate assumed a reduction in gross domestic product of timber-related industries and did not consider substitutability [10].

In addition, our analysis highlights the importance of borers, which were consistently the most costly insect feeding guild. The best estimate of costs to local governments, integrating across the uncertainty distribution, was $1.7 billion per year. Despite the presence of substantial uncertainty, even lower bound estimates revealed considerable costs. For instance, phloem and wood borer damage is expected to cost local governments at least $1.1 billion per year but could cost as much as $1.9 billion per year. Indeed, the effect of the borer guild on local governments dwarfed all other cost categories – most by over an order of magnitude.

The relatively high cost of borers in general may not be surprising given that several borers can cause mortality of their host trees. But this finding is particularly troubling because of the dramatic increase in new detections of established borer species in the last 30 years, coinciding with increased use of wood packaging material, which can transport these organisms [17,23]. Borers accounted for 56% of forest insect invaders detected from 1980–2006, compared to just under 11% before 1930 [17]. Put another way, by integrating the results from our study, there is a 32% risk that a new borer that is as damaging as or more costly than the emerald ash borer will invade in the next 10 years [calculated as $(E = 1 - (1 - p)^{(Y \cdot R)})$, where p is the probability of each introduced pest being more costly than emerald ash borer, Y is the number of years and R is the annual rate of introduction; YR is the expected number of borers introduced. See Appendix S1, eq. 13]. However, if recent international standards which target pathways of introduction such as wood packaging materials (e.g. ISPM-15) are effectively implemented, this introduction rate may be reduced [24]. Although our calculations address the probability that a new poster pest will become established, theoretically, our approach can be extrapolated to any damage level of interest to researchers or policy makers.

The similarity of federal government expenditures for borers and foliage feeders was notable, given that borers (e.g. emerald ash borer and Asian longhorned beetle) generally cause more tree mortality than foliage feeders (e.g. gypsy moth). Some sap feeders also cause localized host mortality (e.g. hemlock woolly adelgid and balsam woolly adelgid), but federal expenditures were almost seven-fold lower for this guild of pests than for defoliators or borers. Federal allocations may reflect factors such as the temporal or spatial extent of a pest, its impacts, the availability of regulatory and management options and external pressures from stakeholders. Hemlock woolly adelgid, for example, may have lower federal costs because it is not regulated. Cost benefit analyses have demonstrated the economic value of efforts such as the gypsy moth "Slow the Spread" program, where government activities prevent or defer costs that would otherwise likely be borne by property owners or municipalities [25,26]. Similar analyses for damaging borers and sap feeders may be appropriate, given their current and projected costs, and the need to optimize spending allocations given current declining budgets.

Our framework can incorporate new information as it becomes available, including explicit cost estimates for additional species and cost categories (Appendix S1). For instance, by causing tree mortality, defoliation and reduced growth of their hosts, non-native forest insects can have important direct and cascading effects on non-market ecosystem services such as water and air quality, nutrient cycling, climate regulation, disease control, and recreation and cultural services [20,27,28]. Furthermore, non-native forest pests threaten native species and entire ecosystems such as the Fraser fir forests of the southern U.S. Appalachian Mountains, the rare Carolina hemlock trees, and redbay trees in the southeastern coastal plains [29,30,31,32]. As data become available for these types of damages caused by exemplary "poster" pests, our framework can be used to estimate guild-wide ecosystem services losses.

Our study provides the most comprehensive estimates of costs of forest invaders currently available for the United States, the probability of future costs and, therefore, the benefits of reducing the rate of invasion. We identify the insect guilds most responsible for, and the societal elements most affected by, these damages, and we provide insight into the introduction pathways that could be targeted by management actions. Our work can be used in quantitative cost-benefit analysis of the preventative measures that are widely regarded to be the best option for addressing invasive species [13,33]. For example, targeted import taxes or fees have been proposed as a means of generating funds to pay for practices to reduce introductions or to eradicate or control invasive pests that have already established [34]. Development, implementation and justification of such policies will require these estimates of nationwide economic damages and the sectors affected by invasive pests. Our analytical framework can be used in any country where data are available and can be easily adapted for estimating costs in a variety of natural resource sectors in addition to invasive species, including fire, disease, and water quality, at scales from municipalities to nations.

Supporting Information

Appendix S1   Detailed economic and modeling methods. Part I describes the economic methods and data sources. Part II describes the database of non-native forest pests and the classification of their impacts. Part III describes the Bayesian model used to estimate total costs.

Figure S1   Study area of U.S. counties infested at the beginning of the study period (shaded green) and at the end of the study period (shaded yellow) by the emerald ash borer (top), the gypsy moth (middle), and the hemlock wooly adelgid (bottom).

Figure S2   Sensitivity of total cost estimate to the lower threshold for each guild and cost category combination. Low and high represent posterior values obtained using a lower threshold one order of magnitude below and above (respectively) expert opinion (medium). Mean and 90% Bayesian credible intervals illustrated.

Figure S3   Sensitivity to the lower threshold of the probability of a new pest as damaging or more damaging than the poster pest for each guild and cost category combination. Low and high represent posterior values obtained using a lower threshold one order of magnitude below and above (respectively) expert opinion (medium). Mean and 90% Bayesian credible intervals illustrated for each guild and cost category combination.

Table S1   Ten-year time horizon for calculating poster pest damages.

Table S2   Host tree density on developed land for the study areas corresponding to the emerald ash borer and hemlock woolly adelgid.

Table S3 Ash density by land use and diameter class for the city of Chicago.

Table S4 Management costs for homeowners and community managers.

Table S5 Parameters for the model of timber losses to forest landowners.

Table S6 Lower economic threshold of damages by damage category.

Acknowledgments

We would like to thank E. Sills for input on gypsy moth costs and T. Stohlgren, E. Fleishman, F. Lowenstein, A. Nuding and J. Turner for valuable discussions. We thank F. Homans, J. Fargione, L. Westphal, F. Lowenstein, and anonymous reviewers for comments on the manuscript.

Author Contributions

Conceived and designed the experiments: JEA BL KK CC KOB JE SJF RGH TPH AML DGM BVH. Performed the experiments: JEA BL KK CC RGH TPH AML DGM BVH. Analyzed the data: JEA BL KK CC RGH TPH AML DGM BVH. Wrote the paper: JEA BL KK CC RGH TPH AML DGM.

References

1. Pimentel D, McNair S, Janecka J, Wightman J, Simmonds C, et al. (2001) Economic and environmental threats of alien plant, animal, and microbe invasions. Agr Ecosyst Environ 84: 1–20.
2. Simberloff D (2000) Nonindigenous species: a global threat to biodiversity and stability. In: Raven P, Williams T, eds. Nature and human society: The quest for a sustainable world. Washington, DC: National Academy Press. pp 325–336.
3. Wilcove DS, Chen LY (1998) Management costs for endangered species. Conserv Biol 12: 1405–1407.
4. Bright C (1998) Life out of Bounds: Bioinvasion in a Borderless World. New YorkW.W.: Norton and Company. pp 287.
5. Everett RA (2000) Patterns and pathways of biological invasions. Trends Ecol Evol 15: 177–178.
6. Perrings C, Dehnen-Schmutz K, Touza J, Williamson M (2005) How to manage biological invasions under globalization. Trends Ecol Evol 20: 212–215.
7. Pimentel D, Lach L, Zuniga R, Morrison D (2000) Environmental and economic costs of nonindigenous species in the United States. Bioscience 50: 53–65.
8. Colautti RI, Bailey SA, van Overdijk CDA, Amundsen K, MacIsaac HJ (2006) Characterised and projected costs of nonindigenous species in Canada. Biol Invasions 8: 45–59.
9. Holmes TP, Aukema JE, Von Holle B, Liebhold A, Sills E (2009) Economic Impacts of Invasive Species in Forests Past, Present, and Future. Year in Ecology and Conservation Biology 2009 1162: 18–38.
10. Born W, Rauschmayer F, Brauer I (2005) Economic evaluation of biological invasions - a survey. Ecol Econ 55: 321–336.
11. Gren IM, Isacs L, Carlsson M (2009) Costs of Alien Invasive Species in Sweden. Ambio 38: 135–140.
12. Williamson M, Fitter A (1996) The varying success of invaders. Ecology 77: 1661–1666.
13. Chornesky EA, Bartuska AM, Aplet GH, Britton KO, Cummings-Carlson J, et al. (2005) Science priorities for reducing the threat of invasive species to sustainable forestry. Bioscience 55: 335–348.
14. Peterson GD, Carpenter SR, Brock WA (2003) Uncertainty and the management of multistate ecosystems: An apparently rational route to collapse. Ecology 84: 1403–1411.
15. Ludwig D (1999) Is it meaningful to estimate a probability of extinction? Ecology 80: 298–310.
16. Lovett GM, Canham CD, Arthur MA, Weathers KC, Fitzhugh RD (2006) Forest ecosystem responses to exotic pests and pathogens in eastern North America. BioScience 56: 395–405.
17. Aukema JE, McCullough DG, Von Holle B, Liebhold AM, Britton K, et al. (2010) Historical accumulation of nonindigenous forest pests in the continental US BioScience 60: 886–897.
18. Simon HA, Ando A (1961) Aggregation of variables in dynamic systems. Econometrica 29: 111–138.
19. Evans M, Hastings N, Peacock B (1993) Statistical Distributions. New York, NY: John Wiley & Sons, Inc.
20. Kenis M, Rabitsch W, Auger-Rozenberg MA, Roques A (2007) How can alien species inventories and interception data help us prevent insect invasions? B Entomol Res 97: 489–502.
21. Holmes TP (1991) Price and welfare effects of catastrophic forest damage from southern pine beetle epidemics. Forest Sci 37: 500–516.
22. Prestemon JP, Holmes TP (2000) Timber price dynamics following a natural catastrophe. Am J Agr Econ 82: 145–160.
23. Haack RA (2006) Exotic bark- and wood-boring Coleoptera in the United States: recent establishments and interceptions. Can J Forest Res 36: 269–288.
24. FAO (2006) ISPM No. 15. Guidelines for regulating wood packaging material in international trade. In: International Standards for Phytosanitary Measures 1 to 27. Rome: Food and Agriculture Orgnization of the United Nations. pp 189–200.
25. Leuschner WA, Young JA, Walden SA, Ravlin FW (1996) Potential benefits of slowing the gypsy moth's spread. South J Appl For 20: 65–73.
26. Sharov AA, Liebhold AM (1998) Bioeconomics of managing the spread of exotic pest species with barrier zones. Ecol Appl 8: 833–845.
27. Jones CG, Ostfeld RS, Richard MP, Schauber EM, Wolff JO (1998) Chain reactions linking acorns to gypsy moth outbreaks and Lyme disease risk. Science 279: 1023–1026.
28. Lovett GM, Christenson LM, Groffman PM, Jones CG, Hart JE, and, et al. (2002) Insect defoliation and nitrogen cycling in forests. BioScience 52: 335–341.
29. Wagner DL, Van Driesche RG (2010) Threats posed to rare or endangered insects by invasions of nonnative species. Annu Rev Entomol 55: 547–568.
30. Gandhi KJK, Herms DA (2010) Direct and indirect effects of alien insect herbivores on ecological processes and interactions in forests of eastern North America. Biol Invasions 12: 389–405.
31. Ward JD, Mistretta PA (2002) Impact of pests on forest health. In: Wear DN, Greis JG, eds. Southern forest resource assessment Gen Tech Rep SRS-53. Asheville, NC, U.S.: Department of Agriculture, Forest Service, Southern Research Station. 635 p.
32. Fraedrich SW, Harrington TC, Rabaglia RJ, Ulyshen MD, Mayfield AE, III, et al. (2008) A fungal symbiont of the redbay ambrosia beetle causes a lethal wilt in redbay and other Lauraceae in the southeastern United States. Plant Dis 92: 215–224.
33. Lodge DM, Williams S, MacIsaac HJ, Hayes KR, Leung B, et al. (2006) Biological invasions: Recommendations for US policy and management. Ecol Appl 16: 2035–2054.
34. Jenkins PT (2002) Paying for protection from invasive species. Issues Sci Technol 19: 67–72.

3

Spatial and Temporal Patterns of Deforestation in Rio Cajarí Extrative Reserve, Amapá, Brazil

Claudia Funi[1,2]**, Adriana Paese**[1,3]*

1 Programa de Pós Graduação em Biodiversidade Tropical, Universidade Federal do Amapá, Macapá, Brazil, **2** Instituto Estadual de Pesquisas do Amapá, Macapá, Brazil, **3** Conservação Internacional, Belo Horizonte, Brazil

Abstract

The Rio Cajarí Extractive Reserve (RCER) is a sustainable use protected area located in Southern Amapá state, Brazil. This protected area is home to traditional agro-extractive families, but has been increasingly invaded by commercial agriculture producers. In this work, we test the hypothesis that the RCER implementation has distinctly affected spatial patterns of deforestation and rates of bare soil and secondary forest formation by the social groups occupying the protected area and its surrounding area. Detailed maps of vegetation cover and deforestation were elaborated, based on Landsat TM images from 1991, 1998, 2007 and 2008 and Linear Spectral Mixture Models. Based on an extensive fieldwork, patches were classified according to the agents causing deforestation and characterized with ten explanatory variables. A discriminant function analysis was used to identify homogeneous groups based on the data. Results show increased rates and distinct spatial patterns of deforestation by three groups: extractivists, non traditional commercial agriculture producers, and a less representative group constituted of miners, cattle and timber producers. In all analyzed dates, clearings by the extrativist community presented the highest total area and smaller average sizes and were located in close proximity to villages. Deforestation patches by the non-traditional group were exclusively associated with ombrophilous forests; these presented higher average sizes and proximity indexes, and showed increased aggregation and large cluster formation. No significant differences were observed in deforestation patterns by the three groups inside or outside the reserve.

Editor: Kimberly Patraw Van Niel, University of Western Australia, Australia

Funding: The International Institute of Education (IEB) supported this work through the (BECA – IEB/Gordon and Betty Moore Foundation Program) by providing a scholarship (number B/2007/01/$BMP/05) for the first author. The funders had no role in study design, data collection and analysis, decision to publish, or preparation of the manuscript.

Competing Interests: The authors have declared that no competing interests exist.

* E-mail: adripaese@gmail.com

Introduction

As new targets for reducing deforestation in tropical forests are being set and as policy and incentives for reducing deforestation and forest degradation in developing countries are being discussed [1,2,3], a clearer understanding of the drivers of deforestation inside and outside protected areas is still needed. Moreover, there is still a need to quantify the effects of improved forest management on carbon retention [4].

Predictive models of the effects of infrastructure development and conservation policies in the Brazilian Amazon have demonstrated the contribution of the existing protected areas to regional climate regulation, [5] and their potential role in reducing CO_2 emissions from deforestation [6] and post clearing agricultural practices [7]. Although coarse-scale global and regional assessments have demonstrated the effectiveness of protected areas in reducing the clearing of tropical forests [8,9,10], refinements to these analyses demonstrate a broader range of efficacy [11]. These refinements have incorporated biophysical and socioeconomic factors that affect the location of protected areas [12] or were conducted at temporal scales that match the patterns of human-caused disturbances [11].

Empirical evidence also shows that the projected scenarios may not be as optimistic as they may appear. Deforestation is known to be pervasive, and such pervasive deforestation is a result of diverging interests of different social and political groups. Small scale decisions by these groups affect land use inside and outside protected areas, hindering the implementation of the areas affected [13,14].

Brazils Amazonian protected area network is constituted of municipal, state and federal sustainable use and strictly protected areas. These areas are unified under the National System of Protected Areas [15]. Protected areas in the Brazilian Amazon increased by 709,000 km^2 between 2002 and 2009 [6]. Over the past 24 years, the average rate of forest clearing in the Brazilian Amazon was 16,341.71 km^2 yr^{-1} [16]. In 2007, sustainable use areas and strictly protected areas represented 61% and 39%, respectively, of the total protected area in the region [17]. Sustainable use areas (e.g. extractive reserves, national forests, environmental protection areas) allow local people to develop subsistence agricultural activities, and partially restrict deforestation. In strict protection areas (e.g. parks, biological reserves, ecological stations), the exploitation of natural resources by human populations is prohibited [15]. Between 2000 and 2006, the average deforestation rate within protected areas was 1,520 km^2 yr^{-1}. The overall protected area network has helped maintain intact 98.6% of the forest cover inside the protected areas [11]. However, analyses of different categories or individual protected areas have indicated a broad range of efficacy. If clearings occur along their perimeters, extractive reserves are more vulnerable to forest cutting than other categories, including strictly protected areas and indigenous lands [9,11].

Assessments of deforestation rates in the Brazilian Amazon are based primarily on the data from the PRODES Project (*Projeto de Monitoramento do Desmatamento na Amazônia Legal*) [6,16,9,11]. The monitoring activities conducted by this project serve to detect deforestation patches above the 6.25 ha threshold. Although data from PRODES have been proved to be useful for detecting deforestation on agricultural frontiers, refinements to this methodology are necessary to detect land use change in more conservative scenarios. These more conservative settings include those involving low human population densities, those associated with vegetation other than upland forests (i.e., upland savannas and inundated savannas), or those associated with forest degradation (e.g. surface fire disturbance and selective logging) [18]. Evidence also demonstrates that protected areas in geographical proximity experiencing similar development pressure may show very different rates of forest clearing [11]. This information underscores the need for a clearer understanding of the social drivers of deforestation and occupation history in protected areas and their surrounding areas.

In this study, we performed a fine-grained analysis of the drivers of spatial and temporal patterns of deforestation by social groups in the Rio Cajarí Extractive Reserve (RCER) and its surrounding area. The RCER is a sustainable use protected area located in southern Amapá State, Brazil. Amapá is the Brazilian state with the highest proportion of native vegetation cover and the lowest average deforestation rate between 1998 and 2011 (58.77 ha) [19]. Approximately 70% of the area of this state is currently protected through designation as strictly protected and sustainable use areas, or as indigenous lands. The RCER is the sustainable use protected area (IUCN category VI protected area) with the greatest average deforestation rate in Amapá State [19]. The RCER was created, as were other extractive reserves in Brazil, with the primary objective of resolving land tenure issues and providing greater tenure security for local extractivist communities [20]. The RCER is home to traditional agroextractivist families, primarily Brazil-nut collectors, but it has been increasingly invaded by commercial agriculture producers.

This study analyzes a satellite image classification that allows the detection and mapping of finer scale spatial and temporal patterns of deforestation in a broad range of native vegetation covers. The pressures and drivers of deforestation are also examined in detail inside the RCER and in its immediate surroundings. We test the hypotheses that the RCER implementation has distinctly affected the spatial patterns of deforestation and rates of bare soil and secondary forest formation by the social groups occupying the protected area and its surrounding area.

Materials and Methods

Study Area

In order to account for potential threats to the reserve, our study area was defined as the Rio Cajarí Extractive Reserve area (RCER) and its immediate surroundings, including a buffer zone extending 5 km from the boundaries of the protected area. This definition of the buffer zone was selected to include all deforestation by extractivists settled in the RCER and its surrounding area.

The RCER was created through Federal Decree number 99,145, on March 12th, 1990 [20], and modified on September 30th, 1997 [21]. As an IUCN category VI protected area, the main goal of the RCER is to reconcile economic development with long-term environmental conservation, while improving the well-being of traditional populations.

The total extent of our study area is 679,421.8 ha, of which 503,448.6 ha (74%) constitute the RCER area and 175,973.2 ha (26%) the reserves immediate surroundings. The study area is located from 0° 15′ S to 52° 25′ W and from 1° S to 51° 31′ W in the municipalities of Laranjal do Jari, Vitória do Jari and Mazagão, in Amapá State, Brazil. The RCER and its buffer zone are covered by dryland and flooded forests, upland savannas and inundated savannas [22]. The climate is classified as tropical wet, with an annual average temperature above 25°C and an average rainfall of 2,300 mm concentrated between December and June [23]. The elevation of the area ranges from 1 to 357 m [21]. The Amazon River forms the eastern boundary of the RCER. The northwestern portion of the reserve is crossed by an unpaved federal highway (BR 156). In absolute terms, the reserve shows the highest deforestation rate among all protected areas in Amapá State [19] (Figure 1).

In 1967, a large-scale enterprise, the Jari Project, was established in southern Amapá State. This project introduced large scale cultivation of exotic species (primarily *Eucalyptus sp*) for pulp production; cattle and buffalo husbandry, and included timber and mineral extraction [24]. As a result of the Jari Project implementation, large tracts of land were deforested, including areas long occupied by extractivist communities [25]. Intensified conflicts between the local communities and the contractors of the Jari Project resulted in the creation of extractivist settlements for the local communities through the agrarian reform agency (*Instituto Nacional de Colonização e Reforma Agraria* - INCRA) in 1987 (Decree No. 627 of 7/30/87), and in the creation of the Rio Cajari Extractive Reserve in 1990 [26]. Today, approximately 3,050 residents, distributed in 552 households, are located in the RCER in close proximity to roads and with access to water corridors [27]. In the RCER, the land is owned by the Federal Government. The extractivist communities hold joint usufruct rights over the extractive reserve. These rights may be transferred by inheritance. The RCER is entailed to the Chico Mendes Institute for Biodiversity Conservation (ICMBio), the governmental agency responsible for all federal protected areas in the country. The governance is decentralized and is conducted by the reserves Steering Committee (*Conselho Deliberativo*), composed of 23 representatives of the local, state and federal governments and local community associations.

A utilization plan, specifying the ways in which the community manages its resources was defined in 1997 as part of the land concession [15]. The utilization plan is a temporary instrument based on the residents knowledge and experience. This plan will remain valid until a management plan is developed [28]. The RCER utilization plan includes the prohibition of commercial hunting and the recognition of common areas, such as rivers, lakes, pathways, banks and jointly managed areas in the reserve. The extraction of timber for commercial purposes is prohibited. According to the RCER utilization plan, each family is allowed to deforest areas no greater than 15 ha for the cultivation of subsistence crops (primarily manioc).

The main economic activity of the extractivist families living in the reserve is the exploration of non-timber forest products, primarily Brazil nuts (*Bertholletia excelsa*), the fruit and palm of açaí trees (*Euterpe oleracea*), and andiroba (*Carapa guianensis*). The rights of families to harvest non-timber forest products are dictated by the spatial pattern of the resource and the rules of the community. Families are allowed to explore areas known as *castanhais* or *colocações*, which contain high densities of Brazil nut trees and are informally regarded as the familys property. The rights to the

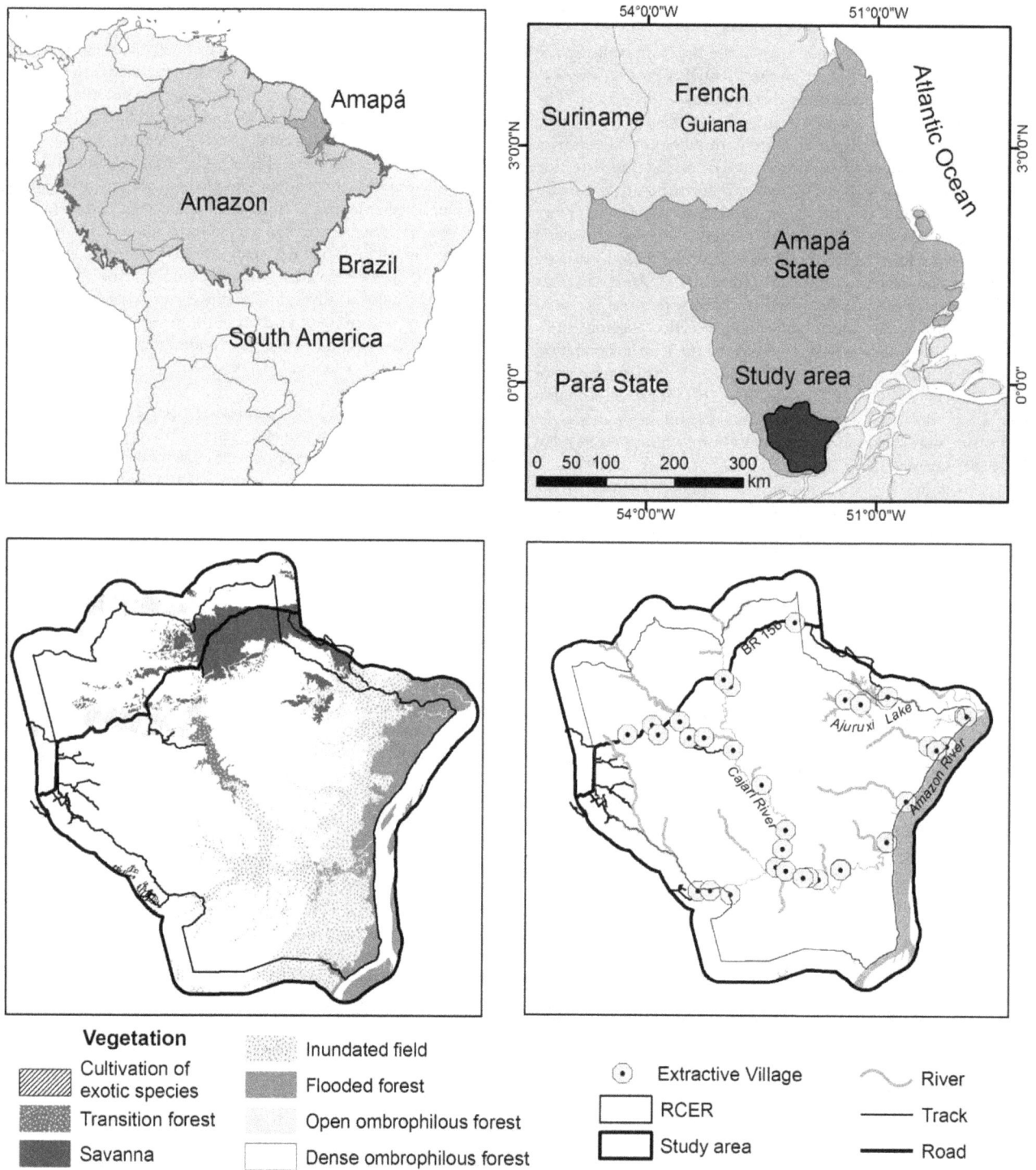

Vegetation

- Cultivation of exotic species
- Transition forest
- Savanna
- Inundated field
- Flooded forest
- Open ombrophilous forest
- Dense ombrophilous forest

- ⊙ Extractive Village
- RCER
- Study area
- ～ River
- — Track
- ▬ Road

Figure 1. Location of the Amazon river basin and Amapá State in South America and Brazil (a); Location of the studied area in Amapá (b); Main vegetation cover in the Rio Cajarí extractive reserve and its immediate surrounding area(c); Location of extractivists' villages terrestrial and fluvial accesses in the study area (d).

exploration of *castanhais* are transferred to family members through inheritance or family division (marriage of a son). This process tends to reduce the sizes of exploration sites and to decrease families opportunities to thrive based solely on non-timber forest production.

Deforestation Mapping

A high-resolution map of the vegetation cover was constructed based on Landsat TM5 satellite images from 1991 and linear spectral mixture models (LSMM) [29,30]. Vegetation patches and classes were validated with the existing coarse-scale vegetation maps (IBGE) and 68 days of on-the-ground and aerial observations. Detailed maps of deforestation were generated for four dates with Landsat Thematic Mapper five images from 1991, 1998, 2007 and 2008. Landsat TM images were chosen for this analysis because of the temporal extent of cloud-free images available in the National Institute for Space Research (INPE) catalog [31]. We wanted to depict land use immediately after the reserve creation (1991), and at intermediate (1998) and recent dates (2007–2008). Amapá is located in the Intertropical Convergence Zone, where increased precipitation and cloud cover are common throughout the year. A mosaic of images from 2007–2008 was created to depict the recent pattern of land use because completely cloud-free scenes were not available for 2008. The deforestation maps were generated using linear spectral mixture models (LSMM) [29,30] and on-screen digitizing. Vegetation, soil and shadow fraction images resulting from the LSMM allowed the identification of deforestation patches in distinct vegetation types.

Agents Causing Deforestation

Deforestation polygons were classified as (a) agroextractive/traditional (AE), (b) non-traditional (NT), and (c) other occupants (OT) according to the agent causing deforestation. The AE deforestation polygons resulted from traditional community activities.

AE patches are deforestation patches opened by agroextractive families. Most of these families have been in the reserve area since the rubber boom in the late 19th century [32]- and remain involved with the extraction of timber and non-timber forest products and subsistence agricultural production. They possess substantial knowledge of the local natural resources and biological cycles. They practice non-mechanized agriculture and may use fire in these activities. Family members may serve as the labor force for family agricultural activities. The reserve utilization plan prohibits the use of contract employees for natural resources exploration or agricultural activities.

NT deforestation patches result from non-traditional land use practices. Non-traditional producers do not extract non timber forest products to complement their income. They practice intensive commercial agriculture and may have employees. The deforestation patches classified as Others (OT) include a less representative, but more heterogeneous group. This group consists of small cultivation sites, abandoned mining sites, or small patches opened by cattle and timber producers. These clearings are located on islands in the Amazon river in the area surrounding the RCER. The OT patches include clearings associated with mining activities in the northern part of the reserve, deforestation associated with the village of Jarilândia (located south of the reserve in the municipality of Vitória do Jari), or clearings associated with higher income farmers. Higher-income farmers are located exclusively in the proximity to the BR-156 road, outside the RCER. Inside the RCER, deforestation associated with the OT group is primarily located relatively far from agroextractive villages.

Temporal Patterns of Deforestation

The analysis of fractional images also allowed the classification of deforestation patches as bare soil or secondary forest. The bare soil class represents very recently cleared land, plantations with less than one year or sites cultivated uninterruptedly for no longer than three years, pastures or villages, whereas the secondary forest class represents abandoned areas or shifting-cultivation fallows that follow the development of manioc plantations. According to interviews with residents, fallow cycles may last, on average, 5.5 years in the study area, but may vary from three to 25 years.

The classification of deforested areas was validated with false color composites of Landsat TM 5 bands 3, 4, and 5 and with extensive field work. Ground-truthing activities were conducted during eight visits to the reserve. These visits included a total of 68 days between May 2007 and May 2008. Land use data were collected in meetings with the residents of agroextractive villages, and representatives of Chico Mendes Institute for Biodiversity Conservation (ICMBio). No specific permits were required for the field studies because all visits to the reserve were conducted in the company of ICMBio staff. Visits to plantations and secondary forests were also accompanied by at least one resident who provided data on the history of land use, the time since occupation, the crops cultivated, the average annual size of the plantations, the fallow cycles, the estimated age of the secondary forest patches, and the fire management of cleared lands. Satellite images were also shown to the residents. This way, it was possible to locate deforested areas that were not visited. Visits to NT patches were conducted during a demographic census by the ICMBio staff. The field studies did not involve endangered or protected species. The data from interviews were double-checked through comparisons with satellite images and with low- altitude flights. Every deforestation patch for which the land cover classification was initially doubtful was visited on the ground.

Spatial Patterns of Deforestation

Deforestation patches were classified according to elevation [33] and according to their linear distances to: water bodies (rivers and streams), main roads, trails or secondary access roads (ramais), municipalities heads, and agroextractive villages, in all analyzed data. Four metrics calculated with Fragstats [34] (number, area, nearest neighbor distance and proximity index) were also used to quantify the spatial patterns of deforestation by the groups. The variables were analyzed with a multivariate discriminant function. This method serves to evaluate the null hypothesis that there are no real groups in the data (i. e., the explanatory variables show similar behavior between the a priori groups) [35]. This analysis assumes that the explanatory variables are normally distributed. Variables that did not follow a normal distribution were log transformed. The variables that still did not follow a normal distribution after transformation were omitted from the discriminant function analysis, but included in the description of the visual patterns. To avoid input data redundancy, a tolerance value of 0.5 was used as the lower threshold for variable acceptance. A significance value of $p \leq 0.001$ was used for the analysis.

Results

Deforestation Rates

The analyses of deforestation maps, in 1991, 1998 and 2007/2008 showed that by 2008 human activities had replaced 9,179.2 ha of the original vegetation cover in the RCER and 4,938.5 ha in the area surrounding the reserve. These values represent 1.82% of the total area of the RCER and 2.81% of the surrounding area (Figure 2).

A clearer understanding of the dynamics of forest replacement in the study area was possible through the detailed mapping and classification of deforestation patches as bare soil or 'secondary forest'. Bare soil represents very recently cleared land, plantations established for less than one year or sites cultivated uninterruptedly for no longer than three years, pastures or villages. In contrast, the 'secondary forest' represents abandoned areas or shifting cultivation fallows following manioc plantations.

The rate of bare soil formation was higher than that of secondary forest formation in the area surrounding RCER, during

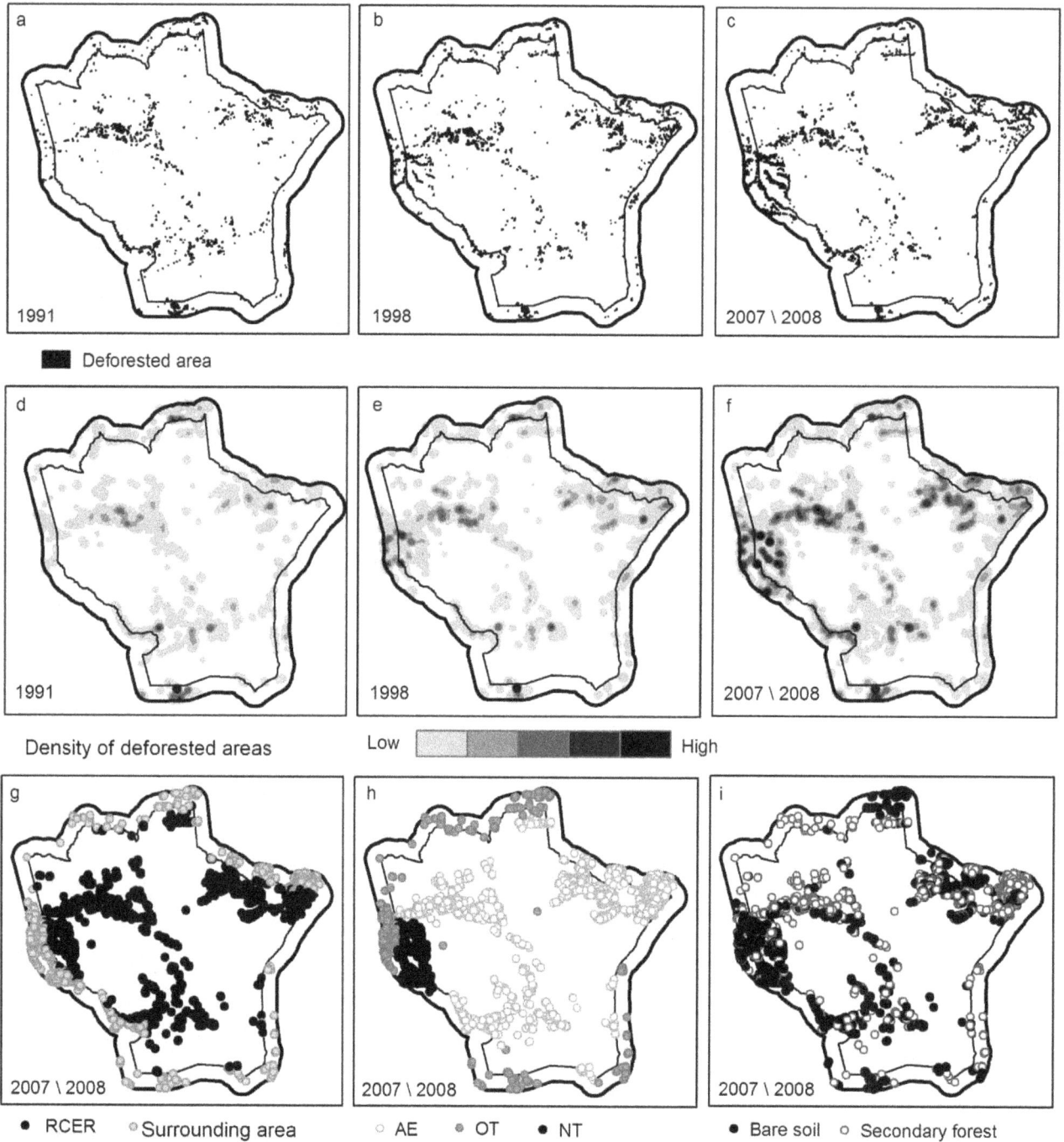

Figure 2. Deforestation in the RCER and its surrounding area in 1991(a), 1998 (b) and 2007/2008 (c). Density of deforested areas in 1991 (d), 1998 (e) and 2007/2008 (f). Deforested areas located in the RCER or in the RCER surrounding area (g). Deforestation patches by causing agents (h). Location of bare soil and secondary forest patches (i).

1991–2008. Between 1991 and 1998, the rate of bare soil formation was 0.028% of the area surrounding the RCER per year, whereas the corresponding rate of secondary forest formation was of 0.017% yr^{-1}. During 1998–2008, bare soil in the RCER surrounding area increased at a rate of 0.052% yr^{-1}, whereas secondary forests increased at a rate of 0.028% yr^{-1} (Figure 3a). Within the RCER, the rate of formation of secondary forests (0.031% yr^{-1}) exceeded the rate of formation of bare soil (0.012% yr^{-1}) between 1991 and 1998. Between 1998 and 2008, bare soil

increased at a higher rate (0.029% yr^{-1}) than secondary forests (0.005% yr^{-1}).

A better understanding of deforestation was also made possible by the analysis of the contribution of the three primary agents causing deforestation (AE, OT and NT) to the percentages of use and to the rates of bare soil and secondary forest formation in the study area (Figures 3a, b). OT was the most representative group in the area surrounding the RCER in 1991 and 1998. AE showed the highest proportions of bare soil and secondary forest in the

(a)

(b)

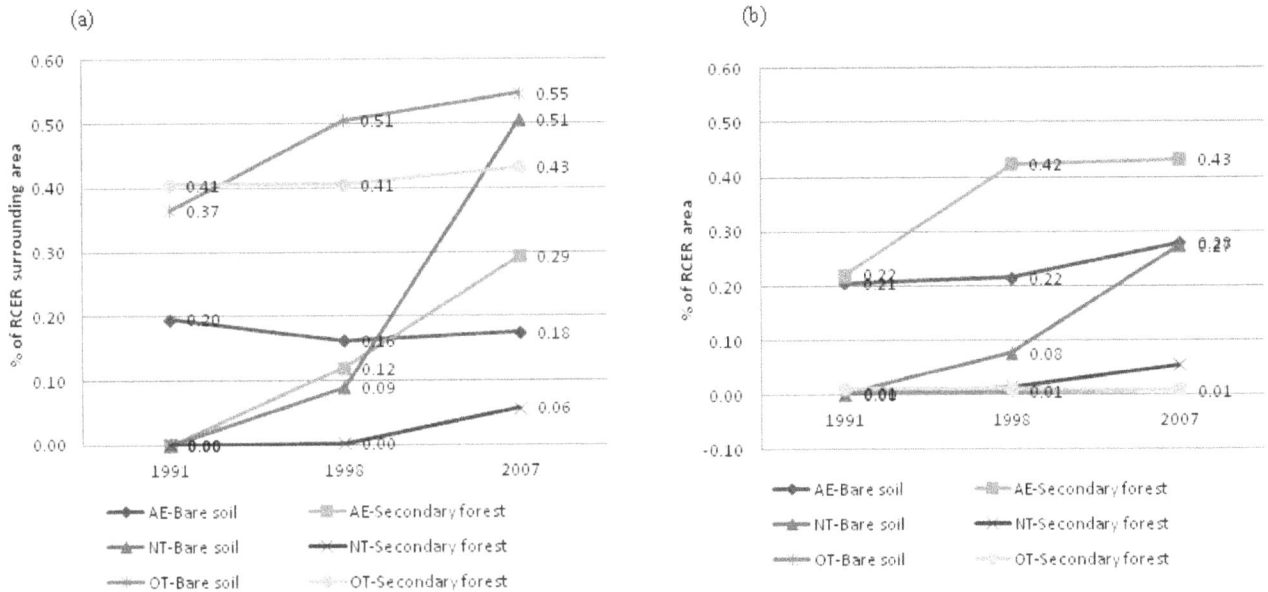

Figure 3. Contributions of the three main agents causing deforestation (non-traditional community (NT), agro-extractive (AE) and others (OT)) to the total area of bare soil and secondary forest in the RCER and in its surrounding area in 1991, 1998 and 2007/2008.

RCER in 1991, 1998 and 2007 (Figures 3a, b). In 2007, the proportion of bare soil associated with NT (0.50%), in the area surrounding the RCER, exceeded the proportion of bare soil associated with all other groups except OT (0.54%). The increasing rate of deforestation in the area surrounding the RCER between 1991 and 2008 was primarily associated with the expressive rate of bare soil formation by the NT group between 1998 and 2008 (0.034% of the area surrounding the RCER per year). The increasing deforestation rates within the RCER were also primarily associated with higher rates of bare soil formation by NT during the 1998–2007 period (0.022% yr^{-1}). To a lesser extent, the increasing deforestation rates were also associated with the formation of bare soil by the AE group at a rate of 0.007% of the RCER area per year. A decreasing rate of secondary forest formation by the AE group in the RCER area (from 0.029% yr^{-1} to 0.001% yr^{-1}) during the second period of the study is also of interest.

Spatial Patterns of Deforestation

Deforestation patches associated with AE were more numerous on the data analyzed than those associated with the other two groups (Table 1). Inside the RCER, the average size of deforestation patches associated with AE was smaller in 1991 and 1998 than the average size of clearings by the other two groups. In 2007/2008, the average size of clearings by the OT group was smaller than the average size of clearings by the AE group due to the abandonment and reduction of the area of mining sites or to increased cloud cover over the Amazon River islands and in the northern part of the reserve, where most deforestation by the OT is located. According to the data analysis, the number of clearings by the NT group in the area surrounding the RCER was smaller than the number of clearings by the two other groups. Inside the RCER, the number of deforestation patches associated with NT was smaller than the numbers associated with AE and OT in 1991. In 1998 and 2007/2008, OT was the least representative group inside the reserve. The patches of deforestation inside the RCER associated with NT

showed the greatest average sizes and relatively high values of the proximity index. The high values of the proximity index imply that areas cleared by the NT group tend to aggregate to form larger clusters. In the area surrounding the RCER, the NT group was associated with the greatest average size of deforestation patches in 1998 and 2008 (Table 1).

A further analysis of the patterns of deforestation by the three groups showed that the AE group is the least heterogeneous group. A single group associated with deforestation caused by AE inside the RCER and in its surrounding area was identified in 1991, 1998 and in 2007/2008. The same pattern is observed for deforestation caused by NT inside the RCER and in its surrounding area. In contrast, the OT group is the most heterogeneous and least representative group. Descriptive patterns of this group overlap with those of AE and NT (Figures 4a, b, c). The OT group could potentially be divided into other subgroups, and these subgroups would require further investigation. For 1991, the two variables (Axis 1 and 2) in the multivariate discriminant function analysis that best explain the distinctiveness of the deforestation patches associated with the three groups are the distances to any fluvial access and the distance between deforested patches and other vegetation types than dense ombrophilous forest (upland forest) (e. g. flooded forests, inundated savannas and upland forests and open ombrophilous forest). For 1998, the principal variables explaining the distinctiveness of the groups are the distances between deforested patches and other vegetation types and distance to any access (fluvial or terrestrial). For 2007/2008, the main variables explaining the distinctiveness of the groups are the distances to water access, the distances between deforested patches and other vegetation types than upland forests (Table 2).

Clearings for agricultural purposes, by all groups (AE, NT and OT), are located exclusively in dense ombrophilous forest due to the fertile soils associated with this vegetation type and shadow conditions. The distance from deforested patches to other vegetation types than the dense ombrophilous forest is an important variable to explain the distinctiveness among the groups

Table 1. Number of deforestation patches (N), their average size (Size), proximity index (Prox), nearest neighbor distance (ENN).

AE	Location	N	Size (ha)	Prox	ENN (m)
1991	RCER	826	2.5 (±5.2)	6.5 (±21.4)	348.3 (±659.6)
	SA	149	2.1 (±3.2)	3.5 (±4.7)	493.6 (±1715.3)
1998	RCER	955	3.3 (±6.9)	9.2 (±21.8)	251.1 (±526.3)
	SA	161	2.8 (±4.8)	6.8 (±19.7)	245.7 (±312.6)
2008	RCER	836	4.2 (±8.3)	9.2 (±21)	270.5 (±412.4)
	SA	239	3.2 (±5.5)	6.4 (±11.7)	231.3 (±403.1)
NT		**N**	**Size (ha)**	**Prox**	**ENN (m)**
1991	RCER	6	3.7 (±5.8)	0.29 (±0.31)	731.1 (±626.5)
	SA	1	0.12	0	0
1998	RCER	70	7.5 (±17.8)	11.1 (±25.3)	252.8 (±295.8)
	SA	15	6.8 (±18.5)	6.2 (±6.5)	242.5 (±235.9)
2008	RCER	97	18.4 (±46.9)	62.1 (±133)	222.2 (±224.2)
	SA	54	15.6 (±39.8)	17.7 (±35.9)	223.5 (±263.1)
OT		**N**	**Size (ha)**	**Prox**	**ENN (m)**
1991	RCER	39	3.4 (±5.6)	4.2 (±11.7)	1369 (±2120.3)
	SA	204	6.3 (±25.3)	22.9 (±138.1)	443.9 (±764.9)
1998	RCER	29	3.8 (±7.8)	0.4 (±0.6)	1884.6 (±2181.9)
	SA	272	5.7 (±22.4)	9.4 (±24.4)	373.7 (±734)
2008	RCER	27	3.2 (±6.7)	0.38 (±0.8)	2929.9(±5957)
	SA	192	8.7(±27.4)	16.1 (±36)	468 (±1168.4)

because the AE group usually establishes settlements on the border between two or more vegetation types and at relatively lower elevations. For this reason, deforestation patches by AE is located in closer proximity to other vegetation types than NT and OT.

Deforestation associated with the AE is clearly located close to agroextractive villages and far from municipality centers, whereas the opposite pattern is observed for deforestation by the NT group. Patches of deforestation associated with NT are located at higher elevations (from 35 to 171 m) and in close proximity to roads. Clearings by NT are located in upland forests (dense ombrophilous forest) distant from the borders of other vegetation types. The deforestation polygons associated with this group are located far from agroextractive villages and in close proximity to secondary roads and to the municipal center of *Laranjal do Jari* (Table 2).

OT is a more heterogeneous group. This group includes abandoned mining sites north of the reserve, and clearings associated with agriculture and with cattle and timber production on islands in the Amazon river. They are also located near the village of *Jarilândia*, south of the reserve in the municipality of *Vitória do Jari*, and they may be associated with higher-income farmers settled near road BR-156. The deforestation patches associated with OT are located primarily in the area surrounding the RCER, but may also be found in upland forests. They may also be located on islands in the Amazon River, at lower elevations, and in the northern part of the reserve.

Discussion

The total deforested area calculated in this study differs significantly from the estimates developed by the PRODES Project (*Projeto de Monitoramento do Desmatamento na Amazônia Legal*)

[16]. The deforestation value calculated in this study was 3.72 times higher than the official estimates for the RCER over the 17 year time frame of that analysis. In the area surrounding the RCER, the deforestation value calculated in this study was 3.57 times higher than the total deforested area estimated by PRODES for the same period. The PRODES monitoring program has been useful for detecting deforestation on agricultural frontiers, such as those in Mato Grosso state and in eastern Pará state. These results need to be explored further in other areas if protected areas in the Brazilian Amazon are to be included in national programs for reducing emissions from deforestation and forest degradation [11] due to their implications for CO_2 baseline estimates. Refinements to the PRODES methodology, such as the one adopted in this work, are necessary to detect land use change in more conservative scenarios. These scenarios include situations in which human population densities are small. They also include situations in which deforestation patches are small, scattered or associated with vegetation types such as upland savannas and inundated savannas rather than with forests.

Clearings detected by the PRODES methodology are restricted to ombrophilous forests. A low resolution polygon mask used by this Program, with scales varying from 1:500,000 to 1:1,000,000, filters out deforestation in other vegetation types and in areas of contact between two or more vegetation types. In the RCER, deforestation by extractivists occurs primarily on the border of two or more vegetation types or is associated with patches of forest surrounded by other vegetation types and can be smaller than 6.25 ha. Refinements to the PRODES methodology allowed the identification of deforestation patches as small as 0.09 ha, in all vegetation types in the RCER. A classification of forest clearings associated with distinct social groups in the RCER and its surrounding area was only possible as a result of refinements to the PRODES methodology.

The deforestation numbers in the study area (1.82% of the total RCER area and 2.81% of the area surrounding RCER) are similar to the average deforestation associated with protected areas in the Brazilian Amazon [8]. A closer analysis of the contribution of the three principal social groups occupying the study area AE, OT and NT to the formation of bare soil and secondary forest provided additional evidence about the spatial and temporal patterns of deforestation in the protected area and in the area surrounding the RCER. Increasing rates of deforestation in the RCER and its surrounding area are primarily associated with increasing rates of bare soil formation by NT. This group tends to use their cultivated areas uninterruptedly, with no fallows, and the cultivated sites are usually abandoned after the soil becomes impoverished. The patches of deforestation associated with NT show the greatest average sizes. They also show relatively high values of the proximity index and tend to aggregate to form larger clusters. Deforestation associated with NT is located near to access roads, which facilitate the marketing of their production. Approximately 62% of the NT occupants in the RCER live in urban areas [36] and land use and occupation of the RCER by NT is illegal. Deforestation rates in the RCER could be greatly reduced with an effective regulatory enforcement by the public authorities over the settlement of NT inside this protected area.

The spatial patterns of deforestation by agroextractivists clearly differ from those associated with the NT and OT groups. Despite the higher observed proportion of secondary forests associated with AE, the rate of bare soil formation by this group has increased within the RCER. A reduction in the rate of secondary forest formation by this group within the RCER was also observed. One cause of the increasing rates of deforestation and decreasing rates of secondary forest formation by agroextractivists is the lack of

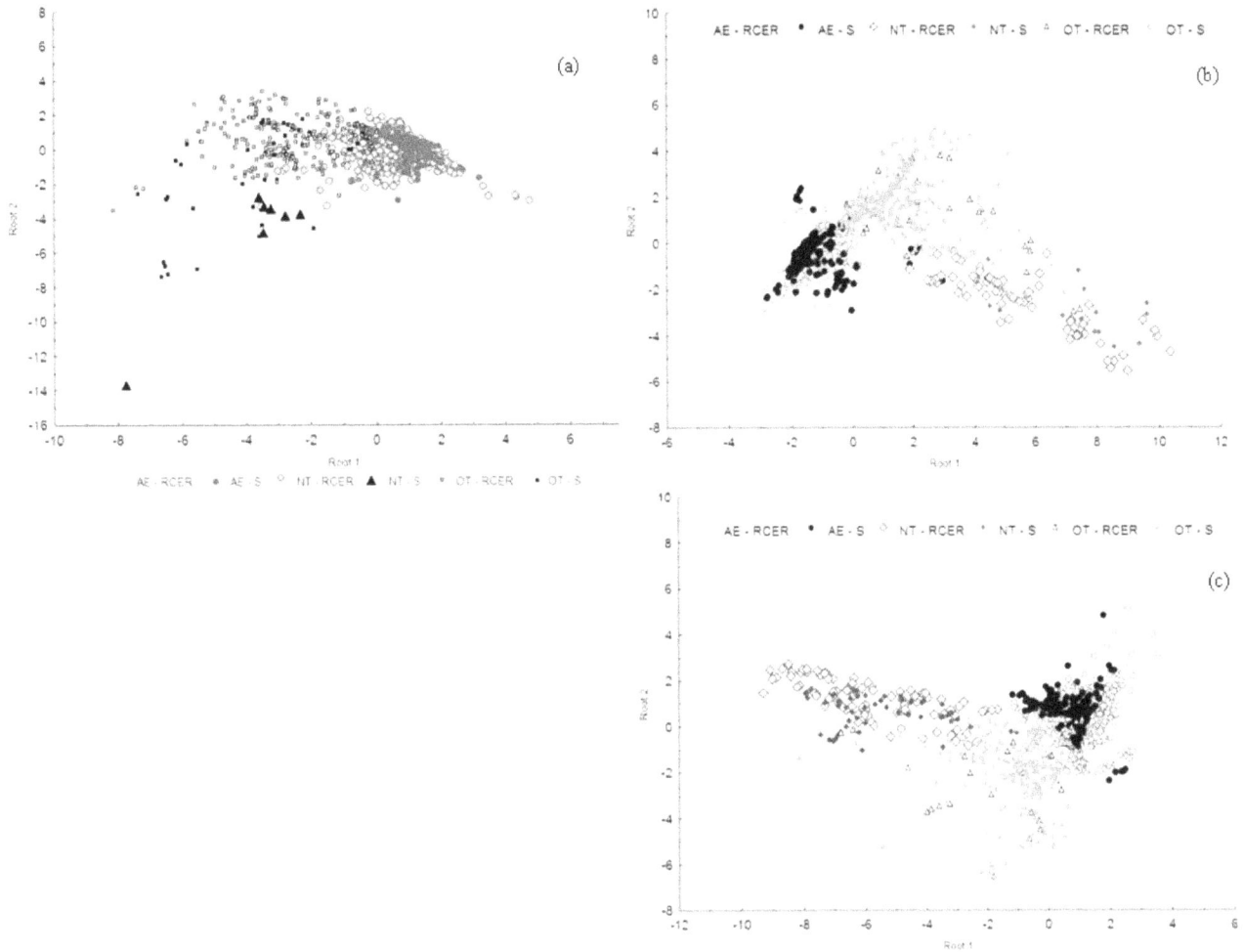

Figure 4. Discriminant function analyses differentiating deforestation spatial patterns by AE, NT and OT inside the RCER and in the RCER surrounding area (S) in 1991 (a), 1998(b) and 2007/2008 (c).

technical assistance for agricultural production. According to the RCER utilization plan, agroextractivists are allowed to increase their crop sizes at the rate of one additional ha yr^{-1}. In interviews, members of the AE group have reported that they abandon cultivation sites after the third cycle of manioc cultivation and return to the same site only after the completion of a five year to six- year fallow cycle. The shifting mosaic of crops and fallow areas associated with the AE group results primarily from slash-and-burn techniques. These techniques impoverish the soil after short periods of land use. The same site is not cultivated by an AE family for more than three years. Improved agricultural techniques could ensure that the same site was cultivated for a longer period; thus, preventing deforestation of new areas. Increasing deforestation rates by the AE group have also resulted from migrations following the provision of public services (power, transportation, health services and education) by the municipal and state governments inside the RCER and from increasing population densities of extractivists in the RCER area [27], and from the low value added that is associated with the Brazil nut production chain. AE residents have employed forest management practices with Brazil nut extraction, which favors the perpetuation of forest stands in the long-term. High densities of Brazil nut trees, as high as 104 individuals per hectare, were observed in cultivation sites by agroextractivists [37]. This fact may be associated with dispersion

of seeds by agoutis who are attracted by the mature crop [38,39], and to the abandonment of sites and protection of secondary forests that are naturally enriched with Brazil nut trees by extractivists who recognize the potential extractive value of these sites.

No significant differences were observed in the spatial patterns of deforestation caused by AE, NT or OT inside or outside the RCER. This result implies that the spatial patterns of deforestation by the three groups are not determined by the protected areas boundaries, but by the land management practices adopted by each group. Although the spatial patterns of deforestation seems to be less determined by the protected areas existing boundaries, the protected area creation has provided greater tenure security for the local extractive communities [20] in a area of low development pressure [8]. Historically, the local extractive communities have employed agricultural management practices that are less detrimental to biodiversity. They have also actively contributed to the RCER implementation. AE patches of deforestation are located near villages and far from areas cultivated by nontraditional groups. This difference reflects the tendency of the members of the AE group to protect their *colocações*. It also prevents the deforestation of the areas they occupy, by the non-traditional group of illegal occupants.

Table 2. Correlation matrix of variables used in the multivariate discriminant function analysis to differentiate deforestation caused by AE, NT and OT in 1991, 1998 and 2007/2008.

Variable	1991		1998		2007/2008	
	Axis 1	Axis 2	Axis 1	Axis 2	Axis 1	Axis 2
Area					−0.109658	0.008756
Proximit Indexy	−0.034651	0.243246	−0.015238	−0.056613	−0.148585	0.165721
Neighboord	−0.138946	−0.453934	0.079136	0.071881	0.012402	−0.139182
distance to road	0.020321	0.066654	−0.098044	0.149326	0.073256	0.06726
distance to secondary roads	0.086162	0.187847	−0.413649	0.238567	0.204088	−0.196119
distance to main rivers	−0.100146	0.293467	−0.000106	0.217821	0.056753	−0.186297
distance to any fluvial access	−0.533715	0.316146	0.232069	0.468568	−0.085927	−0.667105
distance to any access (terrestrial or fluvial)	−0.503684	0.249871	0.164824	0.550004	−0.042391	−0.649891
distance to extractivistš villages.	−0.430331	0.224412	0.357369	0.384927	−0.309962	−0.447015
distance tovegetation types	−0.41678	−0.642048	0.692978	−0.477713	−0.852393	0.248494
distance to municipality heads	−0.045537	0.211869	−0.189318	0.279947	0.338565	−0.156118
elevation	−0.482771	0.025274	0.302827	0.246627	−0.220432	−0.483216
Eigen-	2.242621	0.284196	3.251465	1.056399	4.060145	0.801868
Chi-Sqr.	1953.556	527.781	3589.97	1444.401	3525.974	1234.943
df	44	30	55	40	60	44
p-level	0	0	0	0	0.000000	0.000000

Invasion of protected areas and fine-grained spatial patterns of deforestation are rarely quantified in the Brazilian Amazon [14], but their effects are cumulative and should not be overlooked since they may be an obstacle to the implementation of protected areas in the long-term. Threats to protected areas are conspicuous and may be overlooked or poorly identified in other protected areas of low development pressure due to the lack of detailed maps and analysis of the spatial patterns of deforestation.

Acknowledgments

We are grateful to the residents of The Rio Cajarí Extrativist Reserve. For their help with revisions, we thank Dr. Helenilza Cunha, Dr. Sávio Luis Santos Carmona, Dr. Adriano Paglia. We also thank the International Institute of Education (IEB), the Programa de Pós-graducão em Biodiversidade Tropical of Universidade Federal do Amapá, and the Instituto Chico Mendes de Conservacão da Biodiversidade.

Author Contributions

Conceived and designed the experiments: AP CF. Performed the experiments: CF AP. Analyzed the data: AP CF. Contributed reagents/materials/analysis tools: CF AP. Wrote the paper: AP CF.

References

1. Parker C, Mitchell A, Trivedi M, Mardas N (2008) The Little REDD Book. Oxford, UK: The Global Canopy Programme.
2. Kintisch E (2009) Deforestation moves to the fore in Copenhagen. Science 326: 1465.
3. Nepstad D, Soares-Filho BS, Merry F, Lima A, Moutinho P, et al. (2009) Environment. The end of deforestation in the Brazilian Amazon. Science 326: 1350–1351.
4. Putz FE, Zuidema PA, Pinard MA, Boot RGA, Sayer JA, et al. (2008) Improved tropical forest management for carbon retention. PLoS Biol 6: 166.
5. Walker R, Moore NJ, Arima E, Perz S, Simmons C, et al. (2009) Protecting the Amazon with protected areas. Proc Natl Acad Sci 106: 10582–10586.11.
6. Soares-Filho B, Moutinho P, Nepstad D, Anderson A, Rodrigues H, et al. (2010) Role of Brazilian Amazon protected areas in climate change mitigation. Proc Natl Acad Sci 107: 10821–10826.12.
7. Galford GL, Melillo JM, Kicklighter DW, Cronina TW, Cerri C, et al. (2010) Greenhouse gas emissions from alternative futures of deforestation and agricultural management in the southern Amazon. Proc Natl Acad Sci 107: 19649–19654.13.
8. Bruner AG, Gullison RE, Rice RE, da Fonseca GAB (2001) Effectiveness of parks in protecting tropical biodiversity. Science 291: 125–128.
9. Nepstad D, Schwartzman S, Bamberger B, Santilli M, Ray D, et al. (2006) Inhibition of Amazon deforestation and fire by parks and indigenous lands. Conserv Biol 20: 65–73.
10. Joppa LN, Bane SR, Pimm SL (2008) On the protection of protected areas. Proc Natl Acad Sci U S A 105: 6673–6678.
11. Barber CP, Cochrane MA, Souza Jr C, Verissimo A (2012) Dynamic performance assessment of protected areas. Biol. Conserv 149: 6–14.
12. Andam S, Ferraro P, Pfaff A, Sanchez-Azofeifa G, Robalino J (2008) Measuring the effectiveness of protected area networks in reducing deforestation. Proc Natl Acad Sci USA 105: 16089–1609.10.
13. Fearnside PM (2003) Conservation Policy in the Brazilian Amazonia: understanding the dilemmas. World Dev 32: 757–77914.
14. Ricketts TH, Soares-Filho B, Fonseca GAB, Nepstad D, Pfaff A, et al. (2010) Indigenous Lands, Protected Areas, and Slowing Climate Change. PLoS Biol 8: 3.
15. Instituto Brasileiro dos Recursos Naturais Renováveis Reserva Extrativista do Rio Cajarí- Plano de utilização. Available: http://www.ibama.gov.br/resex/cajari/plano.htm. Accessed 2011 Aug 1.
16. Instituto Nacional de Pesquisas Espaciais (2012) PRODES: Assessment of deforestation in Brazilian Amazonia. Available: http://www.obt.inpe.br/prodes/. Accessed 2012 Jun 6.
17. Borges SH, Iwanaga S, Moreira M, Durigan CC (2007) Uma análise geopolítica do atual sistema de unidades de conservação na Amazônia Brasileira. Política Ambiental 4.
18. Peres CA, Barlow J, Laurance WF (2006) Detecting anthropogenic disturbance in tropical forest. Trends Ecol Evol 21: 227–229.
19. Secretaria de do Meio Ambiente do Estado do Amapá (2011) Relatório Técnico do Desmatamento no Estado do Amapá referente aos anos de 2009 a 2010 Macapá: SEMA.
20. Ioun C, Aoki Y (2007) Brazilian experiences in sustainable reserves. Brasília: Conservação Internacional 94 p.

21. Brasil (1997) Decreto s/n de 30 de setembro de 1997. Diário Oficial da União n° 189.

22. Instituto de Pesquisas Científicas e Tecnológicas do Estado do Amapá (1997) Zoneamento Ecológico e Econômico (ZEE). Escala 1: 1.000.000. Amapá: IEPA.

23. Souza EB, Cunha AC (2010) Climatologia de precipitação no Amapá e mecanismos climáticos de grande escala in Tempo, Clima e Recursos Hídricos. Technical report. Macapá: IEPA.

24. Picanço JRA (2005) Reserva Extrativista do Rio Cajari: verso e reverso da territorialização no sul do Amapá/José Reinaldo Alves Picanço. Master thesis. Univesidade Federal do Rio Grande do Norte. 158 p.

25. Brasil (2008) Atlas de Unidades de Conservação do Estado do Amapá. Macapá: SEMA.

26. Brasil (1990) Decreto n° 99.145, de 12 de março de 1990. Cria a Reserva Extrativista do rio Cajari. Available: http://www.ibama.gov.br/siucweb/mostraDocLegal.php?seq_uc = 673eseq_tp_documento = 3eseq_finaliddoc = 7. Accessed 2006 Nov 13.

27. Picanço JRA (2009) Desenvolvimento, sustentabilidade e conservação da biodiversidade na Amazonia: a produção familiar agroextrativista em áreas protegidas no sul do Amapá. Dissertation. Universidade Federal do Rio Grande do Norte. 381 p.

28. Sistema Nacional de Unidades de Conservação (2000) SNUC.Lei 9985, 18 de julho de 2000.

29. Shimabukuru YE, Novo EM, Ponzoni FJ (1998) Índice de Vegetação e Modelo Linear de Mistura Espectral no Monitoramento da Região do Pantanal. Brasília: Pesquisa Agropecuária Brasileira33: 1729–1739.

30. Shimabukuru YE, Duarte V, Dos Santos JR, Mello EMK, Moreira JC (1999) Levantamento de áreas desflorestadas na Amazônia através de processamento digital de imagens orbitais. Rio de Janeiro: Floresta e Ambiente. 6: 38–44.

31. Instituto Nacional de Pesquisas Espaciais (2008) Classificação de Imagens. Available: http://www.dpi.inpe.br/spring/usuario/c_clapix.htm. Accessed 2009 Jan 10.

32. Lins C (2001) Jarí: 70 anos de história. Rio de Janeiro: Dataforma.

33. Nasa (2005) National Aeronautics and Space Administration. Shuttle Radar Topography Mission (SRTM) Available: http://www2.jpl.nasa.gov/srtm. Accessed 2005 Feb 11.

34. McGarigal K, Cushman SA, Neel MC, Ene E (2002) FRAGSTATS: Spatial Pattern Analysis Program for Categorical Maps. Computer software program produced by the authors at the University of Massachusetts, Amherst. Available: http://www.umass.edu/landeco/research/fragstats/fragstats.html.

35. Manly BJF (2008) Métodos estatísticos multivariados: uma introdução. Porto Alegre: Bookmam. 229 p.

36. Instituto Chico Mendes de Conservação da Biodiversidade (2007) Relatório Técnico da Força Tarefa realizada na parte oeste da Reserva Extrativista do Rio Cajarí. Macapá: ICMBio.

37. Paiva PM, Guedes MC, Funi C (2010) Brazil nut conservation through shifting cultivation. For Ecol Manage 261: 508–514.

38. Cotta JN, Kainer KA, Wadt LO, Staudhammer CL (2008). Shifting cultivation effects on Brazil nut (*Bertholletia excelsa*) regeneration. For Ecol Manage 256: 28–35.

39. Myers GP, Newton AC, Melgarejo O (2000) The influence of canopy gap size on natural regeneration of Brazil nut (*Bertholletia excelsa*) in Bolivia. For Ecol Manage 127: 119–128.

4

Allocating Logging Rights in Peruvian Amazonia—Does It Matter to Be Local?

Matti Salo[1]*, Samuli Helle[1], Tuuli Toivonen[2]

1 Department of Biology, University of Turku, Turku, Finland, **2** Department of Geosciences and Geography, University of Helsinki, Helsinki, Finland

Abstract

Background: The fate of tropical forests is a global concern, yet many far-reaching decisions affecting forest resources are made locally. We explore allocation of logging rights using a case study from Loreto, Peruvian Amazonia, where millions of hectares of tropical rainforest were offered for concession in a competitive tendering process that addressed issues related to locality.

Methodology/Principal Findings: After briefly presenting the study area and the tendering process, we identify and define local and non-local actors taking part in the concession process. We then analyse their tenders, results of the tendering, and attributes of the concession areas. Our results show that there was more offer than demand for concession land in the tendering. The number of tenders the concession areas received was related to their size and geographic location in relation to the major cities, but not to their estimated timber volumes or median distances from transport routes. Small and Loreto-based actors offered lower yearly area-based fees compared to larger ones, but the offers did not significantly affect the results of the tenders. Local experience in the form of logging history or residence near the solicited concession areas, as well as being registered in the region of Loreto, improved the success of the tenders.

Conclusions/Significance: The allocation process left a considerable number of forest areas under the management of small and local actors, and if Peru is to reach its goal of zero deforestation rate by safeguarding 75 per cent of its forests by 2020, the small and the local actors need to be integrated to the forest regime as important constituents of its legitimacy.

Editor: Sharon Gursky-Doyen, Texas A&M University, United States of America

Funding: Alfred Kordelinin Säätiö, Turun Yliopistosäätiö, the Academy of Finland (grant no: 207270 to SH), Oskar Öflunds Stiftelse, Kone Foundation, and Finnish Society of Forest Science financially supported the study. The funders had no role in study design, data collection and analysis, decision to publish, or preparation of the manuscript.

Competing Interests: The authors have declared that no competing interests exist.

* E-mail: mattsal@utu.fi

Introduction

The fate of tropical forests is a global environmental issue [1], yet many far-reaching decisions concerning forest resources are made at local level [2]. Peru possesses the fourth largest tropical rainforests on Earth. The country has recently set the ambitious goals of safeguarding 75 per cent of its forests, and reducing deforestation rate to zero by 2020, through the "National Programme of Forest Conservation for Climate Change Mitigation" [3]. However, the combination of extensive forests, sparsely distributed valuable timber, difficult physical access, high level of poverty, widespread unemployment, and insufficient funding for control and monitoring have made it difficult for Peruvian authorities to efficiently enforce formal rules regulating access, logging activities, and forest management [4–6].

In Peru, all natural resources, including forests, are owned by the state as a part of the national wealth. Up until the early 2000s, the Peruvian access regime to forest resources was in effect based on, and ideally suited for, migratory selective harvest; logging permits were short in duration and small in extent [7]. A new forest law passed in 2000 [8] and implemented during the 2000s pursued to change this habit by e.g. introducing long term forest concessions as the main access mechanism to forests [5,9]. Since then, however, the Peruvian forest regime has been in a constant turmoil; new reforms have been implemented and new forest values, such as ecosystem services, introduced [10]. Renegotiation of power-relations between the state, private, and communal actors has also led to major protests against regime changes perceived unjust by local communities [4–5,9,11]. We argue that lack of locally perceived legitimacy is one of the most important underlying reasons for the troublesome implementation of the recent forest sector reforms in Peru.

In any case, more than 7 million hectares of forest concessions have to date been allocated in separate competitive tenderings in different Amazonian regions of Peru [5,9], and they are likely to remain a central part of any future access regime. Forest concessions are formal contracts between forest owner (in the Peruvian case this is the state) and another party (concession holder, concessionaire), by which the concession holder leases a right to exploit forest resources, accompanied by the obligation to manage them according to legally established principles and methods within a specified area and a specified time-frame [12]. The success or failure of the Peruvian concessions potentially has substantial long-term effects on forest disturbance rates and biological diversity in the region [13], considering that the concession period is 40 years, which, in the moment of expiration, can be renewed by the parties [8].

In this study, we examine the allocation process of logging rights through forest concessions in Loreto, Peruvian Amazonia, where more than 4 million hectares of tropical lowland rainforest were offered for tender in a competitive allocation process (tendering) in the year 2004. We first briefly describe the study area and the concession allocation process in Loreto in 2004. In continuation, we analyse a number of data sets describing the concession areas (hereafter concession units), as well as actors that took part in the tendering, their tenders, and the results of the process. In order to explore the role of locality in the allocation process, we test which attributes of the concession units were related to the number of tenders they received, and then assess the differences between local and non-local actors regarding their economic offers and their success in the tendering. Finally, we discuss the implications of locality on the current and future forest governance and its legitimacy in Amazonia.

Materials and Methods

Study area and allocation process

Loreto is the largest of the 25 Peruvian regions with an extension of 368,900 km^2, equalling the size of Germany, and comprising more than 28 percent of the Peruvian territory (Figure 1). The tropical lowland rainforests covering the area are among the most diverse ecosystems on Earth [14], and harbour a particularly high diversity of trees [15]. Despite its size, Loreto's human population is less than 1 million (<4 percent of all Peruvians), of which c. 500,000 live in and around the region's capital city Iquitos. Loreto is geographically and culturally relatively isolated and has a strong regional identity [16]. It also has a tradition of export-led forestry [17]. Although long distances and the almost complete lack of terrestrial roads form a constant challenge, forestry is one of the most important economic activities in the region. According to Tello Fernández et al. [17], forestry contributes to around 50 per cent of rural jobs, and forms more than 70 per cent of the value of all exports in Loreto.

In Loreto the forest areas eligible for concession were delimited in 2001 [9]. The tendering was opened in June 2002, and suspended the next month [18] mainly due to strong criticism from local timber companies [5]. The companies' concern was based on at least two obvious reasons: first, new competitors would potentially either enter, or emerge within, the region; and second, the cost of forest management and timber extraction would

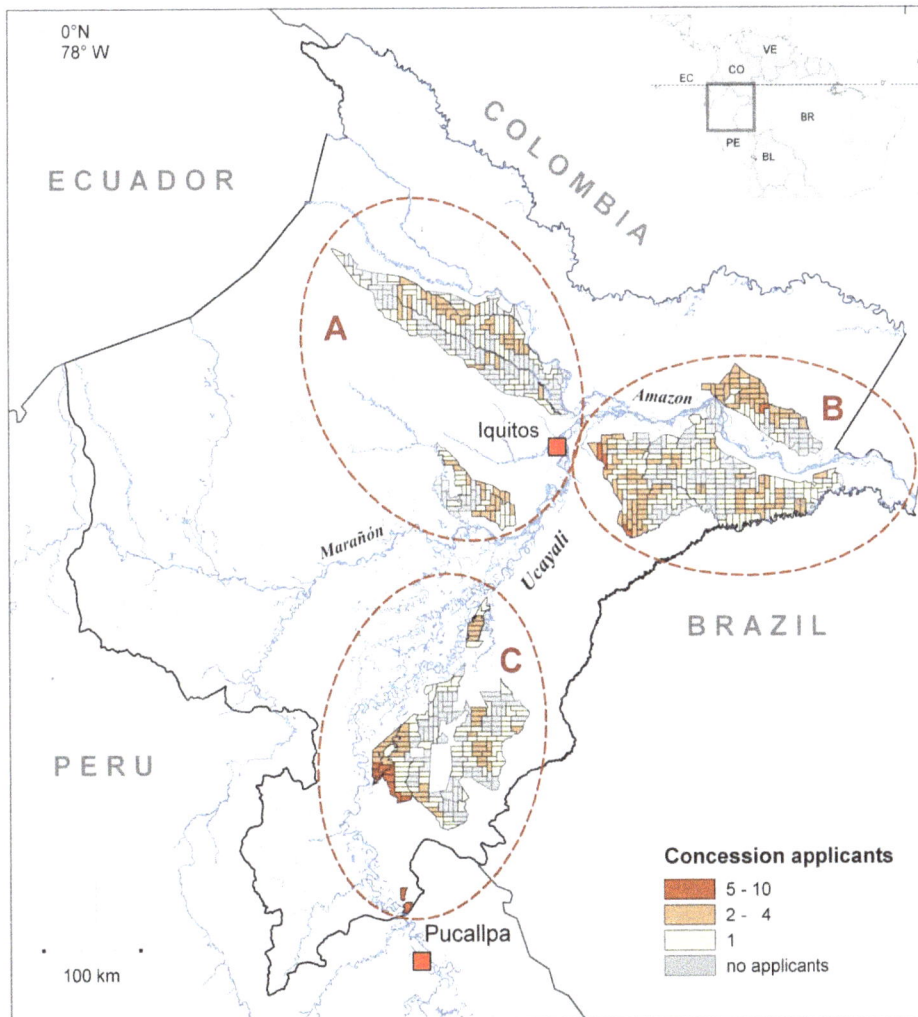

Figure 1. Forest concession units in Loreto, and the number of tenders (concession dapplicants) they received in the allocation process 2004. The cities of Iquitos and Pucallpa are marked with red squares. The concession Blocks are marked with dashed-line ovals indicated with letters A, B, and C.

potentially increase, due to new legal requirements. There was no legal restriction for foreign participation in the concession process, however the tenderings' scoring system, favouring residence or former logging contracts near the solicited concession areas, in addition to tight schedules, hindered international participation. This subtle exclusion of foreigners and the consequent inclusion of the largely informal local small-scale extractors into the formal access-regime was apparently one way to legitimise the reform.

The tendering was reopened in November 2003. Overall, 749 concession units with a total area of 4,644,163 hectares, were offered, the average size of a concession unit being 6,200 hectares (min = 5,000 ha, max = 9,944 ha). The base documents of the tendering were available at a cost of 30 Peruvian nuevos soles (c. 8.50 USD), and a total of 726 base documents were sold. Each applicant submitted a technical tender and an economic offer. The technical tender included information on the applicant's experience and locality, available assets, and work plan, with a maximum score of 100 points. An important feature of the scoring system was that it explicitly favoured local actors; the closer to the solicited concession unit the applicant resided or had past logging contracts, the higher the score obtained.

The economic offer, in turn, was the yearly area-based fee the applicant promised to pay for the total of the concession area. The score of the economic offer was calculated by dividing each offer by the highest bid in the same unit and multiplying the quotient by 100; the maximum score thus being 100. The base value (starting price) of the area-based concession fee was 0.40 USD/ha/year [19], which was uniform in all of the concession units nationwide regardless of their location or production potential.

In order to weight the technical tender in comparison to the economic offer, the technical tender was valued as having a weight of 90 per cent of the total score of the applicant, while the economic offer was worth 10 per cent. This decision was made in order to avoid the tendering turning into an auction, which would arguably discriminate against local actors with limited capital assets. An auction could also potentially form an incentive for unrealistically high offers that had been seen in former allocation processes in Madre de Dios and Ucayali regions, where the first tenderings gave more weight to the economic offer [20].

The total number of actors taking part in the tendering was 328, and 64 per cent (n = 211) of them won a concession. The most numerous group of the participants were individuals (70 per cent, n = 230), whereas companies constituted a quarter of the applicants (25 per cent, n = 82), the remainder being partnerships (associations of two or more individuals committed to form a formal association or company in case of winning a concession; 5 per cent, n = 15) and one Non-Governmental Organisation (NGO). The tendering was concluded in May 2004, and the vast majority, 98 per cent (n = 206), proceeded to sign a concession contract within the next year.

Data and variables used in the analysis of the tendering

We first classified the actors' locality in order to test differences between locals' and outsiders' behaviour and success in the tendering. The distinction between locals and outsiders is common in literature, but as a concept it is often troublesome and vaguely elaborated [21]. We used a combination of two different criteria derived from separate sources. First, we made a distinction between actors that were registered in Loreto and the ones that were registered outside the region; and second, we classified the same actors in two groups according to their direct experience related to the surroundings of the concession units that they solicited. For the analysis, we cross-classified the applicants taking part in the tendering into four locality classes as presented in Table 1.

For the first classification, we used the internet database of the Peruvian taxation authorities, SUNAT (Superintendencia Nacional de Administración Tributaria) [22]. We retrieved the data using the name of the applicant (individual, company, or partnership) as a key word, and as a result we created a data set with the registered office or domicile of all formalised economic actors taking part in the tendering. All applicants with an office or domicile registered by SUNAT in the region of Loreto were labelled 'Loretans' and the ones registered in any other region or outside Peru were labelled 'outsiders'.

The second classification was based on the tendering data provided by INRENA (Instituto Nacional de Recursos Naturales), an institution under the Ministry of Agriculture and at the time of the tendering directly responsible for the management of renewable natural resources. In our classification all participants either residing or with past logging contracts within the watershed where they solicited a concession unit were labelled 'locals', whereas all the rest were labelled 'non-locals'. We refer to "watersheds" as a category used by the forest authorities in order to classify the applicants. We chose the watershed level, as used by INRENA in the tendering scoring system, to be the appropriate spatial resolution for this classification because Loreto lacks a large-scale terrestrial road network and practically all timber logged is transported fluvially. Thus the mouths of the rivers are currently the most feasible, if not the only, locations for systematically controlling the transportation of logged timber.

In the analysis, we first wanted to explore the general popularity of the different concession units by measuring the concession popularity as the number of tenders the units received. In order to study the effects of the geographical location of the units, we attached six attributes to the concession units: concession Block, surface area (in hectares), skidding distance (km), distance to closest city (km), closest city (Iquitos or Pucallpa), and estimated timber volume (m^3) (Table 2). The division of concession Blocks is shown in Figure 1. Skidding distance describes the median Euclidean distance to the nearest river, which we use as a proxy for the cost of transporting felled logs to the primary transport routes formed by rivers. The skidding distance was defined based on a river data set manually digitised from 30-meter resolution Landsat TM imagery by a Finnish-Peruvian environmental cooperation project, Biodamaz [23]. Although river access depends ultimately on water levels in tributary waterways, of which many are too small to be included in our data, our analysis is more realistic and detailed than the analyses using the distance as the crow flies as a proxy for accessibility. The same river network data was used to measure the distances from the concession units to the main timber trade centres, the cities of Iquitos and Pucallpa. The distance from both cities along the river network were calculated using the cost-distance function of ArcGIS 9.2, and the resulting distance to the closest of these cities was stored as the variable 'distance to closest city'. In addition, the variable 'closest city' had a value of either

Table 1. Classification of the locality of the applicants based on INRENA tendering data and SUNAT database [22].

	SUNAT 'Loretans'	SUNAT 'outsiders'
INRENA 'locals'	local Loretans	local outsiders
INRENA 'non-locals'	non-local Loretans	non-local outsiders

Table 2. Attributes of the applicants and the concession units used as explanatory variables in the models.

	Attribute (explanatory variable)	Classes/units
Applicants	locality	local Loretans, non-local Loretans, local outsiders, non-local outsiders
	scale	small extractors, medium-sized actors
	type	individuals, companies, partnerships
	area-based fee	USD/ha/year
Concession units	concession Block	A, B, C (see Figure 1)
	concession unit area	≤6004 ha, >6004* ha (*median)
	skidding distance	≤6.65 km, >6.65* km (*median)
	distance to closest city	<500 km, 500–700 km, >700 km
	closest city	Iquitos, Pucallpa
	estimated timber volume/ha	<80 m^3, 80 m^3

Iquitos or Pucallpa, depending on which of these two cities was closer, via river, to the concession unit in question.

The base documents of the tendering provided a description of forest types found within the concession units, and the area these types covered in each of the units. This information was based on inventories carried out by INRENA [24] and contained volumetric estimations of timber resources technically available in different forest types. The data provided theoretical upper and lower limits for timber volumes, and we used the lower figures to calculate an estimate of minimum timber volume per hectare technically available for each concession unit. We acknowledge limitations in the quality of this data set due to the preliminary nature of its analysis and inaccurate input data used. However, we included the variable 'estimated timber volume' in our models because the data was publicly available for all the applicants, and thus potentially influenced their decisions.

We were also interested in the variables related to the applicants' economic offers, and to their probabilities of winning a concession. Thus, we selected the yearly area-based fee offered and the result of the tender as response variables in our analyses. We decided to study the economic offer, expressed as an area-based fee per year, because it reflects the actor's willingness to pay for the concession rights in the long term. The actor's success, expressed as the probability of winning a concession, was studied because it can be used to assess the tendering system's potential bias towards local actors. 'Area-based fee' was also used as an explanatory variable when the probability of winning was modelled. The variables stored as attributes of the applicants and the concession units are presented in Table 2.

Due to the type and the scale of the actors taking part in the tendering being potentially related to the economic offers they make, we classified the applicants according to scale and type. The participants could either solicit only one concession unit, or more than one. In the former case the applicant belonged to the scale class 'small extractors' and in the latter to the class 'medium-sized actors'. In the tendering scoring system, small extractors were favoured by higher scores. Furthermore, the variable 'type of actor' was classified into three categories: 'individual', 'company', and 'partnership'. The economic offer of the applicant in USD/ha/year was stored as variable 'area-based fee'. Moreover, in addition to the attributes of the applicants, we wanted to capture the possible effects of the geographic location, size, and accessibility of the concession units on the economic offers the participants made, and therefore we used the variables based on the concession units' attributes described above.

Statistical analyses

In order to study the behaviour of the different actors taking part in the tendering, we examined whether the explanatory variables (Table 2) were associated with the number of tenders per concession unit, the area-based fees offered by the applicants, and the applicants' probability in winning the solicited concession unit. The associations between the explanatory variables and the number of tenders the concession units received, and the area-based fee offered by the applicants, were examined with regression models assuming Poisson distributed errors and log link function, since these responses were count variables. The influence of the explanatory variables on whether the applicant won the concession unit or not was examined using logistic regression model, with binomial errors and logit link function. When needed, Pearson's χ^2 was used to rescale the parameter covariance matrix to adjust for any under- or overdispersion.

Prior to the analyses, the variables of 'concession unit area', 'median skidding distance', and 'estimated timber volume' were divided in two categories, based on the variables' median values (to distinguish between large and small, close and remote, and abundant and scarce units). Meanwhile, the variable 'distance to the closest city' was divided in three classes, to detect the possible effect of remoteness (implying less extraction pressure and thus more available timber on the one hand, and more difficult access on the other, which we hypothesised could favour intermediate distances). The breaking points of these classes are presented in the Table 2. All concession units were included in the analysis of popularity, but only the units that received tenders were included in the analyses of economic offers and tendering success. To account for the facts that some applicants made a tender for several concessions, and several concession units had multiple applicants, we applied Generalized Estimating Equations (GEE) to account for the respective correlation structure in the models for the area-based fees offered by the applicants, and for the probability of the applicants for winning the solicited concession unit [25]. Unstructured working correlation matrix, with concession identity nested within applicant identity, was used to accomodate these correlations [25]. Statistical inference was based on Score test [25]. No stepwise model reduction was applied because such methods dramatically increase the rate of type I errors [26], and because our aim here was to obtain the most accurate point estimates and their confidence intervals (CI) [27]. In the case of statistically significant association between categorical variables having more than two levels and the response, statistical interpretation of the group-differences was

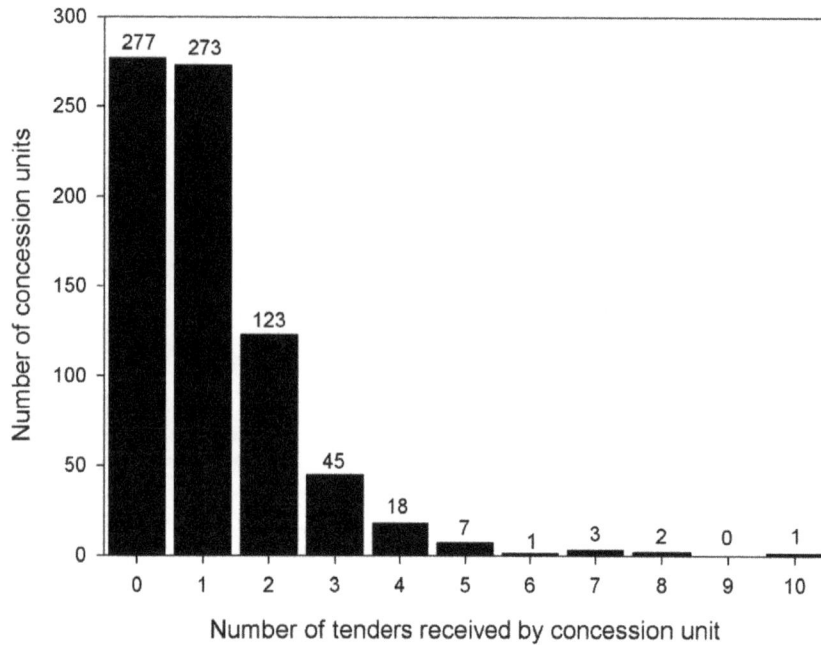

Figure 2. The distribution of the number of tenders received by the concession units (min 0; max 10).

based on the 95% confidence intervals of the means. For example, if the 95% confidence intervals of means overlap half the length of one arm, this corresponds approximately to statistical significance at p = 0.05 [28]. All analyses were conducted with SAS statistical software version 9.2 (SAS Institute Inc, Cary, North Carolina, USA).

Results

Demand and offer of forest land: popularity of concession units

The allocation process revealed that there was more offer than demand for concession units in Loreto; up to 37 per cent (n = 277) of all the units did not receive a tender at all and only 27 per cent (n = 200) of the units were subject to competition between two or more applicants. A mere 4 per cent of the units received tenders from more than 3 applicants, the maximum being 10 tenders per unit (Figures 1 and 2).

According to our analysis, the geographical location and the total area of the concession units had a significant effect on their popularity (Table 3; concession Blocks presented in Figure 1). Units in Block B were the most popular, having on average 1.23 tenders per unit (95% confidence intervals [CIs] = 0.96, 1.59), whereas units in Blocks A and C received on average 0.76 tenders (95% CIs = 0.58, 0.99) and 1.06 (95% CIs = 0.87, 1.29) per unit, respectively. Larger concession units received more tenders than smaller ones (Table 3). That is, large concession units had on average 1.10 tenders (95% CIs = 0.88, 1.37) compared to, an average of 0.91 tenders (95% CIs = 0.74, 1.11) in small units.

The river distance to the closest city (Iquitos or Pucallpa) was also significantly related to the popularity of a given concession unit (Table 3). Concession units at intermediate distances (500–700 km) from cities received the highest number of tenders (on average 1.28 tenders per unit, 95% CIs = 1.02, 1.62), while the units located closer (on average 0.96 tenders per unit, 95% CIs = 0.79, 1.16) or further from the cities (on average 0.81 tenders per unit 95% CIs = 0.60, 1.07) were less popular. This is consistent

with our hypothesis of preference for intermediate distances because of proximity to cities, implying more extraction pressure – and as a consequence less timber left – on the one hand, and growing distance implying more difficult access, on the other. Meanwhile, whether the concession unit was located closer to Iquitos or Pucallpa, estimated timber volumes and median skidding distances of the concession units were not statistically related to the number of tenders they received (Table 3).

Willingness to pay: area-based fees

The highest area-based fee offered was 1.48 USD/ha/year, and the highest winning offer was 1.30 USD/ha/year. The offered area-based fees were related to the locality of the applicants (Table 4). Figure 3 shows that, on average, non-local outsiders offered the highest yearly fees per hectare of concession, contrasting to the non-local Loretans who offered the lowest fees. The economic offers were also related to the scale of actor (Table 4); small extractors offered, on average, lower area-based fees than medium-sized actors (0.52 USD/ha/year [95% CIs = 0.48, 0.57] vs. 0.64 USD/ha/year [95% CIs = 0.60, 0.68], respectively). The type of applicant was also associated with the area-based fees offered (Table 4): partnerships

Table 3. The effect of explanatory variables on the number of tenders received by the concession units.

Predictor	$df_{num,den}$	F	P
Distance to the closest city	2, 739	6.15	0.0023
Closest city	1, 739	0.07	0.80
Concession Block	2, 739	10.14	<0.0001
Concession unit area	1, 739	5.12	0.024
Estimated timber volume/ha	1, 739	0.35	0.55
Median skidding distance	1, 739	0.60	0.44

Table 4. The effect of explanatory variables on the area-based fees (USD/ha/year) offered by the applicants.

Predictor	df	χ^2	P
Locality	3	52.8	<0.0001
Type of actor	2	7.85	0.02
Scale of actor	1	30.0	<0.0001
Closest city	1	6.62	0.01
Concession Block	2	17.2	0.0002
Concession unit area	1	1.69	0.19
Median skidding distance	1	0.41	0.52
Estimated timber volume/ha	1	4.78	0.029
Distance to the closest city	2	8.87	0.012

(0.60 USD/ha/year [95% CIs = 0.55, 0.65] and individuals (0.59 USD/ha/year [95% CIs = 0.55, 0.63] made almost equally high offers, whereas companies offered less (0.55 USD/ha/year [95% CIs = 0.50, 0.59].

Concession units in certain Blocks received higher offers than units in other Blocks (Table 4). The units in Block C received the highest offers; an average of 0.65 USD/ha/year (95% CIs = 0.62, 0.72, whereas units in Blocks A and B received on average offers of 0.52 (95% CIs = 0.47, 0.58) and 0.55 USD/ha/year (95% CIs = 0.50, 0.61), respectively. Distance to the closest city also had a significant effect on the area-based fees offered (Table 4); concession units with short (0.60 USD/ha/year [95% CIs = 0.56, 0.64] and intermediate (0.60 USD/ha/year [95% CIs = 0.56, 0.65] distances from the closest city received, on average, higher offers than those located further (0.53 USD/ha/year [95% CIs = 0.48, 0.59]). Whether the closest city was Iquitos or Pucallpa influenced area-based fees as well, since in the case of Iquitos the area-based fees were, on average, lower (0.53 USD/ha/year

[95% CIs = 0.51, 0.55]) than in the case of Pucallpa (on average 0.63 USD/ha/year [95% CIs = 0.55, 0.71]). Moreover, higher estimated timber volumes per hectare attracted somewhat higher offers than lower ones (0.60 [95% CIs = 0.57, 0.64] USD/ha/year vs. 0.55 [95% CIs = 0.50, 0.61]). Neither the distance to the river network (median skidding distance) nor the concession unit area were statistically related to the area-based fees offered.

Winners and losers: success of tenders

Table 5 shows that Local Loretans won more than half (1.56 million ha) of all the concession land that was finally allocated in the tendering (2.58 million ha). There was a statistically significant relationship between the locality and the success of the applicants (Table 6). Local Loretans had the highest probability of winning, compared to non-local Loretans, local outsiders, and non-local outsiders (Figure 4). It thus seems that in addition to applicants having previous ties to the forest areas they solicited, Loretans in general benefitted from the tendering's scoring system. Applicants registered outside Loreto, whether or not they were considered locals, seemed to have lower probabilities of winning a concession.

Although medium-sized actors won the majority of the concession land (1.84 million ha), a considerable area (0.82 million ha) was also allocated to those small-scale extractors only soliciting one concession unit (Table 5). Neither the area-based fees offered by the applicants, nor the type or scale of actor, were statistically related to the applicants' success (Table 6).

Discussion

Our analysis revealed three general tendencies in the forest concession allocation process of Loreto. First, there was little true competition in the tendering; second, the area-based fees offered did not significantly affect the results of the tenders; and third, the Loreto-based applicants had a higher probability of winning a concession unit than those based outside Loreto.

In the end, only a quarter of all the offered concession units were subject to two or more competing tenders. However, without

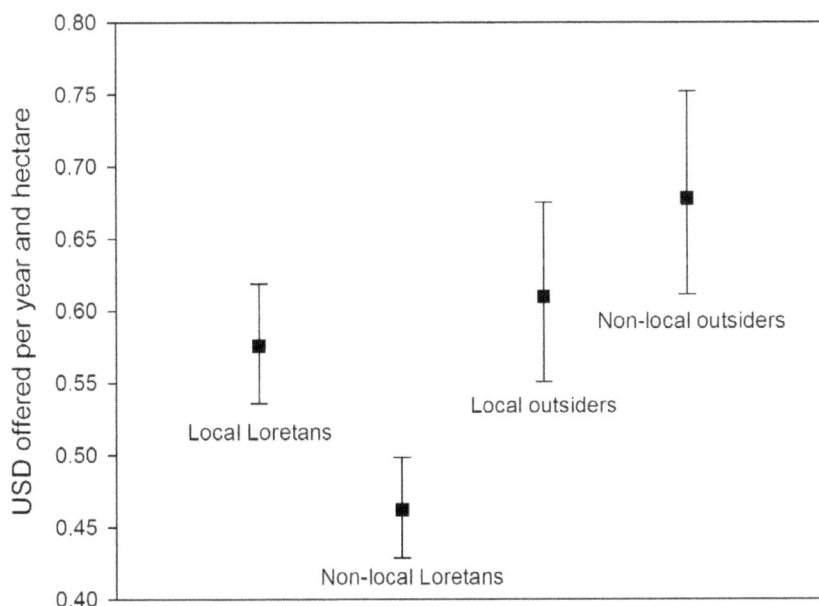

Figure 3. The differences between the offered area-based fees (USD/ha/year) of the four locality groups studied. Squares represent estimated marginal means and error bars their 95% confidence limits.

Table 5. The number of concession units and concession area (hectares) applied and won by applicants representing different locality and scale classes.

	Units applied	Units won	Win%	Hectares applied	Hectares won	Win%
Locality						
Local Loretans	380	250	66	2.376,520	1.564,015	66
Non-local Loretans	156	89	57	973,099	536,962	55
Local outsiders	100	42	42	668,699	276,101	41
Non-local outsiders	131	32	24	827,845	206,731	25
Total	*767*	*413*	*54*	*4.846,163*	*2.583,809*	*53*
Scale						
Small	227	126	56	1.483,923	821,621	55
Medium-scale	591	299	51	3.680,719	1.840,677	50
Total	*818*	*425*	*52*	*5.164,642*	*2.662,298*	*52*

the rules of the game clearly favouring the small and the local, the process could arguably have lost its legitimacy before it was even initiated. Previous examples show that it would have also been possible to restrict the participation in the process only to small actors, as was the case in two tenderings organized in the Madre de Dios and Ucayali regions in 2003 [20]. In Loreto, however, the allocation system turned out to emphasise the local experience of the participants rather than their size. Actors registered in Loreto and with local experience in the form of previous logging contracts, or residence near the solicited concession units, achieved the largest share of the units and concession area. This indicates that the allocation method in general, and the scoring system in particular, were successfully designed to serve the purpose of favouring applicants that had already established ties with the offered forest areas.

While this was one way in which the process succeeded in achieving at least some degree of legitimacy within the region, the tendering simultaneously failed to fundamentally change the power-relations regulating access to timber resources in Peruvian Amazonia [4–5,29–31]. What was important for the forest industrialists was to guarantee a steady flow of timber, and local small-scale extractors depending on locally organized chains of trade are vital for this kind of supply. What these small-scale extractors needed was to bolster their direct access to the forest. Although the allocation process left large areas of forest in the hands of small actors, it is not certain to which degree things have changed in the field. According to a recent report published by the Environmental Investigation Agency [4], with the concessions in function for several years now, it is still commonplace in Loreto that urban timber merchants equip small-scale extractors by advance payments which frequently feed a circle of debt and impoverishment.

Another anxiety that was frequently voiced before the tendering was that of forest inventories, operative plans, and management planning required by the forest law, proving to be prohibitively costly and thus making concessions not attainable for small extractors [32]. This viewpoint cannot be ruled out as an explanation for the low level of competition in the tendering. Furthermore, the possibility of small and local actors being used only as legal representatives of larger timber merchants or companies cannot be straightforwardly ruled out, but according to our analysis, the fear of large national or foreign capital overwhelming small and local applicants – a common concern before the tendering [33] – proved to be unfounded. While the

concession rights are transferable, i.e. the contracts can be further traded, the commitment of local actors can be reinforced through new options based on a wider variety of forest values. New approaches embedded in the future forest regime could contribute positively to the Peruvian efforts to halt forest degradation in and around logging areas. There are several experiences that can be studied to identify such approaches.

Policy implications

Most Amazonian countries face problems similar to those of Peru, and many have undergone forest regime reforms during the last 15 years. In Ecuador and Colombia, the forest regime is in need of reform, but the lack of detailed analyses of their particular characteristics hinders comparisons to other Amazonian countries' forest sector reforms. Neither of these two countries currently applies long-term forest concessions as a major administrative arrangement. In Colombia, a new forest law decreed in 2006 introduced a system based on forest concessions, but the law was declared unconstitutional and revoked in 2008 because its preparation did not adequately address issues related to consultation of local and indigenous communities [34]. In Ecuador, the forest policy development has recently been described as unpredictable [35]. In Bolivia and Brazil, large scale forest concessions have been implemented, and their reforms have received more attention internationally [35–36], but comparative studies between Amazonian countries are still largely lacking.

Recently, major efforts to reshape the forest sector in Peru have been a consequence of international agreements binding the Peruvian government to reform forest legislation while also implementing and enforcing the current rules more efficiently [4,10,37–38]. Particularly the free trade agreement between Peru

Table 6. The effect of explanatory variables on whether or not the applicant's tender won the race for a concession.

Predictor	df	χ^2	P
Locality	3	19.55	0.0002
Type of actor	2	0.11	0.95
Scale of actor	1	2.45	0.12
Area-based fee	1	0.37	0.55

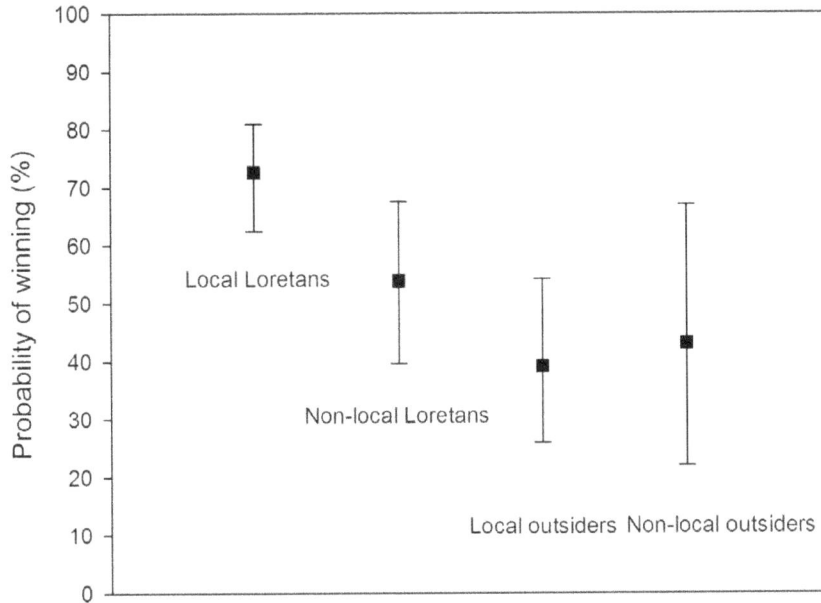

Figure 4. The differences between the probabilities (%) of the applicants for winning a concession unit in the four locality groups studied. Squares represent estimated marginal means and error bars their 95% confidence limits.

and the US has driven changes in the Peruvian legislation [4]. A package of laws related to the free trade agreement, including a new forest law, was approved in 2008. The new law ('Legislative decree 1090') included mechanisms aimed at opening Amazonia for new investment, such as forest concessions solicited through private initiative, and it also enabled drastic changes in land-use designations. In Amazonia, and particularly among the region's indigenous population, these changes were commonly seen as intended to facilitate the privatisation of indigenous peoples' traditional lands. Consequently, protests culminating in tragic acts of violence in northern Peru between state agents and indigenous protesters in June 2009 forced the government to revoke the most controversial of these laws, including the forest law [10].

Currently, a draft of a new forest law is being discussed in a process claimed by the government to be more participatory than the previous one [4,39], yet controversies still remain as to the implementation of the reform, not least regarding the indigenous and local communities' land rights issues and the concessionaires' ability and willingness to follow the law [40]. As a part of any future regime, Peru is in need of new approaches integrating wider environmental values also to the forest concession mechanism. Today, fortunately, there is more diversity than ever of possible additional values that can be directly linked to sustainable forest management. Peruvian forest concessions are intended to be a long-term commitment to forest management; in addition to the timber and non-timber resources that the forests under concession contain, they also entail vital, albeit hard-to-value, ecosystem services [10,41].

Forest concessions will most likely form an important part of the National Programme of Forest Conservation for Climate Change Mitigation [3], promoted by the Peruvian government. The kind of arrangements that will be applied as a part of the Peruvian programme in areas surrounding the forest concessions will certainly affect their feasibility, and vice versa. The Peruvian programme aims at safeguarding the forest cover in an area of 54 million hectares, representing 75 per cent of Peruvian forests. The programme is also planned to include direct area-based payments to its main beneficiaries for conserving forests in their possession. The beneficiaries are defined as "[…] the entitled native and rural communities and population that lives in and around the tropical Amazonian and dry forests of the country" [3]. It remains to be seen how this will be achieved, and what the role of local small scale forest concessionaires will be in this deal.

Acknowledgments

We would like to thank Jukka Salo, Risto Kalliola, Anders Sirén, and Sanna-Kaisa Juvonen, and Aili Pyhälä for their important comments on different versions of the manuscript.

Author Contributions

Conceived and designed the experiments: MS. Analyzed the data: MS SH TT. Wrote the paper: MS. Collected the primary data: MS. Carried out the analysis: SH TT MS.

References

1. Malhi Y, Roberts J, Betts R, Killeen T, Li W, et al. (2008) Climate change, deforestation, and the fate of the Amazon. Science 319: 169–172.
2. Persha L, Agrawal A, Chhatre A (2011) Social and Ecological Synergy: Local Rulemaking, Forest Livelihoods, and Biodiversity Conservation. Science 331: 1606–1608.
3. Peru (2010) Decreto Supremo 008–2010–MINAM.
4. Environmental Investigation Agency (2010) Peru's Forest Sector: Ready for the New International Landscape? Available: http://www.illegal-logging.info/uploads/ReportPeruForestMay10ENG.pdf. Accessed 15 Sep 2010.
5. Smith J, Colan V, Sabogal C, Snook L (2006) Why policy reforms fail to improve logging practices: The role of governance and norms in Peru. Forest Policy Econ 8: 458–469.
6. Chirinos C, Ruiz M (2003) Informe final del estudio: Desarrollo e implementación de lineamientos de control de la extracción ilegal para un manejo forestal sostenible en el Perú. ITTC, Ciudad de Panamá, Panama.
7. Barrantes R, Trivelli C (1996) Bosques y madera. Análisis económico del caso peruano. Lima: Instituto de Estudios Peruanos/Consorcio de Investigación Económica. 121 p.

8. Peru (2001) Aprueban el Reglamento de la Ley Forestal y de Fauna Silvestre, Decreto Supremo 014-2001-AG. Diario Oficial El Peruano XIX, 201105–201146.

9. Salo M, Toivonen T (2009) Tropical timber rush in Peruvian Amazonia: spatial allocation of forest concessions in an uninventoried frontier. Environ Manage 44: 609–623.

10. Hiedanpää J, Kotilainen J, Salo M (2011) Unfolding the organised irresponsibility: Ecosystem approach and the quest for forest biodiversity in Finland, Peru, and Russia. Forest Policy Econ 13: 159–165.

11. Hughes N (2010) Indigenous Protest in Peru: The 'Orchard Dog' Bites Back. Social Movement Studies 9: 85–90.

12. Gray JA (2002) Forest Concession Policies and Revenue Systems: Country Experience and Policy Changes for Sustainable Tropical Forestry. Washington DC: World Bank. 107 p.

13. Oliveira PJC, Asner GP, Knapp DE, Almeyda A, Galvan-Gildemeister R, et al. (2007) Land-use allocation protects the Peruvian Amazon. Science 317: 1233–1236.

14. Bass MS, Finer M, Jenkins CN, Kreft H, Cisneros-Heredia DF, et al. (2010) Global Conservation Significance of Ecuador's Yasuní National Park. PLoS ONE 5(1): e8767.

15. Pitman NCA, Mogollón H, Dávila N, Ríos M, García-Villacorta R, et al. (2008) Tree Community Change across 700 km of Lowland Amazonian Forest from the Andean Foothills to Brazil. Biotropica 40: 525–535.

16. Chirif A (2002) El frente patriótico de Loreto: fortalezas y debilidades. Quehacer 135: 62–73.

17. Tello Fernández H, Quevedo Guevara A, Gasché J (2004) Sistema de incentivos para el manejo de bosques de Loreto: El caso de los recursos forestales maderables. Iquitos: IIAP – CIES. 170 p.

18. Peru (2002) Decreto Supremo 046-2002-AG.

19. Peru (2003) Resolución Jefatural 050-2003-INRENA.

20. Hidalgo J (2003) Estado de la situación forestal en el Perú. Seminario Permanente de Investigación Agraria (SEPIA) X, Lima. 52 p.

21. Moseley C, Reyes Y (2008) Forest restoration and forest communities: Have local communities benefited from Forest Service contracting of ecosystem management? Environ Manage 42: 327–343.

22. SUNAT (2006) Consulta de RUC. Available http://www.sunat.gob.pe/cl-ti-itmrconsruc/jcrS00Alias. Accessed 15 May 2006.

23. IIAP–Biodamaz (2004) Estrategia regional de la diversidad biológica amazónica. Iquitos: Instituto de Investigaciones de la Amazonía Peruana. 82 p.

24. INRENA (2004) Mapificación y evaluación forestal del Bosque de Producción Permanente del Departamento de Loreto – Documento de Trabajo, Lima, Peru.

25. Lipsitz S, Fitzmaurice G (2009) Generalized estimating equations for longitudinal data analysi. In Fitzmaurice G, Davidian M, Verbeke G, Molenberghs G, eds. Longitudinal Data Analysis, Chapman & Hall/CRC, Boca Raton, USA. pp 43–78.

26. Mundry R, Nunn C (2009) Stepwise Model Fitting and Statistical Inference: Turning Noise into Signal Pollution. Am Nat 173: 119–123.

27. Harrell FJ (2001) Regression Modelling Strategies: With Applications to Linear Models, Logistic Regression and Survival Analysis. New York: Springer. 568 p.

28. Cumming G (2009) Inference by eye: Reading the overlap of independent confidence intervals. Stat Med 28: 205–220.

29. Dourojeanni M, Barandiarán A, Dourojeanni D Amazonía peruana en 2021: Explotación de recursos naturales e infraestructuras: ¿Qué está pasando? ¿Qué es lo que significan para el futuro? ProNaturaleza, Lima. 162 p.

30. Defensoría del Pueblo (2010) La Política Forestal y la Amazonía Peruana: Avances y obstáculos en el camino hacia la sostenibilidad. Serie Informes Defensoriales – Informe 151, Lima. 298 p.

31. Derecho Ambiente y Recursos Naturales (DAR) (2011) Informe Anual 2010: Transparencia en el Sector Forestal Peruano, Lima. 68 p.

32. La Región (2002) Costos de los inventarios del bosque, planes operativos y plan de manejo son elevados. Concesiones forestales no están al alcance de pequeños extractors madereros. Iquitos: 22 Apr. 3 p.

33. Iquitos al Día (2004) Las concesiones forestales: Muchas posiciones nada de claridad. Iquitos: 30 Apr. 6 p.

34. Arcila Rueda JL, Velásquez Arredondo HI (2009) El precedente jurisprudencial y los instrumentos de regulación ambiental del sector eléctrico colombiano. Energética 42: 53–61.

35. Ebeling J, Yasué M (2009) The effectiveness of market-based conservation in the tropics: Forest certification in Ecuador and Bolivia. Journal of Environmental Management 90: 1145–1153.

36. Bauch S, Sills E, Estraviz Rodriguez LC, et al. (2009) Forest Policy Reform in Brazil. Journal of Forestry 107: 132–138.

37. Blundell AG (2007) Implementing CITES regulations for timber. Ecol Appl 7: 323–330.

38. Youatt A, Cmar T (2009) The fight for red gold: ending illegal mahogany Trade from Peru. Natural Resources & Environment 23: 19–23.

39. Peru (2010) Proyecto de Ley Forestal y de Fauna Silvestre [Draft of the new forest law submitted for parliamentary evaluation 22.6.2010]. Accessible: http://www.actualidadambiental.pe/documentos/proyecto_ley_FFS.pdf. Accessed 18 Mar 2011.

40. International Tropical Timber Organization (2010) Tropical Timber Market Report 15: 18, 16–30 Sep.

41. Baker TR, Jones JPG, Rendón Thompson OR, Román Cuesta RN, del Castillo D, et al. (2010) How can ecologists help realise the potential of payments for carbon in tropical forest countries? J Appl Ecol, doi: 10.1111/j.1365-2664.2010.01885.x.

5

Relating Demographic Characteristics of a Small Mammal to Remotely Sensed Forest-Stand Condition

Hania Lada*, James R. Thomson, Shaun C. Cunningham, Ralph Mac Nally

School of Biological Sciences, Monash University, Melbourne, Victoria, Australia

Abstract

Many ecological systems around the world are changing rapidly in response to direct (land-use change) and indirect (climate change) human actions. We need tools to assess dynamically, and over appropriate management scales, condition of ecosystems and their responses to potential mitigation of pressures. Using a validated model, we determined whether stand condition of floodplain forests is related to densities of a small mammal (a carnivorous marsupial, *Antechinus flavipes*) in 60 000 ha of extant river red gum (*Eucalyptus camaldulensis*) forests in south-eastern Australia in 2004, 2005 and 2011. Stand condition was assessed remotely using models built from ground assessments of stand condition and satellite-derived reflectance. Other covariates, such as volumes of fallen timber, distances to floods, rainfall and life stages were included in the model. Trapping of animals was conducted at 272 plots (0.25 ha) across the region. Densities of second-year females (i.e. females that had survived to a second breeding year) and of second-year females with suckled teats (i.e. inferred to have been successful mothers) were higher in stands with the highest condition. There was no evidence of a relationship with stand condition for males or all females. These outcomes show that remotely-sensed estimates of stand condition (here floodplain forests) are relatable to some demographic characteristics of a small mammal species, and may provide useful information about the capacity of ecosystems to support animal populations. Over-regulation of large, lowland rivers has led to declines in many facets of floodplain function. If management of water resources continues as it has in recent decades, then our results suggest that there will be further deterioration in stand condition and a decreased capacity for female yellow-footed antechinuses to breed multiple times.

Editor: Francisco Moreira, Institute of Agronomy, University of Lisbon, Portugal

Funding: The Australian Research Council supported this research (DP0984170, DP120100797); http://www.arc.gov.au/. The funders had no role in study design, data collection and analysis, decision to publish, or preparation of the manuscript.

Competing Interests: The authors have declared that no competing interests exist.

* E-mail: hania.lada@monash.edu

Introduction

Fast, extensive change now is the dominant characteristic of ecological systems across the world. This is due to climate change, many direct actions of humans, and myriad indirect effects arising from those actions [1,2,3]. Rates of change are so rapid and extents so large (e.g. forest dieback increased from 45% to 70% in 16 years over 100 000 ha of Murray River floodplains in Australia [4]) that new methods for evaluating and tracking ecosystem change are required to anticipate and potentially to mitigate undesirable ecological outcomes. Traditional methods of field-based surveys require many years to cover large areas but the time-scales of threats (land clearance, fires, dam building) usually are much shorter than the intervals between assessments [5,6].

Remote sensing offers the capacity to represent dynamically, and at appropriate spatial scales, the condition of surrogates (e.g. land use or vegetation condition) and then to project changes in biodiversity as the surrogate itself responds to anthropogenic pressures (e.g. land-use change) or natural processes (e.g. regional climate change, forest senescence). For example, in the Cumberland Mountains in the USA, 200 000 ha were assessed remotely as a potential habitat for cerulean warblers *Setophaga cerulean* [7]. The constructed model then was employed to evaluate the effects of proposed coal surface mining on warbler habitat [7]. At the tens of ha scale, aerial photography and surveys of birds and macroin-

vertebrates were used to monitor decline of seagrass coverage and population crash of seagrass-dependent species in a South African marine reserve [8].

Remotely sensed data have been used to estimate forest stand condition (an indicator of dieback) over >200 000 ha in the southern Murray-Darling Basin in south-eastern Australia [9]. In floodplain forests, changes in stand condition can be assessed in response to water management [4]. Mac Nally et al [10] found that abundance, effective species richness and breeding of birds were related to modeled stand condition; they predicted negative consequences of climate change on stand condition and avifauna on these floodplains.

The yellow-footed antechinus *Antechinus flavipes* Waterhouse is the only native, terrestrial, carnivorous mammal (a marsupial) on these floodplains. This species is most abundant in floodplain forests [11] but has been lost from floodplains of the lower, more arid sections of the Murray River [12]. Understanding of the reasons for changes in antechinus population characteristics (densities, survival) is important for planning management actions. Capture rates of antechinus are related to the occurrence of, and proximity to, floods [13]. Adult male antechinuses die after synchronized breeding in winter (before young are born), and many females die too (after weaning), with only a fraction of females surviving to produce offspring in their second year [14]. The occurrence of second-year females with suckled teats means that they survived to breed again and probably weaned offspring. Lada *et al.* [15] considered capture rates of females and of second-year females with suckled teats as excellent measures of realized habitat quality and thus the probability of population persistence in a given location.

If there are relationships between modeled stand condition and population characteristics of antechinus (e.g. abundance of females), these could provide a tool for rapid, remote assessment of antechinus across vast landscapes.

In this paper we explored the relative importance of modeled stand condition and of in-stand variables in explaining population characteristics of antechinus.

Materials and Methods

Study areas

The study was conducted in seven floodplain forests and woodlands of river red gum (*Eucalyptus camaldulensis* Dehnh.) in south-eastern Australia (Fig. 1 and [11]). We investigated floodplains of the unregulated Ovens River and floodplains of the highly regulated Murray River, including Barmah, Millewa, Koondrook, Gunbower Island, Guttrum and Campbells Island Forests (Table S1 in File S1).

Eucalyptus camaldulensis is a mono-dominant tree species on these floodplains, forming open forests and woodlands (trees 10-30 m tall, 20–45% projective foliage cover) [16]. Groundcover is fallen timber, litter, low shrubs, sedges and grasses [17]. Mean annual rainfall over the study area ranges spatially from 395 (± 115,

temporal SD, based on 30 years) to 624 (± 186, based on 23 years) mm yr^{-1} and mean monthly maximum temperatures range from 12.9°C for the coldest month to 31.8°C for the hottest month during the year [18]. Barmah, Millewa, Koondrook and Gunbower Island Forests are Ramsar-listed wetlands. The Murray River wetlands experienced regular floods prior to river regulation, with extensive floods 45 years per century [19]. Between 1980 and 2011, the frequency of extensive floods was halved compared with pre-regulation frequency [20] due to water management (dams, locks, irrigation channels). The floodplains have experienced grazing, logging and removal of fallen timber for > 130 years [21], along with significant water extractions [22]. Since 2000, small management floods have been used in selected areas of floodplains to improve ecosystem condition [13].

Data sources for *Antechinus flavipes*

Antechinuses were captured in the austral summer, autumn and winter in 2004, 2005 [11] and 2011 at 272 randomly selected square 0.25 ha sites (Table S1 in File S1). Sites visited in 2004 were revisited in 2005 [11] and had been randomly selected in 24 within-forest areas (500 × 500 m) prior to the availability of the model of stand condition (see below). Sites for 2011 surveys were randomly selected with Hawth's Tools (http://spatialecology.com) using an existing map of stand condition [23]. We obtained sites from the full range of stand condition, avoiding spatial clustering of sites with similar stand condition. Sites with representative volumes of fallen timber were chosen because antechinus numbers increase with higher volumes [24]. Sites were > 300 m apart to

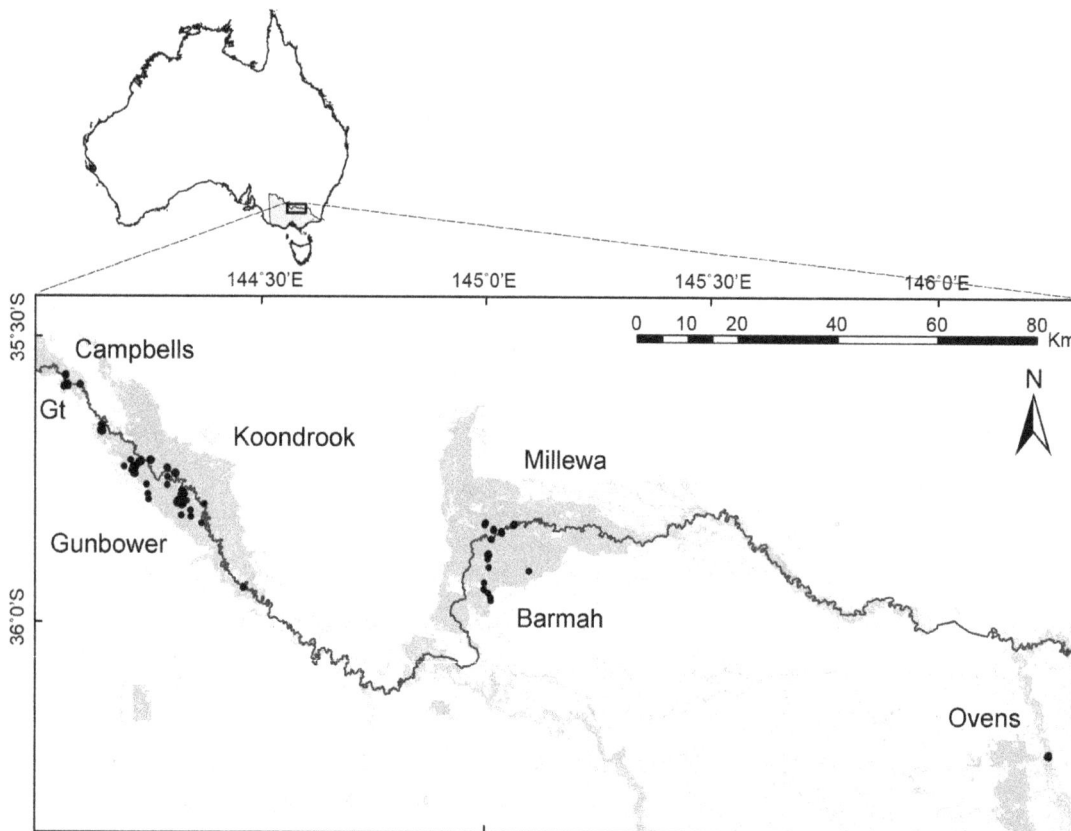

Figure 1. Location of study sites on the middle Murray River floodplain, south-eastern Australia. Grey areas represent extant forests. Black dots represent locations of 1–27 study sites. Extensive forest areas are labeled, with Gt indicating Guttrum forest. Gunbower and Barmah were surveyed in 2004, 2005 and 2011; Campbells, Guttrum, Koondrook, Millewa and Ovens – in 2004 and 2005.

avoid recaptures from other sites. Five or six small-mammal (Elliott) traps were placed along logs and beneath trees at each site for 1–5 nights. Traps were set for at least two nights on 83% of visits to sites. Differences in trapping effort arose because of early collection of traps on sites where individuals were repeatedly recaptured or access was impossible because of logging and management releases of water; differences were accounted for statistically (see below, this section). This trapping effort was sufficient to capture most, if not all, antechinuses present on site because recapture rates were high and animals were seen moving so extensively within sites that they were unlikely to miss locations of traps [11]. We did not trap in spring to avoid possibly stressing lactating females [25]. We used capture rates (= numbers of different individuals captured as a function of trapping effort, hence a binomial response variable) as a measure of density of antechinuses. We analyzed capture rates for all distinct individuals, males, females, second-year females, second-year females with suckled teats.

Ethics statement

This work was carried out under permits BSCI/2011/03, BSCI/2003/02 (Monash University Biological Sciences Animal Ethics Committee, permits 10005842 and 10002325 (Department of Sustainability and Environment). Traps contained lure and bedding, were protected by plastic bags and covered with bark. Animals were released at the point of capture.

Stand-condition scores

Stand-condition scores (see Table 1) were obtained from GIS rasters of modeled stand condition of river red gum forests in 2006 (for 2004 and 2005 trapping) and 2010 (for 2011 trapping) [23,26]. Stand condition was estimated from the three variables: percentage live basal area, plant area index (PAI) and crown extent, which are known to be reliable indicators of condition in *E. camaldulensis* stands (see Methods S in File S1and [27] for further details). Stand condition was calculated at a resolution of 1 ha because this is similar to the home range for yellow-footed antechinus (this home range is inferred from trapping data in our study area in 1999–2005, and is consistent with radio-tracking studies elsewhere by [28]). The stand condition model for 2006 was built from extensive on-ground surveys ($N = 140$ stands) in that year and had high predictive power in an independent, follow-up validation survey in 2007 ($N = 42$ stands, $R^2 = 0.78$). The stand condition model for 2010 was built from on-ground surveys in that year ($N = 175$ stands).

Other environmental variables

Volume of fallen timber (Table 1) was estimated at each 0.25 ha site by considering all logs with a diameter ≥ 10 cm [11]. Distance to floodwaters (for floods occurring anytime between previous breeding season and current trapping) was measured from maps of inundations informed by field observations. The number of large webs of golden orb-weaving spiders *Nephila* sp. were counted within 2 m of each trap line. We considered the number of webs as a proxy for abundances of spiders and possibly of other macroinvertebrates, which are food for antechinus [12]. We tested the suggestion of Lada *et al.* [11] that these spiders may be associated with higher densities of antechinus. Values of annual rainfall of the previous year (weather conditions pre-, post- and during-breeding season of the previous generation and pre-weaning conditions for the current generation) were obtained from Bureau of Meteorology [18].

We included two indicator variables to determine whether capture rates were affected by a life stage: juvenile dispersal variable, which was used to compare capture rates in January (juvenile dispersal) to those in June and July (breeding season); and post-juvenile dispersal variable (capture rates in March to May vs during breeding season).

Statistical analyses

We used logistic regression in WinBUGS v1.4 [29] to examine the relationship between capture rates of antechinus and environmental variables including stand condition. The model was:

$$y_i \sim Binomial(p_i, n_i); \operatorname{logit}(p_i) = \alpha_{year(i)} + \sum_j \beta_j x_{ij} + \varepsilon_{forest(i)} + \varepsilon_{site(i)}$$

Here, y_i is the total number of individuals (i.e. males and females) captured over n_i trap-nights (i.e. each trap-night regarded as a Bernoulli trial, so that n_i = the number of available traps) at site i, and p_i is the corresponding capture rate to be estimated (i.e. probability of success for a single trap-night at a given site, in a given year). The αs are year-specific intercepts, logit-transformed mean capture rates for 2004, 2005 and 2011, and β_j is the regression coefficient for the jth predictor variable. The εs are random effects for forest and site. Forest random effects were assigned exchangeable, normal prior distributions (Gelman 2005) with means 0 and variances, σ_{forest}^2. The site random effects were modeled with a spatial disc function such that the correlation between sites declined nearly linearly with distances up to 500 m (typical home range for antechinus is 1 ha). We assigned uniform priors (0, 5) to the standard deviations of the random effects terms, σ_{site}, and σ_{forest} [30].

We used Bayesian model selection, implemented with the reversible jump MCMC add-on to WinBUGS [31], to identify a subset of the candidate predictor variables that should be included in the best model, or, equivalently, to identify which of the linear coefficients β are non-zero (see Methods S1 in File S1). The posterior probability of a non-zero coefficient $\Pr(\beta_j \neq 0)$ is a measure of the evidence that variable j is a predictor of the response. We considered $\Pr(\beta_j \neq 0) > 0.75$, which is equivalent to a threefold increase in the prior odds (which were unity), to be evidence that a predictor has a strong effect on the response variable [32]. We used exchangeable prior distributions for the coefficients $\beta_j \sim N(0, \sigma_\beta)$; $\sigma_\beta \sim Uniform(0, 2)$. We re-fitted models with a range of plausible upper limits on σ_β (0.3, 1, 2, 4) and obtained similar results.

Models were estimated with three Markov chain Monte Carlo (MCMC) chains run for 100000 iterations following burns-in of 50000 iterations.

We checked the adequacy of the model structure by posterior predictive checks using the χ^2-discrepancy statistic. We also calculated *pseudo-R²* values (proportion of deviance explained divided by maximum) as a measure of model fit. We performed 10-fold cross-validation on models to check that estimated relationships were not spurious and to evaluate likely predictive capacity. The data were split into 10 sets of sites (folds), and each fold, comprising all surveys at each site, served as test data for models built with the remaining sites. Sites were semi-randomly allocated to folds, but sites within 500 m of each other were allocated to the same fold so that predictions were not informed by spatial autocorrelation. We calculated cross-validation *pseudo-R²* [33] for the combined hold-out data, and used random permutation of predictions to calculate the probability of obtaining equal or higher *pseudo-R²* values by chance.

Table 1. Predictor variables used in the analysis of capture rates of the yellow-footed antechinus *Antechinus flavipes* in river red gum woodlands in 2004, 2005 and 2011 in south-eastern Australia.

Predictor variable	Year 2004 Mean ± SD (in *n* forests)	Year 2005 Mean ± SD (in *n* forests)	Year 2011 Mean ± SD (in *n* forests)	Data source
Stand-condition score	7.23 ± 0.72 (7)	7.22 ± 0.74 (7)	6.93 ± 0.97 (2)	GIS rasters of modeled stand condition; 2006 model [23] for 2004 and 2005 trapping; 2010 model [26] for 2011 trapping
Volume of fallen timber (m³/ha)	65.41 ± 36.11 (7)	66.69 ± 36.01 (7)	44.74 ± 29.82 (2)	All logs with diameters ≥ 10 cm on 0.25 ha sites [11]
Distance to floodwaters (km)	3.16 ± 4.14 (7)	3.03 ± 4.11 (7)	0 ± 0 (2)	Maps of inundations and field observations in 2003–2011
Number of orb-weaving spider webs	NA	NA	4 ± 5.1 (2)	Webs counted within 2 m of each trap line
Annual rainfall in previous year (mm)	535.2 ± 114.5	407.7 ± 73.4	657 ± 20.4	Data from three weather stations in 2004 and 2005, two stations in 2011 [18]
Juvenile dispersal				Categorical variable; whether trapping was in January (juvenile dispersal) or in June and July (breeding season)
Post-juvenile dispersal				Categorical variable; whether trapping was in March to May (post-juvenile dispersal) or in June and July (breeding season)

NA = not collected.

Estimates of temporal trends in fractions of stand condition

To track changes in stand condition across these floodplains, a temporal series (1990, 2003, 2006, 2009 and 2010) of condition maps for the region was compiled from previous work (see Methods S1 in File S1).

Results

There were 0–5 distinct individuals of *Antechinus flavipes* captured per site in 2004, and 0–7 individuals in 2005 and 2011. No more than two second-year females were caught on a given site. At least 6% of second-year females failed to produce any offspring (i.e. they did not have any suckled teats). There were 557 captures of antechinus (same-site recaptures discounted) with the average of 1.1 ± 1.4 SD individuals per site. There was strong evidence that capture rates of second-year females and of second-year females with suckled teats increased with increasing stand condition (Table 2). There was strong evidence that the capture rates of total number of individuals increased with the volume of fallen timber, and that capture rates of males were lower during the post-juvenile dispersal period than during the breeding period. There was no evidence that other environmental variables (distance to floods of current and previous year, rainfall previous year, golden orb-weaving spider webs) influenced capture rates (Table 2).

Variable-selection results in cross-validation tests were consistent with values obtained for the full data (Table 2). For all variables that had $Pr(\beta_j \neq 0) > 0.75$ in the full model, $Pr(\beta_j \neq 0)$ values exceeded 0.75 in at least 8 of 10 cross-validation iterations and were never < 0.6 (hence > 0.5, the prior probability). Predictive capacity of models was low ($pseudo\text{-}R^2 \leq 10\%$) but better than random for all response variables [Pr(observed or lower $pseudo\text{-}R^2$ | random) < 0.001].

The best estimate of the percentage of the floodplain forest stands in 'good' condition (SCS > 8, see example in Fig. S1) declined from 57% in 1990 to 25% from 2006 onwards (Fig. 2,

redrawn from [10]), with a very rapid decline between 2003 and 2006 following almost a decade of much-below-average rainfall. Much of the change was into the 'declined' class until 2010, when there was a substantial rise in the percentage of 'poor' condition forest (Fig. 2, and see example in Fig. S1).

Discussion

The first and second generation of *Antechinus flavipes* differed in their responses to modeled stand condition (Table 2). Modeled stand condition was the best predictor for relative abundances of second-year females and second-year females with suckled teats, but not for abundances of males or for total females (first-year and second-year females combined). That the probabilities of capturing second-year females and second-year females with suckled teats were related positively to stand condition, while densities of males were unrelated, is consistent with the species' biology. After weaning in summer, juvenile males disperse from natal areas and, during the breeding season, roam widely searching for females with which to mate [28,34]. Females are much more philopatric and remain near natal areas and with female relatives [35]. The response of second-year females and second-year females with suckled teats to modeled stand condition is of greater demographic importance than is the lack of male response, given that population dynamics effectively are driven by the survival of females and their offspring (i.e. total breeding failure in one year would extirpate a population if there were no subsequent immigration [15]). The lack of response in total females to stand condition suggests that stand condition is a better predictor of past breeding success (females with suckled teats) and survival of females to their second year (second-year females) than of future breeding potential (all females). This idea of differences in past breeding success and future breeding potential in relation to stand condition needs to be explored in other species of vertebrates.

The relationship between densities of second-year females and stand condition may reflect an association between stand condition

Table 2. Results of Bayesian regression analyses [posterior mean regression coefficient, β, and probability of non-zero coefficient, $\Pr(\beta \neq 0)$] of capture rates of the yellow-footed antechinus *Antechinus flavipes* in river red gum woodlands in 2004, 2005 and 2011 in south-eastern Australia with respect to environmental variables and stage of life cycle.

Variable	Total Mean ±SD	Pr	Males Mean±SD	Pr	Females Mean±SD	Pr	F2 Mean±SD	Pr	F2 with teats Mean±SD	Pr
α_{04}	−2.97±0.25		−3.43±0.25		−3.96 ±0.29		−5.99±0.76		−6.88±0.77	
α_{05}	−2.31±0.24		−2.71±0.30		−3.47±0.29		−5.14±0.59		−5.59±0.66	
α_{11}	−2.16±0.41		−2.56±0.51		−3.61±0.46		−5.52±1.01		−6.03±1.01	
Condition	−0.01±0.03	0.22	−0.12±0.11	0.69	0.07±0.10	0.54	**0.45±0.29**	**0.86 (0.85)**	**0.38±0.40**	**0.75 (0.77)**
FloodDist	−0.01±0.04	0.27	0.00±0.03	0.17	0.00±0.04	0.34	0.00±0.03	0.14	0.00±0.05	0.20
Logs (m^3h^{-1})	**0.13±0.08**	**0.87 (0.82)**	0.05±0.07	0.49	0.08±0.09	0.65	0.01±0.04	0.16	0.00±0.03	0.17
RainPrevYr	−0.03±0.09	0.39	0.00±0.16	0.51	−0.04±0.09	0.42	−0.17±0.21	0.63	−0.02±0.09	0.26
Webs	0.00±0.02	0.19	0.00±0.01	0.10	0.00±0.02	0.26	0.00±0.01	0.09	0.00±0.01	0.09
JD	0.06±0.22	0.46	−0.12±0.28	0.46	0.13±0.35	0.56	0.07±0.33	0.34	0.26±0.68	0.51
PostJD	−0.14±0.18	0.61	**−0.46±0.25**	**0.93 (0.93)**	0.01±0.07	0.41	−0.39±0.57	0.53	0.00±0.08	0.27
Pseudo-R^2	0.56 (0.09)		0.42 (0.10)		0.43 (0.05)		0.28 (0.07)		0.33 (0.07)	

$\Pr(\beta \neq 0)$ values in parenthesis are averages of cross-validation fits [shown only for variables with $\Pr(\beta \neq 0) > 0.75$]. Response variables: F2 = second-year females, F2 with teats = second-year females with suckled teats. Covariates: Condition = modeled stand condition at 100 m resolution, Webs = number of webs of golden orb-weaving spiders, FallenTimber = volume of fallen timber, FloodDist = Euclidean distance to flood waters, RainPrev6mon = rainfall over 6 months preceding the month of trapping, RainPrevYr = annual rainfall previous year, JD = whether trapping occurred during juvenile dispersal phase, postJD = whether trapping occurred between juvenile dispersal and breeding stages. Pr = probability that the covariate is a predictor of the response. Mean = regression coefficient. SD = standard deviation of regression coefficient. Pseudo-R^2 is the proportion of the binomial deviance [−2log(likelihood)] explained by the fitted model divided by the maximum possible value, values in parentheses are the corresponding values for 10-fold cross validation.

and resource provision for antechinus, but more likely reflects the positive effects of flooding on both stand condition and floodplain productivity [4,13]. Proximity to flooding is a strong predictor of second-year female densities [13] and females move into the regions closest to floods [35]. Emergence of large-bodied macroinvertebrates (beetles and spiders) after the flood recession almost certainly provides abundant large prey for antechinuses [36]. While stand condition may appear to be a proxy for flooding

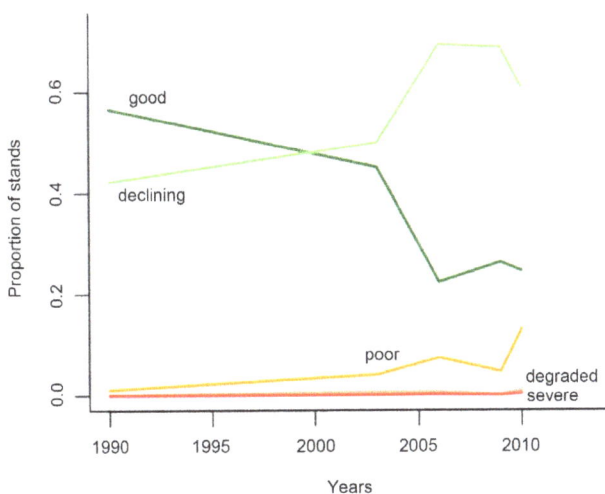

Figure 2. Proportion of floodplain forest in different stand-condition categories along the mid-Murray River between 1990 and 2010. Stand condition was predicted from maps that were built from ground surveys and Landsat imagery [4,23,26]. Key: (1) good, dark green; (2) declined, light green; (3) poor, orange; (4) degraded, brown; (5) severe, red. Redrawn from [10].

frequency, this simple relationship is complicated because the densities of males increase with proximity to flooding [13] but not with higher levels of stand condition (Table 2). To determine why stand condition is a strong predictor of second-year female densities, we need to establish the relationship between stand condition and the availability of food and shelter (on or in trees) for antechinuses. Stand condition might affect ground habitats through shading and the amount of fallen timber.

These floodplain forests have been subjected to extensive logging, grazing and water extractions, which led to changes in forest structure from one dominated by large, spreading trees with mixed-aged patches to one of relatively young, same-age stands with few large trees [20]. Tree recruitment is now rare because of reduced flood frequency, grazing and salinity of soil and groundwater [20]. Sometimes water is returned to floodplain forests to mitigate extensive tree dieback (70% dieback in 2006 [4]) or for safety of towns and upstream dams. Some management floods are detrimental to both trees and to antechinus. Flows during summer support irrigated agriculture but were rare before widespread river regulation [37]. Management floods may be maintained for up to four months, much longer than historical floods [38], which almost certainly limits foraging opportunities for antechinus females carrying young and delays macroinvertebrate emergence. Recruitment of river red gum is highest following winter to early spring floods [39] because these allow development of deep root systems to avoid subsequent water deficits. Given the common need for spring floods of antechinus and of river red gum, stand condition may provide a more integrated assessment of flooding history, and thus productivity, for antechinus on floodplains.

Maps of stand condition over 100 000 ha have been produced annually from on-ground surveys and remotely-sensed data since 2009 [4]. These maps may be used to prioritize and monitor releases of limited amounts of management water onto floodplains at temporal and spatial scales to maximize the positive effects on stand condition, which may also lead to positive effects on survival of antechinus mothers.

There is a lack of information on how changes in land-use and management actions, such as management floods, affect (meta) population dynamics [40]. Typically, at least one of the many parameters in metapopulation viability analyses is poorly known. For antechinus, we have knowledge on effective dispersal from genetic analyses [41]. Unfortunately, capture rates of antechinus (especially second-year females, 0.002 in 2004, 0.006 in 2005, Table S2 in File S1) are too low to accurately estimate the densities of second-year females in patches of habitat of different (modeled) condition. One or two captures can cause large deviation from expected capture rates. For example, simulated captures for second-year females based on actual trapping efforts and the predicted capture probabilities yielded pseudo-R^2 values lower than the observed 7% (values for 10-fold cross validation in Table 2) with probability 0.49. That is, even if predicted capture rates matched the true underlying probabilities of capture perfectly, we would not expect higher pseudo-R^2 values. However, our results provided strong evidence that densities of second-year females generally will increase with increasing stand condition, all else being equal. Of course, there may be other limiting factors, such as macroinvertebrate abundance (food availability for antechinus) and fallen timber volumes (which were not correlated with stand condition) that will cause local variation in antechinus densities independently of stand condition.

Our findings have relevance to management of floodplains in semi-arid areas. River regulation and water diversions are associated with similar tree population declines on floodplains in southwest of the USA [42] and in Spain [43]. In Australia, the widespread decline in river red gum condition across the Murray-Darling Basin has been addressed poorly by very limited allocations of overbank flows across all of the floodplains [44]. The south-eastern Australia was gripped by severe drought (1997–2010) and subject to consistently increasing temperatures [45]. Severe droughts are likely to recur under predicted climate change [46]. During the severe drought, densities of antechinus were higher closer to floods [13], and stands in good condition were restricted to areas closest to the river channel, to permanent wetlands or where management floods had occurred [4]. These results, combined with our results of a positive relationship between stand condition and inferred breeding success of antechinus (females with suckled teats) suggest advantages for increased flood allocations. Depriving floodplains of water most likely will lead to further decreases in stand condition and to decreased ability of antechinus mothers to breed again. Floodplains may offer sanctuaries for antechinus populations during droughts. This is in contrast to box-ironbark forests, where the antechinus population in Chiltern [highest capture rate of any forests in 2004 and 2005; 11] crashed to small numbers in 2011, most likely because of low rainfall in the preceding years [25].

Our results suggest that remote sensing offers a means to assess rapidly habitat quality for one mammal species over large areas. We linked a stand condition model, which originally was developed to assess health of river red gum forests, to population dynamics of antechinus and birds [10]. Remotely-sensed estimates of stand condition and structure have been related to biodiversity measures in many parts of the world (see review by [47]), and the capacity to make larger-scale estimates of biodiversity status is an important advance. Our results, when coupled with strong positive responses of birds to variation in stand condition in the same floodplain forests [10], contribute to this growing body of knowledge that allows upscaling from plot-scale measurements to landscape and regional scale estimates of ecosystem condition and biodiversity status. We present an approach that is potentially transferable to other species and ecosystems, and that may be a means to dynamically link both anthropogenic and natural environmental change to habitats of species and measures of species' status and trends.

Supporting Information

Figure S1 Examples of stand condition states: good, poor and severe.

File S1 Contains the following files: **Methods S1.** Model selection with reversible jump MCMC. Stand-condition scores. Estimates of temporal trends in fractions of stand condition. **References S1. Table S1** Study area and site information. **Table S2** Mean trapping rates of the yellow-footed antechinus in 2004, 2005 and 2011 over all sampled sites in river red gum forests.

Acknowledgments

We thank Tara Draper, Greg Horrocks and Anna Lada for assistance with fieldwork and Erica Fleishman, Sue Carthew and anonymous reviewers for comments.

Author Contributions

Conceived and designed the experiments: HL JRT RMN SCC. Performed the experiments: HL SCC. Analyzed the data: JRT RMN. Wrote the paper: HL JRT RMN SCC.

References

1. Sala O, Chapin F, Armesto J, Berlow E, Bloomfield J, et al. (2000) Global biodiversity scenarios for the year 2100. Science 287: 1770–1774.

2. Kingsford RT, Watson JEM, Lundquist CJ, Venter O, Hughes L, et al. (2009) Major conservation policy issues for biodiversity in oceania. Conservation Biology 23: 834–840.

3. Mantyka-Pringle CS, Martin TG, Rhodes JR (2011) Interactions between climate and habitat loss effects on biodiversity: A systematic review and meta-analysis. Global Change Biology 18: 1239–1252.

4. Cunningham SC, Thomson JR, Mac Nally R, Read J, Baker PJ (2011) Groundwater change forecasts widespread forest dieback across an extensive floodplain system. Freshwater Biology 56: 1494–1508.

5. Palmer MA, Liermann CAR, Nilsson C, Florke M, Alcamo J, et al. (2008) Climate change and the world's river basins: anticipating management options. Frontiers In Ecology And The Environment 6: 81–89.

6. Klein C, Wilson K, Watts M, Stein J, Berry S, et al. (2009) Incorporating ecological and evolutionary processes into continental-scale conservation planning. Ecological Applications 19: 206–217.

7. Buehler DA, Welton MJ, Beachy TA (2006) Predicting cerulean warbler habitat use in the Cumberland Mountains of Tennessee. Journal of Wildlife Management 70: 1763–1769.

8. Pillay D, Branch GM, Griffiths CL, Williams C, Prinsloo A (2010) Ecosystem change in a South African marine reserve (1960–2009): Role of seagrass loss and anthropogenic disturbance Marine Ecology Progress Series 415: 35–48.

9. Treitz PM, Howarth PJ (1999) Hyperspectral remote sensing for estimating biophysical parameters of forest ecosystems. Progress in Physical Geography 23: 359–390.

10. Mac Nally R, Lada H, Cunningham R, Thomson JR, Fleishman E (2014) Climate-change-driven deterioration of the condition of floodplain forest and the future for the avifauna. Global Ecology and Biogeography 23: 191–202.

11. Lada H, Mac Nally R, Taylor AC (2008) Responses of a carnivorous marsupial (*Antechinus flavipes*) to local habitat factors in two forest types. Journal of Mammalogy 89: 398–407.

12. Menkhorst PW (1995) Mammals of Victoria: distribution, ecology and conservation. Melbourne: Oxford University Press in association with Department of Conservation and Natural Resources. 360 p.

13. Lada H, Thomson JR, Mac Nally R, Horrocks G, Taylor AC (2007) Evaluating simultaneous impacts of three anthropogenic effects on a floodplain-dwelling marsupial *Antechinus flavipes*. Biological Conservation 134: 527–536.
14. Lee AK, Woolley P, Braithwaite R (1982) Life history strategies of Dasyurid marsupials. In: Archer M, editor. Carnivorous marsupials. Sydney: Royal Zoological Society of New South Wales. pp. 1–11.
15. Lada H, Mac Nally R, Taylor AC (2008) Phenotype and gene flow in a marsupial (*Antechinus flavipes*) in contrasting habitats. Biological Journal of the Linnean Society 94: 303–314.
16. Specht RL (1981) Major vegetation formations in Australia. In: Keast A, editor. Ecological Biogeography of Australia. The Hague: Junk. pp. 163–298.
17. Margules & Partners (1990) Riparian Vegetation of the River Murray. Report prepared by Margules and Partners Pty. Ltd., P. & J. Smith Ecological Consultants and Department of Conservation Forests and Lands. Canberra: Murray-Darling Basin Commission.
18. Bureau of Meteorology (2011) Climate Data Online. Australian Government Bureau of Meteorology.
19. Bren LJ, O'Niell IC, Gibbs NL (1987) Flooding in the Barmah Forest and its relationship to flow in the Murray-Edward River system. Australian Forest Research 17: 127–144.
20. Mac Nally R, Cunningham SC, Baker PJ, Horner GJ, Thomson JR (2011) Dynamics of Murray-Darling floodplain forests under multiple stressors: The past, present and future of an Australian icon. Water Resources Research 47: W00G05.
21. Fahey C (1988) The Barmah Forest - a history. Melbourne: Department of Conservation, Forests and Lands. 50 p.
22. Close A (1990) The impact of man on the natural flow regime. In: MacKay N, Eastburn D, editors. The Murray. Canberra: Murray-Darling Basin Commission. pp. 61–74.
23. Cunningham SC, Mac Nally R, Read J, Baker P, White M, et al. (2009) A robust technique for mapping vegetation across a major river system. Ecosystems 12: 207–219.
24. Mac Nally R, Horrocks G (2008) Longer-term responses of a floodplain-dwelling marsupial to experimental manipulation of fallen timber loads. Basic & Applied Ecology 9: 456–465.
25. Lada H, Thomson JR, Cunningham SC, Mac Nally R (2013) Rainfall in prior breeding seasons influences population size of a small marsupial. Austral Ecology 38: 581–591.
26. Cunningham SC, Griffioen P, White M, Mac Nally R (2011) Mapping the Condition of River Red Gum (Eucalyptus camaldulensis Dehnh.) and Black Box (Eucalyptus largiflorens F.Muell.) Stands in The Living Murray Icon Sites. Stand Condition Report 2010. Canberra: Murray-Darling Basin Authority.
27. Cunningham SC, Read J, Baker PJ, Mac Nally R (2007) Quantitative assessment of stand condition and its relationship to physiological stress in stands of *Eucalyptus camaldulensis* (Myrtaceae). Australian Journal of Botany 55: 692–699.
28. Coates T (1995) Reproductive ecology of the yellow-footed antechinus, *Antechinus flavipes* (Waterhouse), in north east Victoria [Ph.D.]. Melbourne: Monash University.
29. Spiegelhalter D, Thomas A, Best N, Lunn D (2003) WinBUGS user manual. Cambridge: Institute of Public Health.
30. Gelman A (2005) Prior distributions for variance parameters in hierarchical models. Bayesian Analysis 1: 1–19.
31. Lunn DJ, Best N, Whittaker JC (2005) Generic reversible jump MCMC using graphical models. pp. EPH-2005-01. London: Department of Epidemiology and Public Health, Imperial College.
32. Jeffreys H (1961) Theory of probability. Oxford: Oxford University Press.
33. Nagelkerke N (1991) A note on a general definition of the coefficient of determination. Biometrika 78: 691–692.
34. Marchesan D, Carthew SM (2004) Autecology of the yellow-footed antechinus (*Antechinus flavipes*) in a fragmented landscape in southern Australia. Wildlife Research 31: 273–282.
35. Lada H, Mac Nally R, Taylor AC (2007) Genetic reconstruction of population dynamics of a carnivorous marsupial (*Antechinus flavipes*) in response to floods. Molecular Ecology 16: 2934–2947.
36. Ballinger A, Mac Nally R, Lake PS (2005) Immediate and longer-term effects of managed flooding on floodplain invertebrate assemblages in south-eastern Australia: generation and maintenance of a mosaic landscape. Freshwater Biology 50: 1190–1205.
37. Page K, Read A, Frazier P, Mount N (2005) The effect of altered flow regime on the frequency and duration of bankfull discharge: Murrumbidgee River, Australia. River Research and Applications 21: 567–578.
38. Mac Nally R, Horrocks G (2007) Inducing whole-assemblage change by experimental manipulation of habitat structure. Journal of Animal Ecology 76: 643–650.
39. Dexter BD (1978) Silviculture of the river red gum forests of the central Murray floodplain. Proceedings of the Royal Society of Victoria 90: 175–194.
40. Chisholm RA, Wintle BA (2007) Incorporating landscape stochasticity into population viability analysis. Ecological Applications 17: 317–322.
41. Lada H, Thomson J, Mac Nally R, Taylor AC (2008) Impacts of massive landscape change on a carnivorous marsupial (Antechinus flavipes) in south-eastern Australia: inferences from landscape genetics analysis. Journal of Applied Ecology 45: 1732–1741.
42. Rood SB, Braatne JH, Hughes FMR (2003) Ecophysiology of riparian cottonwoods: Stream flow dependency, water relations and restoration. Tree Physiology 23: 1113–1124.
43. Gonzalez E, Gonzalez-Sanchis M, Cabezas A, Comin F, Muller E (2010) Recent Changes in the Riparian Forest of a Large Regulated Mediterranean River: Implications for Management. Environmental Management 45: 669–681.
44. VEAC (2008) River red gum forests investigation: Final report. Melbourne.
45. McAlpine C, Syktus J, Ryan J, Deo R, McKeon G, et al. (2009) A continent under stress: interactions, feedbacks and risks associated with impact of modified land cover on Australia's climate. Global Change Biology 15: 2206–2223.
46. Nicholls N (2004) The changing nature of Australian droughts. Climatic Change 63: 323–336.
47. Gillespie TW, Foody GM, Rocchini D, Giorgi AP, Saatchi S (2008) Measuring and modelling biodiversity from space. Progress in Physical Geography 32: 203–221.

6

Reconciling Forest Conservation and Logging in Indonesian Borneo

David L. A. Gaveau[1]*, **Mrigesh Kshatriya**[1], **Douglas Sheil**[1,2,3], **Sean Sloan**[4], **Elis Molidena**[1], **Arief Wijaya**[1], **Serge Wich**[5], **Marc Ancrenaz**[6,7,8], **Matthew Hansen**[9], **Mark Broich**[10], **Manuel R. Guariguata**[1], **Pablo Pacheco**[1], **Peter Potapov**[9], **Svetlana Turubanova**[9], **Erik Meijaard**[1,11,12]

1 Center for International Forestry Research, Bogor, Indonesia, 2 School of Environment, Science and Engineering, Southern Cross University, Lismore, NSW, Australia, 3 Institute of Tropical Forest Conservation (ITFC), Mbarara University of Science and Technology (MUST), Kabale, Uganda, 4 Centre for Tropical Environmental and Sustainability Science, School of Marine & Tropical Biology, James Cook University, Cairns, QLD, Australia, 5 Research Centre in Evolutionary Anthropology and Palaeoecology, School of Natural Sciences and Psychology, Liverpool John Moores University, Liverpool, United Kingdom, 6 Sabah Wildlife Department, Kota Kinabalu, Sabah, Malaysia, 7 HUTAN, Kinabatangan Orang-utan Conservation Programme, Kota Kinabalu, Sa,bah, Malaysia, 8 North England Zoological Society, Chester Zoo, Chester, United Kingdom, 9 Department of Geographical Sciences, University of Maryland, College Park, Maryland, United States of America, 10 The Climate Change Cluster, University of Technology Sydney, NSW, Australia, 11 Borneo Futures Project, People and Nature Consulting International, Ciputat, Jakarta, Indonesia, 12 School of Biological Sciences, University of Queensland, Brisbane, Australia

Abstract

Combining protected areas with natural forest timber concessions may sustain larger forest landscapes than is possible via protected areas alone. However, the role of timber concessions in maintaining natural forest remains poorly characterized. An estimated 57% (303,525 km^2) of Kalimantan's land area (532,100 km^2) was covered by natural forest in 2000. About 14,212 km^2 (4.7%) had been cleared by 2010. Forests in oil palm concessions had been reduced by 5,600 km^2 (14.1%), while the figures for timber concessions are 1,336 km^2 (1.5%), and for protected forests are 1,122 km^2 (1.2%). These deforestation rates explain little about the relative performance of the different land use categories under equivalent conversion risks due to the confounding effects of location. An estimated 25% of lands allocated for timber harvesting in 2000 had their status changed to industrial plantation concessions in 2010. Based on a sample of 3,391 forest plots (1×1 km; 100 ha), and matching statistical analyses, 2000–2010 deforestation was on average 17.6 ha lower (95% C.I.: −22.3 ha−−12.9 ha) in timber concession plots than in oil palm concession plots. When location effects were accounted for, deforestation rates in timber concessions and protected areas were not significantly different (Mean difference: 0.35 ha; 95% C.I.: −0.002 ha–0.7 ha). Natural forest timber concessions in Kalimantan had similar ability as protected areas to maintain forest cover during 2000–2010, provided the former were not reclassified to industrial plantation concessions. Our study indicates the desirability of the Government of Indonesia designating its natural forest timber concessions as protected areas under the IUCN Protected Area Category VI to protect them from reclassification.

Editor: Jason M. Kamilar, Midwestern University & Arizona State University, United States of America

Funding: This work was funded by the Arcus foundation and the CGIAR Research Program on Forests, Trees and Agroforestry. The funders had no role in study design, data collection and analysis, decision to publish, or preparation of the manuscript.

Competing Interests: The authors have declared that no competing interests exist.

* E-mail: d.gaveau@cgiar.org

Introduction

Strictly protected areas are established by governments to conserve biological diversity and sustain other values and functions. Extractive and agricultural activities in protected forests are generally prohibited. Most authorities consider that establishing such strictly protected areas represents the best strategy for conserving tropical forests [1]. However, given economic demands, social pressure on land, and the cost of forest protection [2,3], these areas are unlikely to ever constitute more than a minor part of the tropical landscape, particularly in lowland areas [4,5,6]. Some conservation scientists propose combining protected areas with natural forest timber concessions to sustain larger forest landscapes than otherwise possible via protected areas alone [3,7,8,9,10,11,12]. This strategy has the merit of generating income and employment – arguably making it easier to gain

political and public support for conservation. The integration of natural forest timber concessions in a forest protection strategy makes sense in countries, such as Indonesia, where protected area management remains weak [13,14], where the government seeks economic opportunities for its people, and where the urgency of conservation action is high [15].

Natural forest timber concessions are parcels of natural forest leased out to companies or to communities to harvest timber on a long term basis. When natural forest timber concessions are additional to more strictly protected areas they bring an opportunity to maintain larger and better connected forest landscapes with a greater capacity to maintain low density, large range and high mobility species [16]. Indeed, timber concessions are *de facto* a kind of protected area in most tropical countries, as also indicated by their inclusion in the IUCN protected area categories (as Category VI). Conversion of natural forests to plantations in timber concessions is generally prohibited. Concession managers are legally obliged to maintain permanent natural forest cover [9]. Timber harvesting is supposed to be selective [17]. Concession managers only cut the commercially valuable wood above a certain diameter and leave other trees standing for long term regeneration. In equatorial Asia, between two and twenty stems are typically removed from each hectare of forest, once every few decades [18,19]. Generally, this leaves more than 90% of the trees standing and remaining vegetation recognizably constitutes a forest.

Not only does selective logging maintain a forest structure, a recent global meta-analysis of >100 scientific studies concluded that timber extraction in tropical forests has relatively benign impacts on biodiversity, because 85–100% of mammal, bird, invertebrate, and plant species richness remains in forests that have been harvested once [17]. Thus, a logged tropical forest can remain a biologically rich forest [12]. Not everyone is convinced that natural forest timber concessions should play a major role in tropical forest conservation [20]. Many equate timber harvesting (logging) with forest destruction and loggers with forest destroyers [21,22]. Many concerns relate to the apparently increased likelihood of a forest harvested for timber being further degraded by wildfires or converted to agriculture. Harvested forests appear to have increased vulnerability to fire [23,24,25]. Some governments equate 'logged forests' with 'degraded lands" or "wastelands', and reclassify these forests for conversion to industrial crops such as oil palm [26]. Roads built to extract timber are also of concern. They increase access which may exacerbate and facilitate illegal encroachments and other threats such as hunting [27,28,29,30,31,32,33]. But, any active timber concession requires people on the ground who might in principle at least enforce regulations and deter illegal activities [12] – thus whether being a timber concessions promotes deforestation compared to other forest land classifications remains debatable.

Despite the interest, the role of timber concessions in maintaining natural forest cover remains poorly characterized. One recent study of all protected areas on the Indonesian island of Sumatra revealed that areas allocated for natural timber harvesting resisted conversion to agriculture as well (or, arguably, as badly) as protected areas during the 1990s [34]. We note that high levels of deforestation sometimes occur in protected areas all over the world [14,35,36,37], but no-one would use this to argue against having protected areas, rather most would suggest that greater efforts should be invested in protection.

Here, we focus on natural forest timber concessions in Kalimantan, the 532,100 km^2 Indonesian portion of Borneo. Kalimantan is a globally important region for forest biodiversity [38,39]. Currently, 110,232 km^2 of Kalimantan's forests are under official protection as national parks, nature reserves and other protected areas. Natural forest timber concessions still make up a large share of Kalimantan's forest landscapes (105,945 km^2), and include one-third of the habitat of the endangered Bornean orang-utan (*Pongo pygmaeus*) [40]. But, their long-term existence is in jeopardy. As stated earlier, conversion to plantations is prohibited in Indonesian natural forest timber concessions. However, to compensate for the loss of logging revenues following years of harvesting that depleted commercial timber stocks by the late 1980s, the Indonesian government began reclassifying timber concessions in the 1990s into industrial plantation concessions, like monoculture oil palm (*Elaeis guineensis*) and other tree crops such as *Acacia mangium* [41,42]. Oil palm concessions are parcels of land leased out to companies to establish industrial oil palm plantations. These concessions currently cover 115,500 km^2 of Kalimantan's land area [43]. If undeveloped oil palm concessions contain

natural forests, concession managers are legally obliged to remove these forests to make way for plantations. Usually, the forest is logged first. After all timber resources have been harvested, the remaining trees, shrubs, and debris are often burned. Then, the land is cleared and flattened using heavy machinery to make rows of oil palms. Therefore, reclassification of natural forest timber concessions into oil palm concessions has the immediate effect of legalizing industry-driven deforestation within former timber concessions. During 2000–2010, industrial oil palm plantations in Kalimantan increased from an estimated 8,360 km^2 to 31,640 km^2 [43]. Therefore, considering timber concessions as potential protected areas and maintaining their natural forest status could contain the expansion of oil palm into forested areas, and maintain larger and better connected forest landscape with a greater capacity to conserve endangered forest wildlife.

To inform decision-making about the long-term status of natural forest timber concessions, we assessed transitions between official land use categories in Kalimantan (protected area, natural forest timber concession, and oil palm concessions), and studied the change in natural forest cover in each. We compared the total area (248,305 km^2) set aside for timber harvesting in natural forest (the production zone, or *Hutan Produksi*) by Indonesia's Ministry of Forestry (MoF) in the year 2000 with the area of land allocated for industrial plantations (oil palm and tree crops) and for protection in the year 2010. This production zone includes the 105,945 km^2 active timber concession licenses mentioned earlier, and areas without active timber licenses. This allowed us to estimate timber concession areas reclassified to protected area and for use as plantations (either oil palm or monoculture tree crops). To test whether natural forest timber concessions (that have not been reclassified to another land use) maintain forest cover, we compared 2000–2010 deforestation rates inside timber concessions with rates inside oil palm concessions; and with rates inside protected areas. Because protected areas tend to be in remote locations and deforestation generally increases with accessibility (e.g. topography) and may also be affected by a variety of other factors, a simple comparison of deforestation rates between logging concession and protected areas would misjudge the protection impact of protected areas [44]. We used "propensity score matching" to help control for and thus reduce any such location dependent biases [14,45,46,47].

Methods

Definitions of 'forest' and 'deforestation' and datasets used

To map deforestation, we used a 60 m^2 spatial resolution 'tree cover loss' map from 2000–2010 generated by authors MH, MB, PP and ST using the methods of Broich et al. [48] and Potapov et al. [49]. 'Tree cover' is defined as 60 m^2 tree stands with >25% canopy cover of ≥5 m in height [48]. 'Tree cover loss' is defined as the removal of tree stands. 'Tree cover' encompasses any trees including industrial plantations (e.g. oil palm and acacia), mixed traditional gardens (e.g. rubber, orchards, smallholder oil palm and other agro-forests mixed with forest re-growth), as well as old-growth natural forest. 'Natural forest' refers to lowland, hill and lower montane dipterocarp forests (often mixed with ironwood stands), mountain forests, freshwater and peat swamp forests, heath forest or *kerangas*, and mangrove forests (including Nipah) [50]. Because we are only interested in the loss of natural forests, we excluded from our analysis all 'tree cover loss' pixels (60×60 m) that fell outside of remaining forest areas in year 2000 using a forest cover map generated by Indonesia's Ministry of Forestry (MoF) for year 2000 [51]. The MoF map was created using

Landsat images. We assessed its quality by comparing it to our databases of Landsat images. We found that it was in agreement with our independent visual assessment of what constitutes intact natural forests (Primary forest in MoF classification) as well as natural forests degraded by logging, but where the forest remains recognizably a forest (Secondary forest in MoF classification).

Land use maps

Maps showing the total area set aside for timber harvesting in natural forests (production zone; Hutan Produksi) by the Indonesian government in year 2000 were obtained in 1:250,000 scale from Indonesia's Ministry of Forestry. Maps of natural forest timber concessions (year 2009–2010) and protected areas (national parks, nature reserves, wildlife sanctuaries, recreational and hunting parks, and watershed protection reserves) were obtained in 1:250,000 scale from [40], and originate from Indonesia's Ministry of Forestry. Maps of industrial oil palm concession boundaries (year 2005–2008) were obtained from [43], and originate from the provincial governments of Kalimantan. Protected areas created after 2000, for example the Sebangau National Park, were excluded from the propensity score matching analysis.

Propensity score matching

We tested whether natural forest timber concessions (that were not reclassified to another land use) maintained forest cover during 2000–2010 using propensity score matching. We first generated a sample of homogeneous forest stands, in the form of 100 ha forest plots (1×1 km), which we placed randomly across Kalimantan's 2000 forest cover. Forest plots that were placed within two kilometres of a previously chosen forest plot were rejected. Two kilometres were chosen as a compromise between the need for an adequate sample and the wish to reduce non-independence among observations. From these spatial restrictions, the maximum allowed number of forest plots was n = 6,234 plots. From this sample, only plots that were fully or nearly fully forested (>95 ha in a 100 ha plot) in year 2000 were used to compare deforestation rates between timber concessions, protected areas, and oil palm concessions, to allow the comparison of deforestation amongst plots in number of hectares lost rather than in percentage terms. The final subset retained for this analysis had n = 3,391 plots.

We measured the area of deforestation in each 100 ha plot, with values that ranged from 0–100 ha on a continuous scale, which we considered to be our indicator of effectiveness, and compared the deforestation between plots in timber concession (n = 1,220), in protected areas (n = 1,699), and in oil palm concessions (n = 472).

We used the matching package, MatchIt in R [52] to control for accessibility dependent effects in deforestation rates and in land use allocation between plots in natural forest timber concessions, protected areas, and oil palm concessions. Based on the literature of tropical deforestation, the variables that best characterize accessibility are slope, elevation above sea level, distance (expressed as travel time) to roads, and to cities [53]. Methods used to extract travel times can be found in File S1. In the context of expanding oil palm plantations in Kalimantan we added distance to oil palm mills, and to existing oil palm plantations in year 2000. These six variables were defined as "control variables" (Figure S1 in File S1). A propensity score was defined as the probability of a 100 ha plot being assigned as a timber concession. This probability was obtained from a logistic regression model in which the presence or the absence of a timber concession in the landscape was regressed against the control variables. The nearest neighbor with caliper procedure was implemented in the MatchIt package [52].

For every plot inside timber concessions, MatchIt paired up (matched) a plot inside protected areas (or inside oil palm concessions) that possessed the nearest propensity score. No plot could be matched to more than one other plot (without replacement). Only pairs where the difference in propensity scores did not exceed the caliper width were retained. A narrow caliper width was set to 0.25 times the standard deviation of the propensity scores. This narrow caliper width succeeded in matching more similar sites (e.g. protected area and concession plots of similar elevations and slopes) but with fewer number of pairs, thereby increasing the variance of the estimated treatment effect [54]; i.e. the mean difference in deforestation rate. MatchIt further restricted the matching across the landscape, so that a matched plot inside a protected area (or inside an oil palm concession) fell within the same administration and within the same soil type as the timber concession plot. This step was taken to ensure that pairs possessed similar socio-ecological and soil characteristics by being not too distant from each other. For example, wildfires are an important driver of deforestation in Eastern Kalimantan, but not in Western Kalimantan [25,55]. Therefore, matching within the same administration ensures that a plot inside a protected area (or inside an oil palm concession) from Eastern Kalimantan is not matched with a timber concession plot from Western Kalimantan. Eight different administrative groups ($n = 8$) were considered (Figure S1 in File S1). Peat soils and mineral soils were considered because deforestation patterns differ on peat lands; for example industry-driven deforestation for oil palm tends to avoid peat lands in favour of mineral soils [43]. The performance of our matching procedure was evaluated by investigating whether differences in the control variable between pairs had been eliminated [56]. Kolmogorov–Smirnov test (KS-test) and balance statistics provide a way to assess the quality of the matching method [52]. Both methods provide a measure of the balance between the treated and control group before and after matching. The balance statistic is a measure of the percent improvement in balance and is defined as $100*((|a|-|b|)/|b|)$, where a and b are measures, such as median, mean or maximum, of the original and matched data set respectively [52]. Here, the measures used to compare the un-matched and matched data sets included the empirical quantile median (eQQMedian), mean (eQQ Mean), and maximum (eQQ Max).

Results

The forest cover map generated by Indonesia's Ministry of Forestry indicates that 57% (303,525 km^2) of Kalimantan's area (532,100 km^2) was covered in natural old-growth forests (either intact or logged) in 2000. By 2010, this forested area had decreased by 14,212 km^2, representing a 4.7% loss over the decade. In 2000, the combined area of protected areas and timber concessions contained about 55% (182,185 km^2) of Kalimantan's natural forests (Figure 1A&B). In the subsequent 10-year period, natural forests occurring in protected areas had been reduced by 1,122 km^2, representing a 1.2% loss. Forests in timber concessions had been reduced by 1,336 km^2, representing a 1.5% loss (Table 1). Forests in areas granted to oil palm concessions had been reduced by 5,600 km^2, representing a 14.1% loss.

The total area (248,305 km^2) set aside for timber harvesting in natural forests (production zone; Hutan Produksi) by the Indonesian government in 2000 had shrunk by 25% by 2010 (Figure 2). The production zone includes the active timber concession licenses (the 105,945 km^2 area mentioned in Table 1 and shown in Figure 1A), and areas without active timber licenses. An estimated 63,000 km^2 of the production zone were reclassified to industrial plantation

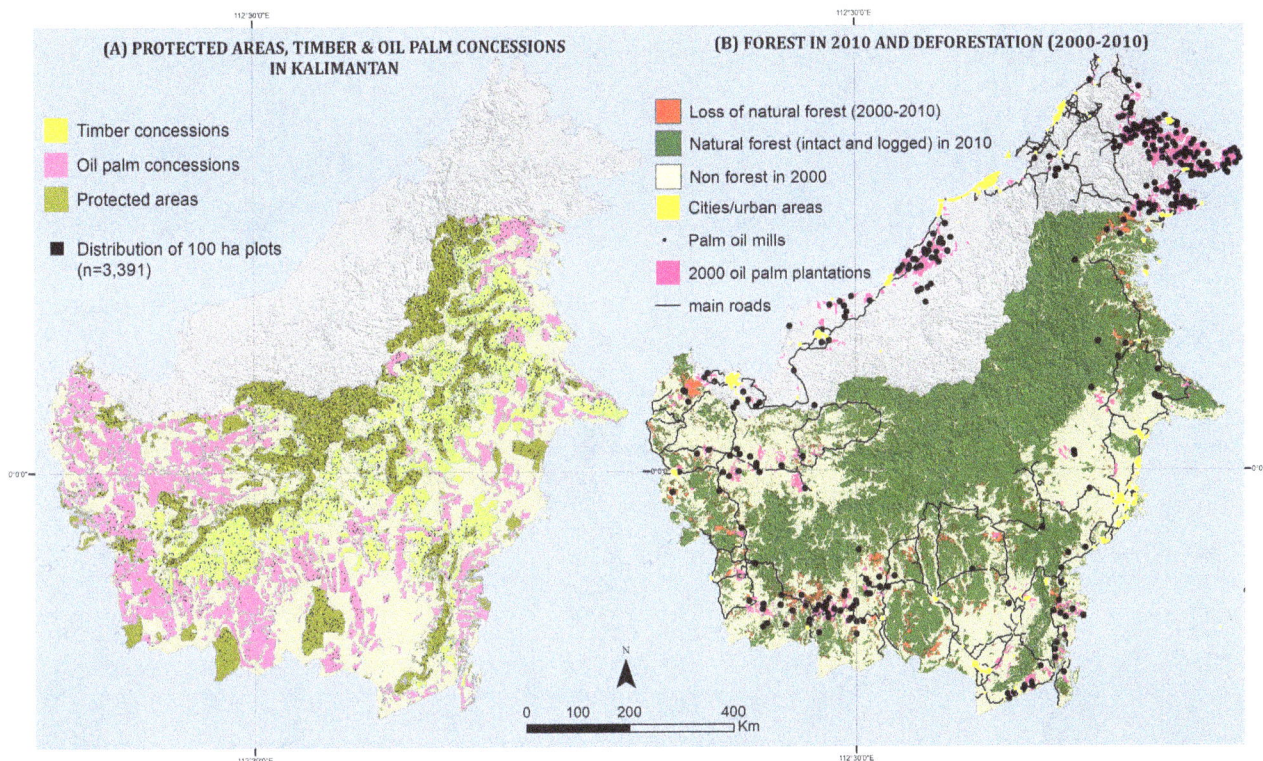

Figure 1. Panel A: protected areas (110,232 km²; brown), timber concessions (105,945 km²; light green), and industrial oil palm plantation concessions (115,500 km²; pink) in 2010 for Kalimantan (532,100 km²), and the spatial distribution of the 3,391 forest plots (100 ha each; black boxes). Panel B: remaining forest in 2010 (dark green), deforestation from 2000–2010 (red), main roads (black lines), realized oil palm plantations in 2000 (purple), urban areas (yellow) and palm oil mills (black dots).

concessions (oil palm and tree crop concessions), while 7,351 km² (3%) were reclassified to protected areas (primarily through the creation of Sebangau National Park). In contrast, less than 1% of protected areas in year 2000 had become reclassified to either natural timber or plantation concessions (Figure 2).

The spatial distribution of our 100 ha plots (n = 3,391) reveals the relative locations of protected areas, timber, and oil palm concessions (Table 2&3). Protected forest plots are typically located in the most remote areas (mean elevation = 636 m; mean slope = 24%; mean travel time to roads, cities, mills and existing plantations >58 hrs; Table 2). Forest plots in oil palm concessions are generally located in the least remote areas (mean elevation = 91 m; mean slope = 4.6%; mean travel time to cities, to mills and existing plantations<18 hrs; Table 3). Forest plots in timber concessions are located in intermediate locations, neither as remote as protected areas or as accessible as oil palm concessions

(mean elevation = 360 m; mean slope = 17%; mean travel time to cities, to mills and existing plantations<44 hrs; Table 2&3).

To control for such location specific effects in our comparison of deforestation rates *Matchit* selected 575 pairs for the logging concession versus protected area analysis and 194 pairs for the logging concession versus oil palm concessions analysis.

The distribution of propensity scores between timber concessions and protected areas differed significantly before matching (KS-test for the "raw" dataset: D = 0.4966, p-value<0.001) and did not differ significantly after matching (KS-test for "matched" dataset: D = 0.0313, p-value = 0.9408, Figure 3A). The distribution of propensity scores between timber concessions and oil palm concessions differed significantly before matching (KS-test for the "raw" dataset: D = 0.6431, p-value<2.2e-16). After matching these differences disappeared (KS-test: D = 0.0309, p-value = 1.00) (Figure 3B).

Table 1. Kalimantan-wide losses in forest cover from 2000–2010.

	Kalimantan	Protected Areas	Timber concessions	Oil palm concessions	Other areas*
Landmass (km²)	532,100	110,232	105,945	115,500	200,423
2000 forest cover (km²)	303,524	93,834	88,351	39,722	81,617
Deforestation (km²)	14,212	1,122	1,336	5,600	6,155
Deforestation (%)	4.7	1.2	1.5	14.1	7.5

*Other areas include areas outside of Timber and oil palm concessions and outside of protected areas.

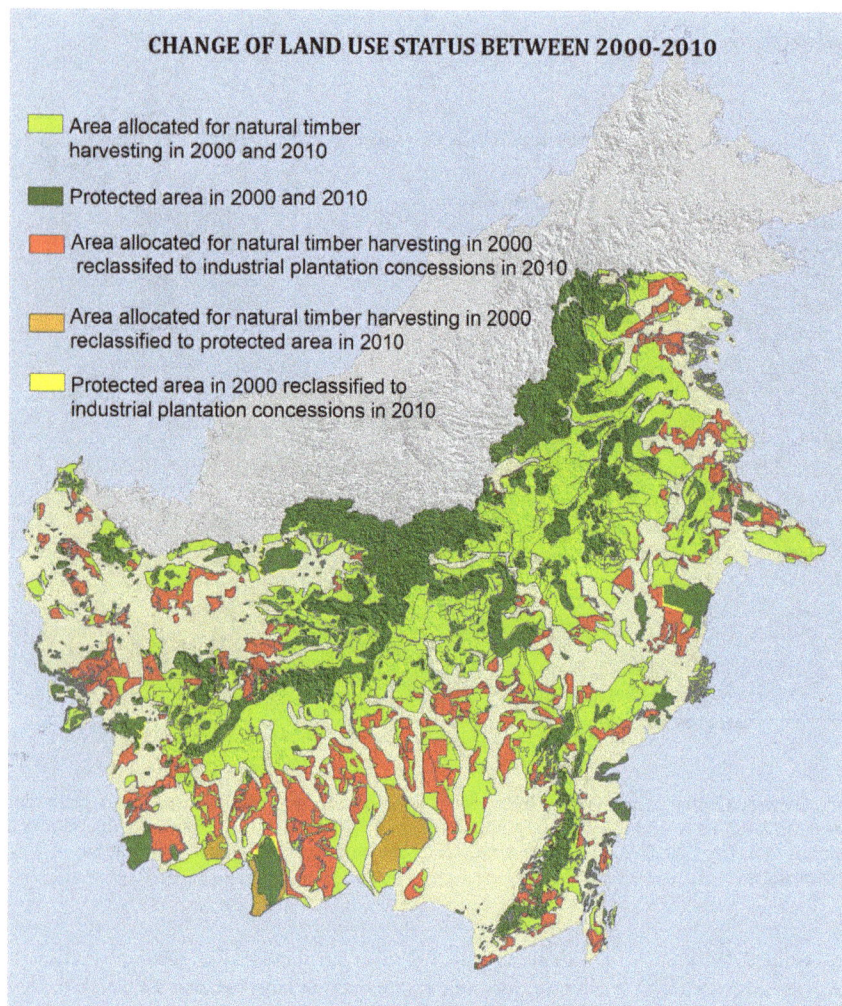

Figure 2. Map showing the change of land use status of area allocated for natural timber harvesting and protected areas during 2000–2010 in Kalimantan. Area allocated for natural timber harvesting in 2000 and 2010 (light green); Protected area in 2000 and 2010 (dark green); Area allocated for natural timber harvesting in 2000 reclassified to industrial plantation concessions in 2010 (red); Area allocated for natural timber harvesting in 2000 reclassified to protected area in 2010 (orange); Protected area in 2000 reclassified to industrial plantation concessions in 2010 (yellow).

For all control variables, the mean difference ("Mean diff") decreased after matching as indicated by the balance indices (Table 2&3). The various measures used to gauge departure from perfect matching, such as the empirical quantile median (eQQMedian), mean (eQQ Mean) and maximum (eQQ Max), showed a common trend. The mean magnitudes of each of these statistics became smaller indicating that the matching had resulted in very similar distributions of all the variables considered. These indicators of good matching give us more confidence that the differences in deforestation we observe among the different land use categories can be attributed to their official status rather than to other factors.

Based on the unmatched sample dataset mean differences in deforestation from 2000–2010 (expressed in hectares lost in 100 ha plots) are all significant (Table 4). After matching, the mean deforestation was still significantly 17.6 ha lower in timber concessions than in oil palm concessions (95% C.I.: −22.3 ha– −12.9 ha; Table 4). Most importantly, any difference in deforestation rates between natural timber concessions and protected areas was smaller than could formally be detected using this method meaning that there is little difference (mean difference: 0.35 ha; 95% C.I.: −0.002 ha–0.7 ha). The spatial distribution of the pairs is shown in Figure 4.

The protected area category included >50% watershed protection forest reserves (*Hutan Lindung*, HL), areas that, except for a few exceptions of locally funded watershed areas, receive neither funds nor are actively managed by governmental agencies. By grouping these HL reserves with protected areas designated for their conservation values (e.g. national parks), the above analysis potentially diluted the protection impact of managed protected areas. However, when HL reserves were excluded from the protected area category, deforestation was still not significantly higher in natural forest timber concessions than in protected areas (mean difference: 0.66 ha; 95% C.I.: −0.11 ha–1.43 ha; Table 4). The distribution of propensity scores between timber concessions and managed protected areas is shown in Figure S2 in File S1. The spatial distribution of the pairs for the timber concession and managed protected areas is shown in Figure S3 in File S1.

Table 2. Summary of balance of the control variables before and after matching for Protected Area (PA) and natural forest Timber Concession (TC) plots.

Variable		Means in TC cells	Means in PA cells	SD Control	Mean Diff	eQQ Med	eQQ Mean	eQQ Max
Travel time to cities (hr)	Before	43.1	61.9	45.8	−18.8	18.1	18.7	37.6
	After	48.9	56.6	41.7	−7.8	7.9	7.8	26.3
	% Balance Improvement[a]				61.9%	61.1%	61.8%	40.6%
Travel time to mills (hr)	Before	42.4	64.6	44.5	−22.2	23.9	22.2	40.0
	After	49.4	55.5	41.4	−6.1	4.2	6.1	37.0
	% Balance Improvement[a]				74.1%	88.0%	74.0%	17.6%
Travel time to roads (hr)	Before	41.7	59.0	42.3	−17.3	18.7	17.2	33.6
	After	48.0	55.1	41.0	−7.1	7.2	7.1	16.8
	% Balance Improvement[a]				62.3%	68.2%	62.2%	53.6%
Travel time to plantations (hr)	Before	39.6	63.6	45.0	−24.0	26.0	24.0	44.3
	After	47.4	54.0	40.5	−6.6	6.3	6.6	22.4
	% Balance Improvement[a]				74.8%	81.6%	74.7%	51.3%
Elevation (m)	Before	359.7	636.4	400.5	−276.6	326.7	282.8	439.4
	After	453.5	505.0	312.5	−51.5	65.6	66.6	371.9
	% Balance Improvement[a]				82.6%	80.8%	77.6%	14.1%
Slope (percent)	Before	17.1	24.2	13.2	−7.1	7.7	7.3	30.9
	After	20.5	21.8	12.0	−1.3	1.1	1.4	11.1
	% Balance Improvement[a]				80.1%	81.3%	79.9%	87.0%

Table 3. Summary of balance of the control variables before and after matching for natural forest Timber Concession (TC) and Oil Palm Concession (OPC) plots.

Variable		Means in OPC cells	Means in TC cells	SD Control	Mean Diff	eQQ Med	eQQ Mean	eQQ Max
Travel time to cities (hr)	Before	17.6	43.1	37.1	−25.5	19.4	26.2	98.1
	After	24.8	28.5	27.9	−3.8	11.6	12.9	99.9
	% Balance Improvement[a]				90.5%	48.5%	51.1%	−4.4%
Travel time to mills (hr)	Before	17.6	42.4	34.7	−24.8	18.1	25.9	93.7
	After	24.9	27.1	25.5	−2.3	9.3	12.2	107.1
	% Balance Improvement[a]				95.3%	51.1%	53.7%	−12.4%
Travel time to roads (hr)	Before	16.4	41.7	37.0	−25.3	18.5	26.0	101.2
	After	22.1	26.9	28.1	−4.8	11.4	12.6	96.0
	% Balance Improvement[a]				86.0%	46.3%	51.6%	2.4%
Travel time to plantations (hr)	Before	13.9	39.6	35.3	−25.7	16.4	26.3	92.3
	After	20.4	24.4	26.1	−4.1	8.2	12.2	94.5
	% Balance Improvement[a]				87.4%	53.0%	53.9%	4.0%
Elevation (m)	Before	90.8	359.7	292.4	−268.9	201.6	269.0	888.8
	After	164.1	167.5	162.3	−3.5	20.0	29.1	235.0
	% Balance Improvement[a]				98.0%	91.0%	89.6%	65.1%
Slope (percent)	Before	4.6	17.1	11.7	−12.4	12.9	12.5	22.6
	After	8.5	8.4	8.5	0.2	0.5	0.8	11.3
	% Balance Improvement[a]				98.9%	97.1%	91.6%	31.9%

Histogram of propensity scores between timber concessions and protected areas

Histogram of propensity scores between timber and oil palm concessions

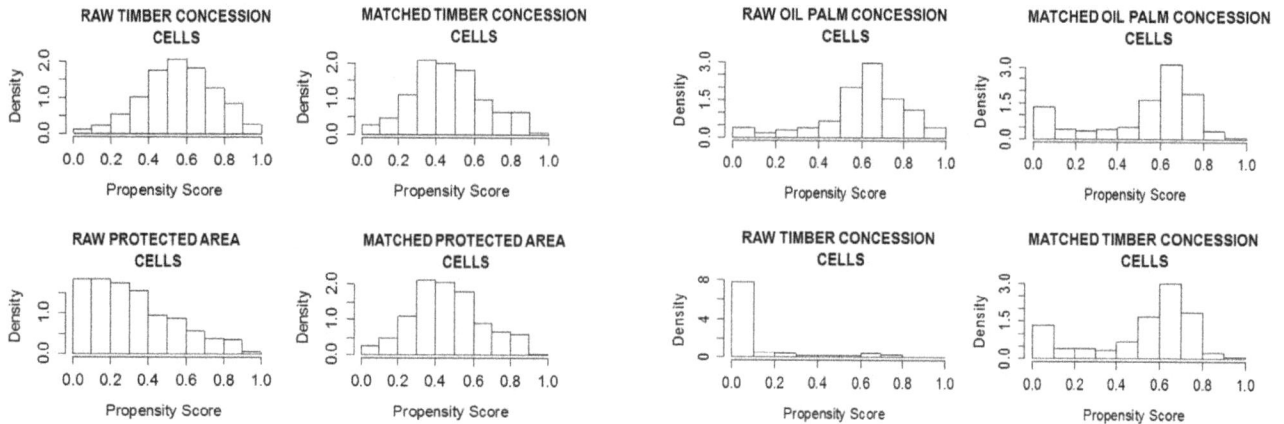

Figure 3. Histogram distribution of propensity scores before and after matching between timber concessions and protected areas (left panel); and between timber and oil palm concessions (right panel).

Discussion

This study reveals that Kalimantan's natural forest timber concessions, i.e. parcels of natural forest leased out to companies to extract timber on a long term basis (>30 years), have as far as we are able to determine with available data and controlling for the influence of location, maintained forest cover just as well as protected areas during the 2000–2010 decade, and have prevented government-sanctioned deforestation; illegal forest conversion to industrial oil palm plantations was marginal within timber concessions. These results corroborate findings in Sumatra where areas allocated for natural timber harvesting (production forest) have been found to resist illegal forest conversion to agriculture as well as protected areas during the 1990s when matched to reduce location specific effects [34]. Thus it appears that timber concessions could be used as a conservation intervention to protect tropical forests. These observations come with caveats.

Firstly, we highlight that our results reflect a statistical conclusion: that is that we cannot detect any significant difference

in the deforestation rates in protected areas and in timber concessions when we account for location. These results do not mean that these rates are equal, only that any differences are relatively small compared with our ability to detect them unambiguously. For example if our null hypothesis was that timber concessions maintained a 50% higher deforestation rate than protected areas under similar spatial contexts we would not have been able to reject that either. So, substantial uncertainties remain. Despite our use of propensity score matching we recognize that these methods are only an approximate solution and that ambiguities remain regarding the variables considered, their measurement, their spatial correlations and the choices made to control for these – this is an area where we would hope to make further methodological investigations in the future in order to improve confidence and better understand how spatial context influences the probability and extent of forest cover loss.

Secondly, as our analysis shows, between 2000 and 2010, the Government of Indonesia reclassified 25% of areas allocated for natural timber harvesting for use as monoculture oil palm and tree crop plantations. In the same period, the government only

Table 4. Comparison of mean differences in deforestation (2000–2010) before and after matching.

	TC *vs* OPC	TC *vs* PA	TC *vs* managed PA
Mean Deforestation rates before matching (ha)	0.91 *vs* 22.21	0.91 *vs* 0.16	0.91 *vs* 0.19
Mean difference before matching (ha)	−21.3	0.75	0.72
(95% C.I.)	−24.8–−18.5	0.43–1.05	0.33–1.11
Number of 100 ha plots	1220 *vs* 472	1220 *vs* 1699	1220 *vs* 594
Mean difference after matching (ha)	−17.6	0.35	0.66
(95% C.I.)	(−22.3–−12.9)	(−0.002–0.7)	(−0.11–1.43)
Number of paired 100 ha plots	194	575	111

These values are expressed in hectares lost in 100 ha plots that were nearly fully forested (>95 ha forest cover) in year 2000. Values ranged from 0 ha lost to 100 ha lost on a continuous scale. Confidence intervals for the unmatched dataset are derived from an independent samples t-test. Confidence intervals for the matched dataset are derived from the matching algorithm, *MatchIt*. The mean difference is between: (i) Timber Concession plots (TC) and Oil Palm Concession plots (OPC); (ii) Timber Concession plots (TC) and Protected Area plots (PA) ; and Timber Concession plots (TC) and managed Protected Area plots (i.e. national parks and nature reserves, but excluding watershed protection forests which are generally not managed).

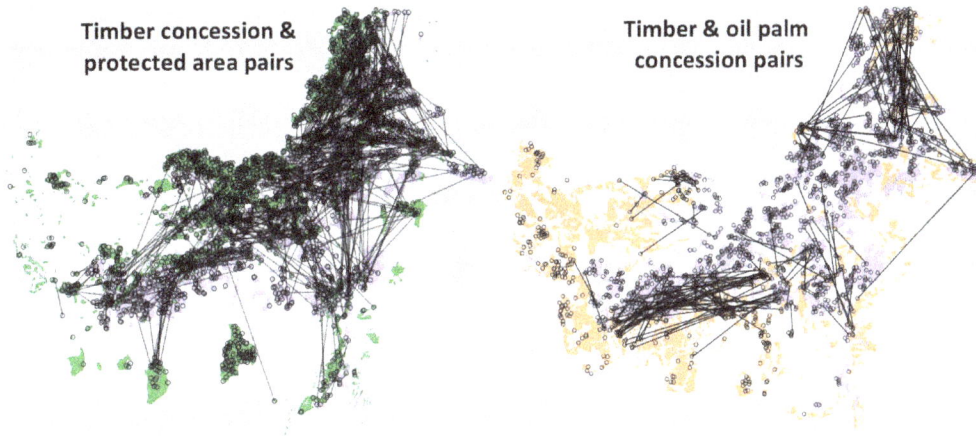

Figure 4. The spatial distribution of the 575 pairs for the natural forest timber concession (purple) *versus* protected area (green) analysis (left panel). The spatial distribution of the 194 pairs for the natural forest timber concession (grey) *versus* oil palm concessions (orange) analysis (right panel).

reclassified 3% of timber concessions to the status of protected area, primarily through the creation of Sebangau National Park in Central Kalimantan. Although timber concessions areas are officially required to keep a permanent forest cover, their classification seems easily changed and reclassification into industrial plantation concessions legalize deforestation. In contrast, less than 1% of protected areas had their status changed to industrial plantation concessions. Thus, compared to protected areas, timber concessions have been more vulnerable to official reclassification that permits forest conversion. We only expect timber concessions to maintain forest cover if they are not reclassified for plantations. This is a crucial point because the Indonesian government tends to equate 'logged' with 'degraded/wasteland,' but as research shows, logged forests can still be extremely valuable habitats for orangutans and other species [16,40,57,58]. The creation of the 5,686 km^2 Sebangau National Park in 2004, an area logged throughout the 1990s, but containing the largest contiguous orangutan population on Borneo [41], indicates that Government of Indonesia is beginning to recognize the value of logged forests for biodiversity conservation.

Despite the legal protection of forests in protected areas and natural forest timber concessions, both land use types lack the management required to prevent all wild fires and illegal agricultural encroachments by small farmers. This situation is not unique to Kalimantan. There is ample evidence that deforestation persists within protected areas because drivers of deforestation, are coupled with a limited protection capacity that largely reflects insufficient management resources [35,59,60,61,62,63,64,65]. Several studies have shown that protected area management in Indonesia is insufficiently effective to abate threats of deforestation, and in particular fire, illegal logging, and illegal encroachment. For example, Kutai National Park in East Kalimantan province was severely damage by prolonged drought and wildfires in 1982–1983 [66]. Gunung Palung National Park in West Kalimantan province was the site of widespread illegal logging during the early 2000s, following an era of breakdown in law and order [13]. Bukit Barisan Selatan National Park, in southern Sumatra suffered massive deforestation through agricultural encroachment by small famers for coffee plantations [60,67]. One reason is insufficient funding. In 2006, Indonesia's terrestrial protected areas received an average USD 1.56/ha in government funding and an estimated USD 0.67/ha in

funding from non-governmental organizations and international donor agencies [68]. This is considerably lower than the average USD 13 spent on protected area management in countries in the Asia-Pacific Region [69,70]. The shortfall in Indonesia's protected area funding – that is the funds needed to achieve what their mandate requires –was estimated at US$ 81.94 million for 2006 [68]. Funding allocation and management choices may have further reduced effectiveness. Data are lacking, but claims have been made that those protected areas involving long-term collaboration between non-governmental organizations (NGOs) and park authorities have been more successful in maintaining forest cover [71].

Our findings indicate that both natural forest timber concessions and protected areas have slowed forest cover loss in Kalimantan in the face of expanding plantations. Timber concessions typically generate a higher per hectare revenues than neighboring protected areas. Timber harvesting in natural forests provides one way in which forest lands can provide income and employment while retaining forest: in simple terms, the forest can pay for its own protection. In addition, studies of the perception of people in Kalimantan about the value of forests for their health, culture, and livelihoods show that logged forests remain important for them [72,73,74,75].

We note that significant forest conservation efforts in Indonesia have been focused on generating and enforcing strictly protected areas. There is little doubt that the reclassification of timber production forest to plantations has been facilitated by the pervasive judgment that equates logged forests with "degraded" or "secondary" undeserving of conservation concern. If we started to pay greater attention to the value of logged forest the protection gains may have been even better. Policy makers, officials and concession staff can all be encouraged to take pride in the value of well managed logged forests and their global conservation values.

Our study indicates the desirability of the Government of Indonesia designating its natural forest timber concessions as protected areas under the IUCN Protected Area Category VI, because they perform as effectively as protected areas in maintaining forest cover and should be protected from reclassification. The World Database of Protected Areas contains many examples of permanent forest reserves where hardwood extraction is one of the activities. Adding Kalimantan's natural forest timber concessions to the protected area network would increase the permanently protected forest in Kalimantan by 248,305 km^2, i.e., the area of production forest that legally should remain forested. Such changes would require a shift in mindset from producers, government, and also conservation groups, especially because government policy presently does not guarantee timber concession permanent status as natural forest. Still, making such a political decision and implementing it accordingly would have long-term benefits for wildlife and the maintenance of ecosystem services from forests, while continuing the generation of income from forests. We note that such changes are required to achieve sustainable forestry practices, which has long been the stated goal of the Ministry of Forestry and such a permanent and inviolate forest estate would certainly also have value under the future of Reducing Emissions from Deforestation and Degradation (REDD) programs in which Indonesia receives payments for reduced forest loss and damage.

Indonesia's government is taking steps towards the long-term maintenance of its natural forests. In recognition of the importance of natural forest timber concessions for biodiversity, economic development, and social aspirations, the government launched the Ecosystem Restoration concept in 2007 [76]. The ecosystem restoration license is granted to companies for a period of 60 years and can be extended once for a further 35 years. The aim of such licenses is to allow heavily harvested forests to recover their potential to produce commercial timber while maintaining a minimum level of ecosystem services, such as biodiversity conservation. The initiative has had a slow start, however, and as of 2012, only 1,005 km^2 in two areas, or about 0.9% of Kalimantan's total concession area, had been granted an ecosystem restoration license [77].

A major impediment to the permanent protection of natural forests in Kalimantan is the high economic potential of oil palm plantations [43]. The returns on plantations are much higher than returns from timber harvesting in natural forests. The conversion of logged forests to plantations makes economic sense. What may be overlooked in the political decision-making regarding such land use conversions are the significant values of natural forests to the well-being of many of Kalimantan's people [72,74,75,78]. This does not only include people living close to these forests, but also the many people in downstream and coastal areas that are affected by the negative environmental impacts (air pollution, temperature increases, changed flooding regimes etc.) from unsustainable land use [72]. For all the benefits that plantations bring to people, poor accounting of negative impacts impairs political decision-making that maximizes the well-being of Kalimantan's people. Therefore, considering the importance of natural forest timber concessions for biodiversity conservation as well as societal aspirations, and the high rate at which these forests are reclassified to plantations, it

seems important that the Government of Indonesia minimize conversion of natural forests to plantations and expand forest restoration opportunities.

Conclusion

Current policies in Indonesia allow logged forests in natural forest timber concessions to be managed for rehabilitation and ecosystem restoration, or to become converted to industrial plantations. The systematic reclassification of timber concessions to plantations should be prevented. Encouraging rehabilitation and restoration, and discouraging conversion of logged forest could play a big role in helping protect forests and wildlife in Indonesia. If Kalimantan's forests are approximately as well protected from illegal encroachments as they are in protected areas, as our analysis shows, the Indonesian government would do well strategically to commit to keep natural forest timber concessions in production over the long term alongside the protected area network to collectively conserve over two-third of Kalimantan's remaining forests, while at the same time providing income and employment. This could be achieved by reclassifying natural forest timber concessions as protected areas under the IUCN Protected Area Category VI. Such a permanent forest estate offers benefits for biodiversity conservation and other environmental benefits as well as for providing a foundation for further investment in sustainable forestry.

Supporting Information

File S1　Supporting information describing how control variables were derived. This file includes Figure S1, Figure S2, and Figure S3.

Acknowledgments

The study is part of a larger set of studies on land use optimization, conservation planning and management, and species ecology ("The Borneo Futures Initiative"). We thank the Government of Indonesia with all the respective forest and wildlife departments and other agencies for supporting our research, as well as Professor Richard Corlett and one anonymous reviewer for their help in improving this study.

Author Contributions

Conceived and designed the experiments: DG E. Meijaard. Performed the experiments: DG MK. Analyzed the data: DG MK E. Molidena. Contributed reagents/materials/analysis tools: MH MB P. Potapov ST AW MA SW. Wrote the paper: DG DS SS E. Meijaard MRG P. Pacheco.

References

1. Chape S, Harrison J, Spalding M, Lysenko I (2005) Measuring the extent and effectiveness of protected areas as an indicator for meeting global biodiversity targets. Philosophical Transactions of the Royal Society B-Biological Sciences 360: 443–455.
2. Carwardine J, Wilson KA, Ceballos G, Ehrlich PR, Naidoo R, et al. (2008) Cost-effective priorities for global mammal conservation. Proceedings of the National Academy of Sciences of the United States of America 105: 11446–11450.
3. Wilson KA, Meijaard E, Drummond S, Grantham HS, Boitani L, et al. (2010) Conserving biodiversity in production landscapes. Ecological Applications 20: 1721–1732.
4. Joppa LN, Loarie SR, Pimm SL (2009) On population growth near protected areas. PLoS ONE 4: e4279. doi:4210.1371/journal.pone.0004279.
5. Rodrigues AS, Akcakaya HR, Andelman SJ, Bakarr MI, Boitani L, et al. (2004) Global gap analysis: priority regions for expanding the global protected-area network. BioScience 54: 1092–1100.
6. Sloan S, Edwards DP, Laurance WF (2012) Does Indonesia's REDD+ moratorium on new concessions spare imminently threatened forests? Conservation Letters 5: 222–231.

7. Billand A, Nasi R (2008) Production dans les forêts de conservation, conservation dans les forêts de production: vers des forêts tropicales durables, à partir du cas de l'Afrique centrale. In: Meral P, Castellanet C, Lapeyre R, editors. La gestion concertée des ressources naturelles: L'épreuve du temps: Gret - Karthala. pp. 201–219.
8. Clark C, Poulsen J, Malonga R, Elkan P Jr (2009) Logging concessions can extend the conservation estate for Central African tropical forests. Conservation Biology 23: 1281–1293.
9. Dickinson M, Dickinson J, Putz F (1996) Natural forest management as a conservation tool in the tropics: divergent views on possibilities and alternatives. Commonwealth Forestry Rev 75.
10. Edwards DP, Laurance WF (2013) Biodiversity despite selective logging. Science 339: 646.
11. Fisher B, Edwards DP, Larsen TH, Ansell FA, Hsu WW, et al. (2011) Cost-effective conservation: calculating biodiversity and logging trade-offs in Southeast Asia. Conservation Letters 4: 443–450.
12. Meijaard E, Sheil D (2007) A logged forest in Borneo is better than none at all. Nature 446: 974.

13. Curran LM, Trigg SN, McDonald AK, Astiani D, Hardiono YM, et al. (2004) Lowland forest loss in protected areas of Indonesian Borneo. Science 303: 1000–1003.

14. Gaveau DLA, Epting J, Lyne O, Linkie M, Kumara I, et al. (2009) Evaluating whether protected areas reduce tropical deforestation in Sumatra. Journal of Biogeography 36: 2165–2175.

15. Sodhi NS, Brooks TM, Koh LP, Acciaioli G, Erb M, et al. (2006) Biodiversity and human livelihood crises in the Malay Archipelago. Conservation Biology 20: 1811–1813.

16. Meijaard E, Sheil D (2008) The persistence and conservation of Borneo's mammals in lowland rain forests managed for timber: observations, overviews and opportunities. Ecological Research 23: 21–34.

17. Putz FE, Zuidema PA, Synnott T, Peña-Claros M, Pinard MA, et al. (2012) Sustaining conservation values in selectively logged tropical forests: the attained and the attainable. Conservation Letters 5: 296–303.

18. Sist P, Dykstra D, Fimbel R (1998) Reduced-impact logging guidelines for lowland and hill dipterocarp forests in Indonesia. CIFOR Occasional Paper.

19. Sist P, Nolan T, Bertault JG, Dykstra D (1998) Harvesting intensity versus sustainability in Indonesia. Forest ecology and management 108: 251–260.

20. Gibson L, Lee TM, Koh LP, Brook BW, Gardner TA, et al. (2011) Primary forests are irreplaceable for sustaining tropical biodiversity. Nature 478: 378–381.

21. Revkin AC (2012) Can logging and conservation coexist? The New York Times. New York.

22. Vanclay JK, Sheil D (2012) Can forest conservation and logging be reconciled? : The Conversation website. Available: https://theconversation.edu.au/can-forest-conservation-and-logging-be-reconciled-7811. Accessed 2013 Apr 22.

23. Cochrane MA (2003) Fire science for rainforests. Nature 421: 913–919.

24. Nepstad DC, Verissimo A, Alencar A, Nobre C, Lima E, et al. (1999) Large-scale impoverishment of Amazonian forests by logging and fire. Nature 398: 505–508.

25. Siegert F, Ruecker G, Hinrichs A, Hoffmann AA (2001) Increased damage from fires in logged forests during drought caused by El Nino. Nature 412: 437–440.

26. Giam X, Clements GR, Aziz SA, Chong KY, Miettinen J (2011) Rethinking the 'back to wilderness' concept for Sundaland's forests. Biological Conservation 144: 3149–3152.

27. Chomitz KM, Gray DA (1996) Roads, land use, and deforestation: a spatial model applied to Belize. World Bank Economic Review: 10 487–512, September 1996.

28. Laurance WF, Albernaz AKM, Schroth G, Fearnside PM, Bergen S, et al. (2002) Predictors of deforestation in the Brazilian Amazon. Journal of Biogeography 29: 737–748.

29. Laurance WF, Croes B, Tchignoumba L, Lahm SA, Alonso A, et al. (2006) Impacts of roads and hunting on Central African rainforest mammals. Conservation Biology 20: 1251–1261.

30. Pfaff A, Robalino J, Walker R, Aldrich S, Caldas M, et al. (2007) Roads and deforestation in the Brazilian Amazon. Journal of Regional Science 47: 109–123.

31. Wilkie D, Shaw E, Rotberg F, Morelli G, Auzel P (2000) Roads, development, and conservation in the Congo Basin. Conservation Biology 14: 1614–1622.

32. Wilson K, Newton A, Echeverria C, Weston C, Burgman M (2005) A vulnerability analysis of the temperate forests of south central Chile. Biological Conservation 122: 9–21.

33. Gaveau DLA, Wich S, Epting J, Juhn D, Kanninen M, et al. (2009) The future of forests and orangutans (Pongo abelii) in Sumatra: predicting impacts of oil palm plantations, road construction, and mechanisms for reducing carbon emissions from deforestation. Environmental Research Letters 4: 34013.

34. Gaveau DLA, Curran L, Paoli G, Carlson K, Wells P, et al. (2012) Examining protected area effectiveness in Sumatra: importance of regulations governing unprotected lands. Conservation Letters 5: 142–148.

35. Curran LM, Trigg SN, McDonald AK, Astiani D, Hardiono YM, et al. (2004) Lowland forest loss in protected areas of Indonesian Borneo. Science 303: 1000–1003.

36. DeFries R, Hansen A, Newton AC, Hansen MC (2005) Increasing isolation of protected areas in tropical forests over the past twenty years. Ecological Applications 15: 19–26.

37. Broich M, Hansen M, Stolle F, Potapov P, Margono BA, et al. (2011) Remotely sensed forest cover loss shows high spatial and temporal variation across Sumatera and Kalimantan, Indonesia 2000–2008. Environmental Research Letters 6: 014010.

38. Whitten T, van Dijk PP, Curran L, Meijaard E, Supriatna J, et al. (2004) Sundaland. In: Mittermeier RA, Gil PR, Hoffmann M, Pilgrim J, Brooks T et al., editors. Hotspots revisited: Another look at Earth's richest and most endangered terrestrial ecoregions. Mexico: Cemex.

39. Kier G, Mutke J, Dinerstein E, Ricketts TH, Küper W, et al. (2005) Global patterns of plant diversity and floristic knowledge. Journal of Biogeography 32: 1107–1116.

40. Wich SA, Gaveau D, Abram N, Ancrenaz M, Baccini A, et al. (2012) Understanding the Impacts of Land-Use Policies on a Threatened Species: Is There a Future for the Bornean Orang-utan? PLoS ONE 7: e49142.

41. Casson A (2000) The hesitant boom: Indonesia's oil palm sub-sector in an era of economic crisis and political change. CIFOR Occasional Paper No. 29. Bogor: Center for International Forestry Research.

42. Kartodiharjo H, Supriono A (2000) The impact of sectoral development on natural forest conversion and degradation: the case of timber and tree crop

43. Carlson KM, Curran LM, Asner GP, Pittman AM, Trigg SN, et al. (2013) Carbon emissions from forest conversion by Kalimantan oil palm plantations. Nature Climate Change 3: 283–287.

44. Joppa LN, Pfaff A (2009) High and Far: Biases in the location of protected areas. PLoS ONE 4: e8273.

45. Andam KS, Ferraro PJ, Pfaff A, Sanchez-Azofeifa GA, Robalino J (2008) Measuring the effectiveness of protected areas network in reducing deforestation. Proceedings of the National Academy of Sciences of the United States of America 105: 16089–16094.

46. Joppa LN, Pfaff A (2010) Global protected area impacts. Proceedings of the Royal Society B. doi: 10.1098/rspb.2010.1713.

47. Nelson A, Chomitz KM (2011) Multiple use protected areas in reducing tropical forest fires: a global analysis using matching methods. PLoS ONE 6: e22722.

48. Broich M, Hansen MC, Potapov P, Adusei B, Lindquist E, et al. (2011) Time-series analysis of multi-resolution optical imagery for quantifying forest cover loss in Sumatra and Kalimantan, Indonesia. International Journal of Applied Earth Observation and Geoinformation 13: 277–291.

49. Potapov PV, Turubanova SA, Hansen MC, Adusei B, Broich M, et al. (2012) Quantifying forest cover loss in Democratic Republic of the Congo, 2000–2010, with Landsat ETM+ data. Remote Sensing of Environment 122: 106–116.

50. McKinnon K (1996) The ecology of Kalimantan: [Hong Kong]: Periplus Editions.

51. Ministry of Forestry (2013) National Forest Monitoring System. Ministry of Forestry, Jakarta.

52. Ho DE, Imai K, King G, Stuart EA (2007) Matching as nonparametric preprocessing for reducing model dependence in parametric causal inference. Political Analysis 15: 199–236.

53. Kaimowitz D, Angelsen A (1998) Economic models of tropical deforestation a review. Bogor: Centre for International Forestry Research.

54. Austin PC (2011) Optimal caliper widths for propensity-score matching when estimating differences in means and differences in proportions in observational studies. Pharmaceutical Statistics 10: 150–161.

55. Wooster M, Perry G, Zoumas A (2012) Fire, drought and El Nino relationships on Borneo(Southeast Asia) in the pre-MODIS era (1980–2000). Biogeosciences 9: 317–340.

56. Rosenbaum PR, Rubin DB (1985) Constructing a Control-Group Using Multivariate Matched Sampling Methods That Incorporate the Propensity Score. American Statistician 39: 33–38.

57. Ancrenaz M, Ambu L, Sunjoto I, Ahmad E, Manokaran K, et al. (2010) Recent surveys in the forests of Ulu Segama Malua, Sabah, Malaysia, show that orang-utans (P. p. morio) can be maintained in slightly logged forests. PLoS ONE 5: e11510.

58. Berry NJ, Phillips OL, Lewis SL, Hill JK, Edwards DP, et al. (2010) The high value of logged tropical forests: lessons from northern Borneo. Biodiversity and Conservation 19: 985–997.

59. Brandon K, Redford KH, Sanderson SE (1998) Parks in peril: people, politics, and protected areas. Washington, D.C.: Island Press. xv, 519 p. p.

60. Gaveau DLA, Linkie M, Suyadi S, Levang P, Leader-Williams N (2009) Three decades of deforestation in southwest Sumatra: effects of coffee prices, law enforcement and rural poverty. Biological Conservation 142: 597–605.

61. Leverington F, Costa KL, Pavese H, Lisle A, Hockings M (2010) A global analysis of protected area Management effectiveness. Environmental Management 46: 685–698.

62. Naughton-Treves L, Alvarez-Berrios N, Brandon K, Bruner A, Buck Holland M, et al. (2006) Expanding protected areas and incorporating human resource use: a study of 15 forest parks in Ecuador and Peru. Sustainability: Science, Practice and Policy 2: 1–13.

63. van Schaik CP, Terborgh J, Dugelby B (1997) The silent crisis: the state of rain forest nature preserves. In: van Schaik C, Johnson J, editors. Last stand: protected areas and the defense of tropical biodiversity. New York: Oxford University Press.

64. Verissimo A, Rolla A, Vedoveto M, de Furtada SM (2011) areas protegidas na Amazonia Brasileira: Avancos e desafios. Imazon/ISA.

65. Soares-Filho B, Moutinhi P, Nepstad D, Anderson A, Rodrigues H, et al. (2010) Role of Brazilian Amazon protected areas in climate change mitigation. Proceedings of the National Academy of Sciences of the United States of America 107: 10821–10826.

66. MacKinnon K, Irving A, Bachruddin MA (1994) A last chance for Kutai National Park-local industry support for conservation. Oryx 28: 191–198.

67. Gaveau DLA, Wandono H, Setiabudi F (2007) Three decades of deforestation in southwest Sumatra: Have protected areas halted forest loss and logging, and promoted re-growth? Biological Conservation 134: 495–504.

68. McQuistan CI, Fahmi Z, Leisher C, Halim A, Adi SW (2006) Protected Area Funding in Indonesia: A study implemented under the Programmes of Work on Protected Areas of the Seventh Meeting of the Conference of Parties on the Convention on Biological Diversity. Jakarta, Indonesia: Ministry of Environment republic of Indonesia, The Nature Conservancy, PHPA, Departemen Kelautan dan Perikanan. 22 p.

69. World Wildlife Fund (2007) Tracking progress in managing protected areas around the world. An analysis of two applications of the Management Effectiveness Tracking Tool developed by WWF and the World Bank. Gland, Switzerland: WWF International.

plantations in Indonesia. CIFOR Occasional Paper No 26 (E). Bogor: Center for International Forestry Research.

70. Emerton L (2001) What are Africa's forests worth? Ecoforum 24: 7.
71. Laurance WF (2013) Does research help to safeguard protected areas? Trends in Ecology & Evolution 28: 261–266.
72. Meijaard E, Mengersen K, Abram N, Pellier AS, Wells J, et al. (in review) Perceptions on the importance of forests for people's livelihoods and health in Borneo. PLoS ONE.
73. Sheil D, Liswanti N, van Heist M, Basuki I, Syaefuddin, et al. (2003) Local priorities and biodiversity in tropical forest landscapes: asking people what matters. Tropical Forest Update 13. Available: http: //www.itto.or.jp/ newsletter/Newsletter.html. Accessed 2012 Dec 01.
74. Padmanaba M, Sheil D (2007) Finding and promoting a local conservation consensus in a globally important tropical forest landscape. Biodiversity and Conservation 16: 137–151.
75. Sheil D, Liswanti N (2006) Scoring the importance of tropical forest landscapes with local people: patterns and insights. Environmental Management 38: 126–136.
76. Ministry of Forestry (2008) A Guide to the Concession Licensing and Management of Ecosystem Restoration. Jakarta, Indonesia: Ministry of Forestry of the Republic of Indonesia and Burung Indonesia.
77. Ministry of Forestry (2011) Forestry Statistics of Indonesia 2011. Jakarta, Indonesia: Direktorat Jenderal Planologi Kehutanan, Kementerian Kehutanan.
78. Sheil D, Puri R, Wan M, Basuki I, Heist Mv, et al. (2006) Recognizing local people's priorities for tropical forest biodiversity. AMBIO: A Journal of the Human Environment 35: 17–24.

7

Comparing Effects of Climate Warming, Fire, and Timber Harvesting on a Boreal Forest Landscape in Northeastern China

Xiaona Li[1], Hong S. He[1,2]*, Zhiwei Wu[1], Yu Liang[1], Jeffrey E. Schneiderman[2]

1 State Key Laboratory of Forest and Soil Ecology, Institute of Applied Ecology, Chinese Academy of Sciences, Shenyang, People's Republic of China, 2 School of Natural Resources, University of Missouri-Columbia, Columbia, Missouri, United States of America

Abstract

Forest management under a changing climate requires assessing the effects of climate warming and disturbance on the composition, age structure, and spatial patterns of tree species. We investigated these effects on a boreal forest in northeastern China using a factorial experimental design and simulation modeling. We used a spatially explicit forest landscape model (LANDIS) to evaluate the effects of three independent variables: climate (current and expected future), fire regime (current and increased fire), and timber harvesting (no harvest and legal harvest). Simulations indicate that this forested landscape would be significantly impacted under a changing climate. Climate warming would significantly increase the abundance of most trees, especially broadleaf species (aspen, poplar, and willow). However, climate warming would have less impact on the abundance of conifers, diversity of forest age structure, and variation in spatial landscape structure than burning and harvesting. Burning was the predominant influence in the abundance of conifers except larch and the abundance of trees in mid-stage. Harvesting impacts were greatest for the abundance of larch and birch, and the abundance of trees during establishment stage (1–40 years), early stage (41–80 years) and old- growth stage (>180 years). Disturbance by timber harvesting and burning may significantly alter forest ecosystem dynamics by increasing forest fragmentation and decreasing forest diversity. Results from the simulations provide insight into the long term management of this boreal forest.

Editor: Dorian Q. Fuller, University College London, United Kingdom

Funding: This research is funded by the Knowledge Innovation Project of Chinese Academy of Sciences (CAS; Grant No.KZCX2-YW-444) and the Natural Science Foundation of China (NSFC; Grant Nos. 41071120 and 2011CB403206). The funders had no role in study design, data collection and analysis, decision to publish, or preparation of the manuscript.

Competing Interests: The authors have declared that no competing interests exist.

* E-mail: HeH@missouri.edu

Introduction

Climate warming has pronounced effects on forests worldwide, particularly in the high latitudes of the boreal forest region. These effects have altered forest productivity [1,2], forest composition [3], and natural disturbance regimes directly and indirectly [4–6], and are expected to continue and intensify in the future [7,8].

Changes in annual and seasonal temperatures and precipitation have directly impacted forest growth rate [9,10] and the establishment of native species and exotic species [11,12]. These changes can also alter competitiveness relations among species [13–15] and lead to shifts in species distributions [16–18]. The resulting alterations in forest composition [3] and distribution are expected to affect the sequestration of carbon by forests at broad spatial scales [1,2].

Climate warming indirectly impacts forest compositions and species' distributional patterns through its effects on natural disturbances such as fires [4,19–21]. In boreal forests, fire is a force that can influence forest succession and structure [22]. Both predictions and observations indicate that fire occurrence and area burned have been projected to increase with longer and warmer growing seasons [5,23–27]. For instance, Stocks et al. [25] projected that the areal extent of extreme fire danger in Russia

and Canada could greatly increase. Flannigan et al. [23] showed that the annual burned area in Canada could increase by 74–118% by the end of this century. Wotton et al. [27] similarly indicated that fire occurrence in the boreal forests of Canada could increase by 75–140% by year 2100. Soja et al. [5] assessed the current situation of boreal ecosystems as they relate to previous predictions of climate-induced ecological change, and indicated that the area burned both in Siberia and North America over recent decades has been steadily increasing. Liu et al. [28] projected that the mean fire occurrence density of a boreal forest in northeast China would increase by 30–230% under climate warming by 2100. Previous studies indicated the effects of increased fires on forest composition and forest productivity may equal or exceed the direct effects of climate warming in the boreal forest region [6,29,30]. For example, Schumacher and Bugmann [30] showed that fire was likely to become almost as important in shaping the forest landscape in the Swiss Alps as the direct effects of climate warming.

Timber harvest is one of the main anthropogenic disturbances to forests. Harvesting alters woody biomass accumulation, forest composition, and patterns of tree distribution across the landscape, and these effects may continue under a climate changing scenario [31,32]. He et al. [32] estimated tree species response to forest harvesting and increased fire due to climate warming in northern Wisconsin forests, and indicated that forest harvesting accelerated the decline of northern hardwood and boreal tree species. Gustafson et al. [31] predicted global change effects on Siberian forests and found that harvesting effects on forest composition in boreal forests in Siberia were more significant than effects of climate warming.

Currently, there is increasing interest in exploring effects of climate warming, burning, and timber harvesting on forest landscapes because quantifying these effects can provide a basis for developing forest management policy under a changing climate. However, predicting the effects of climate warming, burning, and harvesting on forest landscapes is challenging because forests ecosystems involved complex interactions among forest successional trends, natural disturbances (including fire) and anthropogenic disturbances (including timber harvesting).

In this study, we employed a forest landscape model (LANDIS) to simulate forest landscape responses to climate warming, burning, and timber harvesting. LANDIS, a widely used forest landscape disturbance and succession model, independently simulates forest succession, natural disturbances and anthropogenic disturbances [33,34]. LANDIS can incorporate the effects of climate warming on tree species and allow these effects to interact with landscape processes in the simulations [35]. The objectives of this study were to (1) quantify the management, disturbances, and species succession in a boreal forest landscape in northeastern China under a changing climate, (2) design a factorial experiment to assess the relative effects of climate warming as determined by species establishment probabilities and increased fire associated with climate warming, and current harvesting on forest species composition, age structure, and spatial pattern.

Methods

1. Study Area

Our study area included Huzhong Forestry Bureau (770,432 ha) and Huzhong Natural Reserve (166,906 ha), located on the north side of the Great Xing'an Mountains in northeastern China, which covers nearly a million ha ($122°39'30''$ to $124°21'00''$E and $52°25'00''$ to $51°14'40''$N) and is primarily a hilly mountainous region ranging from 450 to 1500 m in elevation. The climate is terrestrial monsoon with long and severe winters (mean January temperature $-25.5°C$) and short, mild summers (mean July temperature $18°C$). Precipitation, which peaks in summer, is 420 mm annually and is unevenly distributed throughout the year.

Vegetation in this region falls within the cool temperate coniferous forests occurring at the southern extension of the eastern Siberian light coniferous forest [36]. Canopy trees include larch (*Larix gmelini*), Mongolian Scots pine (*Pinus sylvestris* var. *mongolica*), Korean spruce (*Picea koraiensis*), birch (*Betula platyphylla*), aspen (*Populus davidiana*), poplar (*Populus suaveolens*), willow (*Chosenia arbutifolia*), and a shrub species *Pinus pumila* (dwarf pine) (Table 1). Larch, comprises about 78.4% of the forest area, and is the dominant and most widely distributed tree species except in some riparian wetlands. Birch, a pioneer species with strong colonization ability, always coexists with larch after burning and harvesting. Mongolian Scots pine, with its limited spatial distribution, is always mixed with larch. Dwarf pine occurs with larch or birch where t elevation is over 700 m. Willow and poplar are confined to along the riverbanks. Because larch and birch are the two most widely distributed species in this region, for convenience we refer to these as major species and the others as minor species to notate conveniently. Both the species composition of vegetation in Huzhong Forestry Bureau and in Huzhong

Natural Reserve is similar. Although forests both in the forestry bureau and the natural reserve have succeeded into old growth stage (200–300 years), forests in the forestry bureau due to timber harvesting mostly are early stage (40–80 years). We divided the area not covered by forest into water and non-forest areas which collectively often serve as fire breaks. We also divide the forest area into four topographic subdivisions (land types: terrace, south slope, north slope, and ridge) to better consider the effects of microclimate, soil, and complex topography on species distributions (Figure 1).

Wildfire was the major disturbance in the Great Xing'an mountains before the forest was exploited by humans, and it is one of the more important factors that drive forest succession [37,38]. Based on Chinese Federal Forest Service data, lightning-caused fires burning 131603.6 ha and 14% of the landscape account for about 94% of the total fires and nearly 99% of the total burned area during the 40-year period from 1965 to 2005 for our study area. The forest in this region has been exploited by humans in various ways since the 1950s, and timber harvesting has significantly impacted forest age structure and natural regeneration [37]. Based on Chinese Federal Forest Service data, larch forests have shifted from late-seral or old-growth stages to a mid-seral stage. Consequently, to maintain older age classes and therefore forest integrity and sustainability, timber harvesting has been restricted by the government since 1999. Hence, the age structure of forests in this region has been shaped primarily by burning and timber harvesting. Mature larch stands, which develop after multiple surface fires have killed ground vegetation and shrubs, allow new cohorts of tree reproduction to establish after each fire, which then form unevenaged stands (≥ 3 age classes). All other stands established after stand-replacing fire or harvest are evenaged.

2. Design of Simulation Experiments

To assess the effects of climate warming, increased frequency of burning, and timber harvesting on forest landscapes under future climate scenarios, we designed a factorial experiment with three independent variables each with two levels: climate (current vs. expected future), fire occurrence (current vs. increased fire), and timber harvesting (no harvest vs. legal harvest). Each treatment combination was simulated at 10-year steps over 300 years (1990–2290) replicated with five times which meets the minimum number for the statistic requirement.

2.1 Climate warming scenarios. Current climate data were obtained from the Northeastern Regional Meteorological Center in China, and daily temperatures and precipitations were compiled for 1961–1990. Average monthly temperature and precipitation were derived from daily temperature and precipitation data. To process climate data for the current and warming scenarios we first used data from six weather stations distributed across the northern Great Xing'an mountains to develop monthly temperature and precipitation values along the longitudes, latitudes and elevations of the mountains. We then calculated monthly temperatures and precipitations based on the developed relationships [39].

Expected future climate parameters were estimated based on projections of the Hadley GCM (UKMO-HadCM3) running under the A2 scenario for 2070–2099. The A2 scenario represents high CO_2 emissions (1250 ppm) related to high human population size and slow technological adaptation (IPCC 2007). We chose the Hadley GCM because it is widely accepted and provides the method currently considered the best for detecting climate warming effects. It predicts warmer and moister summers compared to many other GCMs. We obtained changes (projected

Table 1. Species life-history attributes for canopy species in northeastern China.

Species	Common name	LONG	MTR	SHD	FIRE	EFFD	MAXD	VGP	MVP	CSEPs	WSEPs	DMAX	DMIN
Betula platyphylla	Birch	120	15	1	3	0	2000	1	20	0.153	0.341	3100	600
Populus davidiana	Aspen	100	10	1	1	0	2000	1	15	0.010	0.178	3000	800
Populus suaveolens	Poplar	180	12	1	2	0	2000	1	15	0.013	0.048	1900	400
Chosenia arbutifolia	Willow	250	18	2	1	0	3000	0.9	15	0.018	0.054	2400	600
Larix gmelinii	Larch	300	20	2	5	50	150	0	0	0.288	0.136	1900	400
Picea koraiensis	Spruce	300	30	4	1	100	150	0	0	0.060	0.112	2500	800
Pinus sylvestris var. mongolica	Mongolian Scots pine	250	40	2	4	100	200	0	0	0.175	0.212	2400	700
Pinus pumila	Dwarf pine	250	30	3	3	50	100	0	0	0.225	0.144	1400	300

LONG, longevity of the species (years); MTR, maturity age of the species (years); SHD, shade tolerance value (1–5) (no units), 1 = least tolerant, 5 = most tolerant; FIRE, fire tolerance value (1–5) (no nits), 1 = least tolerant, 5 = most tolerant; EFFD, species effective distance seeding range (m); MAXD, species maximum distance seeding range (m); VGP, probability of vegetative propagation following disturbance (no units); MVP, minimum age of vegetative propagation (years); CSEPs, probability of species establishment under current climate; WSEPs, probability of species establishment under climate warming scenario; DMAX, maximum growing degree day; DMIN, minimum growing degree day each species.

by the Hadley GCM) in monthly average temperature and precipitation over the next 100 years from ClimateWizard [40], and recorded the data as a gridded dataset with a 0.5×0.5° resolution. The Hadley GCM projected that annual average temperature and annual precipitation would increase by 5°C and ~35%, respectively by the 2080s. We modified current climate means and monthly weather records by adding the change in temperature (°C) and precipitation (mm) between the Hadley projection for the periods 1961–1990 and 2070–2099. During the 21st century, we assumed that precipitation and temperature trends are linear and that variability in temperature and precipitation would not change [41].

2.2 Fire regimes under two climate scenarios. The historical fire regimes (including fire ignition probability, mean return interval, mean fire size, and standard deviation of fire size)

for our simulations were parameterized from a fire database from 1990 to 2000. According to Chinese Federal Forest Service data (1990–2000), mean return interval is 238 year, mean fire size is 1884 ha, and fire ignition probability is 0.00402 for the historical fire regimes. Under the A2 scenario, both the fire ignition probability and mean fire size was increased by 200% based on historical regimes, and mean return interval was decreased to 1/3 of the historical regimes. Liu et al. [28] predicted that fire occurrence of boreal forest in the northeast China under the A2 climate change scenario projected by the Hadley GCM would increase by 230% in 2100.

2.3 Forest management scenarios. Our study area is divided into three forest management sub-areas: areas where cutting is banned (50.1%), areas with restricted cutting (23.08%), and areas where timber harvesting is permitted (26.82%). Each of

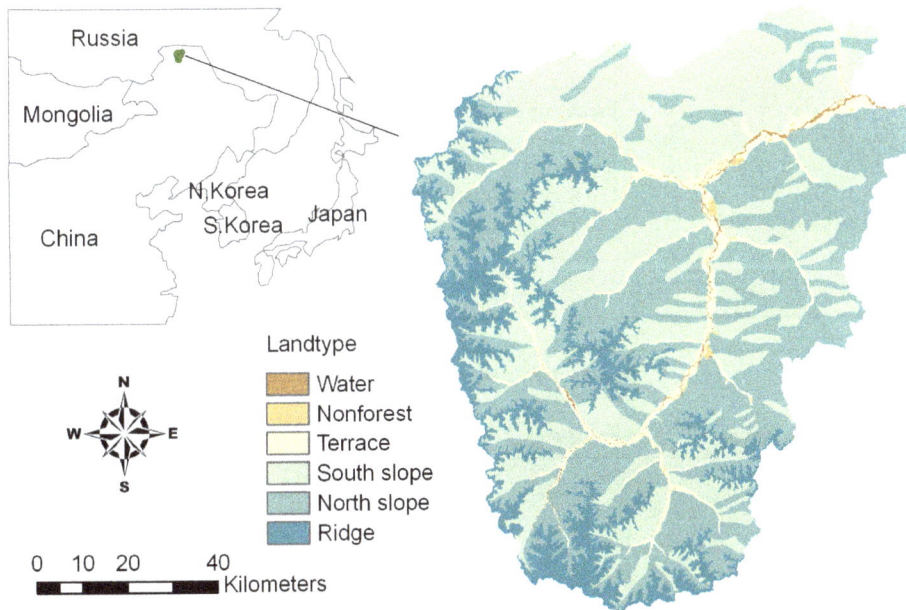

Figure 1. The geographic location of the study area and different land types, among which water and nonforest are not simulated in the model.

these sub-areas is further divided into compartments (stands) with an average size of 20 ha. Harvest units do not exceed 10 ha in size, and adjacent stands are not harvested for at least five years. Current forest managements are different in specific sub-areas for the study area (Table 2). In our study, we assumed that only legal harvesting occurred and that harvesting regimes did not vary. To select stands for harvest, we applied the oldest-first method, in which all stands within a management area were ranked by age and stands of oldest age were harvested first. We also set the minimum stand age for harvest at 40 years, reflecting the current harvest practices. All species except dwarf pine and willow were harvested. Dwarf pine is not harvested because it is a key shrub species for maintaining habitats ≥ 1000 m elevation; similarly willow is crucial to maintaining riparian habits.

3. Model Simulation

We used a forest landscape model (LANDIS) to simulate forest landscape dynamics under different climate, fire, and harvesting scenarios. LANDIS can be used to simulate forest landscape changes related to forest succession and disturbances at large heterogeneous spatial extents (10^3–10^7 ha) over long time spans (10–1000 years) based on raster data in which each cell contains unique information relating to specific species, age cohorts, and time since last disturbance.

We used LANDIS modules (forest succession and seed dispersal, fire disturbances, and harvesting) to simulate forest landscape changes of the study area under different climate scenarios. LANDIS simplifies within-stand processes and individual tree information, and tracks the presence or absence of species age cohorts to simulate succession. Succession at each stand is a competitive process driven by species life history attributes such as longevity, age of sexual maturity, shade tolerance class, fire tolerance class, the minimum age of vegetation sprouting, sprouting probability, and effective and maximum seeding distance. Succession at the landscape scale involves seed spatial dispersal among cells and the different capability for species establishment on each land type [34,42].

LANDIS simulates seed dispersal using an exponential distribution in which the effective and maximum dispersal distances of specific species control seedling distribution [42]. When seeds successfully arrive at a site, the shade-tolerance rank of the seedling relative to the species existing on the site determines the recruitment of the seedling. Whether the seedling successfully establishes and survives is determined by the specie establishment probability (SEP, a value ranging from 0–1). The species establishment probability quantifies how a species establishes in different environmental conditions. Species with high establishment probability have higher probabilities of establishment, and are as responses of tree species to climate in LANDIS. SEPs as input to LANDIS are estimated based on existing experimental data [43] or derived from the simulation results of a gap model [35] such as LINKAGES.

In our study, we employed LINKAGES (a derivative of the JABOWA/FORET class of gap models) [44] to simulate the physiological response of tree species to both current and warming climate within each land type. The physiological response was quantified as individual species biomass, and was used to estimate the SEP for specific species. Individual species biomass was determined by simulating the interactions of climate, soil properties (derived from soil survey data in the study area), and species biological traits (compiled based on previous studies in this area) with ecological processes. The climate properties utilized included monthly temperature and precipitation (Table S1 in Appendix S1). The soil properties included field water capacity, wilting point, total nitrogen, and total carbon. The species biological traits included longevity, maturity, shade and drought tolerance, and seedling capability [36,37]. The ecological processes simulated were competition, succession, and water and nutrient cycling.

To examine variations in species establishment by land type we simulated one species at a time in LINKAGES planting the same number of trees (200 saplings/ha) for each land type [35]. We converted the simulated biomass for all land types under both current and warming climate to two sets of SEPs (Table S2 in Appendix S1) using an empirical method [35] (S1 in Appendix S1).

To simulate a gradual change in climate under the currently expected scenario, SEPs for the climate treatment initially assumed current values and then were modified by a 10-year-interval linear interpolation of values calculated from simulated year 0 to year 100. After 100 years, model values were held constant. Our estimates are probably conservative, because warmer conditions are expected to become more pronounced after 2090 [7].

In LANDIS, fire is simulated as a stochastic process based on the ignition probability distribution, mean return interval and mean size characterized for various land types [34,45]. The fire module simulates temporal patterns of fire regimes using a hierarchical fire frequency model, which divides a fire occurrence into two consecutive events: fire ignition and fire initiation. Fire ignition is generated from the Poisson distribution based on the fire ignition density defined in fire regimes. Whether a fire ignition can result in fire initiation is dependent on the fuel loading, fuel

Table 2. Parameters of harvest scenario.

| Species | Age range (year) | Cutting method | % Area harvested (of each management area per decade | | Regeneration |
			Harvest area	Restricted cutting area	
Larch	120–300	clearcut	0.5%	0.3%	natural
Mongolian Scots Pine	90–250	clearcut	0.5%	0.3%	natural
Spruce	120–300	clearcut	0.5%	0.3%	natural
Birch	60–150	clearcut	0.5%	0.3%	natural
Aspen	40–120	clearcut	0.5%	0.3%	natural
Poplar	50–180	clearcut	0.5%	0.3%	natural

The harvest scenario was adopted from current forest management of Huzhong Forest Bureau and was parameterized in LANDIS harvest module.

arrangement, and fuel moisture content. A fire initiation event starts with the ignition and is successful when an area equal to the cell size is burned. For each fire initiation, LANDIS simulates fire spread using a modified percolation method spread from a burning cell to forested cells in the cardinal directions (north, northeast, east, southeast, south, southwest, west, northwest). These cells are randomly entered into a priority queue. The fire will spread by randomly selecting a fire size using a log-normal distribution based on mean fire size and standard deviation of fire size. Fire intensity is determined by the time since the last fire on the site, as well as the amount of fuel present within each cell. In LANDIS, small or young trees are more susceptible to fire than large or older trees. LANDIS simulates fire intensity from low intensity ground fires to high intensity crown fires as five levels. Correspondingly, trees are sorted into five-fire tolerance classes. Fire severity is the interaction of species fire tolerance, species age cohorts, and fire intensity [34].

In the harvest module of LANDIS, timber harvests are simulated within a specific hierarchical management structure. The overall landscape is divided into forest management areas, each to be treated by specific harvest regimes at specific intensities. Each management area is divided into stands of various forest types. Each stand includes a group of grid cells being populated with a specific combination of species and age cohorts. Within each management area, harvests are implemented by removing specific cohorts of specific species on sites selected for harvest based on harvest regimes. The harvest regimes prescribe the harvest rules (e.g., how to harvest a stand, how to allocate a harvest based on stand ranking which in turn is based on ecological or economic criteria, and how to harvest age cohorts of tree species such as shelterwood, selection, and clear cutting) [33]. The harvest regime is controlled by users based on the targets of managers, and it is determined by the combination of temporal, spatial, and species composition [33].

Various components and processes have been described elsewhere [32–34,42]. The effectiveness of the LANDIS model in boreal forest ecosystems in northeastern China has been demonstrated in previous studies [46–48]. The uncertainty analysis on model parameterization and result variations has been previously done by Xu et al. [48,49]. Their research showed that the uncertainty was low at the beginning of the simulation, increased with simulation year, and finally reached an equilibrium state, where the uncertainties of input parameters had little effect on the simulation results (species percent area and their spatial patterns) at the landscape level. To simulate landscape processes, LANDIS requires setting variable parameters and creating maps for model initialization. Maps delineate forest composition, land types, and type of forest management permitted. Parameters include species vital attributes (Table 1), species establishment probabilities (SEPs), harvest regime attributes, and fire regime attributes. The data for parameterization of LANDIS include the forestry inventory taken in 1990 in the study area, two Landsat TM scenes taken in 1990, fire records from 1990 to 2000, and a Digital Elevation Model (DEM) generated from the contour lines delineated in 1971. These parameters and maps have been specified in previous studies [48,50].

4. Data Analysis

Landscape responses to climate warming, burning, and timber harvesting were quantified in related to species composition, tree age structure, and forest landscape pattern. Model outputs primarily were derived from maps showing the effects of the three independent variables on species composition. Maps were produced for simulations at each 10-year time step. We chose a 10-year time step for mapping the dominant species and the maximum cohort age of all species combined. Species composition was expressed as the proportion of the landscape dominated by each species. We examined forest age structure using five age-based stages: 1–40 years (establishment stage), 41–100 years (early-stage), 101–140 years (mid- stage), 141–180 years (late-stage), and >180 years (old-growth stage) (Gustafson et al., 2010). To quantify spatial pattern of seral stage and the major species (larch and birch) we used an aggregation index (AI) which reflects the tendency of like cells to be adjacent [51], and Shannon diversity index (SHDI) which reflects the heterogeneity of a landscape [52].

The effects of climate warming, forest burning, and timber harvesting were analyzed using multiple analyses of variance (MANOVA). With this method, we tested (for the significance of each independent variable) effect on the dependent variable. We ran separate analyses for each response variable (species composition, tree age structure, and spatial pattern). We chose a subset of representative variables to reduce multi-collinearity within each analysis (Table 3). Because the response variables varied through time, we chose simulation-years 150 and 300 (actual years 2140 and 2290) as representative of the varying response (Table 3). MANOVA models used the error sums of squares and cross products (residual) matrix, and the results were evaluated using Type I sums of squares. The relative effects of every treatment were quantified as the percent of the total variation attributed to each effects and significance was judged conservatively using at $\alpha = 0.01$. Our explanations focused on trends rather than statistical significance because random noise in the tightly controlled simulations was minimal.

5. Simulation Result Verification

Simulation result validation requires long term spatial and temporal data, which are not available. This is especially true for the climate warming scenario. To gain assurance of the simulation results we compared species composition and age composition of simulation years 200–300 to those from the natural reserve in our study area. Forests in the natural reserve have reached the old growth stage with ages ranging from 200 to 300 years. Thus, results from simulation year 200–300 provide legitimate comparisons for result verification.

Results

1. Results Verification

Simulation of the current climate and disturbance regimes (fire and harvesting) showed that the mean of the proportions for most species and age classes were similar to the observed from Huzhong Natural Reserve (Table 3). The ranges of simulated proportions for most species and age classed were coincident. Only the ranges of simulated the proportions for larch and birch were discrepant.

2. Species Composition Responses to Climate and Fire Under the Current Harvest Regime

The abundance of willow, poplar, and aspen for a given future climate and fire regime can be expected to significantly increase compared to the abundance of these species under current climate and fire regimes. Moreover, the simulations showed that the abundance of conifers such as spruce and the two pines greatly decreased whereas birch abundance decreased more slowly. The simulations also indicated that the predominant tree species in this region may be expected to change from predominantly conifers to broadleaf species (Figure 2).

Table 3. Comparison of forest composition outside the natural reserve under the current climate to observed value in Huzhong natural reserve.

	Huzhong natural reserve	Outside the natural reserve		
	%Observed	%Initial conditions	% Range (years 200–300)	%Mean (years 200–300)
Species composition				
Aspen	1.1	4.5	2.0–2.1	2.1
Birch	33.1	34.1	29.0–33.3	32.4
Poplar	1.1	0.5	0.2–0.6	0.4
Willow	1.2	0.7	1.1–1.3	1.2
Larch	52.9	44.7	45.9–51.0	47.3
Spruce	1.0	1.3	2.6–3.4	3.0
Mongolian Scots pine	1.4	5.0	3.3–3.5	3.4
Dwarf pine	7.2	9.3	10.2–10.4	10.4
Age composition				
Establishment (1–40 yr)	3.4	12.9	3.7–6.0	4.5
Early-stage (41–100 yr)	9.6	42.8	8.8–13.0	10.4
Mid-stage (101–140 yr)	10.5	32.5	10.9–14.6	11.9
Late-stage (141–180 yr)	20.2	11.5	24.7–29.8	19.1
Old-growth (>180 yr)	56.3	0.3	47.8–62.5	54.2

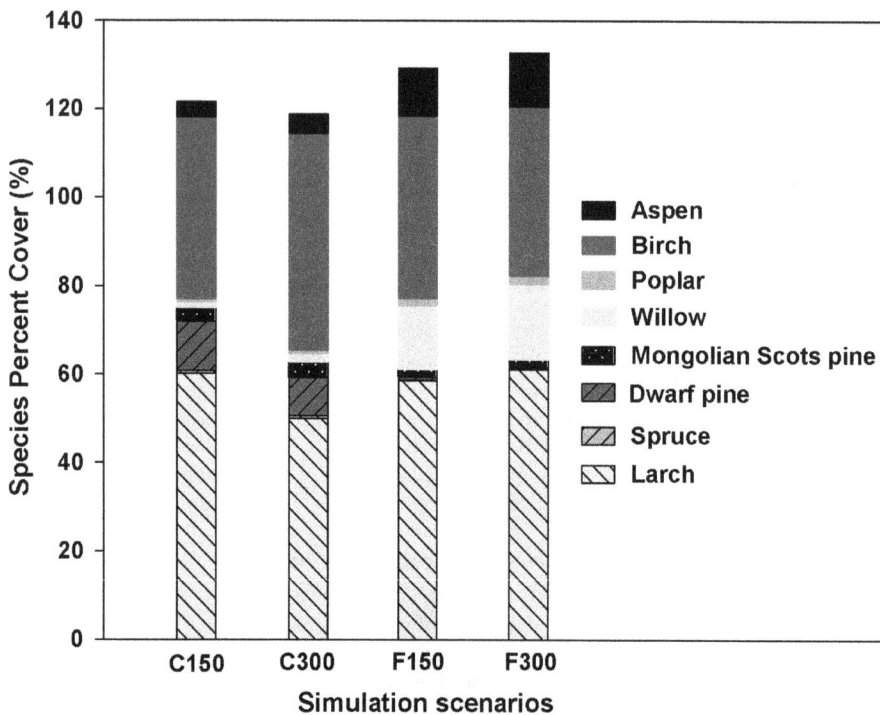

Figure 2. Responses of species composition to climate and fire under current harvest regime. C150 represents the average of the simulated species abundance under current climate and fire regimes in 2000–2140 years; C300 represents the average of the simulated species abundance under current climate and fire regimes in 2150–2290 years; F150 represents the average of the simulated species abundance under future climate and fire regimes in 2000–2140 years; and F300 represents the average of the simulated species abundance under future climate and fire regimes in 2150–2290 years.

3. Effects of Climate Warming, Burning, and Harvesting on Forest Species Composition

The simulations show that tree species composition is be expected to be strongly impacted by climate warming, fire, and harvest (Table 4). Simulated climate warming significantly increased the abundance of most species, and its effects on most species were greater in year 2290 than in year 2140. The effects of climate warming on the abundance of aspen, poplar, and willow were major compared to the main effects of fire and timber harvest. However, effects of climate warming on the abundance of birch, larch, spruce, and pines were much smaller than effects of fire and timber harvest. Burning mostly reduced the abundance of spruce and pines in both year 2140 and year 2290, and its effects were greater in year 2140 than in year 2290. Increased fire strongly increased larch abundance and decreased birch abundance in year 2290. Birch and larch, the major species in this region were mostly impacted by harvest. The harvest treatment increased birch abundance and decreased larch abundance in both years, and its effects were markedly smaller in year 2290 than in year 2140.

4. Effects of Climate Warming, Burning, and Harvesting on Forest Age Structure

Forest age structure was impacted more strongly by fire and harvest rather than climate warming, although direct climate warming effects would increase the SEPs of most species (Table 5). Increased fire had positive effects on the abundance of tree establishment in both year 2140 and year 2290 and had strong negative effects on their abundance during mid-stage and late-stage in both years, and these effects were greater in year 2140 than in year 2290. Harvesting strongly increased tree abundance in the early stage and decreased its abundance in the late stage and old-growth stage, and these effects were markedly greater in year 2290 than in year 2140.

5. Effects of Climate Warming, Burning, and Harvesting on Forest Landscape Pattern

The spatial pattern of the aggregation index (AI) and Shannon's diversity index (SHDI) responded strongly to increased fire and harvest (Table 6). Climate warming significantly affected forest fragmentation and forest diversity. However, its effects were markedly smaller than the effects of increased fire and harvesting. Both increased fire and timber harvesting would increase forest fragmentation and decrease forest diversity. However, increased fire was the predominant influence in both the fragmentation and the diversity of age classes; an exception was the diversity of age classes in year 2290. Harvesting impacts were greater for the pattern-of-response variables for birch and larch. The arithmetic sign and relative strength of these effects were not always consistent through time (e.g., AI-larch and SHDI-birch for timber harvesting effects Table 6). For example, larch was the major tree species that is harvested in this region and was also the dominant species in the early years of simulations. Harvest effects on the aggregation index of larch (AI-larch) in year 2140 were significant (Table 6). However, the abundance of larch rapidly decreased through time due to the action of various disturbances and succession. Thus, harvesting effect on AI-larch was greatly reduced by year 2290.

Table 4. MANOVA results for species composition variables.

Simulation years	Species (%)	Climate effect[‡] Variation explained (%)	t	Fire effect[£] Variation explained (%)	t	Harvest effect[§] Variation explained (%)	t	Fire×harvest Variation explained (%)	R^2
150 (2140)	Aspen	79.8**	**26.3**	3.0**	**4.5**	13.0**	**8.4**	0.2	0.96
	Birch	0.3	**4.9**	1.3**	**17.9**	96.0**	**77.7**	2.0**	1.00
	Poplar	61.3**	**12.7**	12.1**	**5.1**	12.5**	**5.1**	0.9	0.85
	Willow	90.3**	**26**	0.06	0.63	4.6**	**5.3**	0.3	0.95
	Larch	5.3**	**16.5**	24.8**	**−30.2**	68.1**	**−46.7**	1.0**	0.99
	Spruce	5.8**	**6.9**	88.4**	**−20.6**	0.9**	**−3.6**	0.6	0.95
	Mongolian Scots pine	0.3	1.4	78.8**	**−17**	1.8**	0.34	1.3**	0.94
	Dwarf pine	0.02	**−3.7**	99.2**	**−159.1**	0.4	**19.5**	0.3	1.00
300 (2290)	Aspen	92.9**	**49.5**	2.1**	**−5.4**	3.7**	**6.9**	0.001	0.99
	Birch	1.7**	**−7.7**	58.6**	**−30.8**	38.6**	**25.9**	0.02	0.99
	Poplar	82.4**	**13.3**	0.3	−0.64	1.0**	0.91	0.01**	0.82
	Willow	81.0**	**17.5**	7.9**	**−3.3**	1.7**	2.3	0.1	0.90
	Larch	0.7**	**3.7**	37.1**	**15**	58.4**	**−29.5**	2.2**	0.98
	Spruce	10.9**	**6.2**	78.4**	**−12.7**	0.6	**−1.9**	0.4	0.89
	Mongolian Scots pine	39.4**	**13**	51.7**	**−17.8**	11.0**	**3.8**	0.3	0.89
	Dwarf pine	0.01	−1.6	97.7**	**−118.8**	1.3**	**28.2**	1.0**	1.00

The t values test the hypothesis that the response between levels of main effects are equal, and significant ($\alpha = 0.01$) differences are indicated in boldface. All three main effects were significant in both years. Only the fire × harvest interaction was always significant and was included in the model. Significant interactions are indicated by asterisks.
**$P < 0.01$.
[‡]Positive t value means that response variable increases as climate warms.
[£]Positive t value means that response variable increases as fires increases.
[§]Positive t value means that response variable increases when harvest is added.

Table 5. MANOVA results for forest age structure variables.

Simulation years	Seral stage (%)	Climate effect[‡] Variation explained (%)	t	Fire effect[£] Variation explained (%)	t	Harvest effect[§] Variation explained (%)	t	Fire×harvest Variation explained (%)	R²
150 (2140)	Establishment	0.4**	**12.2**	40.8**	**77.9**	58.6**	**94.1**	0.1**	1.00
	Early-stage	7.4**	**15.2**	5.3**	**12.2**	85.5**	**39.6**	0.6**	0.99
	Mid-stage	0.7**	**22.1**	97.3**	**−155.9**	0.08	**19.7**	1.8**	1.00
	Late-stage	0.002	1.5	77.0**	**−239.6**	22.2**	**−116.4**	0.8**	1.00
	Old-growth	0	0.77	5.1**	**−106.2**	94.8**	**−419**	0.09**	1.00
300 (2290)	Establishment	0.2**	**−4.3**	8.6**	**12**	89.4**	**57**	1.4**	1.00
	Early-stage	2.0**	**10.9**	6.6**	**−14.1**	90.8**	**51.6**	0	0.99
	Mid-stage	28.2**	**9.6**	52.2**	**−7.6**	7.3**	**−1.9**	1.5	0.88
	Late-stage	5.6**	**14.9**	15.5**	**−6.5**	71.9**	**−26.7**	6.1**	0.99
	Old-growth	0.2**	**16.4**	0.09**	**8.4**	99.2**	**−220.3**	0.4**	1.00

The t values test the hypothesis that the response between levels of main effects are equal, and significant (α=0.01) differences are indicated in boldface. All three main effects were significant in both years. Only the fire × harvest interaction was always significant and was included in the model. Significant interactions are indicated by asterisks.
**P<0.01.
[‡]Positive t value means that response variable increases as climate warms.
[£]Positive t value means that response variable increases as fires increases.
[§]Positive t value means that response variable increases when harvest is added.

Discussion

We estimated the relative effects of climate warming, burning, and timber harvesting on forest landscapes in northeastern China. The results showed that every treatment (climate, fire, and timber harvest) had strong effects on forest composition and forest spatial pattern. The effects of climate warming on tree species composition were significant but had a lag time (Table 4). However, forest age structure was mostly impacted by forest disturbance rather than direct climate changes (Table 5). This effect is likely related to the direct effects of climate on the abundance birch, larch, spruce, and two pines, which were smaller than the effects of

Table 6. MANOVA results for spatial pattern variables.

Simulation years	Pattern index	Climate effect[‡] Variation explained (%)	T	Fire effect[£] Variation explained (%)	T	Harvest effect[§] Variation explained (%)	t	Fire×harvest Variation explained (%)	R²
150 (2140)	AI-seral stage	2.4**	**14**	52.3**	**−40.8**	44.2**	**−37.1**	0.7**	1.00
	AI-birch	0.2	**−10**	0.2	**20.2**	95.9**	**−113.4**	3.5**	1.00
	AI-larch	4.6**	**10**	26.3**	**−15.5**	67.3**	**−25.7**	0.2	0.98
	SHDI-seral stage	0.9**	**20**	52.4**	**−70.1**	40.3**	**−56.9**	6.3**	1.00
	SHDI-birch	0.5**	**6.6**	4.8	**28.2**	90.4**	**77.8**	3.9**	1.00
	SHDI-larch	2.3**	**−18.7**	45.9**	**80.3**	45.9**	**80.3**	5.8**	1.00
300 (2290)	AI-seral stage	8.0**	**41.9**	83.1**	**−89.8**	8.4**	**−24.9**	0.3**	1.00
	AI-birch	1.5**	**−10**	0.03	**21.1**	83.5**	**−31**	14.5**	0.99
	AI-larch	33.5**	**8.5**	5.4**	1	34.0**	−2.6	11.0**	0.82
	SHDI-seral stage	8.9**	**13.2**	1.9**	2.6	82.6**	**−21.5**	4.8**	0.98
	SHDI-birch	0.007	-0.27	19.8**	**−11.3**	76.9**	**19.7**	0.14**	0.96
	SHDI-larch	7.4**	**−14.7**	23.5**	**17.6**	67.9**	**30.6**	0.07	0.99

The t values test the hypothesis that the response between levels of main effects are equal, and significant (α=0.01) differences are indicated in boldface. All three main effects were significant in both years. Only the fire × harvest interaction was always significant and was included in the model. Significant interactions are indicated by asterisks.
**P<0.01.
[‡]Positive t value means that response variable increases as climate warms.
[£]Positive t value means that response variable increases as fires increases.
[§]Positive t value means that response variable increases when harvest is added.
AI is the aggregation index of He et al. (2000) that reflects the tendency of like cells to be adjacent, SHDI is Shannon diversity index that reflects the heterogeneity of landscape.

disturbance. This is consistent with previous studies that have shown that direct effects of climate warming on forest composition are not as great as effects of harvesting and increased fire [29–31].

In our study, fire regime was specified primarily by area burned, fire size, and fire severity (i.e., underground fire, surface fires, and canopy fires). We reproduced the empirical (current) fire regime quite closely, and found that the total area burned fluctuated in a small range after the year 200 simulation when climate warming continued to exacerbate conditions (Figure 3). As expected, fire had strong effects on forest composition. Fire overshadowed the direct effects of climate warming on the abundance of spruce, pines, and larch. Moreover, fire effects on broadleaf species were much smaller than on coniferous species. This is consistent with the results of He et al. [32], which also showed that increased fire can accelerate the decline of shade-tolerant species. However, potential effects of increased fire may be variable because fire events are highly variable in both size and frequency [23,27,28].

Timber harvesting nevertheless can be expected to strongly impact forest composition in regions undergoing climatic change. In our study region, this is largely the result of changes in the abundance of the major species (birch and larch) and the abundance of late-stage and old-growth tree age classes. This is consistent with the results of Gustafson et al. [31] and He et al. [32]. Gustafson et al. [31] showed that harvest effects were more significant than the effects of climate warming in south-central Siberia. The forests in the Great Xing'an mountains have long been exploited by human, and the resulting effects have thus been continuous and complex. However, timber harvesting and other human activities can be potentially controlled by managers. Therefore, we propose that our modeling approach should be useful in evaluating alternative management policies to mitigate at least some of the negative effects associated with climate warming.

In our study, climate warming was significantly and directly related to forest landscape pattern (Table 6). Nevertheless, those effects were overshadowed by fire or harvesting. Similar to the results of Gustafson et al. [31], timber harvesting in our study also increased forest fragmentation. Increased fire altered the direct effect of climate change on forest fragmentation and the diversity of the pattern of tree age classes. Timber harvesting also altered the fragmentation and diversity of the compositional pattern of tree species. In our study, we selected only the aggregation index (AI) and Shannon's diversity index (SHDI) to represent fragmentation and diversity. It would require more detailed analysis of changes on the landscape including tree species composition, patch size, and connectivity relative to specific species' life requirements to arrive at more specific conclusions.

Our results nevertheless suggest that changes in forest landscapes are complex and involve continuous interactions among climate, fire, and timber harvesting effects. The modeling approach we used can be used to evaluate forest management policy options for mitigating negative effects of climate warming. When interpreting our simulation results, it is important to note the following limitations:

(1) We only incorporated the effects of alteration of temperature and precipitation on forests; however, tree growth is impacted by changes in solar irradiation and CO_2 fertilization [53].

(2) Climate warming indirectly influences forest landscapes through tree species migration, shifts in soil nitrogen deposition [54], changes in natural disturbance regimes such as fires [20] and insect outbreaks [55]. However, we only incorporated changes in the fire regimes.

(3) Fire events are highly variable in size and frequency [24,27]. Krawchuk et al. [24] illustrated that area burned would increase 1.9-fold by 2040–2049 and 2.6-fold by 2080–2089 relative to 1975–1985 conditions. However, we initially parameterized future fire regimes by increasing by 2-fold of the current fire occurrence based on Liu et al. [28].

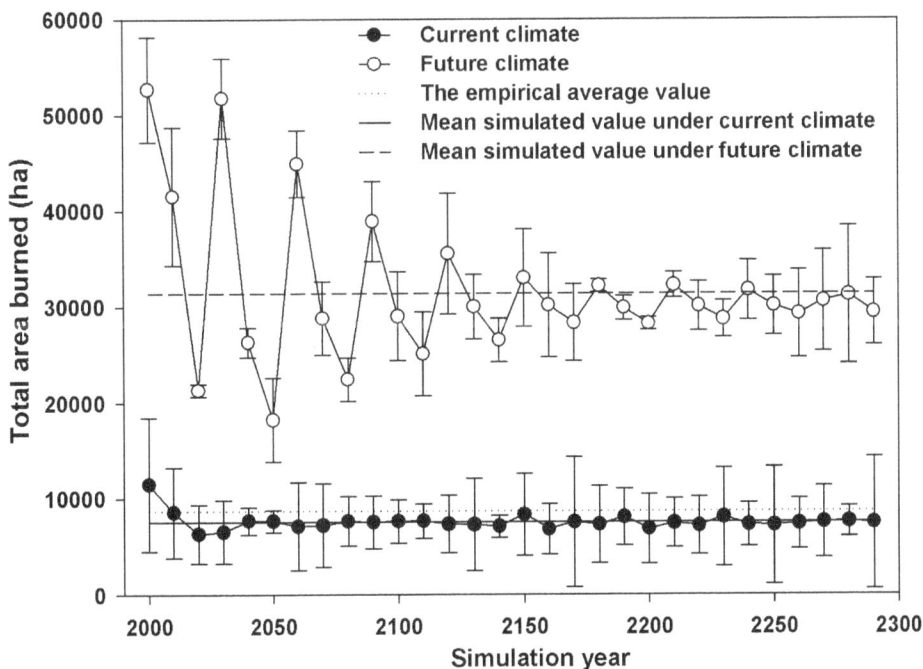

Figure 3. Total area burned per decade and the size distribution of fires (±SD) for 300 simulation years under current climate and future climate.

(4) Spatial pattern dynamics are related to the resolution and grid cell structure used to represent the landscape. The grain size of the initial conditions map may influence the effects of disturbances, because if the grain of a disturbance is the same as the existing pattern, changes in pattern metrics are less likely to be detected.

These limitations provide cautions on concluding that our model makes robust predictions of the future forest dynamics. However, our results should be as reliable as those presented in similar studies.

LANDIS is primarily a process-based model. Its parameters and algorithms are generally accepted as adequate representation of forest dynamics, and that the treatments imposed by modifying parameters related to climate and disturbances should provide useful insight into landscape change. Like other forest landscape models, LANDIS conceptually simulates non-spatial processes and assumes how they interact with landscape processes and with each other. For many processes their behaviors are clearly understood and assumptions are firmly established, but for others less is known about the true behavior of processes. For example, LANDIS simplifies individual tree information and within-stand process and only tracks the presence and absence of species age cohorts for simulating forest succession [42]. The model nevertheless allows large scale questions such as spatial pattern, species distribution, and disturbances to be effectively addressed. When it simulates fires, it performs a Bernoulli trial to address fire ignition, and then randomly select a fire size from a log-normal distribution which is recognized as a distribution useful in simulating fire spread. But fire ignition and spread factually depend upon the combination of weather, fuels, and topography. LANDIS also implements timber harvests within a specific hierarchical management structure, and it removes specific cohorts by specific species on sites selected for timber harvest which are defined by harvest regimes [33]. However, timber harvest and fire also impacts the growth of tree species by affecting the availability of soil nutrients which is a factor not included in LANDIS [56]. Nevertheless our results were reliable, one of the reasons is that LANDIS simulations have been widely used in many simulation studies and its validity reported in other studies in Northeast China [46–48]. Using LANDIS to conduct a controlled simulation experiment allows discovery of general trends in boreal forest responses to climate warming, burning, and harvesting based on our current understanding of the ecological processes that drive forest dynamics.

Conclusions

From our study, we concluded that: (1) the composition of forests of the Great Xing'an mountains is likely to be significantly altered by changing climate, timber harvesting, and burning. (2) The direct effects of climatic change in the study area are not likely to be as important as timber harvesting and the potential for increased burning. (3) Disturbance by burning and harvesting may greatly reduce the abundance of conifer species including larch, spruce, and two pine species. In turn, this may significantly reduce the ecological integrity of these forests by decreasing tree species diversity and increasing forest fragmentation.

Supporting Information

Appendix S1 Estimation of species establishment probability. Table S1 The monthly temperature and precipitation under current climate and future climate. Table S2 Species establishment probability of specific species under two climate scenarios.

Acknowledgments

We thank two anonymous reviewers for their comments which greatly improved earlier versions of this manuscript.

Author Contributions

Designed figures and tables in the manuscript: XL HSH ZW YL. Conceived and designed the experiments: XL HSH YL ZW. Analyzed the data: XL YL JES. Wrote the paper: XL HSH JES.

References

1. Nemani RR, Keeling CD, Hashimoto H, Jolly WM, Piper SC, et al. (2003) Climate-driven increases in global terrestrial net primary production from 1982 to 1999. Science 300: 1560–1563.

2. Zhao MS, Running SW (2010) Drought-Induced Reduction in Global Terrestrial Net Primary Production from 2000 Through 2009. Science 329: 940–943.

3. Johnstone JF, Chapin FS, Hollingsworth TN, Mack MC, Romanovsky V, et al. (2010) Fire, climate change, and forest resilience in interior Alaska. Canadian Journal of Forest Research-Revue Canadienne De Recherche Forestiere 40: 1302–1312.

4. Dale VH, Joyce LA, McNulty S, Neilson RP, Ayres MP, et al. (2001) Climate change and forest disturbances. BioScience 51: 723–734.

5. Soja AJ, Tchebakova NM, French NHF, Flannigan MD, Shugart HH, et al. (2007) Climate-induced boreal forest change: Predictions versus current observations. Global and Planetary Change 56: 274–296.

6. Weber MG, Flannigan MD (1997) Canadian boreal forest ecosystem structure and function in a changing climate: impact on fire regimes. Environmental Reviews 5: 145–166.

7. IPCC (2007) Summary for Policymakers of the Synthesis Report of the IPCC Fourth Assessment Report. Cambridge: UK: Cambridge University Press.

8. Ruckstuhl KE, Johnson EA, Miyanishi K (2008) Introduction. The boreal forest and global change. Philosophical Transactions of the Royal Society B-Biological Sciences 363: 2245–2249.

9. Melillo JM, McGuire AD, Kicklighter DW, Moore B, Vorosmarty CJ, et al. (1993) Global climate change and terrestrial net primary production. Nature 363: 234–240.

10. Wang XC, Zhang YD, Mcrae DJ (2009) Spatial and age-dependent tree-ring growth responses of Larix gmelinii to climate in northeastern China. Trees-Structure and Function 23: 875–885.

11. Hansen AJ, Neilson RR, Dale VH, Flather CH, Iverson LR, et al. (2001) Global change in forests: Responses of species, communities, and biomes. BioScience 51: 765–779.

12. Stueve KM, Isaacs RE, Tyrrell LE, Densmore RV (2011) Spatial variability of biotic and abiotic tree establishment constraints across a treeline ecotone in the Alaska Range. Ecology 92: 496–506.

13. Pan Y, Birdsey R, Hom J, McCullough K, Clark K (2006) Improved Estimates of Net Primary Productivity from Modis Satellite Data at Regional and Local Scales. Ecological Applications 16: 125–132.

14. Peng CH, Apps MJ (1999) Modelling the response of net primary productivity (NPP) of boreal forest ecosystems to changes in climate and fire disturbance regimes. Ecological Modelling 122: 175–193.

15. Xu C, Gertner GZ, Scheller RM (2007) Potential effects of interaction between CO2 and temperature on forest landscape response to global warming. Global Change Biology 13: 1469–1483.

16. Kelly AE, Goulden ML (2008) Rapid shifts in plant distribution with recent climate change. Proceedings of the National Academy of Sciences of the United States of America 105: 11823–11826.

17. Lenoir J, Gegout JC, Marquet PA, de Ruffray P, Brisse H (2008) A significant upward shift in plant species optimum elevation during the 20th century. Science 320: 1768–1771.

18. MacDonald GM, Kremenetski KV, Beilman DW (2008) Climate change and the northern Russian treeline zone. Philosophical Transactions of the Royal Society B-Biological Sciences 363: 2285–2299.

19. Flannigan M, Stocks B, Turetsky M, Wotton M (2009) Impacts of climate change on fire activity and fire management in the circumboreal forest. Global Change Biology 15: 549–560.

20. Flannigan MD, Stocks BJ, Wotton BM (2000) Climate change and forest fires. Science of the Total Environment 262: 221–229.

21. Johnstone JF, Hollingsworth TN, Chapin FS, Mack MC (2010) Changes in fire regime break the legacy lock on successional trajectories in Alaskan boreal forest. Global Change Biology 16: 1281–1295.

22. Johnson EA, Miyanishi K, Weir JMH (1998) Wildfires in the western Canadian boreal forest: Landscape patterns and ecosystem management. Journal of Vegetation Science 9: 603–610.

23. Flannigan MD, Logan KA, Amiro BD, Skinner WR, Stocks BJ (2005) Future area burned in Canada. Climatic Change 72: 1–16.

24. Krawchuk M, Cumming S, Flannigan M (2009) Predicted changes in fire weather suggest increases in lightning fire initiation and future area burned in the mixedwood boreal forest. Climatic Change 92: 83–97.

25. Stocks BJ, Fosberg MA, Lynham TJ, Mearns L, Wotton BM, et al. (1998) Climate change and forest fire potential in Russian and Canadian boreal forests. Climatic Change 38: 1–13.

26. Westerling AL, Hidalgo HG, Cayan DR, Swetnam TW (2006) Warming and earlier spring increase western US forest wildfire activity. Science 313: 940–943.

27. Wotton BM, Nock CA, Flannigan MD (2010) Forest fire occurrence and climate change in Canada. International Journal of Wildland Fire 19: 253–271.

28. Liu ZH, Yang J, Chang Y, Weisberg PJ, He HS (2012) Spatial patterns and drivers of fire occurrence and its future trend under climate change in a boreal forest of Northeast China. Global Change Biology 18: 2041–2056.

29. Kurz WA, Stinson G, Rampley G (2008) Could increased boreal forest ecosystem productivity offset carbon losses from increased disturbances? Philosophical Transactions of The Royal Society Biological Sciences 363: 2259–2268.

30. Schumacher S, Bugmann H (2006) The relative importance of climatic effects, wildfires and management for future forest landscape dynamics in the Swiss Alps. Global Change Biology 12: 1435–1450.

31. Gustafson EJ, Shvidenko AZ, Sturtevant BR, Scheller RM (2010) Predicting global change effects on forest biomass and composition in south-central Siberia. Ecological Applications 20: 700–715.

32. He HS, Mladenoff DJ, Gustafson EJ (2002) Study of landscape change under forest harvesting and climate warming-induced fire disturbance. Forest Ecology and Management 155: 257–270.

33. Gustafson EJ, Shifley SR, Mladenoff DJ, Nimerfro KK, He HS (2000) Spatial simulation of forest succession and timber harvesting using LANDIS. Canadian Journal of Forest Research-Revue Canadienne De Recherche Forestiere 30: 32–43.

34. He HS, Mladenoff DJ (1999) Spatially explicit and stochastic simulation of forest-landscape fire disturbance and succession. Ecology 80: 81–99.

35. He HS, Mladenoff DJ, Crow TR (1999) Linking an ecosystem model and a landscape model to study forest species response to climate warming. Ecological Modelling 114: 213–233.

36. Zhou YL (1991) Vegetation in Great Xing'an Mountains of China. Chinese i, translator. Beijing Science Press. 216 p.

37. Xu HC (1998) Forest in Great Xing'an Mountains of China. Chinese i, translator. Beijing: Science Press. 231 p.

38. Zheng HN, Jia SQ, Hu HQ (1986) Forest fire and forest rehabilitation in the Da Xingan Mountains Journal of Northeast Forestry University 14: 1–7.

39. Fang JY, Lechowicz MJ (2006) Climatic limits for the present distribution of beech (Fagus L.) species in the world. Journal of Biogeography 33: 1804–1819.

40. ClimateWizard (2009): http://www.climatewizard.org/.Accessed 2011 Jun 03.

41. Flato GM, Boer GJ (2001) Warming asymmetry in climate change simulations. Geophys Res Lett 28: 195–198.

42. He HS, Mladenoff DJ, Boeder J (1999) An object-oriented forest landscape model and its representation of tree species. Ecological Modelling 119: 1–19.

43. Shifley SR, Thompson FR, Larsen DR, Dijak WD (2000) Modeling forest landscape change in the Missouri Ozarks under alternative management practices. Computers and Electronics in Agriculture 27: 7–24.

44. Post WM, Pastor J (1996) Linkages - An individual-based forest ecosystem model. Climatic Change 34: 253–261.

45. Yang J, He HS, Gustafson EJ (2004) A hierarchical fire frequency model to simulate temporal patterns of fire regimes in LANDIS. Ecological Modelling 180: 119–133.

46. He HS, Hao ZQ, Mladenoff DJ, Shao GF, Hu YM, et al. (2005) Simulating forest ecosystem response to climate warming incorporating spatial effects in north-eastern China. Journal of Biogeography 32: 2043–2056.

47. Wang XG, He HS, Li XZ, Chang Y, Hu YM, et al. (2006) Simulating the effects of reforestation on a large catastrophic fire burned landscape in Northeastern China. Forest Ecology and Management 225: 82–93.

48. Xu CG, He HS, Hu YM, Chang Y, Larsen DR, et al. (2004) Assessing the effect of cell-level uncertainty on a forest landscape model simulation in northeastern China. Ecological Modelling 180: 57–72.

49. Xu CG, He HS, Hu YM, Chang Y, Li XZ, et al. (2005) Latin hypercube sampling and geostatistical modeling of spatial uncertainty in a spatially explicit forest landscape model simulation. Ecological Modelling 185: 255–269.

50. Chang Y, He HS, Bishop I, Hu Y, Bu R, et al. (2007) Long-term forest landscape responses to fire exclusion in the Great Xing'an Mountains, China. International Journal of Wildland Fire 16: 34–44.

51. He HS, DeZonia BE, Mladenoff DJ (2000) An aggregation index (AI) to quantify spatial patterns of landscapes. Landscape Ecology 15: 591–601.

52. McGarigal K, Marks BJ (1995) FRAGSTATS: spatial pattern analysis program for quantifying landscape structure. Portland: USDA Forest Service, Pacific Northwest Research Station: General Technical Report PNW-GTR-351.

53. Norby RJ, DeLucia EH, Gielen B, Calfapietra C, Giardina CP, et al. (2005) Forest response to elevated CO2 is conserved across a broad range of productivity. Proceedings of the National Academy of Sciences of the United States of America 102: 18052–18056.

54. Euskirchen ES, McGuire AD, Kicklighter DW, Zhuang Q, Clein JS, et al. (2006) Importance of recent shifts in soil thermal dynamics on growing season length, productivity, and carbon sequestration in terrestrial high-latitude ecosystems. Global Change Biology 12: 731–750.

55. Logan JA, Regniere J, Powell JA (2003) Assessing the impacts of global warming on forest pest dynamics. Frontiers in Ecology and the Environment 1: 130–137.

56. Scheller RM, Mladenoff DJ (2005) A spatially interactive simulation of climate change, harvesting, wind, and tree species migration and projected changes to forest composition and biomass in northern Wisconsin, USA. Global Change Biology 11: 307–321.

8

Emulating Natural Disturbances for Declining Late-Successional Species

Than J. Boves[1][*][¤], **David A. Buehler**[1], **James Sheehan**[2], **Petra Bohall Wood**[3], **Amanda D. Rodewald**[4], **Jeffrey L. Larkin**[5], **Patrick D. Keyser**[1], **Felicity L. Newell**[4], **Gregory A. George**[2], **Marja H. Bakermans**[4], **Andrea Evans**[5], **Tiffany A. Beachy**[1], **Molly E. McDermott**[2], **Kelly A. Perkins**[2], **Matthew White**[5], **T. Bently Wigley**[6]

1 Department of Forestry, Wildlife, and Fisheries, University of Tennessee, Knoxville, Tennessee, United States of America, 2 West Virginia Cooperative Fish and Wildlife Research Unit, Division of Forestry and Natural Resources, West Virginia University, Morgantown, West Virginia, United States of America, 3 U.S. Geological Survey, West Virginia Cooperative Fish and Wildlife Research Unit, West Virginia University, Morgantown, West Virginia, United States of America, 4 School of Environment and Natural Resources, Ohio State University, Columbus, Ohio, United States of America, 5 Department of Biology, Indiana University of Pennsylvania, Indiana, Pennsylvania, United States of America, 6 National Council for Air and Stream Improvement, Inc., Clemson, South Carolina, United States of America

Abstract

Forest cover in the eastern United States has increased over the past century and while some late-successional species have benefited from this process as expected, others have experienced population declines. These declines may be in part related to contemporary reductions in small-scale forest interior disturbances such as fire, windthrow, and treefalls. To mitigate the negative impacts of disturbance alteration and suppression on some late-successional species, strategies that emulate natural disturbance regimes are often advocated, but large-scale evaluations of these practices are rare. Here, we assessed the consequences of experimental disturbance (using partial timber harvest) on a severely declining late-successional species, the cerulean warbler (Setophaga cerulea), across the core of its breeding range in the Appalachian Mountains. We measured numerical (density), physiological (body condition), and demographic (age structure and reproduction) responses to three levels of disturbance and explored the potential impacts of disturbance on source-sink dynamics. Breeding densities of warblers increased one to four years after all canopy disturbances (vs. controls) and males occupying territories on treatment plots were in better condition than those on control plots. However, these beneficial effects of disturbance did not correspond to improvements in reproduction; nest success was lower on all treatment plots than on control plots in the southern region and marginally lower on light disturbance plots in the northern region. Our data suggest that only habitats in the southern region acted as sources, and interior disturbances in this region have the potential to create ecological traps at a local scale, but sources when viewed at broader scales. Thus, cerulean warblers would likely benefit from management that strikes a landscape-level balance between emulating natural disturbances in order to attract individuals into areas where current structure is inappropriate, and limiting anthropogenic disturbance in forests that already possess appropriate structural attributes in order to maintain maximum productivity.

Editor: Johan J. Bolhuis, Utrecht University, The Netherlands

Funding: This research was funded and supported by the Department of Forestry, Wildlife, and Fisheries at the University of Tennessee; the United States Geological Survey West Virginia Cooperative Fish and Wildlife Research Unit; the School of the Environment and Natural Resources at Ohio State University; the Department of Biology at Indiana University of Pennsylvania; Tennessee Wildlife Resources Agency; Ohio Division of Wildlife; Kentucky Department of Fish and Wildlife Resources; West Virginia Division of Natural Resources Wildlife Diversity Program; U.S. Fish and Wildlife Service; National Council for Air and Stream Improvement, Inc.; National Fish and Wildlife Foundation (grant numbers 2005-0064-000, 2006-0042-000, 2007-0004-000, and 2008-0009-000); The Nature Conservancy (through a USFWS Habitat Conservation Plan planning grant with the Tennessee Wildlife Resources Agency); MeadWestvaco Corporation; and U.S. Forest Service. The funders had no role in study design, data collection and analysis, decision to publish, or preparation of the manuscript.

Competing Interests: MeadWestvaco Corporation provided some funding, however they did not ask to review/comment on our results prior to publishing and their financial contribution had no contingencies with respect to the study. Conservation science is often funded via public-private partnerships such as this.

* E-mail: tboves@illinois.edu

¤ Current address: Department of Natural Resources and Environmental Science, University of Illinois, Urbana, Illinois, United States of America

Introduction

Ecologists have long appreciated the fundamental role of disturbance in maintaining biodiversity in many ecosystems (e.g., intermediate disturbance hypothesis [1]). This understanding has led to the development of management practices that seek to emulate natural disturbance regimes (hereafter, ENDR), particularly in systems where disturbances have been suppressed or altered, in order to restore biodiversity and improve habitat conditions for vulnerable species [2]. ENDR strategies have been relatively well-established as a method of improving conditions for many declining early successional species [3,4], however, it is relatively unknown how declining late-successional species may respond to such practices.

Although severe disturbances (e.g., intense fires, volcanic eruptions) within mature forests are known to return entire systems to early successional stages at large scales, less intense disturbances such as wind-throw, tree senescence, and low-intensity fires, have the ability to create more subtle micro-conditions within forests that some late-successional forest species may respond to favorably. One region where interior forest disturbance regimes have been suppressed or altered is the eastern United States. Prior to European colonization, old-growth forests in the eastern U.S. were regularly disturbed by natural events such as windthrow, tree senescence, and fire [5–7]. However, since the early 1900s when forests in this region were almost completely cleared for timber and subsequent agricultural opportunities [8], much of the region has regenerated as second-growth forest and interior disturbances are now rare. Fire has become virtually non-existent because of suppression [7], and because <1% of forests are currently in old-growth condition [9], disturbances caused by treefalls (via senescence and wind) occur less frequently and have less impact [10]. Reduction of fire and other natural disturbances has been linked to a number of negative vegetative responses in eastern forests: declines in disturbance-adapted tree species such as white oak (*Quercus alba*) [11], reduction in canopy heterogeneity [12], proliferation of invasive species [13], and a reduction in tree diversity [14]. Concurrently, a number of forest-dependent animal species have undergone steep population declines during this era. These include vulnerable species such as the Indiana bat (*Myotis sodalis*), West Virginia northern flying squirrel (*Glaucomys sabrinus fuscus*), and cerulean warbler (*Setophaga cerulea*) [15–17]. Population declines of these species are likely multi-faceted (particularly for migratory species), but some vulnerable late-successional species may require the specific conditions that small-scale disturbances create and may thus be adversely affected by a lack of perturbations in contemporary second-growth forests [17–21]. Hence ENDR, via timber harvesting or prescribed fire, has been suggested as a strategy to restore natural patterns to forest environments that were historically shaped by periodic disruptions and to potentially restore habitat conditions required by these species [6,19,22].

Birds are an ideal group to use when evaluating how forest succession and the reduction of natural disturbances during the last century has affected wildlife in the eastern U.S., in part because of long-term monitoring programs such as the Breeding Bird Survey (BBS) [23]. Based on BBS data, the regrowth of eastern forests over the past century has been, expectedly, correlated with increasing populations of some avian forest species, such as northern parula (*S. americana*) and blackburnian warblers (*S. fusca*). However, the successional process has also been, seemingly paradoxically, negatively related to population trends of other species that would seem to benefit from what appears to be an increase in breeding habitat, such as the eastern wood-pewee (*Contopus virens*) and Canada warbler (*Cardellina canadensis*) [24]. Perhaps the most notable declining avian species of eastern forests is the cerulean warbler. The cerulean warbler is a Neotropical-Nearctic migratory species that breeds solely in the canopies of deciduous forests in eastern North America and has long been considered to be a prototypical late-successional species [25,26]. However, despite recent increases in their putative breeding habitat, cerulean warblers are one of the fastest declining passerines in North America; populations declined 3.2%/yr from 1966 to 2003 and the trend has recently worsened to a decline of 4.6%/yr [24]. They are currently listed as a species of conservation concern by the US government [16] and are considered 'vulnerable to extinction' by BirdLife International [27]. Contrary to the long-standing paradigm that their preferred habitat is closed-canopy forest, recent evidence suggests that the cerulean warbler's decline may actually be related to a lack of small-scale, interior forest disturbances in their eastern U.S. breeding grounds [21,28,29], particularly in the Appalachians, where an estimated 70% of their remaining population breeds [30]. Consequently, ENDR has been suggested as a method of mitigating degraded forest conditions and restoring habitat for cerulean warblers [21,31]. However the effectiveness of this strategy, as well as the ideal scale and intensity of the disturbances to be emulated, is not known.

Many studies have documented numerical responses of populations (i.e. abundance or density) to anthropogenic disturbance via forest management [e.g., 32,33]. However, our understanding of the mechanisms responsible for numerical responses to environmental perturbations is much more limited. These mechanisms may begin with individual changes in habitat selection, physiology, breeding behavior, and dispersal [e.g., 34,35] and then may be scaled up to population changes in reproductive rates, annual survival rates, and age structure [e.g., 36,37,38]. Evaluating more than numerical responses is essential because simple use of, or even preference for, a habitat does not necessarily indicate the quality of that habitat [39,40]. Mismatches between habitat selection and individual fitness have been identified in several taxa, particularly those inhabiting human-modified habitats where ecological processes have been altered recently and rapidly [e.g., 41,42,43]. Thus, before considering ENDR to be an appropriate strategy for restoring conditions for declining forest species, detailed studies of individual and population-level responses to disturbance are needed to ensure that our actions do not create such a situation.

In this study, we investigated the consequences of emulating natural disturbances for a late-successional avian species, the cerulean warbler. To do so, we experimentally disturbed mature forest stands at various intensities, spanning the range of local disruptions that could occur naturally in mature forests, across the core of the warbler's breeding range in the Appalachian Mountains. We then assessed short-term responses (up to four years) to these manipulations in terms of breeding density, body condition, age structure, and reproductive output. In addition, we explored regional variation in these responses and the potential impacts of emulating disturbance on the source-sink dynamics of cerulean warblers in the Appalachian region using a deterministic population model. Finally, we discuss the implications of our results for cerulean warbler conservation and management.

Methods

Study sites

We conducted this study at seven sites in the Appalachian Mountains (Figure 1), all within the Central Hardwoods' mixed-mesophytic forest region [44], which also corresponds to the core of the cerulean warbler breeding range. These sites were: Royal Blue Wildlife Management Area, TN (RB), Sundquist Forest, TN (SQ), Raccoon Ecological Management Area, OH (REMA), Daniel Boone National Forest, KY (DB), Lewis Wetzel Wildlife Management Area, WV (LW), Wyoming County, WV (WYO), and Monongahela National Forest, WV (MON). The two most southern sites (RB and SQ) were both located in the Cumberland Mountains, an ecophysiographically distinct section of the Appalachian chain [45,46] that has previously been identified as a critical breeding locale for the species [47,48]. Thus, we refer to these two sites hereafter as the "southern region" and the other five study sites as the "northern region." Because cerulean warblers often require large tracts of contiguous forest [26], we selected sites embedded within a matrix of mature forest; mean percent forest cover within 10 km of the site center was 83.2 ± 2.8 [SE] % (range = 74–95%, 2001 NLCD). Plant composition differed slightly among sites, but common overstory tree species included tulip poplar (*Liriodendron tulipifera*); sugar maple (*Acer saccharum*); northern red, white, and chestnut oak (*Quercus rubra, Q. alba, and Q. prinus*); and various hickory spp. (*Carya* spp.).

Disturbance treatments

We randomly assigned treatments to four plots at each field site: light, intermediate, and heavy canopy disturbance, as well as an undisturbed control plot. Disturbance plots were 10 ha and control plots were 20 ha in size (Figure 2). We used larger undisturbed control plots because territory density was lower and nests more difficult to locate in these habitats. Each plot was located >200 m from all other plots to maintain independence. At the periphery of each disturbance treatment were two 5-ha plots of undisturbed forest that we designated as "buffers" (see Figure 2 for plot design). We included buffers to examine potential edge effects of disturbances. Buffers were not spatially independent from disturbed treatments, so we compared them to controls in separate but identical analyses.

Disturbances were designed to emulate natural processes that spanned the range of potential forest disruptions in the Appalachian region and were implemented via timber harvest in the fall of 2006 and spring of 2007. Light treatments (least intense disturbances) mimicked stands disrupted by multiple small tree-fall gaps; we reduced basal area (BA) and overstory canopy cover (CC) on these treatments by approximately 20% (residual $BA = 21\pm1$ [SE] m^2/ha; residual $CC = 61\pm6$ [SE] %). Intermediate treatments mimicked more severe natural disturbances such as fire,

windthrow, or larger tree fall gaps; here we reduced BA and CC by approximately 40% (residual $BA = 14\pm1$ [SE] m^2/ha; residual $CC = 45\pm6$ [SE] %). Heavy treatments emulated the most severe natural disturbances such as more intense fire and windthrow, icestorms, or landslides; we reduced BA and CC by 75% (residual $BA = 6\pm1$ [SE] m^2/ha; residual $CC = 18\pm4$ [SE] %). We left control plots and buffers undisturbed throughout the duration of the study ($BA = 27\pm1$ [SE] m^2/ha; $CC = 73\pm5$ [SE] %). We attempted to apply disturbances uniformly across all treatment plots and overstory tree species composition was largely unchanged after disturbances were implemented [49]. Residual stands on the intermediate and heavy treatments were comprised of dominant and co-dominant trees. Because cerulean warblers prefer productive slopes [29,48], plots were predominantly placed on north- or east-facing slopes to maximize warbler presence and to control for potential interactions between aspect and response.

Territory density response

We used a before-after-control-impact study design to evaluate changes in territory density in response to treatments. We delineated and quantified territories of cerulean warblers using the spot-mapping technique. Because male warblers sing often and are easily detectable in all habitat types, spot-mapping is an ideal form of estimating density for this species. We performed eight morning censuses (from sunrise to 1030) per plot during the height of each breeding season (1 May to 15 June), 2005–2010 (two years pre-disturbance and four years post-disturbance). On gridded maps, we recorded all locations of male vocalizations including all instances of counter-singing among neighboring males, as well as any other territorial behaviors. We defined territories as geographic clusters of two or more registrations from different spot-mapping sessions and used counter-singing or other territorial behavior when available to help separate adjacent territorial

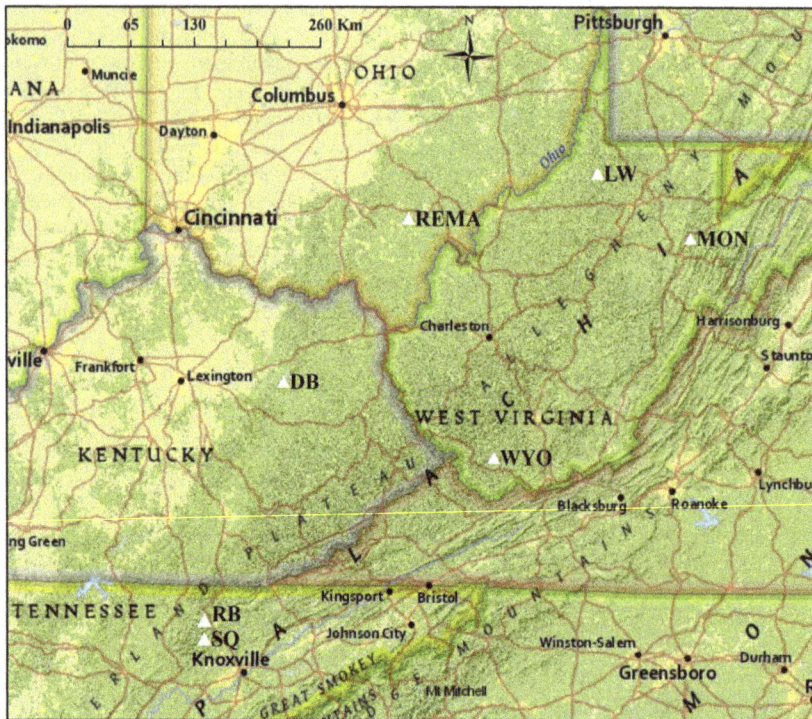

Figure 1. Map displaying locations of seven study sites in the Appalachian Mountains. All sites (white triangles) are located within the core of the cerulean warbler breeding range.

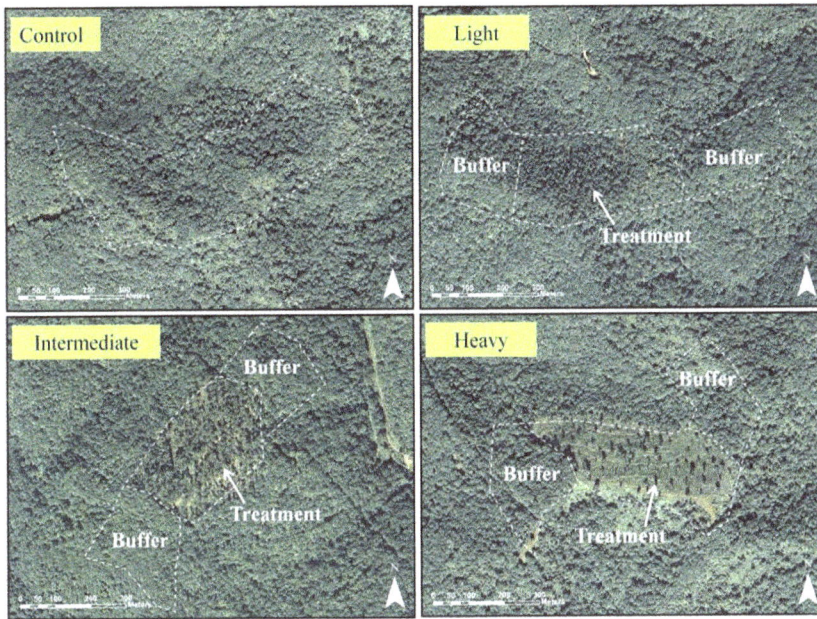

Figure 2. Aerial photos from a study site (LW) depicting treatment plot design and intensity of disturbances. Each field site consisted of three 10-ha treatment plots of various disturbance intensity (created via partial timber harvest) and one 20-ha control plot (undisturbed). Ten ha of undisturbed forest outside the borders of each treatment plot (buffers) allowed for examination of edge effects of the disturbances.

individuals [50]. We also used nest and banding data (see below) to refine spot-mapping data and to validate delineation and estimation of territory numbers. We assigned fractions of territories to individuals whose territories only partially occurred within the borders of a plot (based on the proportion of registrations that fell within the plot).

We estimated baseline territory density on plots by calculating the mean density of pre-disturbance spot-mapping data (2005–06). We first compared density between the two pre-disturbance years using repeated-measures ANOVA. We performed pre-disturbance spot-mapping on MON and WYO sites in 2006 only, so these sites were not included in this pre-disturbance analysis. If we found no significant year effects, we used mean pre-disturbance density (of 2005 and 2006) as a starting point for subsequent analyses. We estimated change in territory density from pre- to post-disturbance by calculating

$$Density\ ratio(DR) = Post\ density / Pre\ density$$

where we defined density as the number of territorial males/10 ha. Two plots were unoccupied pre-disturbance so we replaced zero values with 0.25 (the lowest recorded territory density other than zero) to estimate DR; this resulted in more conservative rates of increase than in reality, but had no effect on our inferences. Values of DR were log-transformed to meet parametric assumptions of normality and equal variance.

We analyzed this experiment as a randomized complete block design with sites treated as blocks. We compared log DR among treatments using a repeated measures mixed-model ANOVA with treatment, year, and treatment x year modeled as fixed effects and site and site x year as random effects. Year was modeled as a fixed effect because we were interested in whether treatment effects were contingent on the number of years since disturbance. If we found a main effect of treatment, we performed pairwise contrasts to evaluate differences among treatments and controls. To examine edge effects, we performed a separate, but identical, analysis to compare changes in density in buffers vs. control plots. We found no statistical difference in log DR among buffers of the three treatment types in any year (one-way ANOVA; $P>0.30$ in all years), so we used the mean density of the three buffers in this analysis.

Age structure and body condition

To compare age structure and body condition of individuals occupying territories in each treatment type, we captured male cerulean warblers using mist-nets while broadcasting territorial songs and call notes during the height of the breeding season (May and June) during 2008–2010 (all post-disturbance). We aged males as second-year (SY; first breeding season) or after-second-year (ASY) by molt limits (particularly useful is that SY birds retain brownish juvenile primary coverts and typically two juvenile alula feathers) [51]. We measured wing chord to the nearest 0.5 mm and mass to the nearest 0.1 g. We then assigned each male to a single treatment that best reflected the individual's territory location based on evidence gathered from spot-mapping and nest searching efforts (described below). Birds were captured and aged at REMA, SQ, RB, LW, and WYO and weighed at REMA, SQ, and RB.

We compared age structure of male warblers among treatments against a null hypothesis of no difference in proportion of SY males using Pearson's chi-square tests for all sites pooled and for each region (north or south) separately (to determine if regional variation existed). To evaluate the impact of disturbance in general and to increase power, we also compared the age structure of birds captured in a disturbance of any kind (pooled) with birds captured in controls. To examine edge effects, we performed a separate, but identical, analysis to compare age structure in buffers versus controls. No difference in age structure existed among buffer types (all sites pooled: $\chi^2_2 = 0.69$, $P=0.71$; north: $\chi^2_2 = 0.01$, $P=0.90$;

south: $\chi^2_2 = 1.54$, $P = 0.46$), so we pooled all birds captured in buffers into a single group.

We compared body mass of males occupying territories on differing treatment plots using a two-way mixed generalized linear model (GLM). Individual birds were the sampling units in this instance and we specified age, treatment (light, intermediate, heavy, or control), and site as fixed factors and year as a random factor. We also included all two-way interactions and Julian capture date as a covariate. All two-way interactions were non-significant (all $P > 0.19$), so we removed these terms and re-ran the GLM. If the treatment effect was significant, we subsequently conducted Fisher's LSD tests to determine where differences existed (at $\alpha = 0.05$ and 0.10). We used body mass as an indicator of condition in this analysis because mass is often more closely related to the amount of nutritional reserves than unverified indices [52,53]. However, to be certain this did not affect our inferences, we also calculated wing-mass residuals and found them highly correlated to body mass ($r = 0.93$); we performed analyses with both measures and found no difference. To examine edge effects, we performed a separate, but identical, analysis to compare body condition of males in buffers versus control plots. No difference in body condition existed among individuals occupying different buffer types ($F_{2,20} = 0.60$, $P = 0.56$), so we pooled all birds captured in buffers into a single group. If individuals were captured in more than one season, we randomly selected one capture event to use in the analysis.

Reproduction

We searched for nests during the entire breeding season (late April to late June), 2008–2010. We used female behavioral cues during building and incubation, and to a lesser extent male vocalizations and behavior, to locate the majority of nests. Because we were more efficient at locating nests on disturbed treatment plots, we stratified our search efforts by increasing the time spent searching on controls and buffers (in an attempt to locate an equal proportion of nests on each plot). We were unable to examine the contents of nests until nestlings were approximately 5 d old, and therefore considered nests 'active' when we observed parental activity at the nest that indicated egg or nestling presence (incubation or provisioning). Once active, we monitored nests every 1–3 d until fledging or confirmed nest failure occurred. From nestling day six until fledging, we monitored nests daily whenever possible for ≥30 min using spotting scopes equipped with 20–60× magnification eyepieces to count the number of nestlings present. As cerulean warbler nestlings near fledging age, they become increasingly restless (climbing over each other, begging, and preening incessantly) and are often easily counted, particularly on the steep slopes of our field sites (T.J. Boves pers. obs.). To conclusively determine nest fate and number of fledglings produced, we also attempted to observe fledging events. If we were unable to directly observe these events, we searched the vicinity of nests after putative fledging for parental and juvenile activity and assumed that the number of nestlings present on the last day of the nestling stage (typically day 10) to be equal to the number of fledglings produced. We considered any nest that fledged ≥1 cerulean warbler young to be successful and did not distinguish between initial and re-nesting attempts. Highly concealed nests where nestlings were difficult to count were excluded from fledgling estimates.

We initially compared logistic exposure models in Program MARK to determine the relative influence of spatial and temporal factors and treatment on daily nest survival rates (DSR). This method uses a generalized linear model with binomial distribution for each day (nest fate = 1 if failed, 0 if successful) with a logit link function to assess the influence of covariates on DSR. We compared and ranked models using a corrected version of Akaike's information criterion adjusted for small sample sizes (AIC_c), where the minimum AIC_c indicates the best model (a combination of parsimony and explanatory power) [54]. We first compared models that included the spatial factors of region (southern vs. northern; RGN) and site (SITE). We found strong support for region as the spatial factor that best explained variation in DSR (when compared with region, site $\Delta AIC_c = 6.78$), so we used this spatial factor alone in future models. We then compared all univariate and additive combinations of RGN, year (YEAR), and treatment (TRT), as well as YEAR×TRT and RGN×TRT interactions to test for temporal and spatial variation in treatment effects. We also included a constant survival model (NULL) for a total of 14 candidate models. We found only one nest at MON, so this site was not included in this analysis.

After this initial evaluation, we made post-hoc comparisons of nest survival rates among treatments and controls partitioned by factors determined to be influential (i.e., included in top models). We calculated cumulative survival rates for the entire nesting period by raising covariate-specific DSR to a power equal to the average length of the nest cycle (25 d) and used Program CONTRAST to determine statistical significance [55]. We approximated entire nest success variance and standard errors using the delta method following Powell [56]. We report these cumulative survival rates (hereafter, 'nest success') throughout the remainder of this paper for ease of interpretation. We conducted an identical analysis comparing controls and buffers to examine potential edge effects on nest success. There were no differences in reproductive success among buffers of different treatment plots in either region (north: $\chi^2_2 = 0.30$, $P = 0.89$; south: $\chi^2_2 = 2.39$, $P = 0.30$), so nests found in any buffer were combined into a single group.

We compared the number of fledglings produced per successful nest among treatments and controls using a mixed model ANOVA with treatment and region specified as fixed factors and year as a random factor. We again conducted an identical analysis comparing controls and buffers to examine edge effects. We used Program MARK (v6.1), JMP (v9.0), and SAS (v9.2) statistical software packages for analyses. For all statistical tests, we considered differences to be significant at $P \le 0.05$ and marginally significant at $0.05 < P \le 0.10$. We report means ± 1 SE.

Source-sink modeling

We employed a deterministic population model, following Buehler et al. [47], to explore how the reproductive consequences of our treatments may affect regional source-sink dynamics. Input parameters included regionally and treatment-specific nest success and number of young produced/successful nest (as we detected regional variability in reproductive output, see results) derived from this study, as well as external estimates of after-hatch-year (AHY) and hatch-year (HY) survival, proportion of individuals that attempt to re-nest after failing, and number of re-nesting attempts. Because we were specifically interested in assessing how the reproductive consequences of disturbance may impact source-sink dynamics, we assumed equal annual survival rates, proportion of re-nesting, and number of re-nesting attempts across treatments and regions. We were unable to obtain reliable adult survival estimates from our study, likely because of high dispersal rates between breeding seasons [57], so we compared two published adult annual survival rates: 54% from Ontario [58] and 65% from Venezuela on their wintering grounds [59]. No data exist for cerulean warbler HY survival, so we assumed HY to be half of AHY survival, as has been used in previous models and has been found empirically in other passerines [60,61]. We recognize that variation in breeding habitat may lead to differential carry-over effects on migratory or winter survival rates [62], however, we observed within-breeding season survival to be nearly 100%, and parents and offspring often dispersed from their chosen breeding habitat soon after fledging occurred (T.J. Boves, *unpub. data* and *pers. obs.*). Thus, it is likely that variation in breeding habitat had a greater impact on reproduction than on these other parameters (which were likely more highly influenced by post-breeding habitat decisions).

Results

Territory density

We found no significant year effects ($F_{1,16} = 0.05$; $P = 0.41$) or year x plot interaction ($F_{3,16} = 0.16$; $P = 0.49$) on pre-disturbance densities, so we used mean pre-disturbance density as a single pre-treatment value. After disturbance, we found a main treatment effect on log DR ($F_{3,18} = 4.96$, $P = 0.01$) and also a treatment x year effect ($F_{9,72} = 2.79$, $P = 0.007$), so we performed contrasts to evaluate differences for each year independently. In 2007 (first year post-disturbance), log DR was significantly greater on intermediate treatment plots than on all other treatment and control plots, and marginally greater on light treatment plots when contrasted with heavy (Figure 3, Table 1). In 2008, log DR remained significantly greater on intermediate treatment plots than on control and heavy treatment plots, and was marginally greater on light treatment plots than on control plots (Figure 3, Table 1). In 2009, log DR was significantly greater on intermediate treatment plots, and marginally greater on heavy and light treatment plots, when contrasted with controls, but there were no differences among any of the disturbed treatments (Figure 3, Table 1). As of 2010, log DR was significantly greater on intermediate treatment plots than on control and light treatment plots, and for the first time, was significantly greater on heavy treatment plots than on control plots (Figure 3, Table 1). Additionally, in 2010 there was no longer a statistical difference between light treatment and control plots and only a marginal difference between heavy and intermediate treatment plots (Figure 3, Table 1). We also found evidence of an edge effect as log DR was significantly greater on treatment plot buffers than on control plots (Table 1); there was no treatment x year effect in this case ($F_{3,36} = 0.88$; $P = 0.46$).

Age structure and body condition

In total, we captured and aged 204 male cerulean warblers; 27% were SY birds, 73% ASY. With all sites pooled, there was no difference in the age structure of males occupying the various treatment and control plots ($\chi^2_3 = 1.03$, $P = 0.79$). There was also no difference in the age structure of males occupying any disturbed treatment plot vs. males occupying control plots ($\chi^2_1 = 0.05$, $P = 0.83$). Assessing each region separately, no difference in age structure existed among treatment and control plots (north: $n = 58$, $\chi^2_3 = 0.64$, $P = 0.89$; south: $n = 67$, $\chi^2_3 = 3.78$, $P = 0.29$) or when all disturbed treatment plots were compared with control plots (north: $\chi^2_1 = 0.09$, $P = 0.93$; south: $\chi^2_1 = 0.05$, $P = 0.82$). No edge effect was observed as age structure of birds occupying buffers did not differ from those occupying control plots when all sites were pooled ($\chi^2_1 = 0.17$, $P = 0.68$), or within regions (north: $\chi^2_1 = 1.18$, $P = 0.28$; south: $\chi^2_1 = 0.36$, $P = 0.55$).

Controlling for site, age, and year effects, body condition of male warblers differed by treatment ($F_{3,56} = 3.41$, $P = 0.02$, Figure 4). Males occupying territories on light and intermediate treatment plots were in significantly better condition than those

occupying control plots (Fisher's LSD, $P \leq 0.05$; Figure 4) and males occupying light treatment plots were in marginally better condition than those occupying heavy treatment plots (Fisher's LSD, $P \leq 0.10$). Body condition also differed by age (SY males = 9.21 ± 0.07, $n = 17$; ASY males = 9.52 ± 0.04, $n = 49$; $F_{1,56} = 12.19$, $P = 0.001$) but did not differ by site ($F_{2,56} = 0.82$, $P = 0.45$). No edge effect was detected as body condition of males occupying buffers did not differ from those on control plots (Controls = 9.26 ± 0.84, $n = 21$; Buffers = 9.18 ± 0.72, $n = 29$; $F_{1,42} = 0.49$, $P = 0.49$).

Reproduction

We found and monitored 413 nests for a total of 6,384 exposure days. All four of the top models included treatment (as well as region) and the top model (RGN+YEAR+TRT) was 96× more supported than the simpler model that did not include treatment (RGN+YEAR; Table 2). There was some support for a region x treatment interaction as it was included in the third- and fourth-ranked models, but virtually no support existed for a year x treatment interaction as it was not included until the seventh-ranked model ($\Delta AIC_C = 11.97$). Confidence intervals (95%) of β coefficients from the top model for the northern region (negative slope), control treatment (positive slope), light treatment (negative slope), and for 2009 (negative slope) did not include zero, which suggests their importance in explaining variation in DSR (Table 3).

Cumulative nest success differed among all sites ($\chi^2_5 = 27.56$, $P < 0.0001$) but did not differ among sites within regions (North: $\chi^2_3 = 1.61$, $P = 0.66$; South $\chi^2_1 = 1.44$, $P = 0.23$). Thus, we pooled nests from respective regions to further assess treatment effects on nest success. In the southern region, cumulative annual nest success varied from 0.48 ± 0.06 in 2009 to 0.67 ± 0.05 in 2010. When pooling nests from all three years (Figure 5), nest success in this region was greater on control plots than on light ($\chi^2_1 = 15.02$, $P < 0.0001$), intermediate ($\chi^2_1 = 4.41$, $P = 0.04$), and heavy treatment plots ($\chi^2_1 = 15.02$, $P < 0.0001$). Nests on intermediate treatment plots were more successful than those on light treatment plots ($\chi^2_1 = 4.38$, $P = 0.04$). There was no evidence of an edge effect on nest success as controls and buffers did not differ ($\chi^2_1 = 1.89$, $P = 0.17$). Annually, nest success was greater on control plots than heavy treatment plots in 2009 ($\chi^2_1 = 26.07$, $P < 0.0001$) and greater than light treatment plots during 2009 ($\chi^2_1 = 33.73$, $P < 0.0001$) and 2010 ($\chi^2_1 = 5.64$, $P = 0.02$).

In the northern region, annual nest success ranged from 0.22 ± 0.04 (2009) to 0.40 ± 0.06 (2010). When pooling nests from all three years (Figure 5), nest success was marginally greater on control plots than on light treatment plots ($\chi^2_1 = 3.50$, $P = 0.06$), but did not differ among any other pairwise combination of treatments and controls. There was marginal evidence of an edge effect as nests on control plots were marginally more successful than those on buffer plots ($\chi^2_1 = 3.12$, $P = 0.08$). On an annual basis, nest success did not differ between control or any treatment or buffers (all $P > 0.10$), however small sample sizes hampered our ability to detect statistical differences annually.

The number of fledglings produced/successful nest differed by region; warblers in the south produced more fledglings/successful nest ($\bar{x} = 3.33 \pm 0.07$) than in the north ($\bar{x} = 2.28 \pm 0.14$; $F_{1,97} = 33.98$, $P < 0.0001$; see Figure 6). However, there was no effect of treatment ($F_{3,42.95} = 0.64$, $P = 0.60$). Comparing controls with buffers, nests in the south again produced more young ($F_{1,73.85} = 19.04$, $P < 0.0001$), but there was no evidence of an edge effect on fledglings produced ($F_{1, 73.05} = 0.05$, $P = 0.82$).

The cause of nest failure was directly observed or inferred from evidence at only 36 (of 174 failed) nests. Predation was the main cause of nest failure ($n = 22$ of these 36 nests) followed by disease or

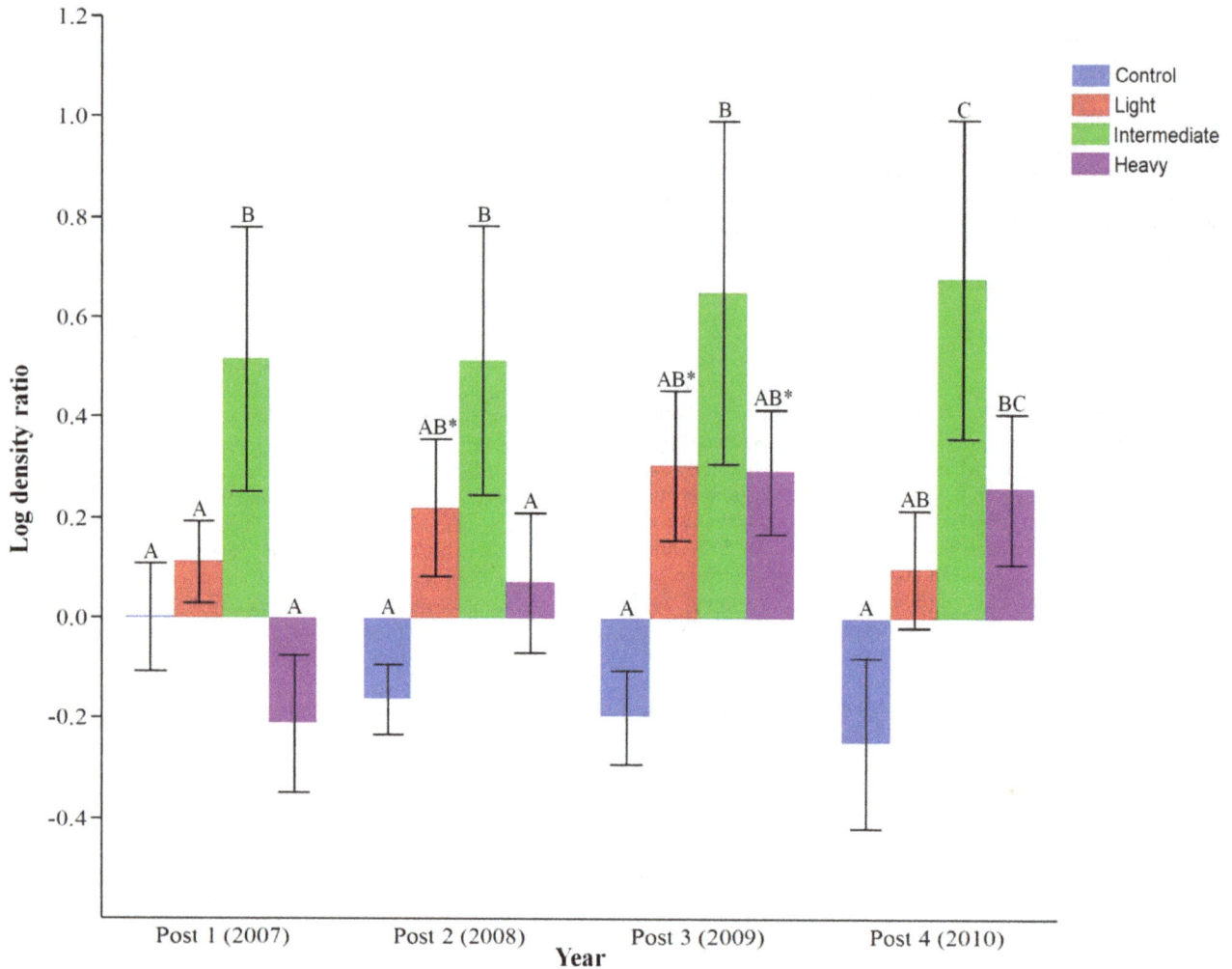

Figure 3. Breeding density ratio (post/pre-disturbance, log-transformed) of cerulean warblers on plots disturbed by various intensities of timber harvest. Log density ratio = 0 reflects no change in density; all values above 0 indicate increased density, all values below indicate density reduction. Different letters indicate significant differences ($P \leq 0.05$) among respective treatments for a given year, based on independent contrasts. Asterisks indicate marginal differences ($0.05 < P \leq 0.10$) between respective treatment and control for a given year. Error bars represent $\pm$ 1 SE.

starvation ($n = 6$). The majority of failed nests were abandoned suddenly for unknown reasons, suggesting that predation was most likely, but nest desertion subsequent to brown-headed cowbird (*Molothrus ater*) parasitism cannot be ruled out.

Source-sink dynamics

Our graphical model shows that given an AHY annual survival rate of 54%, only control plots in the southern region had levels of reproduction sufficient to maintain a stable (or source) population (Figure 6). If annual survival was increased to 65%, all treatment plots in the southern region would act as sources ($\lambda > 1$). We found no treatment plot in the northern region, including controls, that could maintain a stable population given either of these two survival rates; all require either greater annual survival or reproductive output, immigration from other locations, or an adjustment in model assumptions to persist.

Discussion

We hypothesized that existing second-growth forest in the eastern United States may not provide quality habitat for some late-successional species, especially if those species are adapted to small-scale natural disturbances that have been altered or suppressed within contemporary forests. Accordingly, we documented attraction to emulated disturbances of various intensities by a declining species typically considered to be late-successional, the cerulean warbler, in highly-forested ecosystems in the Appalachian Mountains. The density response we observed is congruent with recent correlative studies that found cerulean warblers associated with canopy disturbances within mature forests [21,29]. In our study, attraction was greatest after intermediate and heavy disturbances, suggesting that the species is adapted to fire, intense windthrow, landslides, or other moderate interior natural disturbances, rather than smaller single tree-fall gaps caused by tree senescence, for instance. Density increases after intermediate disturbances on some sites were unexpectedly strong and immediate (e.g., 0.25 territories pre-disturbance to 8.5

Table 1. Density of cerulean warbler territories (± 1 SE) and results of independent contrasts comparing log density ratio (post/pre-density) of treatment plots with controls for each given year.

Treatment	Year	Density	Df	F	P
Control	Pre-disturbance	4.82±1.59			
	2007	4.70±1.20			
	2008	3.43±1.27			
	2009	4.16±1.84			
	2010	4.52±1.89			
Light	Pre-disturbance	5.52±1.92			
	2007	7.14±2.40	1,18	0.41	0.53
	2008	**7.89±2.07**	**1,18**	**3.25**	**0.09**
	2009	**9.11±2.70**	**1,18**	**3.96**	**0.06**
	2010	6.93±2.56	1,18	2.11	0.16
Intermediate	Pre-disturbance	4.95±2.34			
	2007	**7.43±2.18**	**1,18**	**8.93**	**0.008**
	2008	**8.07±2.06**	**1,18**	**10.16**	**0.005**
	2009	**11.43±3.43**	**1,18**	**11.25**	**0.003**
	2010	**10.57±3.02**	**1,18**	**15.03**	**0.001**
Heavy	Pre-disturbance	2.34±1.13			
	2007	1.82±1.00	1,18	1.53	0.23
	2008	3.29±1.53	1,18	1.20	0.29
	2009	**4.75±1.98**	**1,18**	**3.76**	**0.07**
	2010	**5.21±2.66**	**1,18**	**4.50**	**0.05**
Buffers	Pre-disturbance	4.81±1.33			
	2007–2010	**5.11±0.58**	**1,6**	**6.08**	**0.05**

Densities displayed are untransformed no. of territories/10 ha. Significant ($P \leq 0.05$) or marginal ($0.05 < P \leq 0.10$) results are in bold. Buffers and controls were compared in a separate analysis with no significant treatment x year interaction, so individual annual contrasts were not performed.

territories in the first breeding season post-disturbance at LW); on other sites increases were more modest, perhaps because of pre-disturbance saturation. At RB, pre-disturbance density was at a (likely) near-saturation level of 17 territories/10 ha. Density increased on this plot post-disturbance, but only to a maximum of 20.5 territories in 2010. At such great pre-disturbance densities, it would seem unlikely that many more birds could occupy the area, no matter how attractive the habitat became. Densities increased more gradually after heavy disturbances (and actually decreased in the first year post-disturbance). This suggests that some physiognomic cue important for habitat selection required multiple growing seasons to develop after these more severe disturbances, and could be related to temporal changes in canopy or understory structure [63]. The edge effect that we detected (i.e., density increases in undisturbed buffers surrounding disturbed plots) was primarily related to an increase in birds establishing territories that overlapped both the treatment plot and buffers (J. Sheehan, *unpub. data*).

As disturbances attracted warblers at higher densities, the lack of difference in age structure among treatments runs counter to the expectation that older birds should out-compete inexperienced males and settle in preferred habitat more often [64,65]. However, we did find that males occupying light and intermediate treatment plots, regardless of age, were in better condition than those inhabiting controls. We do not know if this difference reflects a settlement bias (e.g., if individuals on disturbed treatments were in better condition on arrival or of higher quality), if disturbances allowed individuals to improve their condition (e.g., by virtue of

increased insect availability after disturbances), or if a combination of the two was responsible for this pattern. Canopy gaps can alter the composition of arthropod communities [66] and cerulean warblers may be better adapted for foraging on invertebrate species inhabiting broken canopies. Indeed, George [49] found that warblers increased their use of aerial foraging maneuvers after partial timber harvests occurred. However, it is not known if this behavioral alteration results in improved condition; future studies that monitor settlement patterns and individual changes in body mass across habitat types would help tease these possibilities apart.

Despite the density increases and improved body condition of individuals occupying treatment plots, per capita reproductive output was lower on many of the treatment plots compared to local control plots. Reproductive differences were most obvious in the southern region, where disparities in nest success between control and treatment plots were statistically apparent in all cases. In the northern region, factors seemingly unrelated to the manipulations reduced overall reproductive success to where disturbance had less influence, and low sample sizes made it difficult to detect statistical differences in some instances (e.g., $n = 5$ nests on heavy treatment plots). However, nest success was marginally greater on control plots than on light treatment plots (and buffers) in this region as well. Thus it appears that individuals, particularly in the southern region, often chose to breed in habitats where they failed to maximize reproduction.

There are numerous potential explanations to this seeming contradiction [see 67 for an exhaustive list]. One possibility is that by breeding in disturbed habitats, individuals increased their

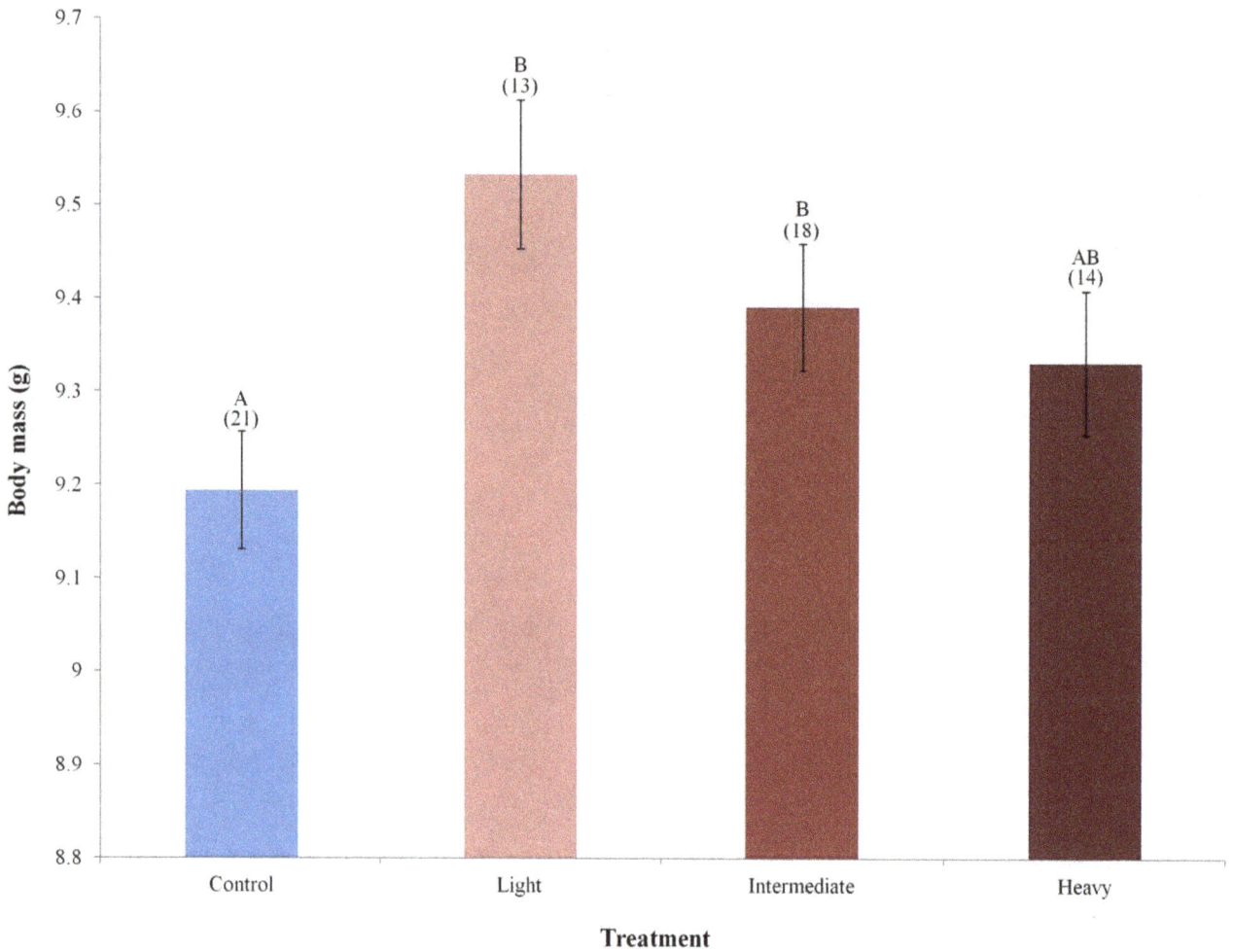

Figure 4. Body mass of male cerulean warblers by treatment after controlling for age, year, and site, 2008–10. Different letters indicate significant differences ($P \leq 0.05$) between respective treatments. Error bars represent ± 1 SE and numbers above bars indicate sample size.

Table 2. Model selection results for factors influencing daily survival rate of cerulean warbler nests.

Model	k	AIC$_c$	ΔAIC$_c$	w
S(RGN+YEAR+TRT)	7	1142.28	0	0.535
S(RGN+TRT)	5	1143.24	0.96	0.331
S(RGN+YEAR+TRT+RGN*TRT)	13	1146.23	3.95	0.074
S(RGN+TRT+RGN*TRT)	11	1147.01	4.73	0.050
S(RGN+YEAR)	4	1151.42	9.14	0.006
S(RGN)	2	1153.74	11.46	0.002
S(RGN+TRT+YEAR+TRT*YEAR)	18	1154.25	11.97	0.001
S(TRT+YEAR)	6	1156.39	14.11	0.001
S(SITE)	6	1157.43	15.15	0.000
S(RGN+TRT+YEAR+TRT*YEAR+RGN*TRT)	25	1157.80	15.52	0.000
S(TRT)	3	1159.81	17.53	0.000
S(YEAR)	3	1165.55	23.27	0.000
S(TRT+YEAR+TRT*YEAR)	16	1168.18	25.90	0.000
S(NULL)	1	1168.62	26.34	0.000

Models with a lower ΔAIC and a greater AIC$_c$ weight have greater support. Model weight (w) and number of estimated parameters (k) are indicated.

lifetime fitness (despite reductions to current reproductive output) by improving their chances of surviving to the next breeding season or by improving their offspring's chances of survival during the dangerous post-fledging period. Increased annual survival of cerulean warblers after canopy disturbances may be possible by virtue of the potential carry-over effects of improved body

Table 3. Parameter estimates (on logit-link scale), standard errors (SE), and 95% confidence intervals (CI) from top-ranked model (RGN+YEAR+TRT) estimating daily survival rate of cerulean warbler nests.

Parameter	β estimate	SE	Lower 95% CI	Upper 95% CI
Intercept	3.7735	0.2383	3.3064	4.2407
RGN$_{north}$	−0.7191	0.1823	−1.0764	−0.3618
TRT$_{control}$	0.7873	0.3372	0.1263	1.4482
TRT$_{light}$	−0.5395	0.1949	−0.9216	−0.1574
TRT$_{intermediate}$	0.3610	0.2682	−0.1646	0.8866
YEAR$_{2008}$	−0.2521	0.2260	−0.6950	0.1908
YEAR$_{2009}$	−0.4299	0.1989	−0.8197	−0.0400

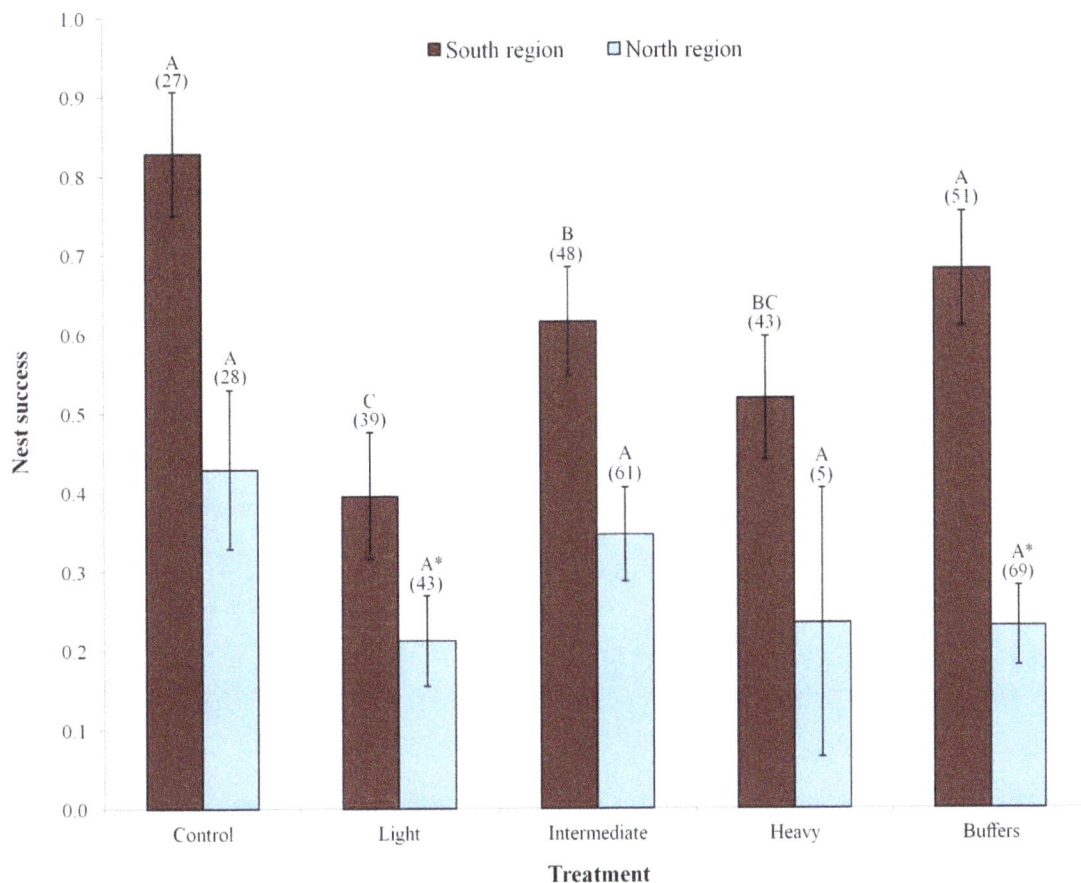

Figure 5. Cerulean warbler nest success by treatment and region, 2008–10. Different letters indicate significant differences ($P \leq 0.05$) between respective treatments within a region (based on CONTRAST χ^2 test). Asterisks indicate marginal differences ($0.05 < P \leq 0.10$) between respective treatments and controls within a region. Markings above buffer columns refer only to their relationship with controls. Error bars represent ± 1 SE and numbers above bars indicate nest sample sizes.

condition on migratory or winter survival [62,68,69], and post-fledging survival rates may be greater because of the abundance of concealing understory vegetation on intermediate and heavy treatment plots [70,71]. However, as alluded to previously, the influence of breeding habitat on these future components of fitness may be relatively indirect and is currently unclear, while the influence of breeding habitat on nest success and fledgling production is direct and obvious. A second possibility is that density was not an accurate reflection of habitat preference and individuals were forced into disturbed habitats via competitive exclusion by more dominant individuals [39,65]. The evidence does not support this possibility however, as we documented no age differences among individuals occupying treatment and control plots, and those individuals that did occupy territories in disturbed habitats were, in fact, in better condition than those in undisturbed control plots.

A third possibility is that individuals may have made maladaptive decisions when selecting disturbed habitats (i.e., disturbed interior forest stands may act as "ecological traps" [72]), particularly when choosing among habitats at the local scale. Under evolutionarily-relevant historical conditions, canopy disturbances in old-growth forests caused by fire or natural treefalls may have created habitats where warblers were able to achieve relatively high levels of fitness. After emulated natural disturbances, environmental cues associated with high fitness may still elicit

the same habitat selection behavior, however other conditions, contemporary in nature, may have also been altered, thereby potentially decoupling the habitat cues from historically high levels of reproduction. If broad-scale factors (such as landscape-scale fragmentation) [73] are responsible for altering the ecological pressures that are at play, then the source of disturbance may be unimportant as even natural disturbances may result in maladaptive behavior. In response to a natural disturbance event, Jones et al. [74] reported a decrease in cerulean warbler nest success a year after an ice storm in Ontario, Canada. However densities also declined in that case, likely producing a sink rather than a trap. Thus, despite our best intentions, forests disturbed by human activity may only resemble naturally disturbed forests, but may differ in terms of tree age-class distribution [75], increased soil disturbance [76], a lack of standing dead trees or snags [77], or in spatial scale and canopy structural complexity [6]. These artificial modifications may result in differing predation pressures, arthropod composition [78], or other factors that may make it difficult for warblers to correctly assess habitat quality. Potential ecological traps created by timber harvests have recently been identified for other declining species including olive-sided flycatchers (*Contopus cooperi*) breeding in selectively logged forests in Montana [43] and rusty blackbirds (*Euphagus carolinus*) breeding in regenerating clear-cuts in northern New England [79]. In the future, research evaluating survival during the post-fledging period across distur-

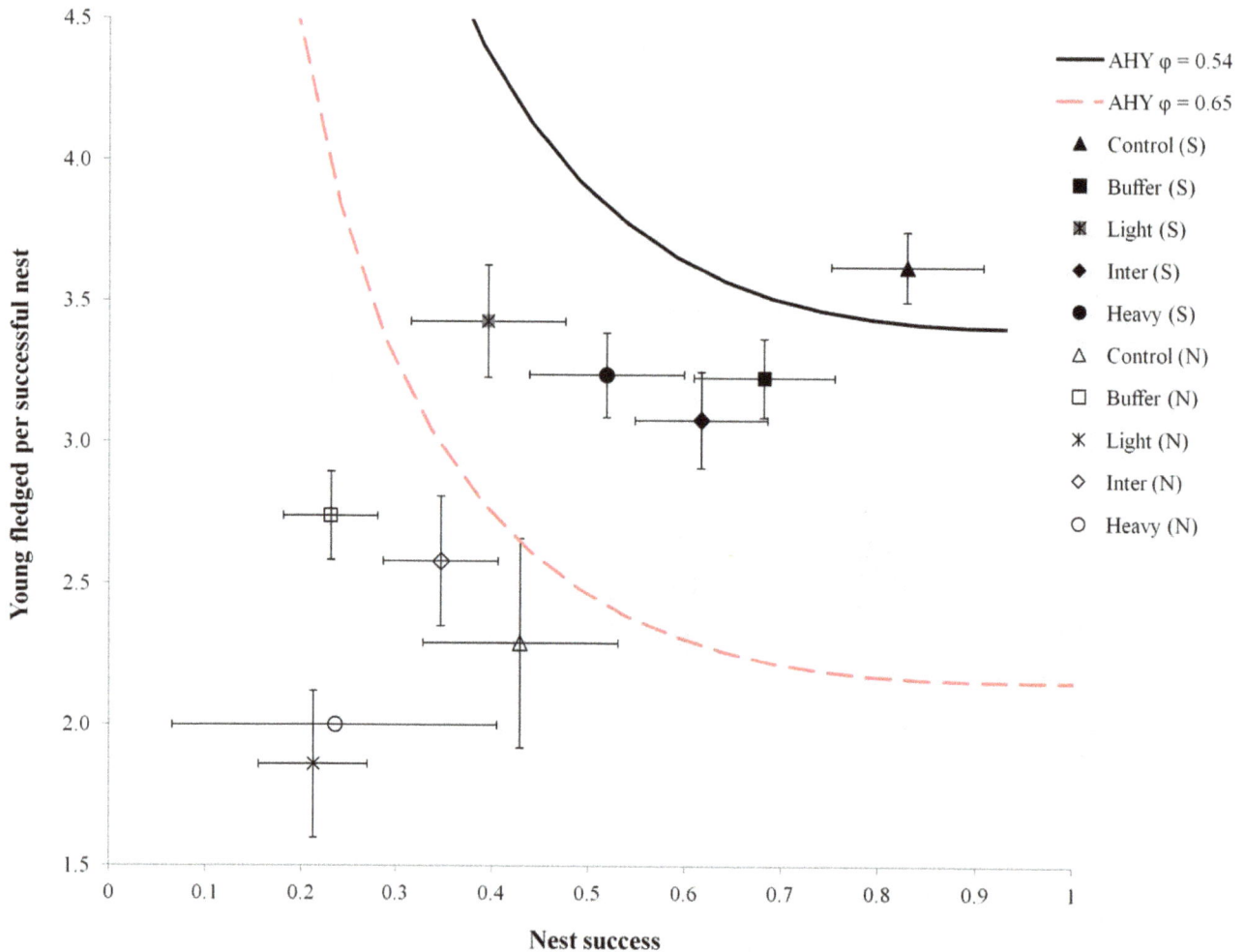

Figure 6. Graphical model of cerulean warbler source-sink dynamics in relation to regional reproductive consequences of emulated disturbances. We used point estimates of nest success and mean number of young fledged/successful nest on various treatments from the southern (S) and northern (N) regions, 2008–10. Error bars indicate ±1 SE. Two possible lambda threshold curves are displayed, each based on a published annual survival rate for cerulean warblers: (1) from Ontario (54% AHY survival), and (2) from Venezuela (65% AHY survival). Points to the left of (or below) the threshold curve, for each given survival rate, represent decreasing, or sink populations, and points to the right of (or above) the curve represent increasing, or source populations. HY survival was considered to be 0.5 of the AHY rate and three re-nesting attempts were assumed to occur for all failed nests.

bance gradients is warranted for cerulean warblers (and other canopy nesting species), although this work will be challenging because of difficulties in capturing nestlings and fledglings. In addition, comparisons of selective pressures in natural versus emulated disturbances and 24-hour video surveillance of nests, will improve our understanding of the causes of nest failure and adaptive nature of habitat selection behavior.

An important caveat of our study is that we measured responses that were short-term in nature (1–4 years), and responses may vary over time. We may have even observed an adjustment in habitat selection behavior in 2010, only four years post-disturbance. While densities increased in 2008 and 2009 on the light treatment plots, by 2010 the density response to light treatments was no longer statistically different than the response to controls. Birds may track variation in breeding success and adjust their habitat selection decisions to match local conditions [80,81]. If habitat selection behavior is dynamic, and relatively low levels of nest success persist on disturbed treatments, densities on light (and possibly other) treatment plots may eventually drop below densities on the control

plots, but this hypothesis will require further study. An alternative explanation is that some canopy closure had already occurred on the light treatment plots [e.g., 82], and attraction to the resulting structural features of the vegetation had begun to wane. Continued monitoring of these field sites to assess the persistence of the trends we have observed would be very useful.

Conservation and Management Implications

The conservation and management implications of our results are complicated by the spatial variability of the impact of disturbances on reproduction, and regional variation in reproductive output in general. In previous studies that have documented putative maladaptive habitat selection, preference has only been considered at local scales (e.g., between adjacent habitats; [43,83–85]). However, for migratory or highly dispersive species, habitat selection behavior also occurs at broader scales (e.g., the decision to breed in the northern or southern portion of the range) [86]. Thus, simply comparing choices made during the final stages of habitat selection greatly simplifies, and possibly misrepresents, this complex behavioral process. In the case of cerulean warblers, although our results suggest that preference for disturbed forest may be maladaptive at the local scale in the southern region, selection for disturbed habitat in this region could actually be adaptive if the alternative option was to migrate further north to breed, or to not reproduce at all. Therefore, a fundamental question that affects our interpretations, as well as those of any study of habitat selection that assesses the adaptive nature of this behavior, is: what alternative breeding locations do birds forego to breed in attractive habitat types? As cerulean warblers appear to regularly engage in long-distance dispersal (putatively searching for recently disturbed forest habitat), the creation of attractive habitats in the southern region (the Cumberland Mountains) may actually be beneficial to the overall sustainability of the global population because it could provide additional breeding opportunities in this highly productive region. However, for this management strategy to be successful, it requires that birds attracted to disturbances in the Cumberlands to have otherwise attempted to breed in less productive regions (e.g., the northern region), or not at all (i.e., 'floaters'), rather than breeding in local undisturbed forest.

In the northern region, emulating disturbances did not always result in major declines in local reproductive success and thus doing so may not create traditional ecological traps. However, if newly created disturbances in this region attract birds from distant locations where fecundity may have been greater (e.g., Cumberlands), a broader-scale trap could be created. Again, if individuals attracted to disturbances in the north would have otherwise failed to reproduce at all, even these northern disturbances with relatively low per capita productivity could have a positive population effect. These contingencies demonstrate how the true impact of putative ecological traps may be quite complex and difficult to assess when viewed in isolation.

Despite those complexities, our study provides evidence that increasing, or even maintaining, populations of cerulean warblers, and potentially other disturbance-adapted late-successional species, into the future will likely require a cooperative, landscape-scale approach to managing forests. The challenge for conserva-tion will be to determine the appropriate locations for implementing disturbances on the landscape in order to provide habitat for a maximum number of breeding pairs while maintaining maximum individual productivity. Accordingly, a conservative approach to management is warranted which would involve emulating disturbances similar in scale and intensity to our intermediate treatments in locations where existing forest structure is unsuitable and breeding densities are low, while limiting disturbance in areas where forest structure is currently appropriate and breeding densities are higher. Determining where appropriate forest structure currently exists may be accomplished by performing systematic bird surveys (to directly assess density) or by applying predictive models which use vegetative and topographic measurements [similar to 21,48]. Future studies examining annual survivorship and long-distance dispersal patterns of cerulean warblers inhabiting various disturbed treatments in multiple regions could help inform this situation further. Finally, it is important to note that we found only minimal impacts of disturbance, beneficial or otherwise, extending beyond the borders of the area treated (i.e., buffers), which suggests that the consequences of any of the forest management practices evaluated here will mostly apply only to the harvested stands themselves.

Acknowledgments

We thank the 50 hard-working field assistants who made this research successful. We thank D.A. Buckley, T.M. Freeberg, and two anonymous reviewers for helpful comments on earlier versions of this manuscript. Banding in Tennessee was conducted under U.S. Geological Survey banding permit no. 22585, in Ohio under permit no. 22272, and in West Virginia under permit no. 23412. Use of trade names does not imply endorsement by the Federal Government.

Author Contributions

Conceived and designed the experiments: TJB DAB JS PBW ADR JLL PDK TAB FLN MHB AE GAG MEM KAP TBW. Performed the experiments: TJB DAB JS PBW ADR JLL TAB FLN MHB AE GAG MEM KAP. Analyzed the data: TJB DAB. Wrote the paper: TJB DAB PDK ADR PBW TBW JLL.

References

1. Connell JH (1978) Diversity in tropical rain forests and coral reefs. Science 199: 1302–1310.
2. Perera AH, Buse LJ, Weber MG (2004) Emulating natural forest landscape disturbances: concepts and applications. New York, NY, , USA: Columbia University Press.
3. Askins RA (1999) History of grassland birds in eastern North America. Studies in Avian Biology 19: 60–71.
4. Vickery PC, Tubaro PL, Cardoza Da Silva JM, Peterjohn BG, Herkert JR, et al. (1999) Conservation of grassland birds in the Western Hemisphere. Studies in Avian Biology 19: 2–26.
5. Lorimer CG (1980) Age structure and disturbance history of a southern Appalachian virgin forest. Ecology 61: 1169–1184.
6. Seymour RS, White AS, deMaynadier PG (2002) Natural disturbance regimes in northeastern North America - evaluating silvicultural systems using natural scales and frequencies. Forest Ecology and Management 155: 357–367.
7. Van Lear DH, Waldrop TA (1989) History, uses, and effects of fire in the Appalachians. Southeastern Forest Experiment Station. Asheville, NC, USA. U.S. Department of Agriculture Forest Service. General Technical Report SE-54.
8. Williams M (1989) Americans and their forests: a historical geography. Cambridge, UK: Cambridge University Press.
9. Parker GR (1989) Old-growth forests of the Central Hardwood Region, USA. Natural Areas Journal 9: 5–11.
10. Bormann FH, Likens GE (1979) Pattern and process in a forested ecosystem. New York, NY, , USA: Springer Verlag.
11. Abrams MD (2003) Where has all the White Oak gone? Bioscience 53: 927–939.
12. Lorimer CG, Frelich LE (1994) Natural disturbance regimes in old-growth northern hardwoods - implications for restoration efforts. Journal of Forestry 92: 33–38.
13. MacDougall AS, Turkington R (2005) Are invasive species the drivers or passengers of change in degraded ecosystems? Ecology 86: 42–55.
14. Strong TF, Teclaw RM, Zasada JC (1997) Monitoring the effects of partial cutting and gap size on microclimate and vegetation responses in northern hardwood forests in Wisconsin. St. Paul, MN, USA. U.S. Department of Agriculture Forest Service. General Technical Report NE-238.
15. U.S. Fish and Wildlife Service (1985) Final ruling for listing Carolina flying squirrel and Virginia flying squirrel as endangered. 50FR27002.
16. U.S. Fish and Wildlife Service (2008) Birds of conservation concern. Arlington, Virginia, , USA: U.S. Department of Interior, Fish and Wildlife Service, Division of Migratory Bird Management.
17. U.S. Fish and Wildlife Service (2007) Indiana bat (Myotis sodalis) draft recovery plan. Fort Snelling, MN, USA.
18. Artman VL, Hutchinson TF, Brawn JD (2005) Fire ecology and bird populations in eastern deciduous forests. Studies in Avian Biology 30: 127–138.
19. Brawn JD, Robinson SK, Thompson FR (2001) The role of disturbance in the ecology and conservation of birds. Annual Review of Ecology and Systematics 32: 251–276.
20. Menzel JM, Ford WM (1985) Nest tree use by the endangered Virginia northern flying squirrel in the Central Appalachian Mountains. American Midland Naturalist 151: 355–368.
21. Bakermans MH, Rodewald AD (2009) Think globally, manage locally: the importance of steady-state forest features for a declining songbird. Forest Ecology and Management 258: 224–232.
22. Long JN (2009) Emulating natural disturbance regimes as a basis for forest management: a North American view. Forest Ecology and Management 257: 1868–1873.
23. Sauer JR, Hines JE, Fallon J (2008) The North American Breeding Bird Survey, Results and Analysis, version 5.15.2008. Patuxent Wildlife Research Center, Laurel, MD, USA. U.S. Geological Survey.
24. Ziolkowski DJ, Jr., Pardieck KL, Sauer JR (2010) The 2003–2008 summary of the North American breeding bird survey. Bird Populations 10: 90–109.

25. Wilson A (1811) American Ornithology. Vol. 3. Philadelphia, PA, , USA: Bradford and Inskeep.

26. Hamel PB (2000) Cerulean Warbler (*Dendroica cerulea*). In: Poole A, editor. The Birds of North America Online. Ithaca, NY, , USA. Cornell Laboratory of Ornithology. http://bna.birds.cornell.edu/bna/species/511.

27. BirdLife International (2010) *Setophaga cerulea*. IUCN Red List of Threatened Species. Version 2010.4. www.iucnredlist.org.

28. Oliarnyk CJ, Robertson RJ (1996) Breeding behavior and reproductive success of Cerulean Warblers in southeastern Ontario. Wilson Bulletin 108: 673–684.

29. Wood PB, Bosworth SB, Dettmers R (2006) Cerulean Warbler abundance and occurrence relative to large-scale edge and habitat characteristics. Condor 108: 154–165.

30. Hamel PB, Rosenberg KV (2007) Developing management guidelines for Cerulean Warbler breeding habitat. In: Buckley DS, Clatterbuck WK, editors; Proceedings of the 15th Central Hardwoods Conference; Southern Research Station, Asheville, NC, USA. US Department of Agriculture Forest Service. General Technical Report SRS-101. Pp. 364–374.

31. Wood PB, Duguay JP, Nichols JV (2005) Cerulean warbler use of regenerated clearcut and two-age harvests. Wildlife Society Bulletin 33: 851–858.

32. Holmes SB, Pitt DG (2007) Response of bird communities to selection harvesting in a northern tolerant hardwood forest. Forest Ecology and Management 238: 280–292.

33. Vanderwel MC, Malcolm JR, Mills SC (2007) A meta-analysis of bird responses to uniform partial harvesting across North America. Conservation Biology 21: 1230–1240.

34. Liker A, Papp Z, Bokony V, Lendvai AZ (2008) Lean birds in the city: body size and condition of House Sparrows along the urbanization gradient. Journal of Animal Ecology 77: 789–795.

35. Rodewald AD, Shustack DP (2008) Urban flight: understanding individual and population-level responses of Nearctic-Neotropical migratory birds to urbanization. Journal of Animal Ecology 77: 83–91.

36. Evans KL, Gaston KJ, Sharp SP, McGowan A, Simeoni M, et al. (2009) Effects of urbanisation on disease prevalence and age structure in Blackbird *Turdus merula* populations. Oikos 118: 774–782.

37. Gram WK, Porneluzi PA, Clawson RL, Faaborg J, Richter SC (2003) Effects of experimental forest management on density and nesting success of bird species in Missouri Ozark forests. Conservation Biology 17: 1324–1337.

38. Lampila P, Monkkonen M, Desrochers A (2005) Demographic responses by birds to forest fragmentation. Conservation Biology 19: 1537–1546.

39. Van Horne B (1983) Density as a misleading indicator of quality. Journal of Wildlife Management 47: 893–901.

40. Battin J (2004) When good animals love bad habitats: Ecological traps and the conservation of animal populations. Conservation Biology 18: 1482–1491.

41. Dwernychuk LW, Boag DA (1972) Ducks nesting in association with gulls - an ecological trap. Canadian Journal of Zoology 50: 559–563.

42. Pelicice FM, Agostinho AA (2008) Fish-passage facilities as ecological traps in large Neotropical rivers. Conservation Biology 22: 180–188.

43. Robertson BA, Hutto RL (2007) Is selectively harvested forest an ecological trap for Olive-sided Flycatchers? Condor 109: 109–121.

44. Fralish JS (2003) The Central Hardwood Forest: its boundaries and physiographic provinces. In: Sambeek JW, Dawson JO, Ponder J, F., editors; North Central Research Station, St. Paul, MN, USA. US Department of Agriculture Forest Service. General Technical Report NC-234. Pp. 1–20.

45. Braun EL (1942) Forests of the Cumberland Mountains. Ecological Monographs 12: 413–447.

46. Fenneman NM (1938) Physiography of the eastern United States. New York, NY, , USA: McGraw-Hill.

47. Buehler DA, Giocomo JJ, Jones J, Hamel PB, Rogers CM, et al. (2008) Cerulean Warbler reproduction, survival, and models of population decline. Journal of Wildlife Management 72: 646–653.

48. Buehler DA, Welton MJ, Beachy TA (2006) Predicting Cerulean Warbler habitat use in the Cumberland Mountains of Tennessee. Journal of Wildlife Management 70: 1763–1769.

49. George GA (2009) Foraging ecology of male Cerulean Warblers and other Neotropical migrants. Morgantown, WV, , USA: PhD Dissertation. West Virginia University.

50. Bibby CJ, Burgess ND, Hill DA, Mustoe SH (2000) Bird census techniques. Second edition. San Diego, CA, , USA: Academic Press.

51. Pyle P (1997) Identification guide to North American birds. Bolinas, CA, , USA: Slate Creek Press.

52. Labocha M, Hayes J (2011) Morphometric indices of body condition in birds: a review. Journal of Ornithology: 1–22.

53. Schamber JL, Esler D, Flint PL (2009) Evaluating the validity of using unverified indices of body condition. Journal of Avian Biology 40: 49–56.

54. Burnham KP, Anderson DR (2002) Model Selection and Multimodel inference: A Practical Information-theoretic Approach, 2nd ed. New York, NY, , USA.: Springer-Verlag.

55. Hines JE, Sauer JR (1989) Program CONTRAST: a general program for the analysis of several survival or recovery rate estimates. U S Fish and Wildlife Service Fish and Wildlife Technical Report 24. US Fish and Wildlife Service, Washington, DC, USA. Pp 1–7.

56. Powell LA (2007) Approximating variance of demographic parameters using the delta method: A reference for avian biologists. Condor 109: 949–954.

57. Girvan MK, Jones J, Norris DR, Barg JJ, Kyser TK, et al. (2007) Long-distance dispersal patterns of male Cerulean Warblers (*Dendroica cerulea*) measured by stable-hydrogen isotopes. Avian Conservation and Ecology 2: 1–12.

58. Jones J, Barg JJ, Sillett TS, Veit ML, Robertson RJ (2004) Minimum estimates of survival and population growth for cerulean warblers (*Dendroica Cerulea*) breeding in Ontario, Canada. Auk 121: 15–22.

59. Bakermans MH, Vitz AC, Rodewald AD, Rengifo CG (2009) Migratory songbird use of shade coffee in the Venezuelan Andes with implications for conservation of Cerulean Warbler. Biological Conservation 142: 2476–2483.

60. Noon BR, Sauer JR (2001) Population models for passerine birds: structure, parameterization, and analysis. In: McCollough DR, Barrett RH, editors. Wildlife 2001: populations. New York, NY, , USA: Elsevier Applied Science. Pp. 441–464.

61. Gardali T, Barton DC, White JD, Geupel GR (2003) Juvenile and adult survival of Swainson's Thrush (*Catharus ustulatus*) in coastal Califonia: Annual estimates using capture-recapture analyses. Auk 120: 1188–1194.

62. Harrison XA, Blount JD, Inger R, Norris DR, Bearhop S (2011) Carry-over effects as drivers of fitness differences in animals. Journal of Animal Ecology 80: 4–18.

63. Oliver CD (1980) Forest development in North America following major disturbances. Forest Ecology and Management 3: 153–168.

64. Edler AU, Friedl TWP (2010) Individual quality and carotenoid-based plumage ornaments in male Red Bishops (*Euplectes orix*): plumage is not all that counts. Biological Journal of the Linnean Society 99: 384–397.

65. Fretwell SD, Lucas HLJ (1970) On territorial behavior and other factors influencing habitat distribution in birds. Acta Biotheoretica 14: 16–36.

66. Greenberg CH, Forrest TG (2003) Seasonal abundance of ground-occurring macro-arthropods in forest and canopy gaps in the southern Appalachians. Southeastern Naturalist 2: 591–608.

67. Chalfoun AD, Schmidt KA (2012) Adaptive breeding-habitat selection: Is it for the birds? Auk 129: 589–599.

68. Morrison RIG, Davidson NC, Wilson JR (2007) Survival of the fattest: body stores on migration and survival in Red Knots *Calidris canutus islandica*. Journal of Avian Biology 38: 479–487.

69. Newton SF (1993) Body condition of a small passerine bird: ultrasonic assessment and significance in overwinter survival. Journal of Zoology 229: 561–580.

70. Streby HM, Andersen DE (2011) Seasonal productivity in a population of migratory songbirds: why nest data are not enough. Ecosphere 7: 1–15.

71. Vitz AC, Rodewald AD (2011) Influence of condition and habitat use on survival of post-fledging songbirds. Condor 113: 400–411.

72. Schlaepfer MA, Runge MC, Sherman PW (2002) Ecological and evolutionary traps. Trends in Ecology and Evolution 17: 474–480.

73. Stephens SE, Koons DN, Rotella JJ, Willey DW (2003) Effects of habitat fragmentation on avian nesting success: a review of the evidence at multiple spatial scales. Biological Conservation 115: 101–110.

74. Jones J, DeBruyn RD, Barg JJ, Robertson RJ (2001) Assessing the effects of natural disturbance on a neotropical migrant songbird. Ecology 82: 2628–2635.

75. DeLong SC, Tanner D (1996) Managing the pattern of forest harvest: Lessons from wildfire. Biodiversity and Conservation 5: 1191–1205.

76. Spies TA, Ripple WJ, Bradshaw GA (1994) Dynamics and pattern of a managed coniferous forest landscape in Oregon. Ecological Applications 4: 555–568.

77. Hutto RL (1995) Composition of bird communities following stand-replacement fires in the northern Rocky Mountain (USA) conifer forests. Conservation Biology 9: 1041–1058.

78. Short KC, Negrón JF (2003) Arthropod responses: a functional approach. In: Friederici P, editor. Ecological restoration of southwestern Ponderosa Pine forests. Washington, D.C., , USA: Island Press. Pp. 286–305.

79. Powell LL, Hodgman TP, Glanz WE, Osenton JD, Fisher CM (2010) Nest-site selection and nest survival of the Rusty Blackbird: does timber management adjacent to wetlands create ecological traps? Condor 112: 800–809.

80. Doligez B, Berthouly A, Doligez D, Tanner M, Saladin V, et al. (2008) Spatial scale of local breeding habitat quality and adjustment of breeding decisions. Ecology 89: 1436–1444.

81. Reed JM, Boulinier T, Danchin E, Oring LW, Nolan V, Jr. (1999) Informed dispersal: prospecting by birds for breeding sites. Current Ornithology 15: 189–259.

82. Miller GW, Kochenderfer JM, Fekedulegn DB (2006) Influence of individual reserve trees on nearby reproduction in two-aged Appalachian hardwood stands. Forest Ecology and Management 224: 241–251.

83. Zhu X, Srivastava DS, Smith JNM, Martin K (2012) Habitat Selection and Reproductive Success of Lewis' Woodpecker (*Melanerpes lewis*) at Its Northern Limit. PLoS ONE 7: e44346.

84. Hollander FA, Van Dyck H, San Martin G, Titeux N (2011) Maladaptive habitat selection of a migratory passerine bird in a human-modified landscape. PLoS ONE 6: e25703.

85. Weldon AJ, Haddad NM (2005) The effects of patch shape on Indigo Buntings: evidence for an ecological trap. Ecology 86: 1422–1431.

86. Johnson DH (1980) The comparison of usage and availability measurements for evaluating resource preference. Ecology 61: 65–71.

9

Comparing Population Patterns to Processes: Abundance and Survival of a Forest Salamander following Habitat Degradation

Clint R. V. Otto*¤, Gary J. Roloff, Rachael E. Thames

Department of Fisheries and Wildlife, Michigan State University, East Lansing, Michigan, United States of America

Abstract

Habitat degradation resulting from anthropogenic activities poses immediate and prolonged threats to biodiversity, particularly among declining amphibians. Many studies infer amphibian response to habitat degradation by correlating patterns in species occupancy or abundance with environmental effects, often without regard to the demographic processes underlying these patterns. We evaluated how retention of vertical green trees (CANOPY) and coarse woody debris (CWD) influenced terrestrial salamander abundance and apparent survival in recently clearcut forests. Estimated abundance of unmarked salamanders was positively related to CANOPY ($\hat{\beta}_{Canopy} = 0.21$ (0.02–1.19; 95% CI), but not CWD ($\hat{\beta}_{CWD} = 0.11$ (−0.13–0.35) within 3,600 m^2 sites, whereas estimated abundance of unmarked salamanders was not related to CANOPY ($\hat{\beta}_{Canopy} = -0.01$ (−0.21–0.18) or CWD ($\hat{\beta}_{CWD} = -0.02$ (−0.23–0.19) for 9 m^2 enclosures. In contrast, apparent survival of marked salamanders within our enclosures over 1 month was positively influenced by both CANOPY and CWD retention ($\hat{\beta}_{Canopy} = 0.73$ (0.27–1.19; 95% CI) and $\hat{\beta}_{CWD} = 1.01$ (0.53–1.50). Our results indicate that environmental correlates to abundance are scale dependent reflecting habitat selection processes and organism movements after a habitat disturbance event. Our study also provides a cautionary example of how scientific inference is conditional on the response variable(s), and scale(s) of measure chosen by the investigator, which can have important implications for species conservation and management. Our research highlights the need for joint evaluation of population state variables, such as abundance, and population-level process, such as survival, when assessing anthropogenic impacts on forest biodiversity.

Editor: Benedikt R. Schmidt, Universität Zurich, Switzerland

Funding: Support for this project was provided by the MDNR–Wildlife Division with funds from the federal Pittman-Robertson Wildlife Restoration Act grant administered by the United States Fish and Wildlife Service (W-147-R: Michigan's Statewide Wildlife Research and Restoration Program), the Rocky Mountain Goat Foundation, and the Michigan Society of Herpetologists. The funders had no role in study design, data collection and analysis, decision to publish, or preparation of the manuscript.

Competing Interests: The authors have declared that no competing interests exist.

* E-mail: cotto@usgs.gov

¤ Current address: United States Geological Survey, Northern Prairie Wildlife Research Center, Jamestown, North Dakota, United States of America

Introduction

Anthropogenic habitat degradation is a primary threat to global biodiversity [1,2]. For example, greater than 30% of amphibian species worldwide are at risk of extinction from different forms of environmental degradation, with anthropogenic habitat degradation often cited as a leading cause of population declines [3–5]. One form of habitat degradation that negatively impacts forest dependent wildlife like some amphibians is timber harvesting [6–8]. Research shows that timber harvesting negatively affects forest amphibian abundance [9–11], but the population mechanism(s) that lead to these observed patterns in abundance are poorly understood [12,13].

Most observational and experimental studies on amphibians and forestry use indices like species richness (counts of the number of species), relative abundance (counts of individuals within a species), or occurrence (counts of occupied sites) as response variables [14,15]. Although these state variables are useful for inferring broad-scale impacts of environmental perturbations [16,17], they have been criticized for failing to elucidate

mechanisms of demographic change[13,14,18]. Indeed, research shows that patterns in amphibian counts may not reflect amphibian survival estimates [19]. Studies that directly assess the influence of habitat degradation on population vital rates such as survival, reproduction, and movement of organisms should yield greater inferential power than those that solely assess population indices [20–23]. However, demographic studies are often conducted at small spatial scales with limited replication, which may reduce the breadth of inference and applicability to broad-scale management [24]. Ideally, population patterns (counts) and processes (demography) should be jointly evaluated to better understand wildlife response to habitat degradation.

Few studies that assessed the impact of habitat degradation on terrestrial wildlife have combined broad-scale surveys with demographic research. We combined correlative and experimental approaches to investigate red-backed salamander (*Plethodon cinereus*) response to residual forest structure, such as coarse woody debris and vertical green trees, within recently harvested forests. These structures were purposefully retained to potentially ameliorate the negative effects of clearcutting on forest wildlife [25]. First, we studied how patterns in salamander abundance at two spatial scales (3,600 m^2 and 9 m^2) were related to retention of green trees (CANOPY) and coarse woody debris (CWD) within recent clearcuts in a managed, forested landscape. Second, we quantified how salamander apparent survival over 1 month in the summer was influenced by CANOPY and CWD. By focusing our study on the same species, forested areas, and disturbance type we were able to evaluate if abundance measurements collected at two spatial scales, and survival measurements, yielded similar inferences regarding amphibian response to forest management.

Methods

Study Species

Red-backed salamanders are a terrestrial, lung-less amphibian distributed in woodlands throughout eastern North America [26]. Terrestrial salamanders are recognized as critical components of forested ecosystems through their contribution to the detrital food web, forest biomass and may potentiality serve as indicators of forest health [27,28]. Like all plethodontids, respiration in red-backed salamanders occurs cutaneously, which requires moist skin, making them susceptible to desiccation. As a result, the red-backed salamander has been utilized in many forest management studies and its negative response to clearcutting has been well documented [10,11].

Study Area

We conducted our study across a 560,000 ha area in the northwestern Lower Peninsula of Michigan, USA, in 2010-11. Our study occurred on state-owned forest lands that were managed for aspen (*Populus* spp.) production by the Michigan Department of Natural Resources (MDNR). The MDNR issues "Use Permits" for research conducted on state-owned lands. Use permits for our project were filed and approved consistent with MDNR expectations and are currently stored at the Cadillac and Traverse City, MI, field offices of the MDNR. In Michigan, aspen is typically harvested via clearcutting on a 40- to 60-year rotation. Within harvested stands (where a stand is defined as an area with homogenous vegetation and management focus) the MDNR implemented green-tree retention prescriptions to mitigate the negative effects of timber harvesting on wildlife [25]. These prescriptions called for retention of 3–10% of the pre-harvest green-tree basal area (*i.e.*, the cumulative surface area covered by a cross-section of tree stems at ground level), arranged throughout the stand as single leave-trees or aggregated into retention patches [29]. Harvested areas also contained varied amounts of CWD that was unequally distributed. Additional study area details can be found in [30]. For this study we focused on quantity of green-trees and CWD (*i.e.*, how much), as opposed to characteristics of individual pieces (*i.e.*, size class, decay state, species) because quantity is directly linked to the MDNR structural retention guidelines [29].

Large-scale Abundance Data

All state-owned aspen stands within a four county area that were >8 ha in size and between 1 to 5 years post-harvest were potential candidates for sampling. We used a Geographic Information System (GIS; ArcGIS 9.1; Environmental Systems Research Institute, Redlands, CA) to overlay each aspen stand with a 60×60 m (0.36 ha) lattice and orthophotos from the 2010 National Agricultural Imagery Program (NAIP; http://www.mcgi.state.mi.us/mgdl) to digitize canopy cover of all retained green-trees within the sampling lattice for each forest stand. We assigned each cell of the lattice to a canopy cover group (>25%, 10–25%,

and <10%) and randomly selected 40 cells from each group. We ensured that all selected cells were >200 m apart. We also selected 30 cells within 40- to 60-year-old aspen stands that were adjacent to our harvested stands. We eliminated 16 cells (13 harvested, 3 older) after initial field visits because the dominant cover-type was not aspen. Our final sample size for harvested cells was 107, with varying levels of green-tree canopy cover, and 27 for the 40–60 year-old cells. Hereafter, we refer to the subset of 60×60 m cells used for our study as sites.

Within each selected site we identified 33, 20×2 m transects that were oriented north to south and spaced ≥5 m apart. From the 33 transects we randomly selected 3 transects, with replacement, for salamander sampling. Selected transects were treated as spatial replicates for estimating salamander capture probability [31]. We sampled subunits (i.e., transects) with replacement to minimize estimation bias of the state-space models used for analysis [32]. We used spatial, as opposed to temporal, replication for sampling salamanders because it minimized the number of repeated visits to each site and reduced travel between sites. Furthermore, previous work shows that temporally replicated cover object searches often violate the "closure" assumption of the state-space models we used for analysis [30,33]. Each transect was surveyed once unless it was selected with replacement, in which case it was surveyed again 12–16 days later. For each transect survey, one observer searched for salamanders under woody cover objects >4 cm diameter and >15 cm long. All woody cover objects consisted of downed logs from previous timber harvest or blow-down events. Observers tallied the number of woody cover objects they searched along each transect. We only included transects with >4 CWD objects of sufficient size in the analysis to ensure all transects had a minimum level of sampling effort. Site-level surveys were completed on the same day generally within 30 min. To assess variation in counts among transects within a site, we calculated a standard deviation for salamander counts at each site and then averaged the standard deviation across all sites.

We used counts of salamanders collected at each site and binomial mixture models [31] to estimate salamander abundance (N_j) and detection probability (p). We hypothesized that salamander abundance would be lower in 1–5 year-old sites compared to 40–60 year-old sites (i.e., CONTROL covariate; Table 1). This hypothesis has been tested previously, and thus not a primary focus of our study [9,10,34]. We also hypothesized that salamander abundance at harvested sites would positively relate to structural retention, such as CANOPY and CWD. Although we stratified canopy cover into different categories during site selection, we treated CANOPY as a continuous variable (i.e., percent canopy cover) for all analyses. We considered models where salamander detection probability was held constant (p(.)) or varied as a function of CWD count (p(CWD$_j$) along transect j at a site. As an exploratory analysis we included average daily temperature and daily precipitation as covariates on salamander detection probability for our 2 highest-ranking models. Although we standardized our salamander surveys to the spring and early summer when temperature and precipitation were conducive to salamander surface detection, we included these weather covariates to account for their potential effects on detection variation. Additional details regarding hypothesis and model development can be found in Text S1. We note that inferences for our large-scale study are limited to the proportion of salamander populations underneath or inside CWD objects, not the entire population of salamanders in the leaf litter or soil profile [35]. We assume that salamanders distributed underneath CWD on a given day are representative of the total salamander population. Our previous

Table 1. Ranking of candidate N-mixture (abundance = N) and Robust Design (survival = S) models for red-backed salamanders in harvested aspen stands in the northern Lower Peninsula of Michigan, USA, 2010–2011.

Model	ΔAIC_c[a]	w[a]	K[a]	$-2l$[a]	CANOPY[b]	CWD[b]
Large-Scale Abundance (3,600 m² sites)						
N(CANOPY +CONTROL), p(CWD)	0.00	0.36	5	558.8	0.21 (0.03–0.40)	
N(CWD + CANOPY + CONTROL), p(.)	1.18	0.20	5	559.9	0.21 (0.02–0.40)	0.23 (0.06–0.40)
N(CWD + CANOPY + CONTROL), p(CWD)	1.40	0.18	6	558.0	0.21 (0.03–0.40)	0.11 (−0.13–0.35)
N(CONTROL), p(CWD)	2.54	0.10	4	563.5		
Small-Scale Abundance (9 m² enclosures)						
N(CONTROL), p(t)	0.00	0.36	5	440.8		
N(CONTROL), p(t + CWD)	0.94	0.23	6	439.0		
N(CWD + CONTROL), p(t + CWD)	2.50	0.10	7	437.8		0.20 (−0.19–0.58)
N(CWD + CONTROL), p(t)	2.65	0.10	6	440.7		−0.01 (−0.21–0.18)
Survival (9 m² enclosures)						
S(CWD + CANOPY + CONTROL), p(t) = c(t)	0.00	0.56	7	1220.4	0.71 (0.26–1.17)	0.96 (0.50–1.42)
S(CWD + CANOPY + CONTROL), p(t) = c(t) + b	1.65	0.24	8	1219.1	0.67 (0.27–1.07)	0.85 (0.44–1.27)
S(CWD + CANOPY + CONTROL), p(t + CWD) = c(t + CWD)	2.90	0.13	8	1220.3	0.72 (0.26–1.18)	0.94 (0.45–1.43)
S(CWD + CANOPY + CONTROL), p(t + CWD) = c(t + CWD) + b	4.80	0.05	9	1219.1	0.67 (0.27–1.07)	0.86 (0.42–1.29)
S(CONTROL), p(t + CWD) = c(t + CWD) + b	15.48	0.00	5	1241.4		

[a] ΔAIC_c = difference from the Akaike's Information Criterion (AIC) best model, adjusted for small sample size, w = AIC_c model weight, K = no. of parameters, $-2l$ = twice the negative log-likelihood.
[b] Beta estimates for abundance covariates CANOPY and CWD with 95% CI in parentheses.

work suggests that this assumption is supported in recently harvested aspen stands [36].

Small-scale Abundance and Survival Data

To quantify small-scale abundance (N_i) and survival (S_i) we selected 6 harvested stands that were sampled as part of our large-scale abundance study. These stands were selected to represent non-uniform variation in green tree retention levels. We overlaid a 30×30 m lattice and used a stratified (*i.e.*, >25%, 10–25%, and <10% canopy cover) random selection to identify 36 cells for study. Although our primary objective was to relate salamander survival to the structural characteristics of 1–5 year-old stands, we also selected nine cells within two, 40–60 year-old aspen stands that were adjacent to our 1–5 year-old stands as a basis for comparison. All cells were >50 m apart. Our selection of 40–60 year-old stands represents the near-maximum age class of aspen in our study area. We recognize that 40–60 year-old aspen stands do not represent high-quality habitat for terrestrial salamanders or provide ideal reference conditions for studying salamander survival [37]. We selected this age class of aspen because it represents a substantial portion of the deciduous cover type in this landscape and hence, in some areas, may be the only older deciduous cover type available for terrestrial salamanders.

In early May of 2010 and 2011 we erected a 9 m² enclosure at the center of each 30×30 m lattice cell (total = 45 cells). Enclosures were constructed of aluminum flashing 50 cm high and buried 12–15 cm into the ground (Figure 1). The top of each fence was bent inward at $90°$ to prevent salamander escape. We did not attempt to remove salamanders that naturally occurred within the enclosure area (*i.e.*, unmarked salamanders). CWD that extended beyond the enclosure boundary were cut and those portions external to the enclosure were removed prior to fence construction. We tested the effectiveness of the field enclosure design for preventing salamander escape over the top by adding 12 salamanders to a 0.09 m² replica enclosure, placed in a covered plastic container with air holes, over a 3-day period. No salamanders escaped from the replica during this time. In 2010 we also visited field enclosures during warm, rainy nights to observe if salamanders were attempting to scale the wall; we never observed salamanders attempting to get in or out of the enclosures.

In mid-May, we added 10 adult salamanders to each enclosure that were individually marked with visual implant elastomer (VIE; Northwest Marine Technology, Shaw Island, Washington; [38]). These salamanders were captured within 1 km of our study stands and added to the enclosure within 24 h of capture. Release locations within an enclosure were randomly assigned. Marked salamander density within the enclosures (≈ 1.1 m²) was comparable to observed densities in mature forests in northern Michigan [39]. From mid-May to mid-June the average maximum temperature was $23.3°$C (range 11.7 to $32.8°$C).

In mid-June, we searched enclosures for salamanders on three successive visits, separated by 3.0 ± 1.2 days (mean $\pm$ 1 standard deviation). We assumed that unmarked salamanders made habitat selection choices resulting in their occurrence within the enclosure prior to construction. Prior to each search, enclosures were gridded into 1 m² sections to ensure searches were performed

Figure 1. Example of a salamander field enclosure deployed in a harvested aspen stand.

systematically. Observers thoroughly searched their assigned 1 m^2 area by examining leaf litter and underneath and within pieces of CWD for salamanders. When performing searches, observers placed all leaf-litter and CWD in plastic bins that were assigned to each 1 m^2 area until nothing remained inside the enclosure except for rooted herbaceous and woody vegetation, and mineral soil. Observers placed leaf-litter and CWD back into the enclosure after searching was complete and attempted to reconstruct the micro-habitat to pre-sampling conditions. Captured salamanders were held in coolers until the sampling event was complete. Two observers checked each salamander for unique VIE markings and then re-released them into the enclosure at the point of capture. Unmarked salamanders that were captured inside the enclosures were not marked.

We used N-mixture models [31] to estimate abundance (N_i) and detection probabilities (p) of unmarked salamanders within the enclosures. Additional details regarding hypothesis and model development can be found in Text S1. Briefly, we fit models where abundance varied as a function of CANOPY above, and CWD within, each enclosure. Similar to the large-scale analysis, we treated CANOPY as a continuous variable (i.e., percent canopy cover) for all analyses. We also fit models where salamander abundance was allowed to vary between enclosures that were situated in 1–5 or 40–60 year-old forest stands (i.e., CONTROL covariate). Salamander detection probability was either held constant (p(.)), allowed to vary as a function of CWD within the enclosures (p(CWD)), or vary between our three sampling events (p(t)).

We used the Huggins parameterization of the robust design population model [40,41] to estimate individual salamander survival (S_i), initial capture (p_t) and recapture probabilities (c_t) of individually marked salamanders. Here, we used subscript "i" to denote that survival estimates apply to individual salamanders, as opposed to estimates from abundance models that apply to individual sites ("I"). Our robust design framework consisted of two primary periods: a salamander additions period and a capture/recapture period. During the additions period, marked salamanders were added to the enclosure, as described above. After one month, we searched through all enclosures for marked salamanders on three successive visits (i.e., 3 secondary periods during primary period 2). Thus, $\hat{S}$ represents the probability that a marked salamander survived from mid-May (primary period 1)

until mid-June (primary period 2) and was available for capture during primary period 2. Initial capture (p_t) is the probability that a marked salamander was captured for the first time during visit t of the second primary period ($t=1, 2, 3$). Recapture (c_t) is the probability a marked salamander was recaptured during visit t, conditional on it being captured at least once before during a previous visit (note: $c_1=0$).

Survival probability was allowed to vary as a function of CANOPY, CWD, or CONTROL (see Text S1 for additional model details). We explored whether capture and recapture probabilities were equal and constant across time (p(.) = c(.)), varied across our sampling events (p(t) = c(t)), varied as a function of CWD within an enclosure (p(CWD) = c(CWD)), or if recapture probabilities were lower than initial capture probabilities (p(.) = c(.) + b).

Data Analysis

We analyzed our abundance data using R (version 2.12.1, http://www.r-project.org/; R Development Core Team 2011) with the add-in package unmarked [42]. We analyzed survival data using Program MARK (MARK, version 5.1, http://www.cnr.colostate.edu/~gwhite/mark/mark.htm). We used Akaike's Information Criterion, adjusted for small sample size (AIC$_c$), to rank models [43]. We used cumulative AIC weights (w_+) and evaluation of 95% confidence intervals to determine relative importance of covariates and model parameters. We report model averaged estimates and unconditional 95% confidence intervals for all back-transformed parameters. We also conducted a Pearson Correlation Analysis to test for potential density dependent effects between counts of marked and unmarked salamanders within enclosures.

Ethics Statement

Our salamander sampling and handling protocols were approved by the Michigan State University Animal Care and Use Committee (Animal Use Form no. 07/08-118-00).

Results

For the 1 to 5 year-old stands, covariates CANOPY and CWD were weakly, negatively correlated in the large- (df = 105, r = −0.11, R^2 = 0.01) and small-scale (df = 34, r = −0.33, R^2 = 0.11) analyses.

Large-scale Abundance

Salamander capture probability (p) ranged between 0.27–0.49 among all candidate models. Capture probability was positively related to the quantity of CWD along each transect ($\hat{\beta}_{CWD}$ = 0.27 (95% CI: 0.07–0.47) for the top-ranking model; Table 1). Cumulative weight (w_+) for models that included the effect of CWD on capture probability was 0.69 (Table S1). Our exploratory analysis revealed no support for the influence of average daily temperature or daily precipitation on salamander detection probability (Table S1). All weather covariates had confidence intervals that overlapped zero and the estimated effects of CANOPY and CWD on salamander abundance were not influenced by the inclusion of weather covariates on detection probability (Table S1). From here forward we report model results which lack exploratory weather covariates.

The mean standard deviation for salamander counts among 3 transects within a site was 0.36 (range 0.00–2.08), suggesting that variation in salamander counts among transects within a site was relatively low. As predicted, abundance estimates for red-backed salamanders were higher for 40–60 year-old, 3,600 m^2 sites ($\hat{\bar{N}}_{40-60yr\ Sites}$ = 4.8, 95% CI: 2.2–10.1) when compared to 1–5 year-old sites ($\hat{\bar{N}}_{1-5yr\ Sites}$ = 1.7, 0.9–3.4). Mean values of percent canopy cover (CANOPY) and counts of CWD objects within clearcut sites were 15±17 (±1SD) and 69±32, respectively. Cumulative weights for all models that included the effects of CANOPY or CWD on salamander abundance were 0.77 and 0.50, respectively (Table S1). Large-scale salamander abundance was positively correlated with CANOPY and CWD (Table 1); however, the estimated effect sizes were imprecise for both covariates (Figure 2a). CANOPY was included in 3 of 4 of our top-ranking models. The 95% confidence intervals for CANOPY did not overlap zero for any of the top-ranking models (i.e., AICc weight ≥10%) that included CANOPY (Table 1). CWD occurred in two of four top-ranking models and the 95% CI overlapped zero for one of those (Table 1). Collectively, the evidence suggests that salamander abundance was weakly related to the amount of CANOPY and CWD at 3,600 m^2 sites within 1–5 year-old aspen stands.

Small-scale Abundance

Estimated capture probabilities of unmarked salamanders in 9 m^2 enclosures were 0.70 (0.53–0.83), 0.35 (0.26–0.46), and 0.28 (0.20–0.38) for our three sampling events, respectively. Evidence suggests that capture probability was not influenced by CWD count (Table 1). The 95% confidence intervals overlapped zero ($\hat{\beta}_{CWD}$ = −0.21 (−0.51–0.09)) for the top model which included an association between CWD and capture probability.

On average, we detected 1.6 unmarked salamanders per 9 m^2 enclosure, per visit (±2.2; 1 SD). Mean estimated abundance of unmarked salamanders was 5.6 (3.8–8.1; 95% CI) and 3.2 (2.6–4.2) for 40–60 and 1–5 year-old sites, respectively. Mean values of percent canopy cover (CANOPY) and counts of CWD objects within clearcut sites were 28±17 (±1SD) and 29±28, respectively. Cumulative weights for models that included the effects of structural retention covariates were 0.20 and 0.24 for CANOPY and CWD, respectively (Table S1). For all models, the 95% CIs overlapped zero for CANOPY and CWD (Table 1). Small-scale abundance of unmarked salamanders was not positively correlated with CANOPY or CWD within the enclosures (Figure 2b).

Salamander Apparent Survival

Observers averaged 83 min (range = 47–184) to search an enclosure during a sampling visit. Observers had 270 captures of marked salamanders during the three visits. Of these captures, 265 were made while searching underneath leaf-litter and CWD, and five were from salamanders found in the bottom of the plastic bins at the end of the sampling event. Within each enclosure, there was no relationship between the total number of unmarked salamanders captured and the total number of marked salamanders recaptured in mid-June (R^2 = 0.02).

Initial capture probabilities of marked salamanders were approximately 50% lower in the second and third site visits compared to the first visit (Figure 3). The estimated probability of capturing a marked salamander at least once during the 3 visits was 0.78 (i.e., $1 - \Pi_{t=1}^{3}(1 - p_t)$). Model-averaged recapture probabilities were slightly lower than initial capture probabilities (Figure 3), but the effect was not strong as models with the capture probability structure $p(t) = c(t) + b$ garnered only 0.29 of the cumulative model weight (Table S1). Capture and recapture probabilities were not influence by the amount of CWD within the enclosures (Table 1; $\hat{\beta}_{CWD}$ = 0.02 (−0.15–0.19) for top model which included CWD).

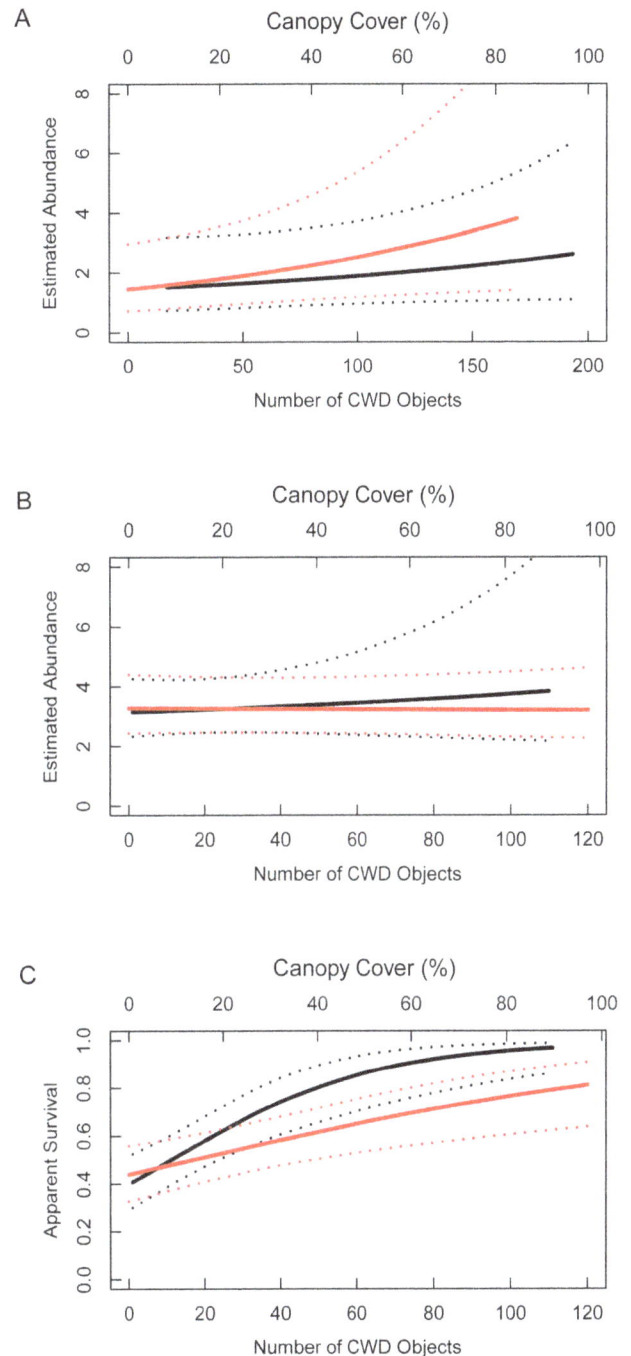

Figure 2. Salamander abundance and survival estimates in harvested forests. Abundance and apparent survival estimates for red-backed salamanders in 1–5 year-old clearcut aspen stands in the northern Lower Peninsula of Michigan, 2010–2011, explained as a function of the amount of green tree canopy retention (Canopy Cover = red line), and the number of coarse woody debris (CWD = black line) objects at each site. Dotted lines are 95% confidence intervals. A) Abundance estimates of unmarked salamanders at 3,600 m^2 sites, B) abundance estimates of unmarked salamanders at 9 m^2 sites, and C) apparent survival estimates of marked salamanders at 9 m^2 sites. All estimates were generated using model averaging.

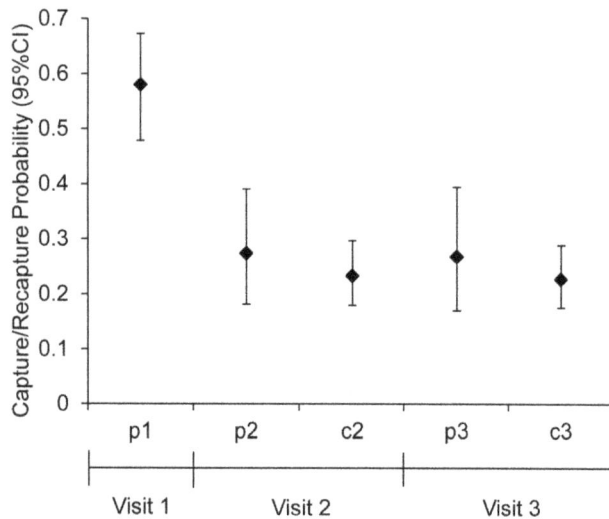

Figure 3. Salamander capture and recapture probabilities within field enclosures. Model-averaged estimates of initial capture (p) or recapture (c) probability of salamanders during three sampling events (visits) in mid-June, northern Lower Peninsula of Michigan, 2010–2011. Initial capture (p_t) is the probability that a marked salamander is captured for the first time during visit t, conditional on it surviving and being available for capture. Recapture (c_t) is the probability a marked salamander is recaptured during visit t, conditional on it being captured at least once before during a previous visit.

Excluding the effects of CANOPY and CWD, apparent survival estimates were $\hat{\bar{S}}$ = 0.62 (0.50–0.72) and $\hat{\bar{S}}$ = 0.64 (0.46–0.79) for 1–5 and 40–60 year-old sites, respectively. Substantial evidence ($w_+ > 0.95$ for CANOPY and CWD; Table S1) indicated that CANOPY and CWD positively influenced salamander apparent survival from mid-May until mid-June in 1–5 year-old harvest stands (Figure 2c). For our top-ranked model, S(CANOPY + CWD + CONTROL), p(t) = c(t) beta parameter estimates were 0.71 (95% CI: 0.26–1.17) and 0.96 (0.50–1.42) for CANOPY and CWD, respectively (Table 1). Estimated apparent survival probabilities were >0.80 for 1–5 year-old sites that contained high levels of canopy cover or CWD (Figure 2c).

Discussion

Ultimately, population processes such as survival, reproduction, and movement determine patterns in species occupancy, richness, and abundance [21]. As such, occupancy and abundance are indirect representations of demographic processes and may offer limited insight into faunal response to habitat degradation [20,21,44]. An appealing characteristic of occupancy and abundance studies is that they can be conducted over broader spatial and shorter temporal scales, thereby increasing the spatial breadth of inference and potential applicability to management. Our inferences regarding salamander response to structural retention in aspen clearcuts were largely conditional on our population parameter of interest. Whereas we found limited support for an influence of structural retention on salamander abundance at two nested spatial scales, we found strong evidence for a positive effect of structural retention on apparent survival probability over 1 month in the summer.

Does Structural Retention Promote Abundance?

At 3,600 m^2 and 9 m^2 we observed higher salamander abundances at 40–60 year-old sites compared to 1–5 year-old sites. This finding is consistent with other studies that show patterns in amphibian counts are positively correlated with time since timber harvest [9,10,34]. In our study, estimated salamander abundance was positively but weakly related to CANOPY and CWD quantity at the large scale (i.e., 3,600 m^2), but not at the small scale (i.e., 9 m^2). At the large spatial scale abundance estimates for 1–5 year-old sites were consistently >20% lower than 40–60 year-old sites, even among sites with high levels of structural retention. Other studies have shown that structural enhancement of young clearcuts can benefit amphibian occurrence and abundance under certain conditions [24]. Our multi-scaled abundance results suggest that the benefits are more uncertain than previously realized.

By sampling across a broad spatial extent (560,000 ha) and over 90 different timber harvest stands, our study represents a realistic range of forest stand conditions typically found in young aspen of the northwestern Lower Peninsula of Michigan. This is important because the spatial scales of past amphibian-forestry studies that assessed occupancy or abundance have often been limited to few experimental forest stands [19,34,45,46]. Inferring broad-scale impacts based on limited spatial extent and site replication is a common limitation in research on the effects of timber harvest on forest biodiversity [24,37], even though the potential pitfalls of doing so in amphibian-habitat research have been discussed [18].

Does Structural Retention Promote Survival?

In contrast to our abundance results, our demographic study showed that leaving structural elements, such as vertical green trees and horizontal CWD within young harvest stands may ameliorate the negative impacts of clearcutting on amphibian survival over 1 month. Past research has shown that clearcutting negatively impacts amphibian survival [19,47] (but see [48]). However, the role of structural retention for influencing demographic parameters is less clear [49]. An enclosure study by Rittenhouse et al. [50] showed that amphibian survival over 30 hours in clearcuts was higher when enclosures were deployed in brushpiles compared to enclosures deployed in open areas with no microhabitat refugia. Our study differs from Rittenhouse et al. [50] in both the area encompassed by the enclosure (0.07 m^2; Rittenhouse et al. [50], 9 m^2; our study) and the time over which survival was estimated (30 h; Rittenhouse et al. [50], $\approx$1 month; our study). Our survival study is temporally limited to 1 month during a relatively mild portion of the summer. It is unclear if salamander survival probability would continue to show a strong positive relationship with CANOPY and CWD during the hottest portions of the summer, July and August. Nonetheless, our research shows that various combinations of green-tree and CWD retention can be used to achieve high apparent survival probabilities for salamanders inhabiting aspen clearcuts in early summer.

Our survival analysis is based on the assumption of no temporary emigration. Although our enclosures prevented horizontal emigration, we were unable to account for potential vertical emigration of salamanders into the soil profile. Live salamanders that migrated into the soil profile before our first resampling event and remained there throughout our three replicate visits were effectively unavailable for capture. These salamanders would likely be recorded as false absences, which can lead to negative bias in survival estimates and potentially bias covariate effects if emigration rates were influenced by covariates that also influence survival [51]. Capture and recapture estimates from our survival study showed that salamanders within the enclosure were less likely to be detected during surveys two and three. This suggests that salamanders in our study had the ability to temporarily emigrate from the sampling area into the soil profile. Thus, our survival estimates likely possess some negative bias and should be considered apparent survival, where S is the product of true survival probability and the probability of remaining in the above ground, sampling area. However salamander capture and recapture probabilities did not depend on the quantity of CWD within the enclosures (Table S1). This suggests that the availability of CWD within the enclosure did not influence the probability that salamanders retreated underground and became unavailable for capture. Although our survival estimates may be biased low, the estimated effects of CWD on apparent survival are likely unbiased.

Population Patterns vs. Process

Single pattern- or process-based response variables are often used to assess the impacts of anthropogenic disturbance on biological diversity; seldom are they jointly evaluated in the same ecological system [19,52]. Inferential power of studies that directly assess demographic processes, such as survival, is generally greater than studies that assess demographic indices, such as occupancy or abundance; however, strength typically comes at the expense of inferential breadth [48]. Conducting detailed studies of demographic processes and population patterns is important for understanding broad-scale anthropogenic impacts and the results of management actions. Our study addresses these concerns through a joint evaluation of multi-scaled abundance patterns and small-scale demographic processes within an actively-managed forested landscape.

Our research shows that structural retention harvest, when done at relatively small spatial scales, can reduce the negative impacts of clearcutting on terrestrial salamanders by lowering apparent mortality rates in 1–5 year-old clearcuts during the summer. This provides resource managers with direct evidence that alternative forest management practices can positively influence salamander population dynamics. However, when viewed at larger spatial scales, we found minimal evidence that structural retention had a strong, positive influence on abundance at sites that were 1–5 years post-harvest. One potential explanation for these seemingly conflicting results is the difference in the temporal scales of abundance and survival studies. Our abundance studies represent snapshots of salamander populations that were subjected to anthropogenic disturbance events that occurred 1–5 years previous, whereas our survival study was conducted over one month. Thus, although structural retention may increase salamander survival over one month, it is unclear if this benefit is negated over a longer temporal period.

An alternative explanation for our conflicting results is that patterns in salamander abundance within clearcuts may be largely influenced by movement dynamics, rather than mortality. Some amphibians will emigrate from recently harvested forest stands (*i.e.*, evacuation hypothesis; [23,53]). Thus, lower abundance estimates for our 1–5 year-old stands could reflect behavioral avoidance of salamanders to clearcuts, regardless of the local availability of structural retention. Indeed, salamander abundances at both spatial scales were not strongly correlated with CANOPY or CWD. This suggests the distribution of free-ranging salamanders across young forest stands was relatively uniform with respect to our habitat covariates. Similar to other past amphibian-forestry

experiments, our survival study did not allow salamanders to choose particular treatments through habitat selection [48,50]. In our survival study we released naïve salamanders into enclosures, thereby eliminating their opportunity to seek refuge in adjacent forest stands and forcing them to make habitat selection choices within a 9 m^2 area. Although our enclosure experiment shows that salamanders can survive over a single season in clearcuts when habitat refugia are provided, it is unclear if these salamanders would choose to remain in clearcuts, if given the option [53,54]. This explanation is consistent with the observed weak correlation between salamander abundance at two spatial scales and structural retention, and the strong effect of structural retention on salamander apparent survival.

Conclusions

Regardless of potential mechanism(s), our study suggests that comprehensive conservation goals for terrestrial salamanders, and other forest-obligate species, will not be accomplished by simply retaining structure within individual harvest units. Our research shows that structural retention can be used to influence population vital rates of terrestrial salamanders, over a relatively short time span. However, our abundance analyses suggest that broad-scale conservation efforts should consider factors that likely extend beyond localized patches of retention. For example, if movement dynamics are truly important to the persistence of forest floor obligates, these species may also require provision of high quality habitats like late-successional forests with structurally complex understories in the managed landscape. Habitat degradation will continue as a dominant force in the global biodiversity and sustainability crisis for the foreseeable future [3,14]. Hence, developing research and monitoring programs that assess broad-scale changes in population patterns, and the demographic processes underlying these changes, should be a shared goal of ecologists and resource managers alike.

Supporting Information

Text S1 Complete description of model development.

Table S1 Complete model set for Table 1. Ranking of candidate N-mixture (abundance) and Robust Design (survival) models for red-backed salamanders in harvested aspen stands in the northern Lower Peninsula of Michigan, USA, 2010–2011.

Acknowledgments

We thank Dave Burt, Andrew Coleman, Joseph and Jonathan Lautenbach, D.J. McNeil, Rachelle Sterling, Shayna Wieferich, and numerous volunteers for their assistance in the collection data. Mike Donovan, Kerry Fitzpatrick, and Larissa Bailey helped develop project objectives and provided guidance on the analysis. A.J. Kroll, Blake Hossack, Larissa Bailey, and 2 anonymous reviewers provided thoughtful comments on an earlier version of this manuscript.

Author Contributions

Conceived and designed the experiments: CRVO GJR RET. Performed the experiments: CRVO RET. Analyzed the data: CRVO. Contributed reagents/materials/analysis tools: CRVO GJR RET. Wrote the paper: CRVO GJR.

References

1. Foley JA, DeFries R, Asner GP, Barford C, Bonan G, et al. (2005) Global consequences of land use. Science 309: 570–574.

2. Fischer J, Lindenmayer DB (2007) Landscape modification and habitat fragmentation: a synthesis. Global Ecology and Biogeography 16: 265–280.

3. Stuart SN, Chanson JS, Cox NA, Young BE, Rodrigues ASL, et al. (2004) Status and trends of amphibian declines and extinctions worldwide. Science 306: 1783–1786.

4. Wake DB, Vredenburg VT (2008) Are we in the midst of the sixth mass extinction? A view from the world of amphibians. Proceedings of the National Academy of Sciences of the United States of America 105: 11466–11473.

5. Hof C, Araujo MB, Jetz W, Rahbek C (2011) Additive threats from pathogens, climate and land-use change for global amphibian diversity. Nature 480: 516–U137.

6. Ferraz G, Nichols JD, Hines JE, Stouffer PC, Bierregaard RO, et al. (2007) A large-scale deforestation experiment: Effects of patch area and isolation on Amazon birds. Science 315: 238–241.

7. Thompson ID, Baker JA, Ter-Mikaelian M (2003) A review of the long-term effects of post-harvest silviculture on vertebrate wildlife, and predictive models, with an emphasis on boreal forests in Ontario, Canada. Forest Ecology and Management 177: 441–469.

8. McDermott ME, Wood PB (2009) Short- and long-term implications of clearcut and two-age silviculture for conservation of breeding forest birds in the central Appalachians, USA. Biological Conservation 142: 212–220.

9. Petranka JW, Eldridge ME, Haley KE (1993) Effects of timber harvesting on southern Appalachian salamanders. Conservation Biology 7: 363–377.

10. Knapp SM, Haas CA, Harpole DN, Kirkpatrick RL (2003) Initial effects of clearcutting and alternative silvicultural practices on terrestrial salamander abundance. Conservation Biology 17: 752–762.

11. Homyack JA, Haas CA (2009) Long-term effects of experimental forest harvesting on abundance and reproductive demography of terrestrial salamanders. Biological Conservation 142: 110–121.

12. Schmidt BR, Feldmann R, Schaub M (2005) Demographic processes underlying population growth and decline in Salamandra salamandra. Conservation Biology 19: 1149–1156.

13. Semlitsch RD, Todd BD, Blomquist SM, Calhoun AJK, Gibbons JW, et al. (2009) Effects of timber harvest on amphibian populations: understanding mechanisms from forest experiments. Bioscience 59: 853–862.

14. Cushman SA (2006) Effects of habitat loss and fragmentation on amphibians: A review and prospectus. Biological Conservation 128: 231–240.

15. deMaynadier PG, Hunter ML (1995) The relationship between forest management and amphibian ecology: a review of the North American literature. Environmental Reviews 3: 230–261.

16. Otto CRV, Forester DC, Snodgrass JW (2007) Influences of wetland and landscape characteristics on the distribution of carpenter frogs. Wetlands 27: 261–269.

17. Kroll AJ, Risenhoover K, McBride T, Beach E, Kernohan BJ, et al. (2008) Factors influencing stream occupancy and detection probability parameters of stream-associated amphibians in commercial forests of Oregon and Washington, USA. Forest Ecology and Management 255: 3726–3735.

18. Kroll AJ (2009) Sources of uncertainty in stream-associated amphibian ecology and responses to forest management in the Pacific Northwest, USA: a review. Forest Ecology and Management 257: 1188–1199.

19. Todd BD, Rothermel BB (2006) Assessing quality of clearcut habitats for amphibians: Effects on abundances versus vital rates in the southern toad (Bufo terrestris). Biological Conservation 133: 178–185.

20. Yoccoz NG, Nichols JD, Boulinier T (2001) Monitoring of biological diversity in space and time. Trends in Ecology & Evolution 16: 446–453.

21. Williams BK, Nichols JD, Conroy MJ (2002) Analysis and Management of Animal Populations. San Diego: Academic Press.

22. Hocking DJ, Babbitt KJ, Yamasaki M (2013) Comparison of silvicultural and natural disturbance effects on terrestrial salamanders in northern hardwood forests. Biological Conservation 167: 194–202.

23. Peterman WE, Crawford JA, Semlitsch RD (2011) Effects of even-aged timber harvest on stream salamanders: Support for the evacuation hypothesis. Forest Ecology and Management 262: 2344–2353.

24. Otto CRV, Kroll AJ, McKenny HC (2013) Amphibian response to downed wood volume in managed forests: A prospectus for future biomass harvest in North America. Forest Ecology and Management 304: 275–285.

25. Franklin JF, Rae Berg D, Thornburgh DA, Tappeiner JC (1997) Alternative silvicultural approaches to timber harvesting: variable retention harvest systems. In: Kohm KA, Franklin JF, editors. Creating a forestry for the 21st century: the science of ecosystem management. Washington D.C.: Island Press. pp. 111–138.

26. Petranka JW (1998) Salamanders of the United States and Canada. Washington, D.C.: Smithsonian Institution Press.

27. Welsh HH, Droege S (2001) A case for using plethodontid salamanders for monitoring biodiversity and ecosystem integrity of North American forests. Conservation Biology 15: 558–569.

28. Davic RD, Welsh HH (2004) On the ecological roles of salamanders. Annual Review of Ecology Evolution and Systematics 35: 405–434.

29. Bielecki J, Ferris J, Kintigh K, Koss M, Kurh D, et al. (2006) Within stand retention guidance. Lansing, MI: Michigan Department of Natural Resources.

30. Otto CRV, Roloff GJ (2011) Using multiple methods to assess detection probabilities of forest-floor wildlife. Journal of Wildlife Management 75: 423–431.

31. Royle JA (2004) N-mixture models for estimating population size from spatially replicated counts. Biometrics 60: 108–115.

32. Kendall WL, White GC (2009) A cautionary note on substituting spatial subunits for repeated temporal sampling in studies of site occupancy. Journal of Applied Ecology 46: 1182–1188.

33. Otto CRV, Bailey LL, Roloff GJ (2013) Improving species occupancy estimation when sampling violates the closure assumption. Ecography 36: 1299–1309.

34. Ash AN (1997) Disappearance and return of plethodontid salamanders to clearcut plots in the southern Blue Ridge Mountains. Conservation Biology 11: 983–989.

35. Bailey LL, Simons TR, Pollock KH (2004) Estimating detection probability parameters for plethodon salamanders using the robust capture-recapture design. Journal of Wildlife Management 68: 1–13.

36. Otto CRV, Roloff GJ (2011) Comparing Cover Object and Leaf Litter Surveys for Detecting Red-Backed Salamanders, Plethodon cinereus. Journal of Herpetology 45: 256–260.

37. Bennett LT, Adams MA (2004) Assessment of ecological effects due to forest harvesting: approaches and statistical issues. Journal of Applied Ecology 41: 585–598.

38. Grant EHC (2008) Visual implant elastomer mark retention through metamorphosis in amphibian larvae. Journal of Wildlife Management 72: 1247–1252.

39. Heatwole H (1962) Environmental Factors Influencing Local Distribution and Activity of the Salamander, Plethodon cinereus. Ecology 43: 460–472.

40. Pollock KH (1982) A capture-recapture design robust to unequal probability of capture. Journal of Wildlife Management 46: 757–760.

41. Bailey LL, Simons TR, Pollock KH (2004) Comparing population size estimators for plethodontid salamanders. Journal of Herpetology 38: 370–380.

42. Fiske IJ, Chandler RB (2011) Unmarked: an R package for fitting hierarchical models of wildlife occurrence and abundance. Journal of Statistical Software 43: 1–23.

43. Burnham K, Anderson DR (2002) Model Selection and Multimodel Inference: A Practical Information-Theoretic Approach. New York, USA: Springer-Verlag.

44. Van Horne P (1983) Density as a misleading indicator of habitat quality. Journal of Wildlife Management 47: 893–901.

45. McKenny HC, Keeton WS, Donovan TM (2006) Effects of structural complexity enhancement on eastern red-backed salamander (Plethodon cinereus) populations in northern hardwood forests. Forest Ecology and Management 230: 186–196.

46. Patrick DA, Hunter ML, Calhoun AJK (2006) Effects of experimental forestry treatments on a Maine amphibian community. Forest Ecology and Management 234: 323–332.

47. Rothermel BB, Luhring TM (2005) Burrow availability and desiccation risk of mole salamanders (Ambystoma talpoideum) in harvested versus unharvested forest stands. Journal of Herpetology 39: 619–626.

48. Chazal AC, Niewiarowski PH (1998) Responses of mole salamanders to clearcutting: Using field experiments in forest management. Ecological Applications 8: 1133–1143.

49. Rosenvald R, Lohmus A (2008) For what, when, and where is green-tree retention better than clear-cutting? A review of the biodiversity aspects. Forest Ecology and Management 255: 1–15.

50. Rittenhouse TAG, Harper EB, Rehard LR, Semlitsch RD (2008) The Role of Microhabitats in the Desiccation and Survival of Anurans in Recently Harvested Oak-Hickory Forest. Copeia: 807–814.

51. Kendall WL (1999) Robustness of closed capture-recapture methods to violations of the closure assumption. Ecology 80: 2517–2525.

52. Welsh HH, Pope KL, Wheeler CA (2008) Using multiple metrics to assess the effects of forest succession on population status: A comparative study of two terrestrial salamanders in the US Pacific Northwest. Biological Conservation 141: 1149–1160.

53. Semlitsch RD, Conner CA, Hocking DJ, Rittenhouse TAG, Harper EB (2008) Effects of timber harvesting on pond-breeding amphibian persistence: Testing the evacuation hypothesis. Ecological Applications 18: 283–289.

54. Rittenhouse TAG, Semlitsch RD, Thompson FR (2009) Survival costs associated with wood frog breeding migrations: effects of timber harvest and drought. Ecology 90: 1620–1630.

10

Encouraging Family Forest Owners to Create Early Successional Wildlife Habitat in Southern New England

Bill Buffum[1]*, Christopher Modisette[2], Scott R. McWilliams[3]

1 Department of Natural Resources Management, University of Rhode Island, Kingston, Rhode Island, United States of America, **2** Ecological Science, United Stated Department of Agriculture, Natural Resources Conservation Service, Warwick, Rhode Island, United States of America, **3** Department of Natural Resources Management, University of Rhode Island, Kingston, Rhode Island, United States of America

Abstract

Encouraging family forest owners to create early successional habitat is a high priority for wildlife conservation agencies in the northeastern USA, where most forest land is privately owned. Many studies have linked regional declines in wildlife populations to the loss of early successional habitat. The government provides financial incentives to create early successional habitat, but the number of family forest owners who actively manage their forests remains low. Several studies have analyzed participation of family forest owners in federal forestry programs, but no study to date has focused specifically on creation of wildlife habitat. The objective of our study was to analyze the experience of a group of wildlife-oriented family forest owners who were trained to create early successional habitat. This type of family forest owners represents a small portion of the total population of family forest owners, but we believe they can play an important role in creating wildlife habitat, so it is important to understand how outreach programs can best reach them. The respondents shared some characteristics but differed in terms of forest holdings, forestry experience and interest in earning forestry income. Despite their strong interest in wildlife, awareness about the importance of early successional habitat was low. Financial support from the federal government appeared to be important in motivating respondents to follow up after the training with activities on their own properties: 84% of respondents who had implemented activities received federal financial support and 47% would not have implemented the activities without financial assistance. In order to mobilize greater numbers of wildlife-oriented family forest owners to create early successional habitat we recommend focusing outreach efforts on increasing awareness about the importance of early successional habitat and the availability of technical and financial assistance.

Editor: Han Y. H. Chen, Lakehead University, Canada

Funding: This research was conducted as part of a University of Rhode Island (URI) project entitled "Forest Management for Wildlife Habitat in Rhode Island" with funding from URI and the United States Department of Agriculture through a McIntire-Stennis Cooperative Research Grant (RIAES-MS975). This is contribution #5354 of the University of Rhode Island Agricultural Experiment Station. The funders had no role in study design, data collection and analysis, decision to publish, or preparation of the manuscript.

Competing Interests: The authors have declared that no competing interests exist.

* E-mail: buffum@uri.edu

Introduction

Encouraging family forest owners (FFO) to manage their forests to create early successional habitat (ESH) is a high wildlife conservation priority in the northeastern United States [1]. Many studies have linked recent declines in early successional wildlife populations in the region to the loss of ESH [2–5]. FFOs, defined as families, individuals, estates, trusts, family partnerships, and other unincorporated groups of individuals owning at least 0.4 ha of forest land, control 55% of all forest land in Southern New England and a similar percentage in the 20 states in the northern United States [6], and thus could play an important role in creating ESH. The Natural Resources Conservation Service (NRCS) of the United States Department of Agriculture encourages FFOs to create ESH through financial and technical support programs such as the Wildlife Habitat Incentives Program and the Environmental Quality Incentives Program [7,8]. However, NRCS funding to support forestry activities has not been fully utilized in recent years in states such as Rhode Island (RI); a missed opportunity for wildlife-oriented FFOs and wildlife conservation efforts in the state. Nationally only 6% of FFOs have participated in federal forestry financial assistance programs [9]. The number of landowners actively managing their forests in some states is decreasing [10]: with increasingly urban backgrounds and lifestyles, many forest owners see forestry as "irrelevant to their landowning objectives and immediate concerns" [11].

Several studies have analyzed participation of FFOs in NRCS forestry programs. None of these studies focused specifically on ESH, but some of the findings are relevant. For example, most participants in federal forestry programs were more concerned about "doing the right thing" than maximizing profits, but the financial incentives provided by the programs increased the number of acres treated [12]. However, landowners who were less financially dependent on their land appeared to be less interested in the financial incentives from the forestry programs [13,14]. The likelihood of FFOs participating in forestry programs increased with the size of their forest holdings [9] and the number of years of ownership [15]. Many FFOs found the federal programs difficult to access and inflexible [16]. The likelihood of active forest management increased if the forest owners lived close to their forests [17].

The objective of our study was to analyze the experience of a group of FFOs who had a strong interest in wildlife and who had been trained to create ESH on their own properties. We conducted the study in RI, one of three states in the Southern New England region of the United States. RI is experiencing ongoing loss of ESH [18], and federal, state and private conservation groups are actively promoting forest management to create more habitat (AFA, 2010; Oehler, 2003; TNC, 2010; USFWS, 2008). RI is representative of Southern New England in terms of forest cover, forest land-use dynamics and forest ownership patterns: for example, FFOs own 57% of RI forests with average holdings of 2.4 ha, compared to 53% and 2.4 ha respectively in Massachusetts, and 50% and 3.6 ha in Connecticut [19].

In the current study we surveyed FFOs who participated in the RI Coverts Program, which has trained FFOs to create wildlife habitat on their own properties since 2008. These FFOs represent a small portion of the total population of FFOs, but we believe they can play an important role in creating wildlife habitat, and that it is important to understand how outreach programs can best reach them. Our study addressed three research questions. (a) What characteristics were shared by this group of FFOs (e.g.

amount of forest holdings, prior experience with forestry, interest in earning income from their land)? (b) How important was technical and financial assistance in motivating these FFOs to create ESH? (c) What other factors made some of these FFOs more likely than others to follow up after the training with activities on their own land? To our knowledge, this was the first study that specifically examined the experience of FFOs in creating ESH. Even though our study had a small sample size and was limited to a subset of FFOs with a strong interest in wildlife conservation, our hope is that the results of the study can be used to strengthen forestry outreach programs in the region and encourage greater numbers of wildlife-oriented FFOs to create ESH on their own properties.

Methods

Ethics Statement

The University of Rhode Island (URI) Institutional Review Board (IRB) reviews research projects conducted at URI that involve human subjects to ensure that two broad standards are upheld: first, that subjects are not placed at undue risk; second, that they give uncoerced, informed consent to their participation. The application for the current study (Project Title: (239239-2) Study of forestry activities by private landowners in Rhode Island), was approved by the IRB on 6 September 2011. As per the

Table 1. Attributes of Family Forest Owners (FFO) who attended RI Coverts Program (N = 34).

	Percent of FFOs
Area of forest ownership: 0.4–3.6 ha	13%
Area of forest ownership: 4–19 ha	38%
Area of forest ownership: 20–39 ha	19%
Area of forest ownership: 40+ ha	31%
Duration of forest ownership: <10 years	15%
Duration of forest ownership: 10–24 years	46%
Duration of forest ownership: 25–50 years	31%
Duration of forest ownership: >50 years	8%
Have some university education	100%
Interest in earning forestry income: Moderate/strong	65%
Interest in earning forestry income: Slight/none	35%
Interest in wildlife: Observe wildlife	94%
Interest in wildlife: Identify species of wild plants or animals	88%
Interest in wildlife: Hunt/fish on own property	47%
Participate in forest certification program	69%
Participate in forest easement program	31%
Before attending Coverts Program: Received forestry advice	71%
Before attending Coverts Program: Started to prepare management plan	62%
Before attending Coverts Program: Hired a forester	62%
Before attending Coverts Program: Implemented some forestry activity	74%
Before attending Coverts Program: Hired a logger	32%
Before attending Coverts Program: Logged without paid help	24%
Before attending Coverts Program: Sold timber or firewood	32%
Before attending Coverts Program: Harvested timber or firewood for personal use	50%
Before attending Coverts Program: Created openings for ESH	38%
Involvement in management of other land: Conservation organizations	38%
Involvement in management of other land: Land of friends and relatives	22%

approved IRB proposal, we requested the participants to formally consent to participating in the study. This was done by either checking a box on an on-line survey, or by consenting in writing if they decided to submit a paper version of the survey rather than the on-line version. We retained the records for all participants.

Data Collection

We surveyed all FFOs who attended the three day RI Coverts Program from 2008 to 2012. A total of 79 persons representing 54 households attended from 2008 to 2012. We sent one survey per household, and did not include program participants who were not FFOs, such as representatives of conservation agencies. Before distributing the 54 surveys, we announced the study in a newsletter to FFOs distributed by NRCS. We then contacted the program participants by email or regular mail, using the contact information they provided when they registered for the program. We gave the respondents the option of filling out paper or online versions of the survey. The survey included 41 questions that were a mix of multiple-choice questions to generate quantitative data as well as open ended questions to generate qualitative data. The survey included three components: (a) general demographics and interest in forestry and wildlife, (b) experience with forestry activities before attending the program, and (c) experience with forestry activities after the training. In the summer of 2011 we surveyed all FFOs who participated in the program between 2008 and 2011. In the summer of 2012 we distributed a shorter version of the survey to the participants of the 2012 program - this version included the first two components of the original survey, but not the section on experience with activities after the training.

Data Analysis

Our small sample size (N = 34) limited the options for statistical analysis, but we used SPSS v. 20 to calculate the Likelihood Ratio Statistic (LX^2) when our data met the requirements for expected frequencies.

Results

The total number of respondents was 34, of whom 19 had implemented activities after attending the program. The overall response rate was 63%. Ninety four per cent of the respondents filled out an on-line survey, and 6% filled out printed surveys. The quantitative results of the survey are presented in Tables 1 and 2, and the qualitative results from open ended questions are presented in the narrative.

Forest Ownership

Fifty percent of our respondents owned more than 20 ha of forest (Table 1). The respondents mentioned several advantages of having larger forest holdings when creating ESH: greater flexibility in site selection for forest management, fewer conflicts with neighbors who resented clearcuts near their property boundaries, and greater ability to engage loggers who preferred larger jobs. Thirteen percent of the respondents owned less than 4 ha of forest. The respondents who owned less than 20 ha were less likely than respondents with larger forest holdings to have a moderate or strong interest in earning income from their forests $(LX^2(1) = 7.197$, n = 34, p<0.01) and to have sold timber or firewood before attending the Coverts Workshop $(LX^2(1) = 6.983$, n = 34, p<0.01). However, there were no significant differences in their likelihood of having engaged in forest management activities before attending the Coverts Program (hiring a forester, preparing a management plan, or harvesting timber or firewood for their personal use) or having followed up after attending the Coverts

Program with forest management activities on their own land $(LX^2(1) = 0.125$, $LX^2(1) = 0.125$, $(LX^2(1) = 0.118$, LX^2 $(1) = 0.000$ respectively; p>0.5 in all cases).

Networking with Other Forest Owners

Before attending the Coverts Program, 71% of the respondents had already received some forestry advice from friends or relatives and 62% had hired a professional forester. Sixty nine percent were involved in forest certification programs such as the Rhode Island Tree Farm Program, 31% were involved in forest easement programs, and 62% had started to prepare a forest management plan. In addition to managing their own forests, 60% were involved in planning forest management activities on land that did not belong to them, such as land owned by friends, relatives or conservation organizations.

Knowledge about Early Successional Habitat

All of our respondents expressed a strong interest in wildlife, whether observing wildlife, hunting, or identifying species of wild plants and animals. Sixty eight percent of the respondents who had implemented activities since attending the program could describe some positive impact of their forest management activities, including increased abundance of birds, mammals, amphibians and insects. Awareness about the importance of ESH for wildlife was low: several respondents commented that before attending the program they did not realize that many wildlife species depend on ESH or that this habitat type was declining in New England. Several commented that they viewed clearcutting negatively before visiting other FFOs who had already created ESH. Thirty three percent of the respondents who implemented activities after attending the program said they probably would not have implemented the forestry activities if they had not attended the program.

Forest Owner Follow-up after Attending Training

Eighty three percent of our respondents who had attended training at least six months before the survey had already followed up with forest management activities on their own land, and all of the other respondents were planning to implement activities. The respondents implemented a range of forest management activities after attending the Coverts Program (Table 2). The most common activity was creating forest openings to generate ESH, with opening sizes ranging from 0.2 ha to more than 6 ha per household. None of the personal attributes recorded in the study were significantly related to how quickly the participants initiated activities on their own land. However, 80% of respondents with management plans had implemented additional forest management activities after the training, whereas none without a forest management plan had implemented any activity since the training except for starting to prepare a management plan (which can take up to one year).

Importance of NRCS Financial Support

Our respondents varied in terms of interest in earning income from their forests: 12% were not at all interested in earning forest income, while 24% were slightly interested and 65% were at least moderately interested. However, the financial and technical assistance offered by NRCS programs appeared to be an important motivating factor for our respondents. Eighty four per cent of our respondents who had implemented activities after attending the RI Coverts Program had received support from NRCS, and 47% said they would probably not have implemented the activities without the financial assistance. Several respondents

Table 2. Attributes of Family Forest Owners (FFO) who implemented forestry activities on their properties after attending Coverts Program (N = 19).

	Percent of FFOs
After attending program: Started forest management plan	32%
After attending program: Hired a forester	42%
After attending program: Harvested timber or firewood for personal use	74%
After attending program: Created forest openings	79%
After attending program: Hired a logger	21%
After attending program: Sold timber or firewood	32%
After attending program: Logged without any paid help	42%
After attending program: Received NRCS financial support	84%
Would probably not have implemented activities without financial support	47%
Would probably not have implemented activities without attending the program	33%
Plan to request future NRCS financial support	79%
Feel that the implementation was harder than expected	20%
Feel that the implementation was easier than expected	20%
Feel that they need additional wildlife/forestry training	53%
Can describe positive impact of implemented activities on wildlife	68%

mentioned that the process of obtaining NRCS support was complex and time consuming, and/or that they were not aware about these financial assistance programs before attending the Coverts Program.

Discussion

Our findings demonstrate that this subset of FFOs shared some common attributes such as education level and interest in wildlife, but differed in terms of land holdings, prior experience with implementing forestry activities, and interest in earning income from their forests.

Forest Ownership

The FFOs in our study tended to have large forest holdings: half owned more than 20 ha of forest, whereas the National Woodland Owner Survey (NWOS) conducted by the US Forest Service reported that only 3% of FFOs in Southern New England and 0% in RI owned more than 20 ha of forest [20]. (The NWOS included 33 respondents in RI, but the area-based sampling frame resulted in very high sampling errors for most attributes [21] so we compared our results to the NWOS results for Southern New England, which included 887 respondents). Our respondents mentioned several advantages of having larger forest holdings when creating ESH, such as greater flexibility in site selection, fewer conflicts with neighbors, and greater ability to engage foresters and loggers. Other studies have reported correlations between larger forest holdings and more active forest management [17,20] and greater participation in forestry programs [15,22]. However, half of our respondents had forest holdings under 20 ha, including 13% with holdings under 4 ha as is more typical for the region (90% of NWOS Southern New England respondents). We found that FFOs with forest holdings under 20 ha were less likely than other respondents to be interested in earning income from their forests, but not significantly less likely to have actively managed their forests before or after attending the Coverts Program. Recent forestry outreach experience in RI has confirmed that FFOs with small forest holdings are interested in

forest management: a series of lectures on forest management supported by NRCS in 2013 attracted 65 FFOs, of whom 31% owned no more than 4 ha of forest and 55% owned no more than 8 ha (Sayles, K. 2013, unpublished data). These findings suggest an opportunity for outreach programs in Southern New England to target FFOs with small forest holdings. These FFOs can make a valuable contribution by creating small openings of 0.6 ha or more, which provide suitable habitat for many shrubland bird species [23], and can contribute to creating habitat for species requiring larger habitat patches, such as New England Cottontail, if their properties are close to existing patches of habitat. These FFOs are eligible to participate in NRCS forestry programs, which, unlike some state forestry programs, do not require that participants own at least 4 ha of forest.

Knowledge about Early Successional Habitat

The strong interest in wildlife expressed by the respondents was not surprising since they had volunteered to participate in a three day training on wildlife issues. More unexpected was their limited understanding about the importance of ESH prior to the training. Several respondents commented that before attending the program they were unaware that many wildlife species depend on ESH and that this habitat type is declining in New England. This applied to several respondents who had already prepared forest management plans before attending the training, which suggests that the consulting forester did not stress the importance of creating ESH during the process of plan preparation. Several respondents commented that they felt negatively about forest clearcutting before attending the Coverts Program, an attitude that is common in other Southern New England states [10]. These findings highlight the importance of educating forest owners about ESH. If FFOs with a strong interest in wildlife, such as our respondents, are unaware of the importance of early successional habit, it seems likely that the awareness of most FFOs is even lower.

Networking with Other Forest Owners

Our respondents appeared to be comfortable networking with other FFOs and receiving advice about forest management. Even before attending the Coverts program, most had received forestry advice from friends, relatives or professional foresters, and many had participated in forest certification programs. Other studies have reported that land owners who receive advice about their forests from professionals or friends are more likely to participate in forestry support programs [15,22] and that participation in a forest certification program may allow landowners to become more comfortable with harvesting trees [10]. Several respondents commented that they learned the most during the Coverts Program from field trips to other FFOs who had already implemented forest management activities on their own properties. The benefits of peer-to-peer learning have been emphasized by several authors as an effective approach to motivate forest owners to manage their forests [17,24]. Our findings suggest that outreach programs should take advantage of existing networks of wildlife-oriented FFOs such as conservation organizations and land trusts to identify FFOs who can potentially be motivated to create ESH on their own properties, and that the outreach programs should promote peer-to-peer learning by expanding the number of forestry events located on land of FFOs who have already implemented forest management activities.

Follow Up after the Training

The main factor affecting how quickly the respondents implemented forest management activities after attending the training appeared to be whether or not the respondent had prepared a forest management plan, a process which can take up to a year. Eighty per cent of the respondents with management plans had implemented additional forest management activities after the training, whereas none without a forest management plan had implemented any activity after the training except for preparing a management plan. As such, we endorse the current NRCS program of providing technical and financial support to encourage FFOs to prepare forest management plans, an important first step in creating wildlife habitat.

Importance of NRCS Financial Support

Our respondents varied in terms of interest in earning income from their forests, but the financial and technical assistance offered by NRCS programs appeared to be an important motivating factor for them. Most respondents who had implemented activities after attending the RI Coverts Program received financial assistance from NRCS and would probably not have implemented the activities without this assistance. We did not see any indications that landowners who are less interested in earning income from their forests were less interested in the financial incentives, as was reported in North Carolina [13]. Several respondents commented that they were disappointed with the low payment offered by loggers when creating ESH, which is not surprising as the annual median stumpage value of hardwood firewood in Rhode Island ranged from $5 to $10 per cord during the study period [25]. Our findings agree with Daniels et al. [12], who reported that profit was not the primary objective of many forest owners, but that the financial incentives increased the area of forest the owners were willing to manage.

However, several respondents commented that they did not know about the NRCS financial assistance programs before attending the Coverts Training. Other respondents mentioned that the process of obtaining NRCS support was complex and time consuming, which is a common complaint of landowners about federal financial assistance programs [16]. Clearly, these important technical and financial assistance programs need to be promoted more widely, and training materials need to be developed that describe the application process in a way that is easier for FFOs to understand.

Conclusions

Our study examined the characteristics of FFOs who have a strong interest in wildlife. We found that this subset of FFOs share some characteristics such as education level, but differ in terms of land holdings and interest in earning income from their forests. Most had prior experience with implementing forestry activities, but 26% were totally unengaged in forest management before attending the program. A key finding of our study was that despite their strong interest in wildlife, many respondents had not been aware before attending the Coverts Program of either the importance of ESH or the availability of financial support currently available from NRCS. However, these FFOs were willing to implement forest management activities on their own land once they were provided with adequate training and financial incentives. These FFOs represent a small portion of the total population of FFOs, but we believe they can play an important role in creating wildlife habitat, and we recommend the expansion of outreach programs such as the Coverts Program. The sample size of our study was smaller than we would have liked, and we recommend conducting a comparable study with a larger sample size to confirm our findings. Meanwhile we offer these preliminary recommendations for an outreach strategy to motivate greater numbers of wildlife-oriented FFOs to engage in forest management: (a) use existing networks of wildlife-oriented FFOs such as conservation organizations and land trusts to identify FFOs who are not yet engaged in forest management but who can potentially be motivated to create ESH on their own properties; (b) focus the content of outreach efforts on the importance of creating ESH and the availability of NRCS technical and financial assistance; (c) encourage peer-to-peer learning by providing more opportunities for FFOs to visit FFOs who have already created ESH on their own land; (d) develop simpler descriptions of NRCS forestry programs and application procedures to make them more easily understood by FFOs; (e) encourage FFOs with smaller forest holdings (4 ha or less) to manage their forests: these FFOs can make a valuable contribution by creating small patches of ESH which provide critical habitat for many species of shrubland birds [23]; and (f) encourage consultant foresters to educate FFOs about the need for ESH by providing refresher trainings for consultant foresters on wildlife issues such as the minimum opening size required by different wildlife species.

Author Contributions

Conceived and designed the experiments: BB CM SW. Performed the experiments: BB CM. Analyzed the data: BB. Contributed reagents/materials/analysis tools: BB. Wrote the paper: BB CM SW.

References

1. DeGraaf RM, Yamasaki M, Leak WB, Lester AM (2006) Technical guide to forest wildlife habitat management in New England. Hanover: University Press of New England. xiii, 305 p.

2. DeGraaf RM, Yamasaki M (2003) Options for managing early-successional forest and shrubland bird habitats in the northeastern United States. Forest Ecology and Management 185: 179–191.

3. Foster DR, Aber JD (2004) Forests in time: the environmental consequences of 1,000 years of change in New England. New Haven: Yale University Press.

4. Litvaitis JA (2001) Importance of early successional habitats to mammals in eastern forests. Wildlife Society Bulletin 24: 466–473.

5. Schlossberg S, King DI (2007) Ecology and management of scrub-shrub birds in New England: a comprehensive review. Beltsville, Maryland, USA: Natural Resources Conservation Service, Resource Inventory and Assessment Division. 0022-541X 0022-541X.

6. Butler BJ, Ma Z (2011) Family Forest Owner Trends in the Northern United States. Northern Journal of Applied Forestry 28: 13–18.

7. Oehler JD (2003) State efforts to promote early-successional habitats on public and private lands in the northeastern United States. Forest Ecology and Management 185: 169–177.

8. Gray RL, Benjamin SJ, Rewa CA (2005) Fish and Wildlife Benefits of the Wildlife Habitat Incentives Program, Paper 98. Lincoln, Nebraska: USDA Forest Service/University Nebraska - Lincoln.

9. Butler BJ (2008) Family Forest Owners of the United States, 2006. Gen. Tech. Rep. NRS-27. Newtown Square, PA: U.S. Department of Agriculture, Forest Service, Northern Research Station. 72 p.

10. Berlick MM, Kittredge DB, Foster DR (2002) The Illusion of Preservation - a Global Environmental Argument for the Local Production of Natural Resources. Harvard Forest Paper 26. Petersham, Massachusetts: Harvard Forest, Harvard University.

11. Sampson N, DeCoster L (2000) Forest Fragmentation: Implications for Sustainable Private Forests. Journal of Forestry 98: 4–8.

12. Daniels SE, Kilgore MA, Jacobson MG, Greene JL, Straka TJ (2010) Examining the Compatibility between Forestry Incentive Programs in the US and the Practice of Sustainable Forest Management. Forests 1: 49–64.

13. Daley SS, Cobb DT, Bromley PT, Sorenson CE (2004) Landowner attitudes regarding wildlife management on private land in North Carolina. Wildlife Society Bulletin 32: 209–219.

14. Kammin LA, Hubert PD, Warner RE, Mankin PC (2009) Private Lands Programs and Lessons Learned in Illinois. The Journal of Wildlife Management 73: 973–979.

15. Ma Z, Butler BJ, Kittredge DB, Catanzaro P (2012) Factors associated with landowner involvement in forest conservation programs in the U.S.: Implications for policy design and outreach. Land Use Policy 29: 53–61.

16. Kilgore MA, Greene JL, Jacobson MG, Straka TJ, Daniels SE (2007) The Influence of Financial Incentive Programs in Promoting Sustainable Forestry on the Nation's Family Forests. Journal of Forestry 105: 184–191.

17. Rickenbach M, Kittredge D (2009) Time and Distance: Comparing Motivations Among Forest Landowners in New England, USA. Small-Scale Forestry 8: 95–108.

18. Buffum B, McWilliams SR, August PV (2011) A spatial analysis of forest management and its contribution to maintaining the extent of shrubland habitat in southern New England, United States. Forest Ecology and Management 262: 1775–1785.

19. Butler BJ, Barnett CJ, Crocker SJ, Domke GM, Gormanson D, et al. (2011) The Forests of Southern New England, 2007: A report on the forest resources of Connecticut, Massachusetts, and Rhode Island. NRS. 55 Newtown Square, PA: : U.S. Department of Agriculture, Forest Service, Northern Research Station. 48 p.

20. USFS (2006) National Woodland Owner Survey Table Maker Ver 1.01. United States Forest Service, Forest Inventory and Analysis, United States Department of Agriculture, Arlington, VA. http://apps.fs.fed.us/fia/nwos/tablemaker.jsp (Accessed 10 September 2012).

21. Butler BJ, Leatherberry EC, Williams MS (2005) Design, implementation, and analysis methods for the National Woodland Owner Survey. Newtown Square, PA: U.S. Department of Agriculture, Forest Service, Northeastern Research Station. 43 p.

22. Poudyal NC, Hodges DG (2009) Factors Influencing Landowner Interest in Managing Wildlife and Avian Habitat on Private Forestland. Human Dimensions of Wildlife 14: 240–250.

23. Askins RA, Zuckerberg B, Novak L (2007) Do the size and landscape context of forest openings influence the abundance and breeding success of shrubland songbirds in southern New England? Forest Ecology and Management 250: 137–147.

24. Ma Z, Kittredge D, Catanzaro P (2012) Challenging the Traditional Forestry Extension Model: Insights from the Woods Forum Program in Massachusetts. Small-Scale Forestry 11: 87–100.

25. Abbot T, Kittredge D (2013) Southern New England Stumpage Price Report. MassWoods Forest Conservation Program.

11

Short-Term Forest Management Effects on a Long-Lived Ectotherm

Andrea F. Currylow*, Brian J. MacGowan, Rod N. Williams

Department of Forestry and Natural Resources, Purdue University, West Lafayette, Indiana, United States of America

Abstract

Timber harvesting has been shown to have both positive and negative effects on forest dwelling species. We examined the immediate effects of timber harvests (clearcuts and group selection openings) on ectotherm behavior, using the eastern box turtle as a model. We monitored the movement and thermal ecology of 50 adult box turtles using radiotelemetry from May–October for two years prior to, and two years following scheduled timber harvests in the Central Hardwoods Region of the U.S. Annual home ranges (7.45 ha, 100% MCP) did not differ in any year or in response to timber harvests, but were 33% larger than previous estimates (range 0.47–187.67 ha). Distance of daily movements decreased post-harvest (from 22 m$\pm$1.2 m to 15 m$\pm$0.9 m) whereas thermal optima increased (from 23$\pm$1°C to 25$\pm$1°C). Microclimatic conditions varied by habitat type, but monthly average temperatures were warmer in harvested areas by as much as 13°C. Animals that used harvest openings were exposed to extreme monthly average temperatures ($\sim$40°C). As a result, the animals made shorter and more frequent movements in and out of the harvest areas while maintaining 9% higher body temperatures. This experimental design coupled with radiotelemetry and behavioral observation of a wild ectotherm population prior to and in response to anthropogenic habitat alteration is the first of its kind. Our results indicate that even in a relatively contiguous forested landscape with small-scale timber harvests, there are local effects on the thermal ecology of ectotherms. Ultimately, the results of this research can benefit the conservation and management of temperature-dependent species by informing effects of timber management across landscapes amid changing climates.

Editor: Csaba Moskát, Hungarian Natural History Museum and Eotvos University, Hungary

Funding: Funding for the project was provided by the Indiana Division of Forestry Grant #E-9-6-A558 and IDNR Division of Fish and Wildlife, Wildlife Diversity Section, State Wildlife Improvement Grant #E2-08-WDS15. The funders had no role in study design, data collection and analysis, decision to publish, or preparation of the manuscript.

Competing Interests: The authors have declared that no competing interests exist.

* E-mail: a.currylow@gmail.com

Introduction

Study of habitat alteration through direct and indirect anthropogenic episodes such as reduction of forest habitats and changing climate is becoming increasingly frequent. The understanding of how these changes affect the physiology and behavior of native fauna is vital to the preservation of diversity. Timber harvesting is likely one of the most prominent land uses affecting forest wildlife [1–5]. Forest management practices change the vegetative structure and local temperature, which may affect community structure and function [6]. Environmental flux also has a greater effect on movements and behavior of poikilotherms than for homeothermic species [7,8]. In response, timber harvests have been implicated as a possible cause for worldwide herpetofaunal declines [9–11]. As a result, management of our eastern hardwood forests has become a balancing act between timber production and ecological conservation.

While some data suggest that heavily logged areas are associated with moderate increases in bird and reptile diversity [3], it is not clear whether this can be considered a general trend for all taxa. Timber harvesting has the potential to affect multiple facets of how ectotherms utilize available habitat both directly and indirectly. Canopy openings may create basking sites or allow herbaceous mass to flourish and provide basilar food sources [12]. Edge effects of openings and access roads have been shown to influence habitat

resources into the forest interior at varying distances [13–15]. Because variation in resources such as vegetation and invertebrate prey occur, daily movements and annual home range sizes may readily expand, contract, or shift in response to this variation. Moreover, the behavior, physiology, and even fitness of ectotherms are strongly affected by temperature fluctuations [16,17]. Temperature dictates ectothermic habitat use based on the animals thermal optima (i.e., the temperature at which movement activity is maximal; [17] which in turn alters behavior [18,19].

Recent attempts to assess the effects of timber harvests on many ectothermic species often suffer from the lack of replication or comparable pre-harvest data (e.g., [20,21]). Furthermore, the majority of these herpetofaunal studies have focused on the harvest effects on amphibian populations (e.g., [5,22–25]), while relatively little is known about the impacts on reptile populations. However, the existing data suggest reptiles are not only sensitive to habitat perturbations, but that the impacts are more pervasive and severe than for amphibians [11]. Negative impacts to reproductive adult reptiles, such as long-lived, K-selected turtles, can devastate entire populations [26,27]. Box turtles, which are among the longest lived of all reptiles, are geographically widespread throughout the eastern forests, yet they are sensitive to environmental disturbances that affect local habitat features [28–30]. Widespread population declines have sparked interest in the conservation of this species. While basic data exist on the habitat requirements of certain ectothermic species, many studies were conducted at a single location and did not empirically assess responses to changing habitat or microenvironmental conditions.

The investigation of ecological mechanisms underlying species declines has become paramount in conservation literature. Simply reporting the extirpation of populations without testing mechanistic causes does little to promote conservation management. Herein, we investigated temporal thermal habitat availability, habitat use, thermal behavior, and intersexual differences among eastern box turtles (*Terrapene carolina carolina*) within the framework of a managed forest setting. The overarching goals of this study were to examine ectothermic response to timber harvesting at both the landscape- and local-scale. At the landscape scale, our specific goals were to assess effects of various timber harvest regimes on habitat use, thermal environments, and thermal ecology. At the local level, our specific goals were to investigate edge effects of timber harvests on thermoregulatory behavior, movement parameters (frequency of movement and steplength), and observed behavior.

Methods

Study area

The research was conducted within approximately 35,000 hectares of Morgan-Monroe State Forest (MMSF) and Yellowwood State Forest (YSF) in Morgan, Monroe, and Brown Counties, Indiana (Figure 1a). From the years 1860 through 1910, routine burning and cutting for cattle grazing characterized the forestland. At the turn of the 20th century, the state of Indiana began purchasing the land and establishing these State Forests. Now, MMSF and YSF boundaries are shared, forming a relatively contiguous forested habitat characterized by hills and ravines of hardwood, deciduous forests with scattered gravel access roads. This is an oak-hickory dominated forest, with the majority of canopy species being *Quercus* spp., such as *Q. montnana* (chestnut oak), and *Carya cordiformis* and *C. ovata* (butternut and shagbark hickory; [31]). These State Forests are managed for multiple-uses, including recreation, education, research, and timber harvesting. Research activities on public lands were conducted under the scientific use permits 09-0080 & 10-0083 issued by the Indiana Department of Natural Resources.

Forest management design and sampling

Our research is part of a long-term (100-yr), landscape-scale (spanning 31 linear kilometers and 3,601 hectares) timber and wildlife research collaborative designed for the study of ecological and social impacts of various silvicultural methods typically employed in the Midwest (Hardwood Ecosystem Experiment [32]. In 2007, we identified nine study sites of approximately 400-ha, each assigned to one of three forest management classes in a randomized complete block design (Figure 1b). The management classes included two 2.72–4.43-ha clearcuts, eight 0.15–2.55-ha group selection openings, and forested controls. The timber harvests were implemented on equal numbers of southwest- and northeast-facing slopes over the winter of 2008–09 within the center 90-ha of each study site. The remaining 300+ hectares at each site remained intact to serve as refugia and maintain species diversity.

To determine the effects of timber harvests on *T. c. carolina*, we collected GPS location and habitat use data before timber harvests (pre-harvest; 2007–08) and after harvests (post-harvest; 2009–10). We initially located adult animals by meandering-transect visual encounter surveys. Upon capture, we assigned a unique ID number and marked each animal using a triangle file along the

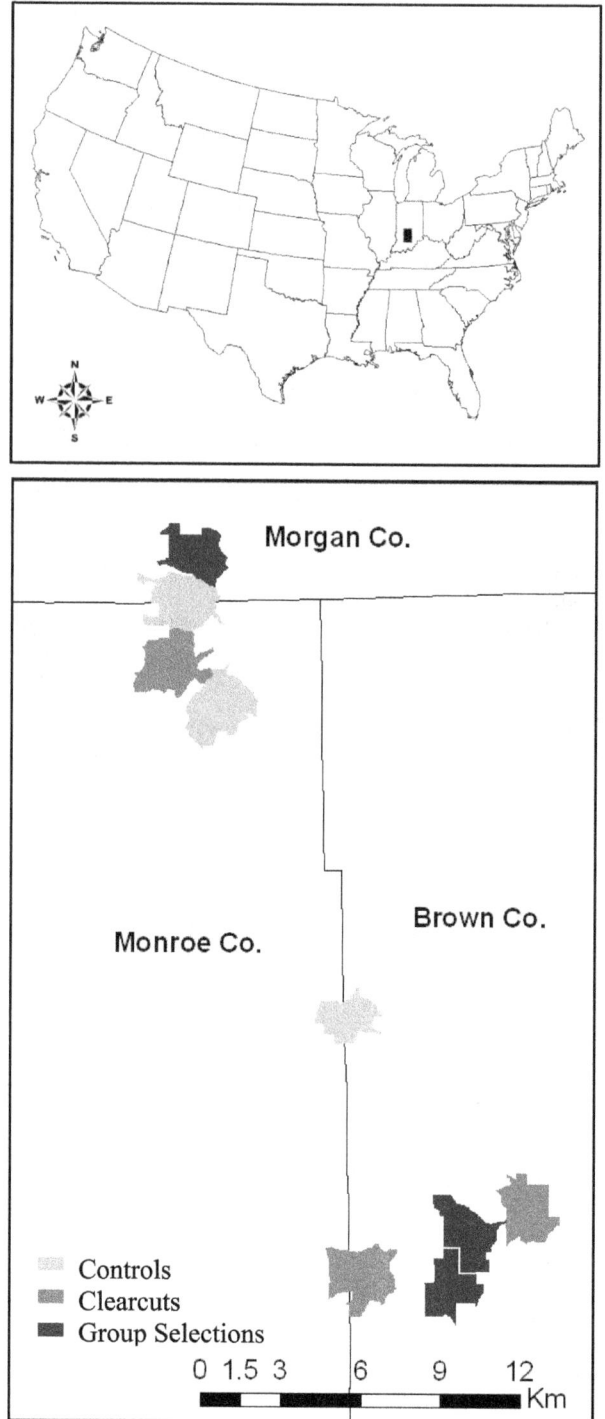

Figure 1. Study Area Maps. Regional and local map of the study area in south-central Indiana. a) The location of the study area in Indiana relative to the continental US. b) The nine study sites spanning Morgan, Monroe, and Brown Counties in IN. Polygon colors represent management classes (controls = light grey, clearcuts = medium grey, group selections = dark grey).

marginal scutes following a modified Cagle scheme [33–35], recorded morphometrics, and affixed a transmitter (model RI-2B Holohil Systems, Ltd., Ontario, Canada) to the carapace. Where possible, we equally divided sex ratios and numbers of the animals among sites and management classes. We subsequently radio-tracked (homing) the animals 2–3 times per week during the active seasons (May through October). For each tracked location, we recorded GPS coordinates, date, ground temperature, elevation, and during the post-harvest years we also recorded observed activity classifications (resting, eating, mating, basking, walking, etc.).

To monitor the thermoregulatory behavior of the animals post-harvest, we affixed iButton temperature dataloggers (model DS1921G-F5#, Maxim Integrated Products, Inc., Sunnyvale, CA) to the carapace of each of the tracked turtles in May 2009. Since carapacial temperature measurements have been shown to correlate well with deep body temperatures [36–39], we used the dataloggers to represent each animal's body temperature (T_b). Temperature datalogger and transmitter weight combined was usually no more than 5% (max 20 g) of the animal's total body weight. Dataloggers recorded temperatures every 45 minutes during the active season (May–October). All animals were handled according to the Purdue Animal Care and Use Protocol 07-037.

To assess the available thermal habitats in harvest areas versus uncut forests, we measured ambient temperature using temperature dataloggers affixed to stakes, 10 cm from soil surface (at approximately *T. c. carolina* carapace height). We randomly placed these 'environmental dataloggers' at four sample locations within each of the nine study sites for a total of 36 individual thermal locations. In each clearcut management site, two environmental dataloggers were randomly deployed inside clearcuts and two in the adjacent forests (between 100 m and 500 m from the nearest harvest edge; harvest-adjacent forest). In each group selection management site, four dataloggers were randomly deployed inside harvest openings. In each control site, we randomly deployed four dataloggers within the forested habitats. This blocked design resulted in equal numbers of environmental dataloggers inside harvest openings (n = 18) and in forested areas (n = 18) representing the four habitat types (clearcut opening, group selection opening, harvest-adjacent forest, and forested control). To eliminate the effect of slope aspect on temperature logged, we used equal numbers of southwest- and northeast -facing slopes. We deployed all temperature loggers from May 2009 to October 2010 for a total of 75 weeks. We programmed dataloggers to record temperatures every 45 minutes to match the carapacial dataloggers described above.

Landscape-scale analyses

Home Range Estimation. We used multiple analyses to examine how various timber-harvesting regimes affect behavior at land-scape- and local-scales. To describe landscape-scale effects of timber harvests, we used all animal location data across all nine study sites throughout the forested landscape. To characterize spatial land use in our population of box turtles, we created a point layer in ArcGIS 9 (version 9.3.1; [40]) using the GPS location data and calculated 100% Minimum Convex Polygons (MCP) with the Hawth's Analysis Tools extension [41] for each turtle in each year, thus creating annual MCP home ranges. We standardized all annual MCP home ranges by the number of GPS locations and log-transformed them for normality.

We used a generalized linear mixed model to test annual MCP home ranges for differences among sites using a crossover design and the PROC GLMMIX command in SAS [42] with a first-order autoregressive covariance structure. We compared all the pre-harvest data then "crossed over" to the post-harvest control comparisons. In our initial model, site, year, and the interaction of site and year were fixed effects and animal ID nested in site was a random effect. By analyzing data in this crossover fashion, we

could verify that control sites were representative of pre-harvest conditions (i.e., site explained very little variation). We grouped sites by management class (clearcut, group selection, and control) for all subsequent analyses and evaluated their effects in the pre- and post-harvest data using a full factorial generalized linear mixed model (GLMM) with unbounded variance components in JMP [43]. We used year, sex, management class, and their interactions as fixed effects and animal ID nested in year as a random effect (to account for repeated measures of individual animals) to find any differences in annual MCP home ranges with relation to harvests. To detect significant differences across effect levels, we used post-hoc Least Squares Means (LSMeans) Tukey-Kramer pairwise comparisons, which adjust significance for multiple comparisons.

Year-to-year variation in movements and habitat use is common (often due to variation in resources such as vegetation and invertebrate prey; [28,44,45], therefore we used biennial (two-year) intervals as indices of longer-term home range sizes and core use areas. These biennial intervals corresponded to the two pre-harvest years and two post-harvest years (hereafter "harvest periods"). To assess differential habitat utilization due to timber harvests, we used biennial MCPs and kernel estimates (ArcGIS Home Range Tools [HRT] extension; [46]) for each animal between harvest periods. We chose to use kernel estimates for further comparisons to other habitat use studies [47] but also continued to calculate MCPs because it has been argued that they better represent herpetofaunal habitat use [48]. We calculated 50%, 90%, and 95% kernel isopleths (percent volume contour) of utilization distributions using the fixed kernel method with least squares cross validation (LSCV) for pre- and post-harvest. For both biennial MCP and kernels, we used a GLMM to test for differences in pre- and post-harvest area measurements (log-transformed) caused by the fixed effect of harvest period (with animal ID nested as a random effect to control for re-sampling error).

Movement and thermal ecology

Animals may not only adjust their annual home ranges in response to harvests, but also vary their movement activity (i.e. move farther distances within their home range or move more frequently). For this analysis, we calculated the Euclidian distance between GPS locations for each animal in ArcGIS using the HRT extension then calculated steplength (average estimated distances by day). To test whether harvest period had an effect on steplength, we log-transformed these data and fitted a full factorial unbounded GLMM with harvest period, sex, management class, and their interactions as fixed effects and animal ID nested in harvest period as a random effect. Then to examine the thermal ecology of *T. c. carolina* in relation to the timber harvests, we tested for correlation between the log-transformed steplength data and ground temperature (T_g; recorded when animals were radio-located). We also used these data to determine the thermal optima (the temperature at which movement activity is maximal) across harvest periods.

Thermal habitats

To test for changes in available thermal habitat, we used differences in ambient temperature among habitat types within sites. We summarized the temperature time series data from each of the 36 environmental dataloggers into three variables; monthly temperature maxima (T_{max}), monthly temperature minima (T_{min}), and monthly temperature mean (T_{mean}) using R [49]. We used unbounded GLMM in JMP to test for significant T_{min}, T_{max}, and T_{mean} differences caused by habitat type, month, and their interaction as fixed effects and individual datalogger ID nested in month as the random effect. We used LSMeans Tukey-Kramer post-hoc comparisons to detect significant differences in T_{min}, T_{max}, and T_{mean} between months.

Local-scale analyses

To determine the thermal effects of harvests on ectotherm behavior, we first characterized the thermoregulatory behavior of our entire turtle population. We examined the max, mean, and min T_b to find the range of selected temperatures for each month. We correlated observed behavior at the time of each GPS location in relation to T_b. We used a GLMM to investigate animal body temperature (T_b) differences explained by the fixed effect of observed behavior category with the random effect of animal ID nested in behavior. Behavior categories included basking, eating, mating, resting, inverted (found upside-down), walking, and buried.

To investigate local-scale harvest edge use and activity, we examined the actual harvest openings and their associated edges in combination with GPS location data. We then created 10- and 50-meter polygon buffers around the harvest boundaries using ArcGIS Analysis Tools. We tested for differences in the percent of animal locations within these three harvest-polygons (inside harvest, 10 m buffer, and within the 50 m buffer) across harvest periods, again controlling for individual effects using an unbounded GLMM as described above. We conducted a similar analysis using the Euclidian distances animals moved within these harvest-polygons to test for differences in activity (frequency of movement or daily distance moved).

To determine the edge effects on thermoregulation, we compared T_b of the animals using the harvests and their edges to the T_b of those same animals when they were located in the forests. To investigate edge effects on movement activity, we used T_b to describe the available thermal habitats in various harvest-polygons. We analyzed harvest edge effects by categorizing harvest proximity polygons (as above) by inside the harvest, 10 m buffer, and 50 m buffer from the nearest harvest opening. We also explored T_b within harvest-polygons by using T_b as the response variable and distance to harvest and month as the fixed effects. We used unbounded GLMMs and controlled for repeated measures using animal ID nested in harvest-polygons as a random effect in each model. We performed post-hoc LSMeans Tukey-Kramer pairwise comparisons to detect significant differences.

Results

Landscape-scale effects

We radio-tracked 23–44 *T. c. carolina* each year (average = 33.5/year), carrying over all that survived each year and were not lost or censored. Losses due to transmitter failure were rare (n = 1). Two animals were separated from their transmitters and censored. Five animals died of various causes including predation (n = 1), severe emaciation (n = 1), suspected disease (n = 2), or failure to emerge from hibernacula (n = 1). Home range MCPs for the remaining animals (n = 50; 23♂, 27♀) with >20 locations per year (avg. = 57.34, SD = 19.10, range = 14–79) were calculated for each year (see supporting information for Table S1).

We found no difference (p-value = 0.418) in the overall size of *T. c. carolina* annual MCP home ranges between all pre-harvest sites and control post-harvest sites, verifying our experiment used true controls. Annual MCP home ranges (4.10 ha to 11.43 ha) did not differ among sex, year (2007–10), management class, or any combination of these factors (all p-values>0.07). The minimum and maximum annual home range sizes were 0.47 ha and 187.67 ha, respectively. The average MCP for all four years was 9.14 ha for males and 5.55 ha for females.

Average pre-harvest biennial MCP home ranges (18.93 ha, SE = 7.51) were generally larger than post-harvest (9.09 ha SE = 5.75; Table 1), however, this difference was not significant ($F_{1,\ 2.435} = 0.018$, $p = 0.90$). There was much variation in kernel areas by sex and harvest period (Table 1) For all three kernel isopleths (50%, 90%, and 95%), the home range areas increased from pre-harvest to post-harvest (all p-values<0.05). No variation in biennial home range area was attributed to harvest type (clearcut or group selection) or sex (all p-values>0.29).

Movements and thermal ecology

Steplength decreased from pre-harvest to post-harvest ($F_{1,\ 66.2} = 33.96$, $p<0.001$) but there were no differences (all p-values>0.13) by sex, management class, or any combination of the three. The percent of steplengths that equaled zero (the percent of time the animals did not change position between GPS locations) was 1.83% pre-harvest and 0.86% post-harvest, meaning the animals moved more often post-harvest. Steplength was positively and significantly correlated with ground temperature ($R^2 = 0.16$, $p<0.001$; Figure 2a & b). Thermal optimum was found at 22–24°C pre-harvest (Figure 2c) and 24–26°C post-harvest (Figure 2d) despite the fact that ground temperatures were higher pre-harvest (mean = 24.5°C) than post-harvest (mean = 22.7°C; $F_{1,\ 7315} = 140.8$, $p<0.001$). Average steplength during the pre-harvest period was 22.08 meters (SE = 1.21) and 15.40 meters (SE = 0.88) during the post-harvest period, with the height of activity varying by month (Figure 3).

Thermal habitats

We processed 388,974 environmental temperatures from 36 locations in four habitat types (clearcut opening, group selection opening, harvest-adjacent forest, and forested control) between May 2009 and October 2010. Available temperatures differed at each level (T_{max}, T_{mean}, and T_{min}) for each habitat type, month, and habitat by month interaction. The interaction term for T_{min} was the only non-significant effect ($F_{33,\ 376.6} = 0.959$, $p = 0.54$) in the model. The strength of the effects varied by month, with the harvest habitat types (clearcut and group selection openings) more similar to one another and forested habitat types (harvest-adjacent forest and forested controls) more similar (Table 2). Habitat type affected T_{max} more strongly than others. Explicitly, the range of temperatures for T_{max} was broader between habitats than for T_{min} or T_{mean} especially during the active period (Figure 4a). Between March and October, T_{max} in both harvest habitat types were significantly warmer (>10°C) than forested habitats (forests $T_{max} = 24.57$°C, SE = 0.73; harvest $T_{max} = 34.43$°C, SE = 0.80; $F_{1,\ 40.25} = 83.56$, $p<0.001$). This difference was most extreme in August when the T_{max} in harvest areas averaged 39.98°C (SE = 0.99) while it was nearly 13°C cooler in forested areas at 27.49°C (SE = 0.88). In contrast, T_{min} and T_{mean} differences remained within 3°C between habitat types, but usually less than 2°C for these months.

Local-scale effects

We recorded and processed 494,548 body temperatures among 50 turtles between May 2009 and October 2010. The maximum, mean, and minimum T_b varied by month (Figure 4b). T_b was highly correlated with T_g ($R^2 = 0.71$, $p<0.001$). Behavioral categories were correlated with T_b over the post-harvest period, but explained very little of the variation ($R^2 = 0.08$, $p<0.001$). Post-hoc analysis revealed significant T_b differences in basking, walking, resting, and being underground behaviors. Behaviors associated

Table 1. Pre-harvest (Pre-harv.; 2007–2008) and post-harvest (Post-harv.; 2009–2010) home ranges of female and male eastern box turtles.

Period	Sex	Mngmnt Class[*]	n	Biennial MCP[†]			Biennial 95% Kernel[‡]		
				Median	Mean	SE	Median	Mean	SE
Pre-harv.	F	Clearcut	5	6.80	15.42	10.20	3.57	32.32	28.97
		Control	4	3.52	4.54	1.86	3.32	3.37	0.73
		GroupSelect	2	10.21	10.21	6.74	4.31	4.31	0.38
	M	Clearcut	5	2.02	4.63	1.75	2.71	4.34	1.07
		Control	4	5.52	83.08	78.41	5.39	14.99	10.63
		GroupSelect	7	3.53	5.70	2.81	3.85	4.35	1.11
Summary	F	All	11	5.27	10.52	4.74	3.94	16.70	13.15
	M	All	16	3.57	24.71	19.62	4.12	7.01	2.72
Totals			27	3.61	18.93	11.70	3.94	10.96	5.52
Post-harv.	F	Clearcut	7	2.57	10.56	5.80	1.45	1.85	0.51
		Control	8	7.96	9.87	2.81	2.28	5.02	2.91
		GroupSelect	9	2.69	5.48	1.89	1.30	1.36	0.23
	M	Clearcut	7	5.98	11.11	6.30	2.22	49.22	46.96
		Control	7	3.65	16.72	13.10	1.59	18.66	17.29
		GroupSelect	8	2.32	2.64	0.51	1.66	1.64	0.21
Summary	F	All	24	4.19	8.42	2.02	1.49	2.72	1.00
	M	All	22	3.02	9.82	4.57	1.75	22.19	15.69
Totals			46	3.02	9.09	2.40	1.57	12.03	7.57

*The associated management class (Mngmnt Class).
[†]biennial Minimum Convex Polygons (MCP) home ranges.
[‡]Only the 95% kernel isopleths areas are listed here, as they are the only relevant comparisons to 100% MCP.

with higher T_b (24–27°C) included basking and mating. Behaviors generally associated with lower T_b (22–23C°) included resting, inverted, walking, and eating, but post-hoc analysis revealed that these were not significantly different than mating. When T_b decreased to an average of 13.8°C, the animals were generally buried underground (near the hibernation season).

We found no significant difference in number of animal locations between the harvest periods within harvest-polygons. While the pre-harvest Euclidian distances within the designated harvest boundaries and their edges did not differ from 2007 to 2008, the averages were significantly different from post-harvest Euclidian distances in each polygon ($F_{1, 516.5} = 32.45$, $p<0.001$). Inside the harvest boundaries, post-harvest Euclidian distances were shorter (11.26 m, SE = 1.66) compared to pre-harvest Euclidian distances of 22.91 m (SE = 2.83). A similar trend was found within edge polygons where post-harvest Euclidian distances (14.45 m, SE = 1.27) were smaller than pre-harvest (23.60 m, SE = 2.10).

Body temperatures did not vary among management classes ($F_{2, 40.72} = 1.624$, $p = 0.21$) but were different among months ($F_{6, 294.7} = 1087.334$, $p<0.001$; Figure 5). However, animals found within the harvest openings maintained 9% higher T_b overall than those found in the forest/harvest edge or forest interior ($F_{2, 73.24} = 8.135$, $p<0.001$). Body temperatures within 50 meters of the harvest edges were lower (21.72°C, SE = 0.35) than farther inside the forest (22.22°C, SE = 0.21) and harvests (23.91°C, SE = 0.44).

Discussion

Recent literature has shown that timber harvesting can have both positive and negative effects on forest dwelling species. Here we investigated the effect of various harvest openings on an ectotherm, the eastern box turtle. Using an experimental design and a variety of approaches, we demonstrate that in a relatively contiguous forested landscape, timber harvests have little effect on the short-term annual behavior of box turtles. However, we did detect a behavioral effect at the local scale where available microenvironmental temperatures were altered. We also offer further evidence that there is much variation in the annual behavior and home ranges of *T. c. carolina* that should be considered when establishing management strategies for forests and this species.

Landscape-scale effects – home ranges and thermal ecology

Ectotherms, such as box turtles, will preferentially use certain types of available habitats for thermoregulation, nesting, and aestivation [50,51]. Home range size likely depends on the quality of available food and other resources within the habitat [28]. Annual MCP home ranges for our adult *T. c. carolina* ranged from 0.47 and 187.67 hectares, the upper extreme being much larger than reports from any other study on this species. Indeed, our average annual home range estimate of 7.45 ha is more than 33% larger than any other published estimates to date (Table 3; [7,45,52–58]). It should be noted that there is a large variance in home range estimates across studies, which is likely associated with study duration, size, and monitoring method. The most likely

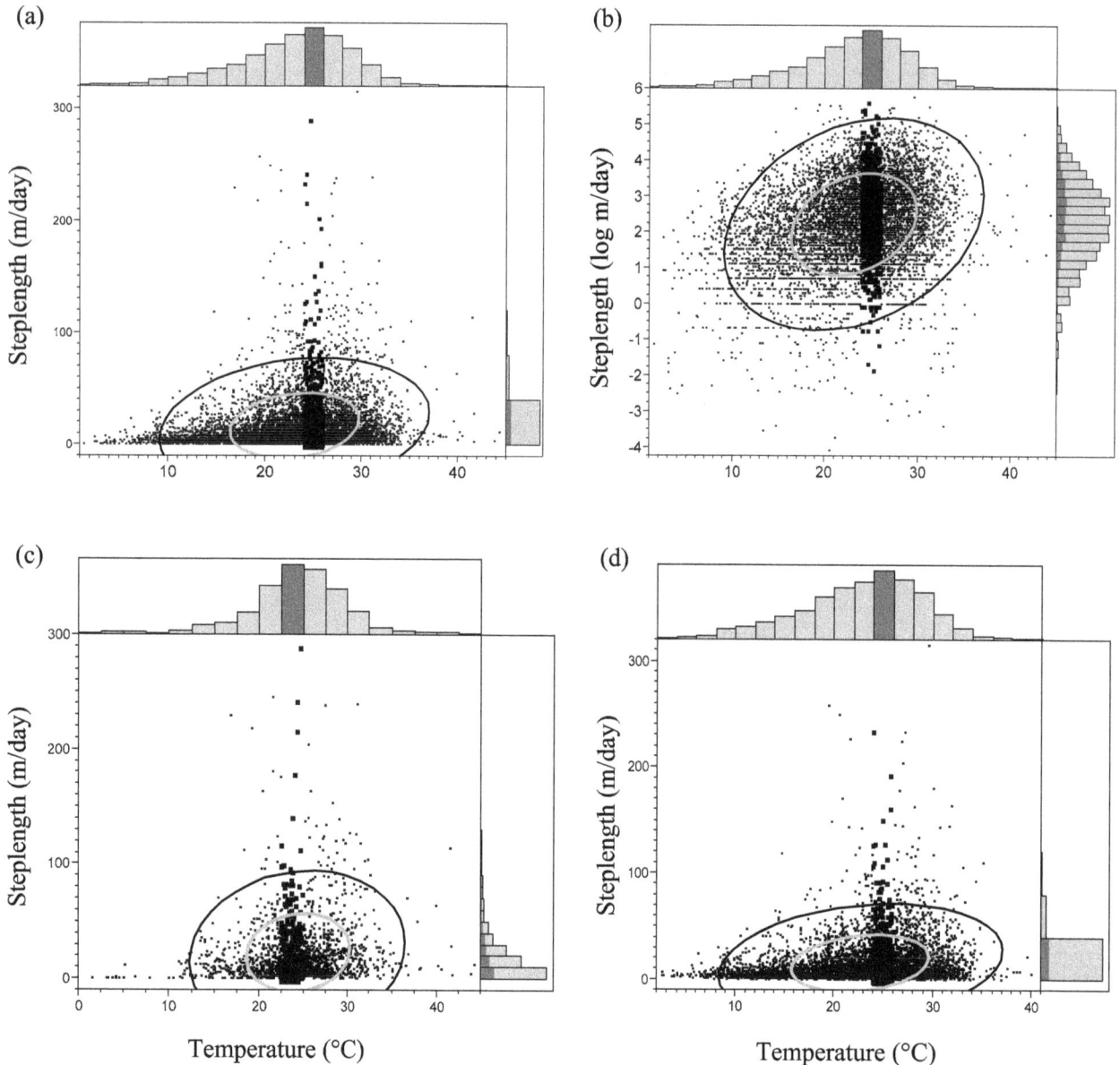

Figure 2. Thermal Optima. Scatter plot of daily distances traveled by eastern box turtles (steplengths; y-axes) by ground temperature (T_g in °C; x-axes). All 2007–10 steplengths (in meters per day) by ground temperature (a.) and the log-transformed steplength by ground temperature (b.). Pre-harvest (2007–08) steplength in meters per day by ground temperature (c.) and post-harvest (2009–10; d.). Plots show 95% (black ellipse) and 50% (grey ellipse) density ellipses around points and histogram densities along plot boarders. Darkened areas represent the peak of activity temperatures (22–26°C; thermal optimum) in these data.

explanation for the large home range size reported here is that our study was conducted within an expansive, relatively contiguous, and undisturbed habitat. Iglay et al. [59] found that turtles in fragmented habitats moved less often than those in contiguous habitats. To this end, many previous studies were conducted within relatively small and fragmented habitats that likely physically limited home ranges (Table 3).

In this study, we found no differences in either annual or biennial home ranges across the landscape in association with any of the three management classes (clearcut opening, group selection opening, or control). This lack of variation was likely due to the fact that the actual timber harvest openings were relatively small (0.15–4.43 ha) in relation to *T. c. carolina* home range size and the

surrounding contiguous forested habitat. Forest species often develop different strategies to cope with habitat perturbations. Some species expand their home ranges in response to forest fragmentation [60] while others inhabit territories that contain only small percentages of preferred habitat [61]. Still other species may gravitate toward mixed-composition habitat [62]. In the current study, the percent of animal locations within harvest edges did not change from pre- to post-harvest, suggesting that no such gravitation occurred. However, the movement parameters we investigated suggested that animals did alter their behavior while in proximity to harvest boundaries.

In pre-harvest years, animals tended to move longer distances (i.e., longer steplengths) than post-harvest years. However, the

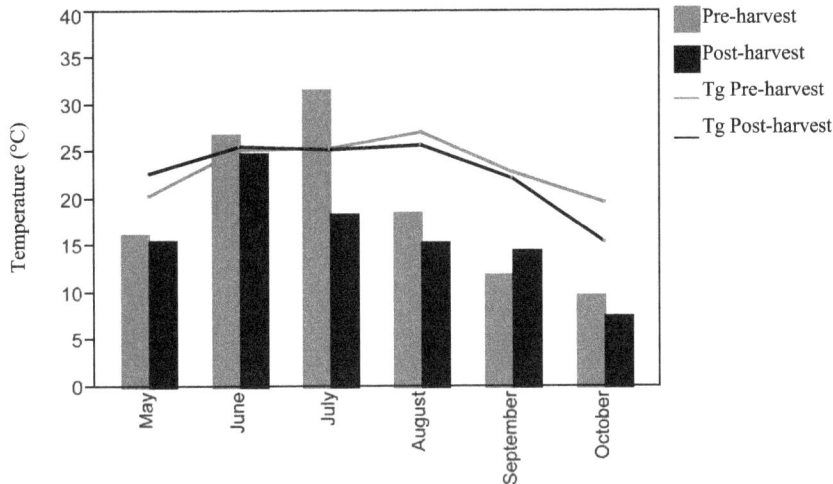

Figure 3. Pre vs. Post Movements. Average steplength (m/day) moved by eastern box turtles each month for both harvest periods (pre-harvest [2007–08] and post-harvest [2009–10]; bars). The average ground temperatures (Tg; °C) recorded at turtle location each harvest period are also plotted (lines).

percent of steplengths that were zero were higher pre-harvest (1.83% vs. 0.86%). This suggests that although the animals moved shorter distances and maintained generally smaller home ranges after the harvests were implemented, they moved more often. These increased short-range movements may be the result of changes in resources. In this altered habitat, animals may need to move frequently for new foraging opportunities as seen with many small mammal and bird species [60,63]. Shorter movements may be a result of downed slash acting as physical barriers or severe climatic conditions (i.e., drought). While it was evident that the animals did reduce movements during drought years, the cumulative effect on our results is minimal because the animals experienced drought years during pre-harvest 2007 and post-harvest 2010. Alternatively, behavioral thermoregulation may explain why the animals regularly moved but remained nearer to the same locations post-harvest.

Studies of fine-scale temperatures over broad spatial expanses are rare, despite the fact that temperature is an important factor in the location and activity of species [16]. A primary effect of the alteration of landscapes is the change in the microclimate of available habitats [64]. We measured these changes temporally across the landscape using temperature dataloggers. Although there was annual variation in ambient temperatures, the microclimatic conditions varied significantly between harvest and forested habitats. The most pronounced period occurred between May and September for T_{max} when differences were often greater than 10°C. These extreme summer temperatures found within harvest areas potentially exclude many plant and animal species. For example, variation in microclimates has been shown to affect the germination of emergent herbaceous and woody species [65]. During periods of highest temperatures, T_{max} within harvest areas was often observed to be near the maximum thermal tolerance for most ectotherms (43°C) effectively reducing the suitability of these areas for *T. carolina* (34.2°C; [66]) and other herpetofauna [67–70]. Although the current study examines a subset of factors affected by timber harvests, the advantage of this approach is the resulting detailed data of mechanisms underlying landscape effects [63]. Our results suggest that population-level responses to small-scale timber harvests (which are typical for the Midwestern U.S.) are minimal.

Local-scale effects – movement and edge effects

Ecotones (either natural or man-made) will influence animal activity differently as surface temperature, air temperature, and canopy cover varies across the landscape [55,71]. Ecotones at the harvest edges may provide cover by fallen logs and downed treetops, increased concentration and variety of forage (soft mast plants and invertebrates), and may facilitate behavioral thermoregulation by providing basking sites. Although we found no significant difference in the relative number of animal locations within the boundary or edges of the harvest areas, we did find differences in the movement parameters that suggest the animals use these areas differently. Prior to the harvests, the animals made longer, scattered movements across would-be harvest areas. Once the harvests were implemented, the movements (Euclidian distances) across the harvests shortened and were concentrated along the edges of the harvests (within edge polygons). Directed movements, although varied, often would alternate from the forest to the harvest edge, and frequently would cross project logging roads to do so. Studies on various turtle species have determined that roads bisecting animal routes were positively correlated with male biased sex ratios [27,72–75], population declines [76], and expanded home range sizes [77]. In this study, two of the sites were bordered by public roads and all sites were adjacent to logging roads, however we did not analyze correlations of the roads to movements or home ranges.

Anthropogenic effects extend beyond the physical boundary of disturbance. In a broader definition of habitat, thermal microclimates limit the use of certain areas both seasonally and spatially. Analyses of the variables that affect ambient temperatures on a microclimate scale will aide in the understanding of habitat requirements of ectotherms [16]. In this study, the animals found inside the harvest areas maintained higher active season body temperatures than those outside the harvests by 10.13%. As expected, basking behavior correlated with higher temperatures. Forested sites located near roads or open areas such as timber harvests, are found to be generally warmer than those further away [16]. However, T_b at timber harvest edges were the lowest during the active period, even lower than in the adjacent forested habitat suggesting that the animals were moving between microhabitats for thermoregulation as seen in other taxa [78,79].

(a)

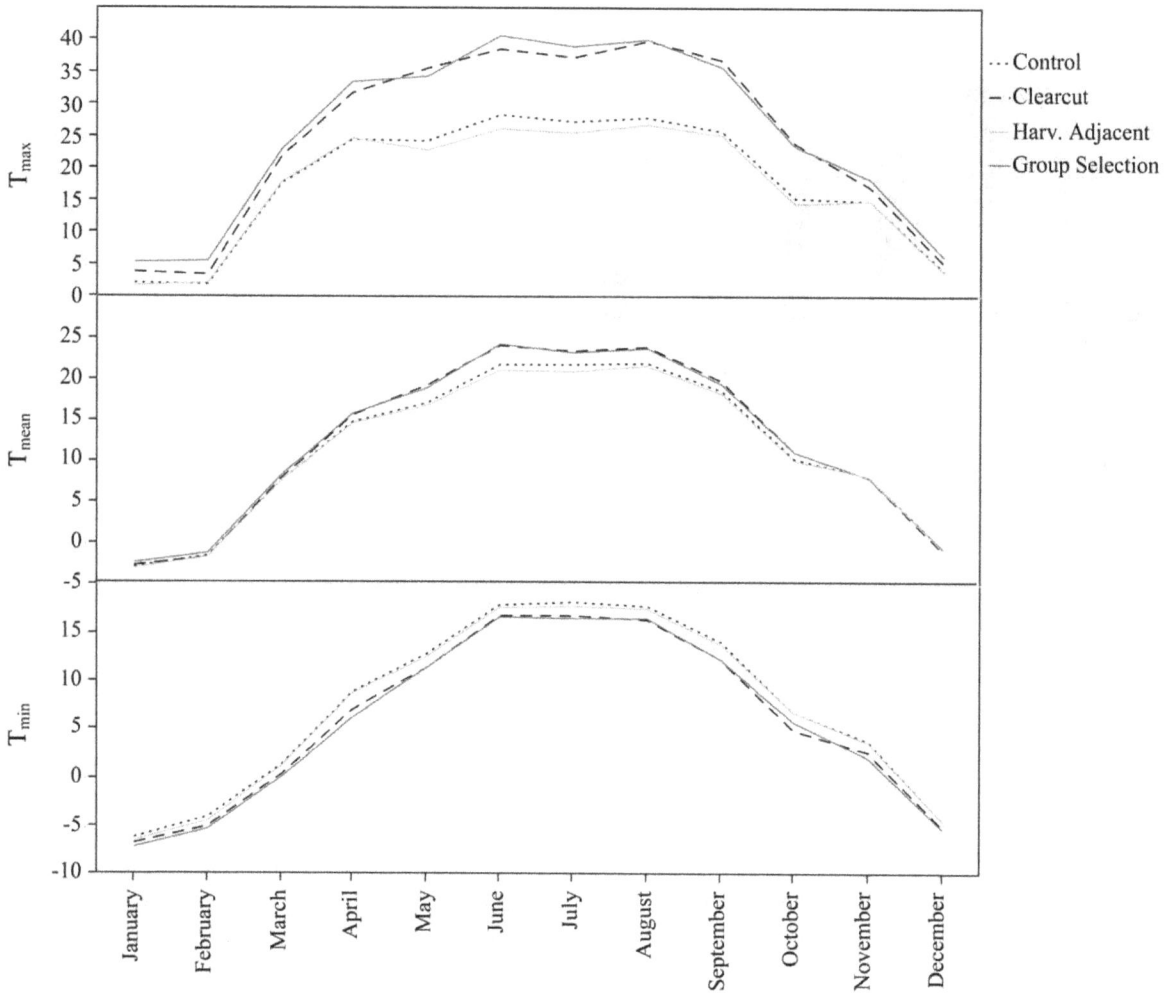

(b)

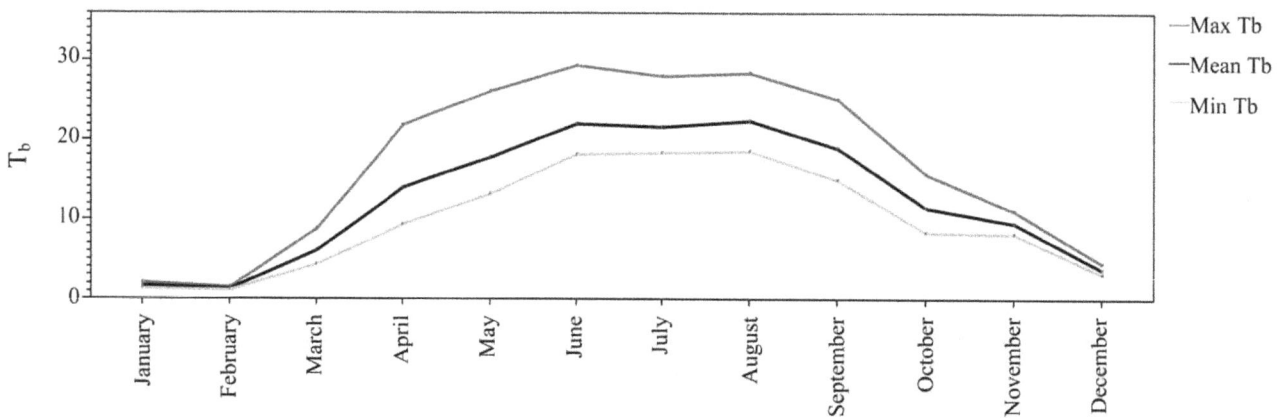

Figure 4. Habitat and Animal Temperature Ranges. Mean monthly temperature maxima (Tmax), mean (Tmean), and minima (Tmin) over two years (2009–2010) by habitat type (clearcut openings, group selection openings, harvest-adjacent forest (Harv. Adjacent), and forested control) (a). Maxima, means, and minima monthly eastern box turtle body temperatures (Tb) for the same period (b).

Table 2. Least Squares Means (LS Mean) Tukey-Kramer post-hoc pairwise comparisons connecting letters report of monthly environmental temperatures (Tmin, Tmax, Tmean) during 2009–2010 within four habitat types (clearcut openings, group selection openings, harvest-adjacent forest, and forested control).

Level	Habitat Type*			LS Mean
T_{mean}	GroupSelection	A		12.4068993
	Clearcut	A		12.3435027
	Control		B	11.4761091
	Harv.Adjacent		B	11.1620106
T_{max}	GroupSelection	A		25.3953822
	Clearcut	A		24.6201529
	Control		B	17.7994578
	Harv.Adjacent		B	17.2141375
T_{min}	Control	A		7.302016
	Harv.Adjacent	A		7.05775033
	Clearcut		B	5.99306883
	GroupSelection		B	5.84889568

*Habitat types at each level not connected by the same letter are significantly different.

The animals within our experimental openings were exposed to a wide range of temperatures. In a laboratory study, the specificity of T_b was investigated between *T. c. carolina* and *T. ornata* with the finding that *T. c. carolina* has less thermal specificity [39]. We routinely found the animals walking while inside the harvests and document that they do have the ability to behaviorally adjust to varying temperatures at a fine scale. These adjustments may play key roles in the physiological requirements of ectotherms throughout ontogeny and in various physical conditions (e.g., in reptiles, gravid females actively adjust to maintain higher body temperatures than males; [80].

Open spaces, such as clearcuts, may have less of an effect on larger-bodied species or those adapted to hot and dry conditions. Canopy cover directly influences light intensity, which is known to be a critical factor for many reptiles during activity periods [81–83]. On the other hand, many reptilian species such as small-bodied snakes are adapted to utilize leaf litter and are likely to be adversely affected by its removal with associated timber harvests [83]. During the active season, *T. c. carolina* use leaf litter to create 'forms' as cover [84]. *T. c. carolina* will use these forms throughout the active period as refuge from the heat, cold, rain, or disturbance [28]. In addition to cover, leaf litter serves as habitat for prey (such as snails, worms, and mushrooms) of box turtles. Immediately following implementation of harvests, leaf litter is degraded, blown from these areas, and often leaves large patches of unsuitable bare ground [85]. Studies have found that the increased soil temperatures and reduced leaf-litter cover (which can take decades to return pre-harvest conditions) in previously cut areas exclude many amphibian species [86,87]. We found that short term effects such as the loss of leaf litter did not cause the animals to abandon the area, but rather continue to use it in a different way (such as for thermoregulation).

Merely reporting species declines without determining their mechanistic causes leaves conservation planners with little recourse. To date, no studies have monitored the response of an ectotherm's movement parameters prior to and after discrete anthropogenic disturbance such as timber harvests. The present study has yielded detailed data on habitat use and spatial ecology of an ectotherm in a managed forest, but has much broader implications on multiple forest-dwelling species in a changing climate. In our study, the timber harvest openings were generally small (<5 ha) and were contained in a relatively contiguous and much larger forest matrix. Our results indicate that in a relatively contiguous forested landscape, small-scale timber harvests have minimal effects on the short-term behavior of these ectotherms. However, temperature fluctuations as seen in the current study

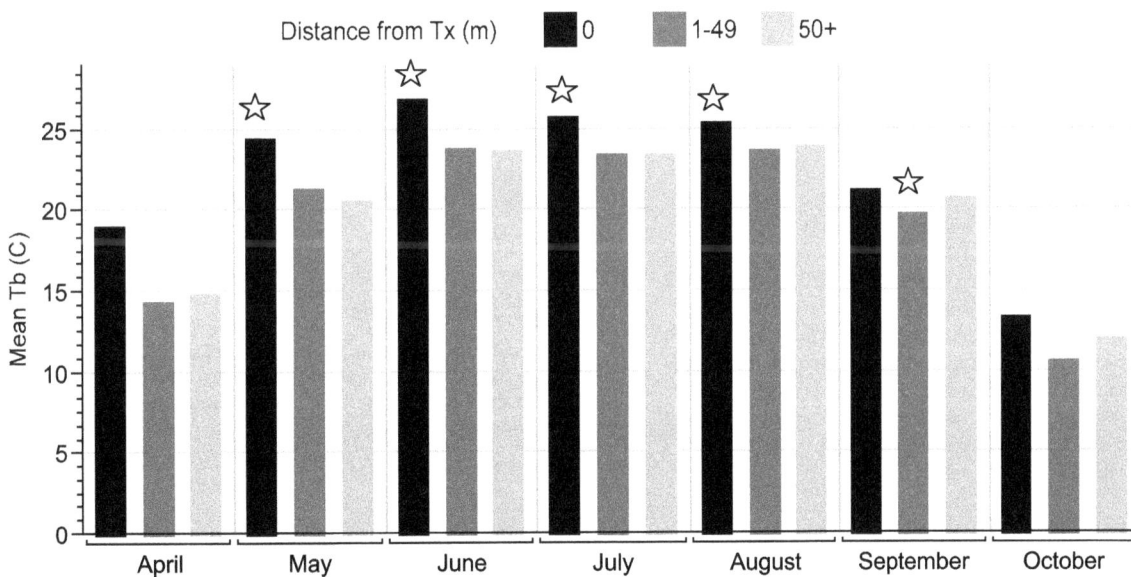

Figure 5. Harvest Proximity Temperatures. Mean eastern box turtle body temperatures (Tb) in degrees Celsius (C) with relation to timber harvest proximity over the active season months for post-harvest years (2009–10 combined). Starred bars represent significantly different mean temperatures during that month.

Table 3. Published studies involving home range estimates from native populations of *T. carolina*.

Authors [Reference]	# of Animals (# of loc/turtle)	Duration of study	Method	Home Range Size Estimate	Location (study size)
Nichols [53]	12 (14)	20 yrs	Mark-recapture	120–200 m diam.	Long Island, NY
Stickel [84]	55 (3+)	3 yrs	Mark-recapture & thread trailing	100 m diameter	Maryland (11 ha)
Dolbeer [54]	'many' of 270 marked	1 yr	Mark-recapture	76.2 m diam.	Tennessee (8.9 ha)
Schwartz & Schwartz [50]	239 (4–18)	8 yrs	Dog capture-recapture & Radiotelemetry	1.9 ha ave. area; (1.2–10.2 ha)	Missouri (22 ha)
Madden [51]	23 (85)	4 yrs	Radiotelemetry	♀ 373 m diam.; ♂ 284 m diam.; 2.12 ha ave. area	New York
Strang [55]	8 (3+)	3 yrs	Thread trailing	167 m diam.	Pennsylvania (29 ha)
Schwartz & Schwartz [44]	37 (11–44)	19 yrs	Dog capture-recapture & Radiotelemetry	5.2 ha ave. area; (0.6–10.7 ha)	Missouri (22 ha)
Bayless [7]	6 (10+)	56 days over 2 yrs	Radiotelemetry & thread trailing	213 m diam.; 1.25 ha	Virginia (49.4 ha)
Williams & Parker [56]	35 (3+)	26 yrs	Mark-recapture	♀ 176 m diam.; ♂ 171 m diam.	Indiana (72.9 ha)
Hallgren -Scaffidi [57]	11 (3+)	2 yrs	Mark-recapture & thread trailing	97 m diam.; 0.2 ha area	Maryland (11.3 ha)
Stickel [45]	103 (3+)	37 yrs	Mark-recapture	145 m diam.; ♀ 1.13 ha area; ♂ 1.2 ha area	Maryland (11 ha)
Quinn [58]	14 (av. 62)	1 yr	Radiotelemetry	♀ 4.0 ha area; ♂ 6.7 ha area; 4.97 ha ave. area	Connecticut
Current Study	50 (av. 34–70/yr)	4 yrs	Radiotelemetry	♀ 5.55 ha area; ♂ 9.14 ha area; 7.45 ha 4-yr ave.	Indiana (35,000 ha)

affect seasonal available habitat for forest-dwelling animals, especially for those with limited dispersal and thermoregulatory capabilities. Microclimates within harvested areas can exclude animals, but they also may create some desired ecotonal habitat. Considerations of habitat requirements and contiguity of surrounding refugia habitat and species ability to recover should be thoroughly considered before timber harvest sizes are determined. These factors are of particular interest when dealing with long-lived species of conservation concern amid a changing climate.

Supporting Information

Table S1   Four-year home range summary. Summary of turtle annual home ranges at all nine study sites from 2007–2010. (PDF). Legend: Summary of the eastern box turtle annual home ranges at nine study sites in south-central Indiana from 2007–10. Year, sex, management class (Mngmt Class), number in group (*n*), and median, mean, and standard errors of annual home range (100% Minimum Convex Polygon; MCP) in hectares (ha). For 2007–08, the management class represents the assigned harvest type prior to harvest implementation.

References

1. Gram WK, Porneluzi PA, Clawson RL, Faaborg J, Richter SC (2003) Effects of experimental forest management on density and nesting success of bird species in Missouri Ozark Forests. Conservation Biology 17: 1324–1337.
2. Gitzen RA, West SD, Maguire CC, Manning T, Halpern CB (2007) Response of terrestrial small mammals to varying amounts and patterns of green-tree retention in Pacific Northwest forests. Forest Ecology and Management 251: 142–155.

Acknowledgments

This paper is a contribution of the Hardwood Ecosystem Experiment, a partnership of the Indiana Department of Natural Resources (IDNR), Purdue University, Ball State University, Indiana State University, Drake University, and the Nature Conservancy. We thank field technicians A. Garcia, A. Hoffman, A. Krainyk, B. Geboy, B. Johnson, B. Tomson, G. Stephens, H. Powell, J. Faller, J. MacNeil, K. Creely, K. Lilly, K. Norris, K. Powers, K. Westerman, L. Keener-Eck, L. Woody, M. Baragona, M. Cook, M. Cross, M. Turnquist, M. Wildnauer, N. Burgmeier, N. Engbrecht, S. Johnson, S. Kimble, S. Ritchie, T. Jedele, and Z. Walker. We also thank members of the Williams lab group for providing helpful comments on previous versions of this manuscript.

Author Contributions

Conceived and designed the experiments: AFC BJM RNW. Performed the experiments: AFC BJM. Analyzed the data: AFC. Contributed reagents/materials/analysis tools: BJM RNW. Wrote the paper: AFC. Provided direction, critical reviews, and suggestions for the manuscript: BJM RNW.

3. Fredericksen TS, Ross BD, Hoffman W, Ross E, Morrison ML, et al. (2000) The impact of logging on wildlife: A study in northeastern Pennsylvania. Journal of Forestry 98: 4–10.
4. Summerville K (2010) Managing the forest for more than the trees: effects of experimental timber harvest on forest Lepidoptera. Ecological Applications.
5. Peterman WE, Semlitsch RD (2009) Efficacy of riparian buffers in mitigating local population declines and the effects of even-aged timber harvest on larval salamanders. Forest Ecology and Management 257: 8–14.

6. Renken RB, Gram WK, Fantz DK, Richter SC, Miller TJ, et al. (2004) Effects of forest management on amphibians and reptiles in Missouri Ozark forests. Conservation Biology 18: 174–188.

7. Bayless JW (1984) Home range studies of the Eastern Box Turtle (*Terrapene carolina carolina*) using radio telemetry. Fairfax: George Mason University. 63 p.

8. Allard HA (1935) The natural history of the box turtle. The Scientific Monthly 41: 325–338.

9. Wake DB (1991) Declining amphibian populations. Science 253: 860.

10. Pechmann JHK, Scott DE, Semlitsch RD, Caldwell JP, Vitt IJ, et al. (1991) Declining amphibian populations - The problem of separating human impacts from natural fluctuations. Science 253: 892–895.

11. Gibbons JW, Scott DE, Ryan TJ, Buhlmann KA, Tuberville TD, et al. (2000) The global decline of reptiles, Deja Vu amphibians. Bioscience 50: 653–666.

12. Perison D, Phelps J, Pavel C, Kellison R (1997) The effects of timber harvest in a South Carolina blackwater bottomland. Forest Ecology and Management 90: 171–185.

13. Donovan TM, Jones PW, Annand EM, Thompson III FR (1997) Variation in local-scale edge effects: mechanisms and landscape context. Ecology 78: 2064–2075.

14. Cadenasso ML, Pickett STA (2001) Effect of edge structure on the flux of species into forest interiors. Conservation Biology 15: 91–97.

15. Delgado García JD, Arévalo JR, Fernández-Palacios JM (2007) Road edge effect on the abundance of the lizard *Gallotia galloti* (Sauria: Lacertidae) in two Canary Islands forests. Biodiversity and Conservation 16: 2949–2963.

16. Cunnington G, Schaefer J, Cebek J, Murray D (2009) Correlations of biotic and abiotic variables with ground surface temperature: an ectothermic perspective. Ecoscience 15: 472–477.

17. Huey RB, Kingsolver JG (1989) Evolution of thermal sensitivity of ectotherm performance. Trends in Ecology & Evolution 4: 131–135.

18. Fox SF, McCoy K, Baird TA (2003) Lizard Social Behavior. Baltimore, MY: The Johns Hopkins University Press. 445 p.

19. Bradshaw WE, Holzapfel CM (2007) Evolution of animal photoperiodism. Annual Review of Ecology, Evolution, and Systematics 38: 1–25.

20. Goldstein MI, Wilkins RN, Lacher TE (2005) Spatiotemporal responses of reptiles and amphibians to timber harvest treatments. Journal of Wildlife Management 69: 525–539.

21. McLeod RF, Gates JE (1998) Response of herpetofaunal communities to forest cutting and burning at Chesapeake Farms, Maryland. American Midland Naturalist 139: 164–177.

22. Rittenhouse TAG, Semlitsch RD (2009) Behavioral response of migrating wood frogs to experimental timber harvest surrounding wetlands. Canadian Journal of Zoology 87: 618–625.

23. Semlitsch RD, Todd BD, Blomquist SM, Calhoun AJK, Gibbons JW, et al. (2009) Effects of timber harvest on amphibian populations: Understanding mechanisms from forest experiments. Bioscience 59: 853–862.

24. Rittenhouse TAG, Semlitsch RD, Thompson FR (2009) Survival costs associated with Wood Frog breeding migrations: Effects of timber harvest and drought. Ecology 90: 1620–1630.

25. Hocking DJ, Semlitsch RD (2008) Effects of experimental clearcut logging on Gray Treefrog (*Hyla versicolor*) tadpole performance. Journal of Herpetology 42: 689–698.

26. Brooks RJ, Brown GP, Galbraith DA (1991) Effects of a sudden increase in natural mortality of adults on a population of the common snapping turtle (*Chelydra serpentina*). Canadian Journal of Zoology 69: 1314–1320.

27. Gibbs JP, Shriver WG (2002) Estimating the effects of road mortality on turtle populations. Conservation Biology 16: 1647–1652.

28. Dodd CK, Jr. (2001) North American Box Turtles: A Natural History. Norman, Oklahoma, USA: University of Oklahoma Press. 231 p.

29. MacGowan BJ, Kingsbury BA, Williams RN (2004) Turtles of Indiana; Brown JW, editor: Purdue University. 63 p.

30. Currylow AF, Zollner PA, MacGowan BJ, Williams RN (2011) A survival estimate of Midwestern adult Eastern Box Turtles using radiotelemetry. The American Midland Naturalist 165: 143–149.

31. Summerville KS, Courard-Hauri D, Dupont MM (2009) The legacy of timber harvest: Do patterns of species dominance suggest recovery of lepidopteran communities in managed hardwood stands? Forest Ecology and Management 259: 8–13.

32. Kalb RA, Mycroft CJ (in press) The Hardwood Ecosystem Experiment: Goals, design, and implementation. In: Swihart RK, Saunders MR, Kalb RA, Haulton S, Michler CH, editors. Northern Forest Experiment Station General Technical Report: U.S. Department of Agriculture, Forest Service.

33. Cagle FR (1939) A system of marking turtles for future identification. Copeia 1939: 170–173.

34. Ernst CH, Hershey MF, Barbour RW (1974) A new coding system for hardshelled turtles. Transactions of the Kentucky Academy of Science 35: 27–28.

35. Ferner JW (2007) A Review of Marking and Individual Recognitions Techniques for Amphibians and Reptiles; Moriarty JJ, editor: Society for the Study of Amphibians and Reptiles.

36. Bernstein NP, Black RW (2005) Thermal environment of overwintering Ornate Box Turtles, *Terrapene ornata ornata*, in Iowa. American Midland Naturalist 153: 370–377.

37. Congdon JD, Gatten RE, Jr., Morreale SJ (1989) Overwintering activity of box turtles (*Terrapene carolina*) in South Carolina. Journal of Herpetology 23: 179–181.

38. Peterson CC (1987) Thermal relations of hibernating painted turtles, *Chrysemys picta*. Journal of Herpetology 21: 16–20.

39. do Amaral JPS, Marvin GA, Hutchison VH (2002) Thermoregulation in the box turtles *Terrapene carolina* and *Terrapene ornata*. Canadian Journal of Zoology 80: 934–943.

40. ESRI (2009) ArcGIS Desktop 9. Redlands, California, USA: Environmental Systems Research Institute, Inc.

41. Hawth's Analysis Tools for ArcGIS. Available: http://www.spatialecology.com/htools. Accessed 2012 Jun 14.

42. SAS Institute Inc (2007) SAS OnlineDoc® 9.2. Cary, NC: SAS Institute Inc.

43. SAS Institute Inc (2008) JMP 8.0. Cary, North Carolina, USA: SAS Institute Inc.

44. Schwartz ER, Schwartz CW, Kiester AR (1984) The Three-toed Box Turtle in central Missouri, Part II: a nineteen-year study on home range, movements and population. Jefferson City, Missouri, USA: Missouri Department of Conservation.

45. Stickel LF (1989) Home range behavior among Box Turtles (*Terrapene c. carolina*) of a bottomland forest in Maryland. Journal of Herpetology 23: 40–44.

46. Rodgers AR, Carr AP, Smith L, Kie4 JG (2005) HRE: The Home Range Extension for ArcView. Thunder Bay, Ontario, Canada: Ontario Ministry of Natural Resources, Centre for Northern Forest Ecosystem Research.

47. Worton BJ (1989) Kernel methods for estimating the utilization distribution in home-range studies. Ecology 70: 164–168.

48. Row JR, Blouin-Demers G (2006) Kernels are not accurate estimators of home-range size for herpetofauna. Copeia: 797–802.

49. R Development Core Team (2009) R: A Language and Environment for Statistical Computing. Version 2.10.1. 2.10.1 ed. Vienna, Austria: R Foundation for Statistical Computing.

50. Schwartz CW, Schwartz ER (1974) The Three-toed Box Turtle in central Missouri: its population, home range, and movements. Jefferson City, Missouri, USA: Missouri Department of Conservation.

51. Madden RC (1975) Home range, movements, and orientation in the Eastern Box Turtle, *Terrapene carolina carolina* [Dissertation]. New York: The City University of New York. 218 p.

52. Donaldson BM, Echternacht AC (2005) Aquatic habitat use relative to home range and seasonal movement of Eastern Box Turtles (*Terrapene carolina carolina*: Emydidae) in eastern Tennessee. Journal of Herpetology 39: 278–284.

53. Nichols J (1939) Range and homing of individual box turtles. Copeia: 125–127.

54. Dolbeer RA (1969) Population density and home range size of the Eastern Box Turtle (*Terrapene c. carolina*) in eastern Tennessee. ASB Bulletin 16.

55. Strang CA (1983) Spatial and temporal activity patterns in two terrestrial turtles. Journal of Herpetology 17: 43–47.

56. Williams EC, Jr., Parker WS (1987) A long-term study of a box turtle (*Terrapene carolina*) population at Allee Memorial Woods, Indiana, with emphasis on survivorship. Herpetologica 43: 328–335.

57. Hallgren-Scaffidi L (1986) Habitat, home range and population study of the Eastern Box Turtle (*Terrapene carolina*): University of Maryland.

58. Quinn DP (2008) A radio-telemetric study of the Eastern Box Turtle (*Terrapene carolina carolina*), home-range, habitat use, and hibernacula selection in Connecticut: Central Connecticut Sate University.

59. Iglay RB, Bowman JL, Nazdrowicz NH (2007) Eastern Box Turtle (*Terrapene carolina carolina*) movements in a fragmented landscape. Journal of Herpetology 41: 102–106.

60. Hansbauer MM, Storch I, Pimentel RG, Metzger JP (2008) Comparative range use by three Atlantic Forest understorey bird species in relation to forest fragmentation. Journal of Tropical Ecology 24: 291–299.

61. Andrén H (1994) Effects of habitat fragmentation on birds and mammals in landscapes with different proportions of suitable habitat: A review. Oikos 71: 355–366.

62. Andrén H (1992) Corvid density and nest predation in relation to forest fragmentation: A landscape perspective. Ecology 73: 794–804.

63. Debinski DM, Holt RD (2000) A survey and overview of habitat fragmentation experiments. Conservation Biology 14: 342–355.

64. Saunders DA, Hobbs RJ, Margules CR (1991) Biological consequences of ecosystem fragmentation: A review. Conservation Biology 5: 18–32.

65. Breshears David D, Nyhan John W, Heil Christopher E, Wilcox Bradford P (1998) Effects of woody plants on microclimate in a semiarid woodland: Soil temperature and evaporation in canopy and intercanopy patches. International Journal of Plant Sciences 159: 1010–1017.

66. Penick DN, Congdon J, Spotila JR, Williams JB (2002) Microclimates and energetics of free-living box turtles, *Terrapene carolina*, in South Carolina. Physiological and Biochemical Zoology 75: 57–65.

67. Hutchison VH, Vinegar A, Kosh RJ (1966) Critical thermal maxima in turtles. Herpetologica 22: 32–41.

68. Blem CR, Ragan CA, Scott LS (1986) The thermal physiology of two sympatric treefrogs Hyla cinerea and Hyla chrysoscelis (Anura; Hylidae). Comparative Biochemistry and Physiology 85A: 563–570.

69. Kroll JC (1973) Comparative physiological ecology of Eastern and Western Hognose Snakes (*Heterodon platyrhinos* and *H. nasicus*). College Station, TX: Texas A & M University.

70. Brattstrom BH (1965) Body temperatures of reptiles. American Midland Naturalist 73: 376–422.

71. Weiss JA (2009) Demographics, activity, and habitat selection of the Eastern Box Turtle (*Terrapene c. carolina*) in West Virginia: Marshall University.

72. Kipp RL (2003) Nesting ecology of the Eastern Box Turtle (*Terrapene carolina carolina*) in a fragmented landscape: University of Delaware.

73. Marchand MN, Litvaitis JA (2004) Effects of habitat features and landscape composition on the population structure of a common aquatic turtle in a region undergoing rapid development. Conservation Biology 18: 758–767.

74. Gibbs JP, Steen DA (2005) Trends in sex ratios of turtles in the United States: Implications of road mortality. Conservation Biology 19: 552–556.

75. Steen DA, Aresco MJ, Beilke SG, Compton BW, Condon EP, et al. (2006) Relative vulnerability of female turtles to road mortality. Animal Conservation 9: 269–273.

76. Shepard DB, Kuhns AR, Dreslik MJ, Phillips CA (2008) Roads as barriers to animal movement in fragmented landscapes. Animal Conservation 11: 288–296.

77. Nieuwolt PM (1996) Movement, activity, and microhabitat selection in the Western Box Turtle, *Terrapene ornata luteola*, in New Mexico. Herpetologica 52: 487–495.

78. Adolph SC (1990) Influence of behavioral thermoregulation on microhabitat Use by two *Sceloporus* lizards. Ecology 71: 315–327.

79. Sepulveda M, Vidal MA, Farina JM, Sabat P (2008) Seasonal and geographic variation in thermal biology of the lizard *Microlophus atacamensis* (Squamata: Tropiduridae). Journal of Thermal Biology 33: 141–148.

80. Tozetti AM, Pontes GMF, Borges-Martins M, Oliveira RB (2010) Temperature preferences of *Xenodon dorbignyi*: field and experimental observations. The Herpetological Journal 20: 277–280.

81. Gould E (1957) Orientation in box turtles, *Terrapene c. carolina*. The Biological Bulletin 112: 336–348.

82. Rose FL, Judd FW (1975) Activity and home range size of the Texas Tortoise, *Gopherus berlandieri*, in south Texas. Herpetologica 31: 448–456.

83. Todd BD, Andrews KM (2008) Response of a reptile guild to forest harvesting. Conservation Biology 22: 753–761.

84. Stickel LF (1950) Populations and Home Range Relationships of the Box Turtle, *Terrapene-c-carolina* (Linnaeus). Ecological Monographs 20: 353–378.

85. Enge KM, Marion WR (1986) Effects of clearcutting and site preparation on herpetofauna of a north Florida flatwoods. Forest Ecology and Management 14: 177–192.

86. Crawford JA, Semlitsch RD (2008) Post-disturbance effects of even-aged timber harvest on stream salamanders in southern Appalachian forests. Animal Conservation 11: 369–376.

87. Petranka JW, Eldridge ME, Haley KE (1993) Effects of timber harvesting on southern Appalachian salamanders. Conservation Biology 7: 363–377.

12

Designing Mixed Species Tree Plantations for the Tropics: Balancing Ecological Attributes of Species with Landholder Preferences in the Philippines

Huong Nguyen[1,2]*, David Lamb[1,3], John Herbohn[1,2], Jennifer Firn[4]

1 School of Agriculture and Food Sciences, The University of Queensland, St Lucia, Australia, 2 Forest Industries Research Centre, The University of the Sunshine Coast, Sippy Downs, Australia, 3 Centre for Mined Land Research, The University of Queensland, St Lucia, Australia, 4 Faculty of Science and Technology, School of Earth, Environmental and Biological Sciences, Queensland University of Technology, Brisbane, Australia

Abstract

A mixed species reforestation program known as the Rainforestation Farming system was undertaken in the Philippines to develop forms of farm forestry more suitable for smallholders than the simple monocultural plantations commonly used then. In this study, we describe the subsequent changes in stand structure and floristic composition of these plantations in order to learn from the experience and develop improved prescriptions for reforestation systems likely to be attractive to smallholders. We investigated stands aged from 6 to 11 years old on three successive occasions over a 6 year period. We found the number of species originally present in the plots as trees >5 cm dbh decreased from an initial total of 76 species to 65 species at the end of study period. But, at the same time, some new species reached the size class threshold and were recruited into the canopy layer. There was a substantial decline in tree density from an estimated stocking of about 5000 trees per ha at the time of planting to 1380 trees per ha at the time of the first measurement; the density declined by a further 4.9% per year. Changes in composition and stand structure were indicated by a marked shift in the Importance Value Index of species. Over six years, shade-intolerant species became less important and the native shade-tolerant species (often Dipterocarps) increased in importance. Based on how the Rainforestation Farming plantations developed in these early years, we suggest that mixed-species plantations elsewhere in the humid tropics should be around 1000 trees per ha or less, that the proportion of fast growing (and hence early maturing) trees should be about 30–40% of this initial density and that any fruit tree component should only be planted on the plantation margin where more light and space are available for crowns to develop.

Editor: Han Y.H. Chen, Lakehead University, Canada

Funding: The study was funded from ACIAR Smallholer Forest projects ASEM/2003/052 and ASEM/2006/091. The funders had no role in study design, data collection and analysis, decision to publish, or preparation of the manuscript.

Competing Interests: The authors have declared that no competing interests exist.

* E-mail: h.nguyen22@uq.edu.au

Introduction

Over the past two decades, there has been rising interest in planting mixtures of tree species to establish plantations that provide multiple services including production and improved nutrient cycling and also to provide more biodiversity at the landscape level [1]. Mixed-species plantations have the potential to generate a variety of forest products, as well as a range of ecosystem services [2–4]. Mixed species plantations are often established using just two or three species but sometimes far more diverse mixtures have been used which include representatives of a variety of successional stages [5].

A major issue facing those wishing to establish mixed species plantations is that knowledge of the silvicultural attributes of most species is usually so limited that it is difficult to predict how these will grow when planted into novel combinations [6]. The growth strategies of species growing in natural forests can provide an indication of the role they might play in a plantation including whether they are shade-intolerant, a canopy dominant species or whether they can grow in sub-canopy strata. The interactions among species will strongly influence the productivity of mixtures

[7,8]. Species attributes that could be used as possible indicators of the performance of species grown in a mixed-species stand include shade tolerance/intolerance, height growth rate, crown structure, foliar phenology, and root depth and phenology [9,10]. Combinations of species with complementary traits can reduce competition and allow for the most efficient use of limiting resources like water, nutrients and light in plant communities [6].

Identifying complementary species is a difficult task, particularly in the tropics when native tree species are preferred, as there is a limited knowledge of growth strategies of most native species. A number of challenges are likely to take place in newly established mixed-species stands because of competition and differences in the species growth rates. Trees grown in mixed species stands can sometimes suffer both intra-specific competition and inter-specific competition. Both of these are influenced by tree density [11–13]. High initial planting density may facilitate early site capture, reduce weed control, improve form and lower stem taper [14]. On the other hand, higher densities mean costs are greater because more seedlings are needed and stand thinning is necessary at an earlier stage of stand development. This problem may not be as great as it seems if the thinned trees have some economic value

Table 1. Site characteristics and planting history of 18 of the mixed-species sites in Leyte Province, the Philippines*.

Site	Site location	Year planted	Area (ha)	Soil type	Topography	No. of plots
02	Marcos, Baybay	1995	0.61	Clay loam	Slightly to moderately rolling	2
03	Catmon, Ormoc	1998	1.4	Clay loam	Flat	12
04	Patag, Baybay	1998	1.0	Clay loam	Slightly rolling	5
05	Cienda, Baybay	1996	0.9707	Clay to clay loam	Flat	9
06	Pomponan, Baybay	1997	0.38	Clay loam	Slightly rolling	2
07	Punta, Baybay	1996	5.442	Limestone	Moderately rolling	9
08	Maitum, Baybay	1996	0.478	Clay loam	Slightly to moderately rolling	2
09	Mailhi, Baybay	1996	3.22	Clay to clay loam	Slightly to moderately rolling	10
10	Vila Solidasidad, Baybay	1995	0.4377	Clay	Flat	4
11	Maitum, Baybay	1996	0.4686	Limestone	Moderately rolling	2
12	Maitum, Baybay	1996	0.9862	Limestone	Slightly rolling	2
13	Maitum, Baybay	1996	0.2518	Clay loam	Moderately rolling	2
14	Pomponan, Baybay	1996	0.9518	Clay loam	Slightly to moderately rolling	2
15	Pomponan, Baybay	1997	0.438	Clay	Moderately rolling	2
16	Pomponan, Baybay	1997	0.41	Clay loam	Moderately rolling	2
17	Pomponan, Baybay	1999	0.8475	Sandy loam	Flat	2
19	Licoma, Ormoc	2000	0.25	Clay loam	Moderately rolling	2
22	Milagro, Ormoc	1996	1.5	Clay loam	Flat	7

*Milan et al. 2004.

and there are usually some fast growing species that can be harvested at an early age.

The Rainforestation Farming plantations in the Philippines present an ideal opportunity to explore basic questions concerning forest dynamics, optimal planting densities and human-use patterns of tropical polycultures [15–19]. We hypothesise that these mixed species plantations could balance the ecological attributes of species with landholder preferences in the Philippines. Here, over three time periods, we measure changes in the structure and species composition of these mixtures. We consider the following questions:

(i) How has the composition and structure of these mixed-species plantings changed over time?

(ii) How have patterns of species loss (due to mortality and harvesting) and recruitment been affected by the design of these species mixtures?

(iii) What are the implications for the future design and management of tropical polycultures used by smallholders?

Materials and Methods

Ethics Statement

The study was permitted by the owners of the lands (for more details see Table 1) to be conducted in 18 sites. It was not possible to sample from all 28 sites that were established under the Rainforestation Farming system because several plantations had been detrimentally affected by fire, harvesting, clearing for other agricultural activities; because access was not granted by the land owners; or did not meet minimum requirements for measurements (e.g. trees greater than 5 cm diameter).

Study Area

The study was conducted in Leyte province, which is one of two provinces located in Leyte Island (Figure S1 in File S1). This is the eighth largest island in the middle of the Philippines archipelago and covers about 800,000 ha [20,21]. Leyte Province has a humid monsoon climate and the average rainfall in the study area for the years 1980−2000 was 2,686 mm with an annual variation of between 1,775 mm in 1987 and 3,697 mm in 1999 [22]. Although there is no pronounced dry season, the region experiences its lowest rainfall of less than 100 mm between March and May [23]. Dry periods of several months duration with rainfall of less than 100 mm can sometimes occur as was the case during the 'El Nino' year of 1993, the year in which the project commenced. The average annual temperature is 27.5°C and ranges from 26.3 to 28.7°C. The relative humidity is always high and the average monthly level for the years 1980−2000 ranged from 75.1% in March to 80.1% in October [22]. The soils are derived from volcanic parent material and were strong acidic with a pH 4.1–4.9 [24]. The natural vegetation in the region is a species-rich, lowland dipterocarp forest but natural forest now only remains on the less accessible slopes of the Leyte cordillera [25].

A unique polyculture reforestation program was started in the Philippines in 1992 called the Rainforestation Farming system [26–29]. It involved 28 small-scale mixed-species plantations on private farms on Leyte Island. These plantings were established to provide smallholders with a form of reforestation that generated income from an early age while also providing the benefits of increased biodiversity in the highly cleared and mainly agricultural landscape. The intent was to create resilient and sustainable forms of reforestation that would also be financially attractive to local farmers [27,28]. The program focussed on native species and used a large number of species to establish mixtures to resemble a few natural forests in the region. In response to demands by landholders some exotic species were also used in the mixtures.

Table 2. Equations to calculate Important value index (IVI) of species.

Index	Equation
Importance value index (IVI)	$\mathrm{Relative\,density + Relative\,frequency + Relative\,dominance}$
Relative density	$\frac{\text{Density of a species}}{\text{Total density of all species}} \times 100$
Relative frequency	$\frac{\text{Frequency of a species}}{\text{Total frequency of all species}} \times 100$
Relative dominance	$\frac{\text{Dominance of a species}}{\text{Total dominance of all species}} \times 100$
Density	$\frac{\text{Number of a species}}{\text{Total area sampled}}$
Frequency	$\frac{\text{Area of plots in which a species occurs}}{\text{Total area sampled}}$
Dominance	$\frac{\text{Total basal area of a species}}{\text{Total area sampled}}$

The system used approximately 100 endemic pioneer and shade-intolerant tree species, longer-lived species of Dipterocarp, fruit trees and a limited number of exotic timber species. These were planted to create a series of small scale plantations in the average area of about 1 ha [29].

At the beginning of the Rainforestation Farming project, species assessed as shade-intolerant pioneers were planted in a spacing of 2 m×2 m to provide an environment for supposedly shade-tolerant timber and fruit tree species to be established in the following year. These were inter-planted at a general spacing of 2 m×1 m [28]. The estimated density at planting time was about 5000 trees/ha at sites. From 7 to 40 species were planted at each farm. The aim was to create a three storied structure with pioneer and shade-intolerant trees in the upper canopy layer, shade-tolerant trees and fruit trees in the second storey and shade-tolerant crops in the lowest layer. Supposedly there would be roughly equal numbers of shade-tolerant and shade-intolerant species. In fact the numbers and identity of species planted at each farm were not consistent but depended upon seedling availability at the time, and the preferences of the local landholder. As a consequence detailed records of plantings at each farm were not kept. Sites were developed over a period of several years from 1995 to 2000. Further details of the Rainforestation Farming methodology are given by Kolb [22].

Data Collection and Analysis

Data were collected from 80 plots distributed across 18 of the mixed-species plantations (Table 1). These plots have also been used in a recent biodiversity-productivity study [18]. The first measurement of species composition and stand structural charac-teristics was undertaken from February to March 2006 when the trees were aged between six and 11 years. Measurements of trees and site properties were collected from randomly located circular plots with a radius of 5 m (78 m^2 area) within the plantations. The number of plots sampled at each farm ranged from 1 to 12 plots on the size of the farm's plantings, with the number being determined by the size of site so that the sampling area occupied at least 5% of plantation [30,31]. All plots and trees within them were permanently marked in the field. In each plot, all trees were counted and identified to the species level and the height (H) and diameter at breast height (dbh) were measured. Each plot contained at least seven trees greater than 5 cm in diameter. All the plots were re-measured during 2008 and again in 2012. On each occasion new recruits were identified and measured and tree deaths were recorded. A distinction was made between deaths and by harvesting (i.e. evidence in the form of stumps) and natural tree deaths.

The species were grouped into two categories based on provenance (i.e. native species and exotic species) and three categories based on ecological types. These included shade-intolerant species (pioneer as well as longer lived secondary forest species), shade-tolerant species (species from the Dipterocarpaceae plus other shade tolerant species) and fruit tree species. All of the shade-tolerant species but especially the Dipterocarps are regarded locally as having highly valued timbers.

Annual rates of mortality and recruitment were calculated as the mean annual proportion of lost trees or new recruits using the total number of stems at the time of the first measurement as the base line.

Table 3. Changes of species composition of canopy trees in the Rainforetation plantations over time.

Category	Remaining number of species		
	2006	2008	2012
Shade-intolerant species	43	38	35
Shade-tolerant species	19	18	19
Fruit-tree species	14	12	11
Native species	57	50	49
Exotic species	19	18	16
Total	76	68	65*

*includes a new species recruited from outside plot.

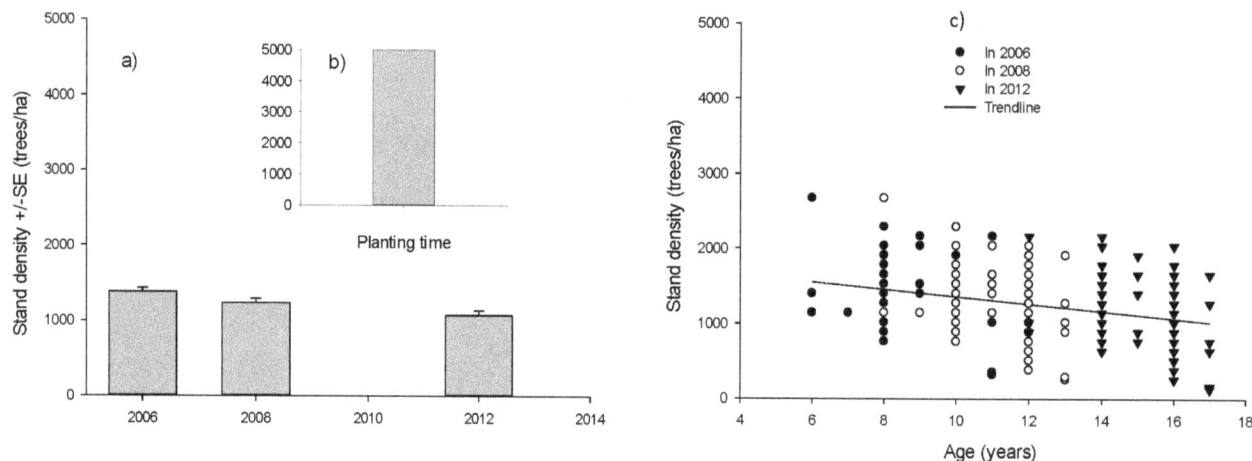

Figure 1. Changes in stand density of Rainforestation plantations over time. Mean stand density at measurements (a), estimated stand density at the planting time (b), and stand density of plots at different ages (c).

The estimated times needed for species growing in Rainforestation stands to reach threshhold sizes of 10 cm and 30 cm were calculated based on relative growth rates of the 32 common species across sites.

The Importance Value Index (IVI; see Table 2) was used to assess the importance of different species in each plantation [32–36].

One-way ANOVA test was used to compare stand densities and the Importance Value Indices between the three measurements. The differences in tree loss and recruitment between species groups were also analysed using a one-way ANOVA combined with LSD post-hoc tests for all pairwise comparisions between group means.

Statistical analysis was undertaken using SPSS 21 and Excel 2010 and data was visually presented using SigmaPlot 12.3.

Results

Species Compositional Changes

A total of 76 canopy species of trees were recorded at 80 plots distributed across the 18 sites at the time of the first measurement in 2006 (Table 3). These represented 33 families and 58 genera. They included 43 shade-intolerant species, 8 shade-tolerant non-Dipterocarp species, 11 Dipterocarp species (all shade-tolerant) and 14 fruit tree species. Of these, 57 are native species and 19 are exotic species.

Each of the different species types were present at most sites, but the proportion of trees represented by the different types of species varied: exotic species represented 36% of all trees, while shade intolerant species represented 78% of all trees. Fruit trees were present in only 21 of the 80 plots and represented 8% of all trees.

Overtime, we found the number of species present decreased from 76 in 2006 to 65 in 2012 (Table 3). The species lost included eight that were shade-intolerant, three fruit tree species and one species of Dipterocarpaceae. On the other hand an additional species (*Strombosia philippinensis*) was recruited from trees growing outside the plots in the period between 2008 and 2012 and grew up to exceed the 5 cm dbh size threshold above which trees were assessed.

Changes in Stand Density Caused by Mortality and Harvesting

The average stand density across all sites decreased from an estimated planting density of 5000 trees ha^{-1} to 1383±50 trees ha^{-1} at the time of first measurement in 2006 when the sites were aged between six and 11 years. From 2006 the average stand density then decreased further to 1145±56 trees ha^{-1} in 2012 (Figure 1). These changes were not uniform and density was constant in 19 plots (24% total plots) during the whole period over which measurements were made. The average site stand density differed significantly over time (F_{18} = 4.720, p = 0.013).

Part of the decline was due to between-tree competition but some was also caused by tree harvesting by farmers. Loss due to competition (i.e. mortality) over the period differed between the various categories of species. Overall, shade-intolerant species and fruit trees lost around 5% of trees each year but only 0.7% of the shade-tolerant trees were lost (Table 4). Around 5.4% and 8.6% of individuals of native and exotic species respectively were lost each year due to competition and harvesting (Table 4, Figure 2). There was no significant difference in the loss of stems between native and exotic species (F_{18} = 0.450, p = 0.507) but a difference between functional groups of shade-intolerant, shade-tolerant and fruit trees (F_{18} = 3.111, p = 0.054) across sites. We found that the shade-tolerant group had lost significantly fewer stems than the other groups (p = 0.030 and 0.043 between shade-tolerant and shade-intolerant, and between shade-tolerant and fruit trees, respectively). We found no significant difference in terms of losing individuals between shade-intolerant species and fruit trees (p = 0.881). Most of the deaths occurred in trees less than 10 cm dbh although a few larger trees had also died by the time of the final measurement (Figure 3). Most of the losses occurred in smaller size classes and were more common in the 2006–2008 period than in the 2008–2012 period (Figure 3).

Evidence of harvesting and the identity of these trees could be seen in the form of old stumps. Harvesting commenced when the trees were in the 5–10 cm dbh class but harvesting was common in trees up to around 25 cm dbh (Figure 3). During the six years, 20.4% of all trees >5 cm dbh recorded in 2006 were harvested. The smaller sized trees harvested belonged to 38 species of all shade-intolerant, shade-intolerant species and fruit trees while the larger trees belonged to only 11 shade-intolerant species (e.g. *Melia dubia, Gmelina arborea, Leucaena leucocephala, Terminala macrocarpa* and

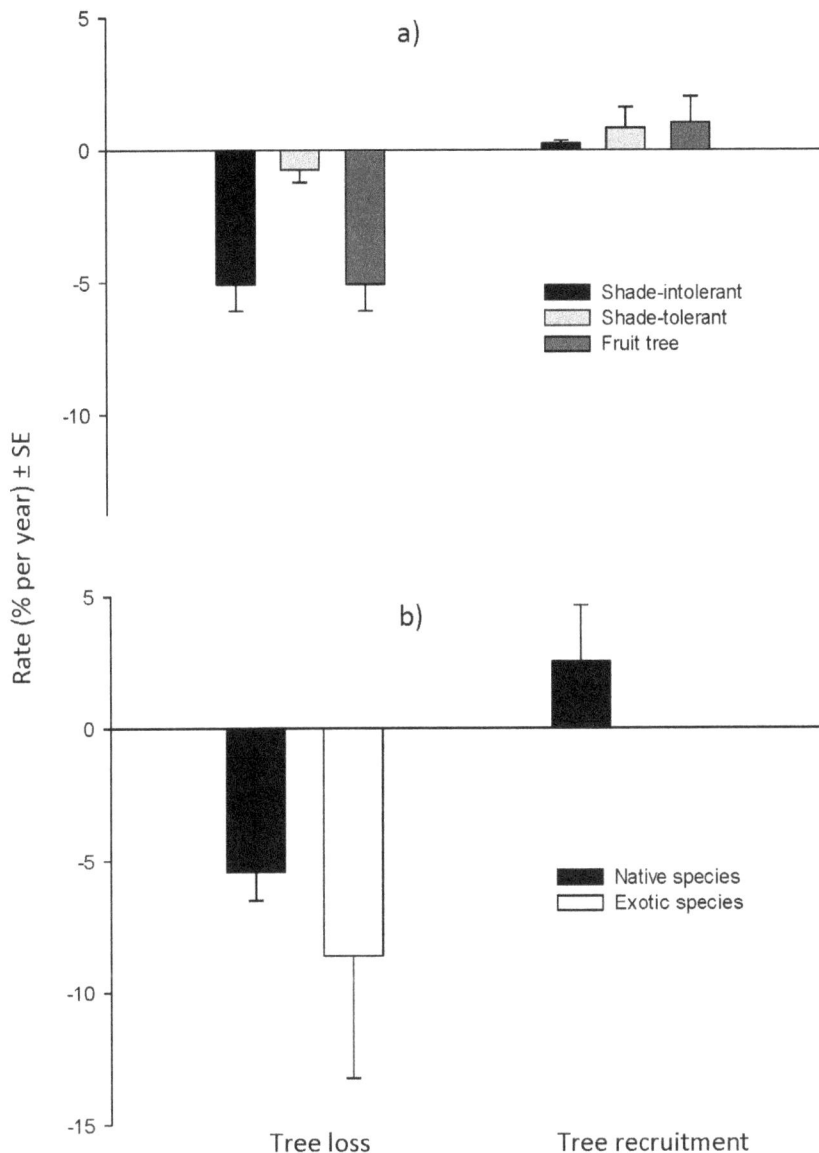

Figure 2. Annual rates of mortality and recruitment at sites during the period of 2006–2012. In categories of species ecology (a) and in categories of species provenance (b).

Swietenia macrophylla) and only one individual of fruit tree (i.e. *Sandroricum koetjape*). Some trees of these species were also harvested at dbh < = 25 cm. The other common species that were harvested at dbh <25 cm included *Vitex parviflora*, *Gymnostoma rumphianum*, *Artocarpus heterophyllus*, *Samanea saman*, *Pterocymbium tinctorium* and *Tectona grandis*.

Most of trees which were died due to competition or disease had dbh smaller than 25 cm, including shade-intolerant, shade-tolerant and fruit trees; only one death of decayed tree of *Melia dubia* had dbh around 30 cm.

In contrast to these losses, a number of trees were also added to the stands as trees grew larger and exceeded the 5 cm dbh size class. These were found in 23 of the 80 plots. At the time of the first measurement in 2008 trees belonging to 9 species had been added to the stands and by 2012 trees from 20 native species were being added. They included Dipterocarp species (e.g. *Parashorea plicata*, *Shorea contorta*, *Hopea malibato*, *Hopea plagata*) and some

individuals of other shade-tolerant, shade-intolerant and fruit trees. There were no exotic species found in the recruitment (Table 4, Figure 2).

Growth Rates

By 2006 about 2% of all trees exceeded 30 cm dbh but this has increased to 6.7% by 2012 (see Figure S2 in File S1). The results showed that three shade-intolerant species (i.e. *Leucaena leucocephala*, *Melia dubia* and *Gmelina arborea*) could reach the dbh threshold of 30 cm by age 10 years. Most these species could achieve 30 cm dbh before 20 years old of planting. The estimated time it takes for particular species to reach a harvestable age is shown in Table 5.

Importance Value Index

Although Importance Value Index of species was not significantly different between measurements across sites for all species ($F_{77} = -1.1E\text{-}13$, p>1), mortality and early harvesting by land-

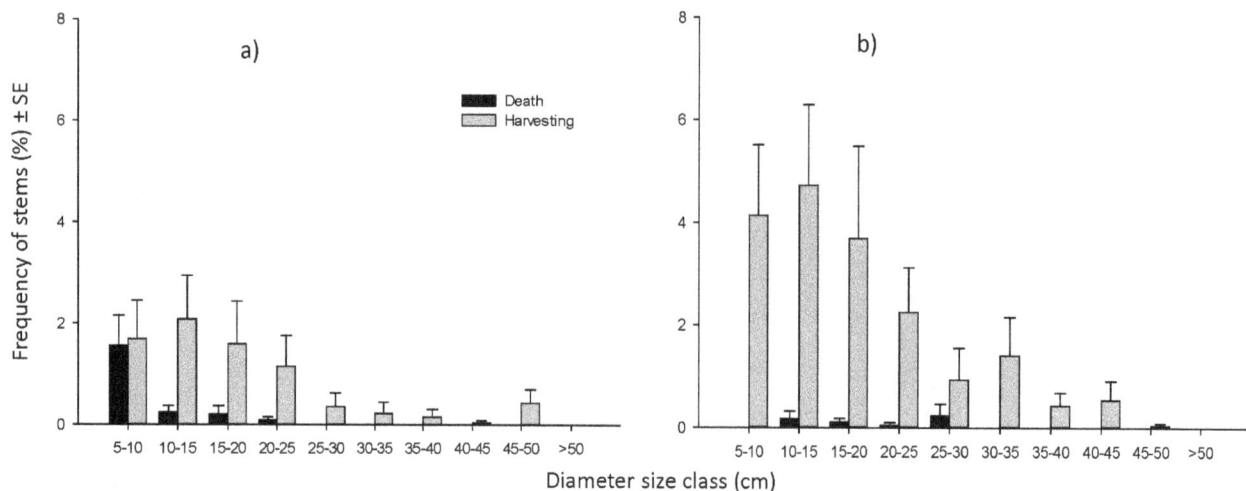

Figure 3. Size class distributions of lost trees (death and harvesting) at the Rainforestation sites. In the periods of 2006–2008 (a) and 2008–2012 (b).

holders caused a change in the Importance Value Index of some species in specific plantations (Figure 4). At the time of the first measurement shade-intolerant species had the highest Importance Value Indices due to their density, relative common and uniform distribution and their faster-growth (e.g. *Swietenia macrophylla*, *Gmelina arborea*, *Terminalia macrocarpa*, *Vitex parviflora* and *Gymnostoma rumphianum*). However, their Importance Values declined because of harvesting and the shade-tolerant species gradually acquired greater Importance Values (e.g. *Gmelina arborea*, *Gymnostoma rumphianum*, *Melia dubia* and *Leucaena leucocephala*) (Figure 4 and Figure S3 in File S1).

Discussion

Our results indicate that the Rainforestation Farming plantations are a highly dynamic system. Changes in stand structure and species composition have been brought about by a combination of anthropogenic factors and natural mortality. It appears that the harvesting of trees by farmers was carried out in an entirely opportunistic way according to their particular circumstances; it was not based on silvicultural prescriptions that specified the timing or intensity of tree removals. This means it is unlikely that production was optimised in the way that managers of industrial

plantations usually seek to achieve. On the other hand, this was exactly the intention of the designers of the Rainforestation Farming system; optimal productivity has been traded-off for the sake of flexibility. Mortality has been hastened by the very high initial planting densities. Not surprisingly this has mostly occurred amongst the shade-intolerant species. The net effect of these two processes has meant that over the six years of the study, slower growing native species have become more dominant while faster growing (mainly shade-intolerant) species have become less dominant. At this point it is not clear how these stands should be now managed because the timing of any future harvesting (i.e. financial benefit to landholders) is difficult to specify. It is likely that farmers will continue to remove trees once they reach some threshold size chosen by the landholder to suit their purpose irrespective of the value of the timbers or market for the logs.

Fruit trees were added to the species mixtures in the expectation that once mature, they would yield a large annual crop of fruit. Experience has shown however that this has not occurred even though some fruit trees are now more than 15 years old. Typically fruit trees require wide spacing and full sun to bear productively [15,35–40]. But in the Rainforestation Farming design they were incorporated into a closely spaced system and were quickly shaded by faster growing pioneer species. As a consequence there appears

Table 4. Mortality and recruitment of species groups in Rainforestation plantations over time during the period of 2006–2012.

Category	Proportion of trees	
	Mortality (% per year)± SE	Recruitment (% per year)± SE
Shade-intolerant species	5.08±1.01	0.24±0.11
Shade-tolerant species	0.73±0.49	0.83±0.78
Fruit-tree species	4.79±2.11	1.03±0.98
Native species	5.43±1.08	2.51±2.13
Exotic species	8.62±4.63	0
New species in plots (but planted species in farms)		0.004±0.004
Total	4.89±1.01	0.49±0.32

Table 5. The estimated time needed for species growing in Rainforestation Farming stands to reach threshold sizes for firewood (10 cm dbh) and lumber (30 cm dbh).

Species	Provenance	Estimated time (years) to reach	
		10 cm dbh (Mean ± SD)	30 cm dbh (Mean ± SD)
Shade-intolerant species:			
Ipil-Ipil (*Leucaena laucocephala*)	Exotic	3.3±1.0	9.8±3.1
Bagalunga (*Melia dubia*)	Native	5.1±2.4	15.2±7.3
Gmelina (*Gmelina arborea*)	Exotic	5.9±3.5	17.6±10.5
Taluto (*Pterocymbium tinctorium*)	Native	7.0±2.6	20.9±7.7
Teak (*Tectona grandis*)	Exotic	7.5±3.6	22.6±10.8
Santol (*Sandoricum koetjape*)	Native	7.7±3.0	23.2±8.9
Mt Agoho (*Gymnostoma rumphianum*)	Native	7.8±2.8	23.5±8.3
Kalumpit (*Terminalia macrocarpa*)	Native	7.9±3.9	23.7±11.6
Raintree (*Samanea saman*)	Exotic	7.9±5.0	23.7±15.1
Dao (*Dracontamelon dao*)	Exotic	8.6±3.7	25.7±11.0
Nangka (*Artocarpus heterophyllus*)	Native	8.9±2.7	26.7±8.2
Thailand acacia (*Senna siamea*)	Exotic	9.3±2.5	28.0±7.4
Mahogany (*Swietenia macrophylla*)	Exotic	9.3±4.2	27.9±12.6
Molave (*Vitex parviflora*)	Native	9.4±3.1	25.2±9.3
Antipolo (*Artocarpus blancoi*)	Native	9.9±4.4	29.8±13.3
Bitanghol sibat (*Calophyllum lancifolium*)	Native	10.4±3.4	31.2±10.3
Narra (*Pterocarpus indicus*)	Native	10.6±3.7	31.7±11.2
Hindang laparan (*Myrica javanica*)	Native	10.9±4.7	32.6±14.2
Lanipga (*Toona ciliate*)	Exotic	10.3±2.8	30.9±8.5
Malakawayan (*Podocarpus rumphii*)	Native	18.2±4.2	54.7±12.5
Shade-tolerant species:			
Mayapis (*Shorea palosapis*)	Native	5.3±1.6	16.0±4.9
Tangeli (*Shorea polysperma*)	Native	6.5±2.7	19.4±8.1
Apitong hagakhak (*Dipterocarpus kunstleri*)	Native	8.0±3.6	24.1±10.8
Marang banguhan (*Artocarpus odoratissimus*)	Native	8.3±3.0	24.8±9.0
Bagtikan (*Parashorea plicata*)	Native	8.7±3.8	26.2±11.4
Rambutan (*Nephelium lappaceum*)	Native	8.8±2.4	26.3±7.2
Cacao (*Theobroma cacao*)	Exotic	8.9±2.6	26.7±7.9
White lauan (*Shorea contorta*)	Native	8.9±3.7	26.8±11.1
Almaciga (*Agathis philippinensis*)	Native	9.5±2.9	28.6±9.5
Durian (*Durio zibethinus*)	Exotic	9.8±3.7	29.5±11.2
Yakal saplungan (*Hopea plagata*)	Native	10.4±3.2	31.2±9.6
Yakal kaliot (*Hopea malibato*)	Native	12.0±2.7	35.9±8.2

to be high mortality of fruit trees. There are reports of some fruit harvesting at some sites by Milan et al. [26] but overall productivity has been very low. A better outcome might have been achieved by plantings fruit trees along plantation edges where they would receive more light and be easier to harvest.

Based on this experience we suggest a modified set of silvicultural prescriptions for smallholder and community tree plantations in this part of the Philippines, along with elsewhere in SE Asia. The suggestions are outlined in Table 6 and show recommended planting densities and constituent species. The planting density is set at around 1100 trees per ha rather than the 5000 trees per ha used in the Rainforestation Farming plantations. This reduces the cost of buying and planting seedlings and, while

there may be slight increase in the time needed for weed control until the seedlings are established, this cost should be modest. The numbers of species used is reduced from that originally established in Rainforestation Farming plantings (at least 76 species assessed in the plots in this study) to between 11–23 species. This is because the type of species (i.e. fast growing and/or exotic versus slow growing/native) drives productivity in these plantings rather than increased biodiversity [18].

The identity of these species has been based on their performance in the field (See Herbohn et al. [15] for the initial assessment) and the prospective markets likely to be available to farmers in this region. They include fast-growing species currently preferred by most farmers as well as slow-growing species likely to

a)

b)

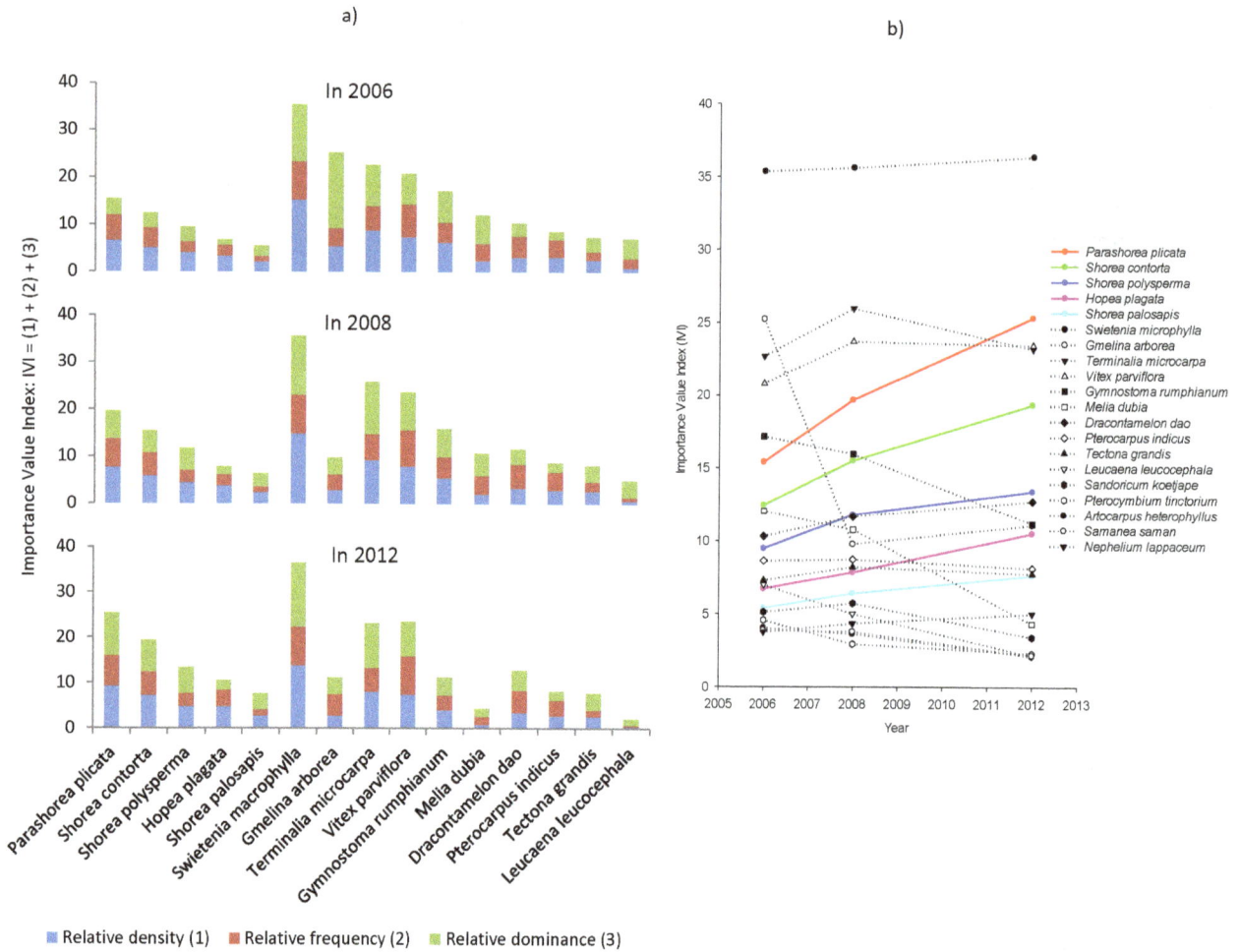

Figure 4. Importance Value Indices (IVIs) of the 15 common species with the highest IVI in rainforestation plantations in period of 2006–2012. IVI at measurements (a) and changing trend of IVI between 2006 and 2012 (b).

generate higher value timbers in the long term [26]. As was the case in the original Rainforestation Farming design, the shade intolerant species will be planted one year before the other species to facilitate their establishment. The intent of these prescriptions is to provide a steady flow of trees of sufficient size and with properties making them suitable for various markets. A significant proportion of the species planted are destined quickly capture the site and then be removed at an early age (6–10 years) in order to generate an income. This harvest will also act as a thinning to prevent the stands stagnating and to hasten the development of larger and more valuable residual trees. Most of these will be faster-growing species such as *Gmelina arborea*, *Casuarina equestifolia* or *Acacia* sp but some slower-growing species having trees with poor form might also be felled at this time as well. *Casuarina equisetifolia* was not found in our plots but observed in surrounding areas, we, therefore, suggest this species for plantations in the future based on our observation and the local farmers and experts during our research periods. At this age these are likely to be mostly used for fuelwood or pulp [41]. A second harvesting period could occur when the stands are 8–12 years old and a number of species are capable of being large enough for a pole market. Based on species growth rates a third harvesting period would develop around 14–18 years leaving a final harvest of slower growing but

higher value trees after 20 years. The selective harvesting could be applied for different species at different times and for different products. Trees of some fast-growing species such as *Leucaena leucocephala*, *Melia dubia* and *Gmelina arborea* could be early harvested for household consumption or firewood while most of Dipterocarp species could be late harvested for high valued products (e.g. lumber). By this time a number of additional seedlings may have regenerated and begun to growing using light from canopy gaps. These seedlings might even be supplemented by enrichment plantings using species favoured by the market. The plantations could then gradually move to become a selectively managed forest or it could be clear-felled and replanted. Throughout this sequence the timbers produced by the plantations are becoming progressively more valuable. Similarly, the market for these timbers is increasing from a largely local market or on-farm use for fuelwood to a regional or national market for higher value sawlogs.

Finally, these mixtures could also achieve the ecological goal of the Rainforestation Farming project i.e. to achieve a planted forest with a physical structure and species composition and succession similar to the original local rainforest ecosystem after around 14 – 16 years. A three-storey structural complexity is now beginning to develop these plantings resembling natural rainforests in the area. The changes in density and species compositions in these

Table 6. Recommended design for smallholder tree plantations in Leyte, Philippines based on performance of Rainforestation Farming plantations.

Product	Time of thinning (yrs)	Number of species	Density (trees/ha)	Typical species
Firewood	6–10	3–5	450	Gmelina (*Gmelina arborea*)
				Bagalunga (*Melia dubia*)
				Ipil-Ipil (*Leucaena laucocephala*)
				Raintree (*Samanea saman*)
				Mt agoho (*Gymnostoma rumphianum*)
				Agoho (*Casuarina equisetifolia*)
				Thailand acacia (*Senna siamea*)
Pole	8–12	2–3	200	Gmelina (*Gmelina arborea*)
				Mahogany (*Swietenia macrophylla*)
				Kalumpit (*Terminalia macrocarpa*)
				Dao (*Dracontamelon dao*)
				Narra (*Pterocarpusindicus*)
				Mt agoho (*Gymnostoma rumphianum*)
Fast-growing timber	14–18	3–5	250	Gmelina (*Gmelina arborea*)
				Mahogany (*Swietenia macrophylla*)
				Teak (*Tectona grandis*)
				Mayapis (*Shorea palosapis*)
				White lauan (*Shorea contorta*)
				Almaciga (*Agathis philippinensis*)
				Tangeli (*Shorea polysperma*)
				Bagtikan (*Parashorea plicata*)
Slower-growing timber	>20	3–10	200	Yakal saplungan (*Hopea plagata*)
				Yakal kaliot (*Hopea malibato*)
				Malakawayan (*Podocarpus rumphii*)
				Molave (*Vitex parviflora*)
				Narra (*Pterocarpus indicus*)
				Malapanau (*Dipterocarpus kerrii*)
				Tangeli (*Shorea polysperma*)
				Apitong (*Dipterocarp grandiflorus*)
				Apitong hagakhak (*Dipterocarpus kunstleri*)
				Dalingdingan (*Hopea dalingdingan*)
				Red lauan (*Shorea negrosensis*)
				Bitanghol (*Calophyllum blancoi*)
				Bitanghol sibat (*Calophyllum lancifolium*)
Total		11–23	1100	Fruit trees: Durian (*Durio zibethinus*), Mango (*Mangifera indica*), Rambutan (*Nephelium lappaceum*), Nangka (*Artocarpus heterophyllus*)

plantings over time indicated a characteristic often found in uneven-aged forests, old-growth forest or natural forest that is a continuous recruitment and mortality in the forest succession [42]. The ratio of Dipterocarp trees to total trees was approximately 1:4 across sites, which is similar to the species proportion in the three canopy strata at Mt. Pangasugan in Leyte, the Philippines [29].

Conclusions

It is clear from our results that the composition and structure of the Rainforestation plantations have changed over time, with the plantations being a highly dynamic system. There has been a decrease in the relative importance of shade-intolerant species, especially exotic species and a corresponding increase in the relative importance of shade-tolerant native species. The design of the species mixtures affected the patterns of species loss in that the very high initial stocking rates resulted in high rates of mortality, mostly among the shade-intolerant species and fruit trees. Other changes in species loss were largely due to ad hoc harvesting decisions by the land owners. We draw on our current results to recommend a modified set of prescriptions for smallholder and community tree plantations. These recommendations include a lower initial stocking rate of 1100 trees per hectare; the proportion of fast growing species should be around 30 to 40 per cent of this initial density; and any fruit trees should only be planted on the plantation margin where more space and light are available for crowns to develop.

Supporting Information

File S1   Figure S1. Map of the Rainforestation sites in Leyte province, the Philippines. The numbers in the map refer to the names of sites. **Figure S2. Size class distributions of species groups in the Rainforestation sites.** Depending on species provenance (i.e. exotic and native) or ecological characteristics of species (i.e. Shade-intolerant, Shade-tolerant and Fruit tree). 1a, 2a & 3a: all size classes of provenance groups; 1b, 2b & 3b: three largest size classes of provenance groups; 4a, 5a & 6a: all size classes of ecological groups; and 4b, 5b & 6b: three largest size classes of ecological groups. **Figure S3. The survival and mortality of the most common species at 80 plots of 18 rainforestation sites in period of 2006–2012.**

Acknowledgments

Thank you to the staff of the Visayas State University, in particular the College of Forestry and Natural Resources and Institute of Tropical Ecology, for supporting the work and also to the people in Leyte province, for allowing us to conduct our study on their land and for their great assistance during the field work. The authors would like to acknowledge to Grant Wardell-Johnson for contributing to the design of the sampling procedure. We thank Jack Baynes for helping with the map of study sites.

Author Contributions

Conceived and designed the experiments: HN JH. Performed the experiments: HN. Analyzed the data: HN DL. Wrote the paper: HN DL JH JF.

References

1. Lamb D, Erskine PD, Parrotta JA (2005) Restoration of degraded tropical forest landscapes. Science 310: 1628.
2. Debell DS, Cole TG, Whitesell CD (1997) Growth, Development, and Yield in Pure and Mixed Stands of Eucalyptus and Albizia. Forest Science 43: 286–298.
3. Montagnini F, Gonzalez E, Porras C, Rheingans R (1995) Mixed and pure forest plantations in the humid neotropics: a comparison of early growth, pest management and establishment costs. Commonwealth Forestry Review 74: 306–314.
4. Piotto D, Víquez E, Montagnini F, Kanninen M (2004) Pure and mixed forest plantations with native species of the dry tropics of Costa Rica: a comparison of growth and productivity. Forest Ecology and Management 190: 359–372.
5. Kelty MJ (2006) The role of species mixtures in plantation forestry. Forest Ecology and Management 233: 195–204.
6. Lamb D (2011) Regreening the bare hills: Tropical forest restoration in the Asia-Pacific region; Palo M, editor. Dordrecht: Springer.
7. Hooper DU, Chapin FS III, Ewel JJ, Hector A, Inchausti P, et al. (2005) Effects of biodiversity on ecosystem functioning: A consensus of current knowledge. Ecological Monographs 75: 3–35.
8. Wormald TJ (1992) Mixed and pure forest plantations in the tropics and subtropics. Rome: FAO. 152 p.
9. Kelty MJ (1992) Comparative productivity of monocultures and mixedspecies stands. In: Kelty MJ, Larson BC, Oliver CD, editors. The Ecology and Silviculture of Mixed-Species Forests. Dordrecht, Boston: Kluwer Academic Publishers. 125–141.
10. Haggar JP, Ewel JJ (1997) Primary Productivity and Resource Partitioning in Model Tropical Ecosystems. Ecology 78: 1211–1221.
11. Bullock BP, Burkhart HE (2005) An evaluation of spatial dependency in juvenile loblolly pine stands using stem diameter. Forest Science 51: 102–108.
12. Grant JC, Nichols JD, Pelletier M-C, Glencross K, Bell R (2006) Five year results from a mixed-species spacing trial with six subtropical rainforest tree species. Forest Ecology and Management 233: 309–314.
13. Kooyman RM (1996) Growing rainforest: rainforest restoration and regeneration - recommendations for the humid sub-tropical region of northern New South Wales and south east Queensland. Brisbane: Greening Australia - Queensland. 79 p.
14. Lamprecht H (1989) Silviculture in the tropics. Tropical forest ecosystems and their tree species - Possibilities and methods for their long-term utilization. Eschborn, Federal Republic of Germany: Technical Co-operation.
15. Herbohn JL, Vanclay J, Nguyen H, Le HD, Baynes J, et al. (2014) Inventory Procedures for Smallholder and Community Woodlots in the Philippines: Methods, Initial Findings and Insights. Small-Scale Forestry 13: 79–100.
16. Le HD, Smith C, Herbohn J (2014) What drives the success of reforestation projects in tropical developing countries? The case of the Philippines. Global Environmental Change, 24:334–348.
17. Le HD, Smith C, Herbohn J, Harrison S (2012) More than just trees: Assessing reforestation success in tropical developing countries. Journal of Rural Studies 28: 5–19.
18. Nguyen H, Herbohn J, Firn J, Lamb D (2012) Biodiversity–productivity relationships in small-scale mixed-species plantations using native species in Leyte province, Philippines. Forest Ecology and Management 274: 81–90.
19. Nguyen HTT (2011) Performance of mixed-species plantaitons using the Rainforestation Farming system in Leyte province, the Philippines [PhD thesis]. The University of Queensland.
20. Milan PP (1997) Appraisal mission of rainforestation farming in Leyte. Baybay, Leyte: Visayas State College of Agriculture (ViSCA). 16 p.
21. Milan PP (1997) Strategy for community involvement in Rainforestation farming. In: Margraf J, Goltenboth F, Milan PP, editors; 1997 3–6 March; Leyte. Visayas College of Agriculture (ViSCA) - GTZ. 336.

22. Kolb M (2003) Silvicultural analysis of "Rainforestation Farming" areas on Leyte island, Philippines [PhD thesis]. Deutschland: Universitat Gottingen. 116 p.
23. Jahn R, Asio VB (2001) Climate, geology, geomorphology and soils of the tropics with special reference to Leyte islands (Philippines). In: Goltenboth F, Asio VB, editors; 2001 9–20 Apr; Baybay, Leyte. Visaya State College of Agroculture. 25–43.
24. Marohn C (2007) Rainforestation farming on Leyte island, Philippines - aspects of soil fertility and carbon sequestration potential [PhD thesis]. Stuttgart: University of Hohenheim.
25. Hussain I (2008) Plant Physiology. 1 ed. New Delhi: Oxford Book Co.
26. Milan PP, Ceniza MJC, Asio VB, Bulayog SB, Napiza MD (2004) Evaluation of silviculltural management, ecological change and market study of products of existing rainforestation demonstration and cooperators' farms. Baybay: Institute of Tropical Ecology.
27. Schulte A (1998) Community-based rainforestation options for the Visayas, Philippines: a review. Hoxter, Germany: GTZ.
28. Schulte A (2002) Rainforestation Farming: Option for rural development and biodiversity conservation in the humid tropics of Southeast Asia. A review of major issues on community-based rehabilitation silviculture and guide to recommended native tree species for the Visayas/Philippines; Goltenboth F, Milan P, editors. Aachen: Shaker. 312 p.
29. Margraf J, Milan P (1996) Ecology of Dipterocarp forests and its relevance for island rehabilitation in Leyte, Philippines. In: Schulte A, Schone D, editors. Dipterocarp forest ecosystems: Towards sustainable management. Berlin: World Scientific. 124–154.
30. Barbour MG (1999) Terrestrial plant ecology. Menlo Park, Calif.: Benjamin/Cummings. 1 v. (various pagings).
31. Barbour MG, Burk JH, Pitts WD (1980) Terrestrial plant ecology. Menlo Park, Calif.: Benjamin/Cummings Pub. Co. xi, 604 p.
32. Arroyo-Rodríguez V, Mandujano S (2006) The Importance of Tropical Rain Forest Fragments to the Conservation of Plant Species Diversity in Los Tuxtlas, Mexico. Biodiversity & Conservation 15: 4159–4179.
33. Dangol DR, Shivakoti GP (2001) Species composition and dominance of plant communities in Westren Chitwan, Nepal. Napal Journal of Science and Technology 3: 69–78.
34. Shrestha R, Karmacharya SB, JHA PK (2000) Vegetation analysis of natural and degraded forests in Chitrepani in Siwalik region of Central Nepal. Tropical Ecology 41: 111–114.
35. Guariguata M, Chazdon R, Denslow J, Dupuy J, Anderson L (1997) Structure and floristics of secondary and old-growth forest stands in lowland Costa Rica. Plant Ecology 132: 107–120.
36. Nebel G, Kvist LP, Vanclay JK, Christensen H, Freitas L, et al. (2001) Structure and floristic composition of flood plain forests in the Peruvian Amazon: I. Overstorey. Forest Ecology and Management 150: 27–57.
37. Crane JH (2004) Selected cultural techniques to improve production of some subtropical and tropical fruit crops. In: Albrigo LG, Sauco VG, editors. Acta Horticulturae; 2004 29 Feb 2004; Toronto. 179–187.
38. Ferguson L, Krueger WH, Reyes H, Metheney P (2002) Effect of mechanical pruning on California black ripe (Olea europea L.) cv. 'Manzanillo' table olive yield. In: Vitagliano C, Martelli GP, editors. Proceedings of the Fourth International Symposium on Olive Growing. Leuven. 281–284.
39. Medina-Urrutia VM, Nunez-Elisea R (1997) Mechanical pruning to control tree size, flowering, and yield of mature "Tommy Atkins" mango trees. In: Lavi U, Degani C, Gazit S, Lahav E, Pesis E et al., editors. 5th International Mango Symposium. Israel. 305–314.

40. Schupp JR, Baugher TA, Miller SS, Harsh RM, Lesser KM (2008) Mechanical thinning of peach and apple trees reduces labor input and increases fruit size. Horttechnology 18: 660–670.

41. Tolentino EL (2008) Restoration of Philippine Native Forest by Smallholder Tree Farmers. In: Snelder DJ, Lasco RD, editors. Smallholder Tree Growing for Rural Development and Environmental Services. Dordrecht; London: Springer. 319–346.

42. Oliver CD, Larson BC (1996) Forest stand dynamics. Updated edition. Wiley, New York.

13

Documenting Biogeographical Patterns of African Timber Species Using Herbarium Records: A Conservation Perspective Based on Native Trees from Angola

Maria M. Romeiras[1,2]*, **Rui Figueira[1,3]**, **Maria Cristina Duarte[1,3]**, **Pedro Beja[3]**, **Iain Darbyshire[4]**

1 Tropical Botanical Garden, Tropical Research Institute (IICT), Lisbon, Portugal, 2 Centre for Biodiversity, Functional and Integrative Genomics (BIOFIG), Faculty of Sciences, University of Lisbon, Lisbon, Portugal, 3 CIBIO - Research Center in Biodiversity and Genetic Resources/InBIO, University of Porto, Vairão, Portugal, 4 Royal Botanic Gardens, Kew. Richmond, United Kingdom

Abstract

In many tropical regions the development of informed conservation strategies is hindered by a dearth of biodiversity information. Biological collections can help to overcome this problem, by providing baseline information to guide research and conservation efforts. This study focuses on the timber trees of Angola, combining herbarium (2670 records) and bibliographic data to identify the main timber species, document biogeographic patterns and identify conservation priorities. The study recognized 18 key species, most of which are threatened or near-threatened globally, or lack formal conservation assessments. Biogeographical analysis reveals three groups of species associated with the enclave of Cabinda and northwest Angola, which occur primarily in Guineo-Congolian rainforests, and evergreen forests and woodlands. The fourth group is widespread across the country, and is mostly associated with dry forests. There is little correspondence between the spatial pattern of species groups and the ecoregions adopted by WWF, suggesting that these may not provide an adequate basis for conservation planning for Angolan timber trees. Eight of the species evaluated should be given high conservation priority since they are of global conservation concern, they have very restricted distributions in Angola, their historical collection localities are largely outside protected areas and they may be under increasing logging pressure. High conservation priority was also attributed to another three species that have a large proportion of their global range concentrated in Angola and that occur in dry forests where deforestation rates are high. Our results suggest that timber tree species in Angola may be under increasing risk, thus calling for efforts to promote their conservation and sustainable exploitation. The study also highlights the importance of studying historic herbarium collections in poorly explored regions of the tropics, though new field surveys remain a priority to update historical information.

Editor: Giovanni G. Vendramin, CNR, Italy

Funding: This work was supported by the Portuguese Foundation for Science and Technology with the FCT/Ciência 2008 to MMR and FCT/Ciência 2007 to RF, EDP Biodiversity Chair (CIBIO) to PB and BioFIG PEst-OE/BIA/UI4046/2011. The funders had no role in study design, data collection and analysis, decision to publish, or preparation of the manuscript.

Competing Interests: The authors have declared that no competing interests exist.

* Email: mromeiras@yahoo.co.uk

Introduction

Legacy data from natural history collections contain invaluable information about biodiversity in the recent past, providing a baseline for detecting change and forecasting future trends [1]. In the case of plants, specimens have accumulated for hundreds of years in herbaria, and these may be used as the basis for identifying threatened or declining species, guiding future research and monitoring programs, and establishing conservation priorities [2]. For instance, the IUCN Sampled Red List Index for plants was driven in its first iteration almost solely by herbarium specimen data [3]. Data from herbaria are particularly important in poorly explored regions of the tropics, where the lack of continuous field-based botanical research has emphasized the pivotal role of herbaria in documenting plant diversity and species distributions [4–6]. The interest in herbaria for undertaking conservation biology research has thus grown in recent years, though less than about 2% of the herbarium specimens have been used to answer biogeographical or environmental questions [6].

Establishing baselines is particularly important for those tropical tree species that are exploited commercially and have come under increasing pressure from the global timber trade [7–8]. Overexploitation has resulted in declining populations of the most valuable timber species and it is one of the foremost causes for the loss and degradation of tropical forests [9], with utmost negative consequences for the conservation of biodiversity and ecosystem services [10–11]. In recent decades, efforts have been made to increase the sustainability of tropical timber exploitation, through for instance the outright ban on or severe restrictions to the trade of endangered species, or the implementation of certification schemes for timber harvested sustainably [8–12]. These approaches face several problems, however, including uncertainties related to the conservation status of many exploited species due to insufficient knowledge of their distribution, abundance and

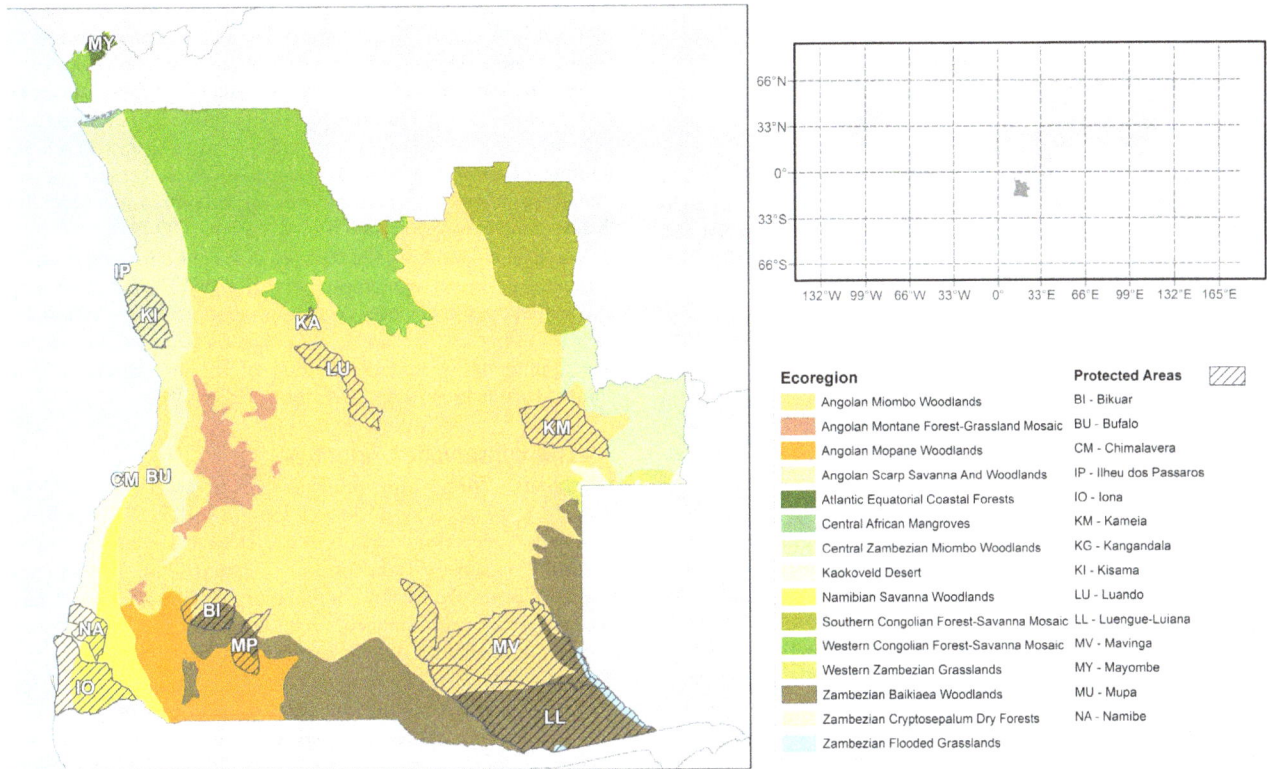

Figure 1. Map of Angola. The 15 WWF ecoregions represented in Angola are displayed together with the network of protected areas (see text for details).

population trends [13–15]. Although this type of information has become increasingly available for tropical forests of Central and South America [16–17] and Asia [18–19], data are still very limited for most African forests [20]. Considering that Africa still holds some of the most important tropical forests in the world [21–22] and that these have been increasingly exploited [23], information on the conservation status of its timber species is urgently required [24].

Angola is one of the African countries for which basic data on timber tree species are most severely lacking, though the country has a forested area of about 40–60 million hectares largely administered by the government [25–26]. Deforestation rates in Angola are among the highest in Sub-Saharan Africa [27], which is likely a consequence of wood extraction for firewood and charcoal, slash-and-burn cultivation, urban expansion, and logging [25]. Illegal logging of valuable timber is considered one of the potential causes of forest degradation, but there is no information on the extent of this problem [25]. Despite some early studies [28–32], botanical data on the forests of Angola are scarce because most of the country was inaccessible to researchers during the war of independence (1961–1974), and the subsequent civil war (1975–2002). Despite increases in safety during the first decade of the 21st century, field biodiversity research has remained very limited, thereby making historic herbarium specimens the main source of data for studying the distribution patterns of tree species exploited commercially in Angola. This information is urgently required because Angola is currently experiencing rapid economic and human population growth, which is likely to place further pressure on its forest resources, with negative consequences for biodiversity, ecosystem services, and ultimately for human well-being [25]. Data on timber trees is also

required to inform ongoing initiatives to improve the protected area network of Angola [33].

The present study focuses on the timber trees of Angola, combining herbarium and bibliographic data to assess biogeographical patterns and conservation priorities, thereby providing baseline information required for their conservation management and sustainable exploitation. Specifically, the study aims (i) to inventory the timber tree species of Angola based on a thorough review of literature and data held in herbaria, (ii) to document biogeographical patterns of the timber species in relation to WWF ecoregions [34]; and (iii) to estimate species conservation priorities based on distribution patterns, representation in protected areas and deforestation rates.

Materials and Methods

Study area

The Republic of Angola (Fig. 1) is the largest country in southern Africa (1.24 million km^2), encompassing a variety of climatic characteristics, which correspond to five climate types by the Köppen–Geiger system [35]. The phytogeographic study of Grandvaux-Barbosa [36] identified 32 vegetation units in the country, ranging from rainforests in the northwest to the desert in the southwest. The global ecoregions map of World Wildlife Fund (WWF) [34] recognises the presence of 15 biogeographic units in Angola (Fig. 1), of which the most widespread are the miombo woodlands of the central plateau, and the western Congolian forest-savanna mosaics in the north. Other important but less widespread forest types include the Atlantic Equatorial coastal forests in Cabinda, the mopane (*Colophospermum mopane*) woodlands, and the Namibian savanna woodlands in the southwest (see Fig. 1). The network of protected areas was mainly established in colonial times to protect large ungulates, and it has been considered too limited to adequately protect most biodiversity components, notably vascular plants [25,33].

Species data

Data on the timber species of Angola were obtained through a combination of bibliographic sources and the study of 2670 herbarium records: 417 of Angolan specimens (see Table S1) and 2253 records from 62 providers available through the GBIF data portal (Tables S2 and S3). First, a thorough literature review was undertaken, focusing on studies of the flora of Angola [36–44], and on studies documenting the use of afro-tropical timber trees [28–32]. Based on this information, we selected for further analysis the subset of timber trees that are: (i) known to be native in Angola; (ii) documented in the Angolan literature to be exploited for timber in colonial times or at present; and (iii) traded in international timber markets. Many of these timber species are important components of the upper forest layer, above 25 m, and all are known for their economic value, thus making them interesting from both ecological and conservation standpoints. For each species selected, we compiled information on their distribution in Angola and across Africa, their habitat, ecology, timber value and characteristics, and their global conservation status based on the IUCN Red List of Threatened Species [45].

Second, a thorough study of herbarium specimen data was undertaken for all timber species selected. The research was concentrated on herbaria holding the largest collections of Angola vascular plants, including LISC (Tropical Research Institute), LISU (University of Lisbon), COI (University of Coimbra), BM (Natural History Museum, London), and K (Royal Botanic Gardens Kew). The collecting locality for each specimen was georeferenced wherever possible, using 1:100,000 cartographic maps and geographic gazetteers [46], and data was compiled in a geographic database prepared in ArcGIS Arcinfo ver. 10.0 [47]. Further information about the global native distribution of each selected timber species in Africa was gathered from the GBIF data portal. Although it is recognised that GBIF does not contain all known records of the species studies, it is deemed adequate to provide a first approximation of their geographic range.

Biogeographic patterns

Patterns of timber tree species distribution in Angola were analysed in relation to the 15 WWF ecoregions identified in the country [34]. We focused on WWF ecoregions because they have been produced mainly as a utility tool for conservation planning [48], and so it was considered important to examine whether they could be used as meaningful spatial units for conservation prioritization and management of Angolan timber tree species.

Analyses were based on a presence/absence matrix, which indicated whether or not each timber tree species had been recorded within each WWF ecoregion. Presence/absence was used instead of the number of records, to reduce the bias associated with geographic variation in sampling effort. Although this approach does not avoid the problem of false absences (i.e. absence due to lack of sampling rather than a true record of absence), we believe that this problem has been minimised by using a small number of spatial units, each covering a large geographic area and encompassing many species records. Hierarchical clustering was then carried out, using the Jaccard index as a measure of similarity between species distribution, and the Ward agglomerative procedure [49]. The Jaccard index was used because it does not consider double absences [49]. Several agglomerative methods were tested (e.g., UPGMA, WPGMA),

but they produced largely similar results. Clusters identified at different levels of the dendrogram were mapped and checked for spatial consistency, i.e., whether each group was associated with a well-defined spatial region, and we selected the number of clusters that maximised spatial interpretability [49]. Quantitative approaches such as the L-Method [50] were also tested but the number of clusters produced was excessively large and with no spatial consistency. Analyses were carried out using 'dist' and 'hclust' functions implemented in R version 3.02 [51].

The spatial distribution of the species groups emerging from the cluster analysis was overlapped with the WWF ecoregions map, and spatial consistency between species groups and ecoregions was visually inspected. A similar investigation was carried out by overlapping the spatial distribution of species groups and the climate classification map of Köppen–Geiger [35].

Species conservation priorities

Estimating conservation priorities from herbaria data is difficult, because a species may no longer exist in localities where it was historically recorded, and because collectors may be biased towards or against certain species or regions [2,5]. To overcome these problems, we used a combination of three relatively coarse criteria, which were judged useful in helping to guide future conservation efforts, despite some potential shortcomings and limitations.

A first approximation for conservation prioritization was obtained by computing the extent of occurrence (EOO) of each species, assuming that the highest priority should be given to species with a small EOO in Angola, and to species with a large proportion of its global EOO concentrated in the country. EOO was computed from the georeferenced locality data for each species, using the minimum convex hull polygon method [52], implemented in GEOCAT [53]. Computations were carried out at the scale of the African Continent and that of Angola, and we calculated Angola's contribution to the overall EOO for each species. Areas offshore from the African continent were calculated using ArcGIS Arcinfo ver. 10.0 [47] and were excluded from the EOO polygon. Although the area of occupancy (AOO) is an important parameter to assess species conservation status [52], it was not estimated because large gaps in species distribution are likely to be due primarily to the lack of comprehensive field surveys or lack of data reporting by herbaria to GBIF, rather than resulting from true species absences.

A second indicator of conservation priority was based on the occurrence of herbarium specimens' locations in national parks and reserves, assuming that a higher conservation risk should be attributed to the species poorly represented within protected areas. We considered both the number of locations recorded within protected areas, and the percentage of the EOO that is included in protected areas. Although we recognise that it is uncertain whether a given species occurs at any particular location within its EOO, we assumed that the overlap between EOO and protected areas could be taken as a coarse approximation of the relative representation of a species within the protected area network. The geographical limits of protected areas were obtained in GIS shape file format from WDPA [54]. New protected areas unavailable in WDPA were digitised in ArcGIS ArcInfo ver. 10.0 [37] from maps published in the official journal of The Republic of Angola (law n° 38/11 of December, 29 2011, p. 6340).

Finally, conservation priority was also evaluated by estimating rates of forest loss between 2000 and 2012 around the georeferenced localities for each species. We assumed that higher conservation priority should be given to species occurring in areas with low forest cover, and where the recent deforestation rate is highest. Forest cover was estimated for each georeferenced specimens location using raster maps provided by Hansen et al. [27], by multiplying the percent tree (crown) cover per pixel and the pixel area (30-m resolution), and then summing across all pixels extracted in a 5-km buffer of the location. Deforestation rate was calculated by estimating the area of pixels showing forest loss, and then expressing it as a percentage of total tree cover in 2000. Similar analyses were carried out using 1, 2.5 and 10-km buffers, but the results were much the same, and so they were not considered further.

Results

From the literature review and the study of herbarium specimens, we identified eighteen native timber species occurring in Angola (Table 1), which have a high commercial value due to the quality of their timber (Table S4). Available herbarium data are rather old, corresponding primarily to specimens collected in 1850–1860, 1910–1920, and 1950–1975 (Fig. S1). Most species (> 80%) belong to the Fabaceae and Meliaceae families, and they are associated with tropical rainforests (11 species), evergreen forests and woodlands (2 species), and mainly with dry forests and savannas (5 species) (Table 1, Table S4). Half the species are either classified as threatened (7) or near-threatened (2) by IUCN at the global scale, whereas the conservation status of eight species has not yet been evaluated.

The cluster analysis of timber trees identified four groups of species (Fig. 2a). The first and second groups have similar spatial patterns, with most occurrences concentrated in the small enclave of Cabinda. The second group, however, is also represented in north-western regions of Angola. For eight of the eleven species in these two groups, Cabinda is the southern limit of wider distributions concentrated in the Guineo-Congolian rainforests (Fig. 3a). The third group includes only two species and it has a distribution concentrated in north-western regions of Angola, though it is absent from Cabinda. The fourth group includes five species, and it occupies most of the Angolan territory, with the exception of Cabinda.

There is a poor match between the spatial distribution of the four species groups and the WWF ecoregions, as each group occurs in several ecoregions (Fig. 2b). Overlay with the climate classification map of Köppen–Geiger suggests a rough association between groups 1–3 and a single climate type (Aw - Equatorial savanna with dry winter), while group 4 occurs in a wide range of climate types (Fig. 2c).

Most of the timber species have a large global extent of occurrence - EOO (Table 2). Smaller EOO values are found for *Entandrophragma spicatum* and *Guibourtia arnoldiana* ($\approx 1 \times 10^5$ km^2), but they are still one order of magnitude above the threshold for species qualifying as threatened under IUCN criterion B (i.e., $> 2 \times 10^4$ km^2) (Table 2). Within Angola, however, there are nine species with a restricted EOO ($< 2 \times 10^4$ km^2) and thus potentially qualifying as threatened at the national level. Three species have more than 15% of their global EOO concentrated in Angola (Fig. 3b–d), reaching >50% in the case of *E. spicatum* and *G. coleosperma* (Table 2). Few of the historical herbarium specimens were collected from within current protected areas, with ≤5 localities for all species evaluated (Table 2). More than 10% of the EOO of eight species overlaps with protected areas, whereas there was no overlap for another six species (Table 2).

Forest cover, in 2000, around the location of collection localities ranged from <20% in the case of *Diospyros mespiliformis, E.*

spicatum and *G. coleosperma*, to >70% in the case of *Entandrophragma candollei, Milicia excelsa, Oxystigma oxyphyllum* and *Terminalia superba* (Table 3). Variation among species in deforestation rate (2000–2012) was less marked, but particularly high values (>10%) were recorded in the occurrence areas of *G. coleosperma* and *Pterocarpus angolensis*.

Discussion

This study recognized the presence of 18 key timber tree species in Angola, most of which are widely used as timber trees elsewhere in Africa [26]. These species are important components of woody vegetation communities and are known for their economic value, thus making them important from both ecological and economic perspectives. Several of these species are highly valued in international timber markets and they have been historically exploited in Angola, including the African mahoganies (*Entandrophragma* spp.), the agba (*G. balsamiferum*), and the tchitola (*O. oxyphyllum*) [43], hence they are under increasing pressure in the country.

Biogeographical patterns

This study revealed striking differences in biogeographic patterns of timber species in Angola, recognizing different groups associated with regions with relatively homogeneous climatic conditions (i.e. tropical rainforests; evergreen forests and woodlands; and dry forests, woodlands and savannas). Three of the four clusters identified were associated with Cabinda and the north-western regions of Angola, showing a close matching with the Aw climate category of Köppen–Geiger and with the Congolian region identified in the recent bioregionalization study of Linder et al. [56]. The timber species included in the first two clusters (ca. 60%) corresponding largely to Guineo-Congolian rainforest species [57], where the rainy season lasts for six months or more, and the relative air humidity is above 80%, with some areas having persistent and dense fogs (locally known as *cacimbo*). In Angola they occur in Cabinda's Maiombe forest (extending through Congo, Democratic Republic of the Congo and Angola), which are dense moist forest formations, with high ecological and floristic diversity [58,59]. Most of the species found in Guineo-Congolian rainforests have their south-western range limit in Cabinda (see Fig. 3a), where they may face drier climatic conditions than within their core range. These peripheral populations may have unique adaptations to specific environmental conditions that are absent from other populations [60], and so may be particularly valuable in a warming scenario due to climate change [61].

The third cluster included just two species (*K. anthotheca* and *P. tinctorius*), and it was restricted to woodlands and evergreen forests of northwest Angola. This region is characterized by a rainy season lasting about six months and relative air humidity of about 75% [36]. Finally, the fourth cluster comprised five species and is widespread throughout the country, albeit little represented in the north-western regions of Angola. Species included in this group occur mainly in dry forests and savannas, and sometimes their distribution reach the semi-arid and arid regions of southern Angola, characterized by xerophytic vegetation. Among the studied timber species, the legumes *G. coleosperma* (see Fig. 3b) and *P. angolensis* (see Fig. 3c) are the most widespread in Angola, and are mainly found in miombo woodlands, which is the dominant forest component of Angola and one of the major dry forest-savanna biomes of the world [34].

The spatial distribution of the four clusters retrieved from the biogeographical analysis showed little concordance with the WWF

Table 1. Geographical distribution in Angola and in the African continent, main types of ecosystems and global conservation status [45] for each species considered in the present study. Nomenclature according to The Plant List [55].

Main types of ecosystems Species	Geographical distribution Angola (district)	Africa	Conservation status and criteria
Tropical Rainforests			
Bobgunnia fistuloides (Harms) J.H. Kirkbr. & Wiersema	Cabinda; Malanje; Zaire	Nigeria, Cameroon, Gabon, Angola, D.R. Congo, Mozambique.	Least Concern
Entandrophragma angolense (Welw.) C. DC.	Cabinda; Cuanza Norte; Cuanza Sul; Malanje	From Guinea to Uganda, Kenya and Angola	Vulnerable A1cd
Entandrophragma candollei Harms	Cabinda	From the Ivory Coast to Angola and D.R. Congo	Vulnerable A1cd
Entandrophragma cylindricum (Sprague) Sprague	Cabinda	From Sierra Leone to Cabinda and Uganda	Vulnerable A1cd
Entandrophragma utile (Dawe & Sprague) Sprague	Cabinda	From the Ivory Coast to Angola, D.R. Congo and Uganda	Vulnerable A1cd
Gossweilerodendron balsamiferum (Vermoesen) Harms	Cabinda	From the south of Nigeria and Cameroon to D.R. Congo and Angola	Endangered A1cd
Guibourtia arnoldiana (De Wild. & T. Durand) J. Léonard	Cabinda; Zaire	Gabon, Congo, Angola (Cabinda), D. R. Congo (Maiombe)	Not Evaluated
Khaya ivorensis A. Chev.	Cabinda	From the Ivory Coast to Angola (Cabinda)	Vulnerable A1cd
Milicia excelsa (Welw.) C.C. Berg	Cabinda; Cuanza Norte	Widely distributed in Africa, from Senegal to Angola, D.R. Congo, East Africa and Mozambique	Near Threatened
Oxystigma oxyphyllum (Harms) J. Léonard	Cabinda	From Nigeria to Angola (Cabinda)	Not Evaluated
Terminalia superba Engl. & Diels	Cabinda	From Guinea to D.R. Congo (Maiombe)	Not Evaluated
Evergreen Forests and Woodlands			
Khaya anthotheca (Welw.) C. DC.	Bengo; Cuanza Norte; Malanje	From Sierra Leone to Uganda and Tanzania, and central Angola, Zambia, Malawi, Mozambique and Zimbabwe; also in the Ivory Coast, the Gold Coast, Nigeria and Cameroon.	Vulnerable A1cd
Pterocarpus tinctorius Welw.	Bengo; Cuanza Norte; Cuanza Sul; Luanda; Malanje; Zaire	Congo, Angola, Zambia, Zimbabwe, Tanzania and Mozambique	Not Evaluated
Dry Forests, Woodlands and Savannas			
Afzelia quanzensis Welw.	Benguela; Bie; Cuando Cubango; Cuanza Norte; Cuanza Sul; Cunene; Huila; Malanje; Namibe	In Angola, Namibia, D.R. Congo, Zambia, Zimbabwe and Botswana, and from Somalia to South Africa	Not Evaluated
Diospyros mespiliformis Hochst. ex A. DC.	Bengo; Cuando Cubango; Cuanza Norte; Cuanza Sul; Cunene; Huila; Luanda; Namibe	From Senegal to Sudan, and southwards to Namibia. It can be found from the mouth of the Zaire river to the Transvaal and South Mozambique, but not in the Guineo-Congolian rainforests	Not Evaluated
Entandrophragma spicatum (C. DC.) Sprague	Benguela; Cunene; Huila; Namibe	Southern Angola and Namibia	Not Evaluated
Guibourtia coleosperma (Benth.) J. Léonard	Bengo; Bie; Cuando Cubango; Cunene; Huambo; Huila; Lunda Norte; Lunda Sul; Moxico	D.R. Congo, Angola, Namibia, Botswana, Zambia, Zimbabwe	Not Evaluated
Pterocarpus angolensis DC.	Benguela; Bie; Cuando Cubango; Cuanza Norte; Cuanza Sul; Cunene; Huambo; Huila; Lunda Norte; Lunda Sul; Malanje; Moxico; Namibe; Uige	From Congo to Namibia and from Tanzania to Swaziland	Near Threatened

ecoregions. The mismatch was particularly notable in the case of the Cabinda forests, which have a very unique set of timber tree species that are not adequately captured by WWF ecoregions. In fact, although one of the two regions dominating the enclave of Cabinda also occurs in a larger area in the northwest of Angola (western Congolian forest-savanna mosaic), most timber species characteristic of the former region were not found elsewhere. Reasons for the mismatches are uncertain, but they are probably related to the operation of climatic and historical factors that are not adequately captured by the WWF ecoregion definitions. Irrespective of the reason, however, these results suggest that

WWF ecoregions may not provide an adequate operational basis for conservation planning exercises targeting timber tree species in Angola.

It is suggested that the biogeographic patterns observed in this study, might be better explained by the climatic classification of Köppen–Geiger [35], which suggests that the native range of these timber species is conditioned by large scale patterns. These regions are broadly similar to those recently proposed by Linder et al. [56], that demonstrate the existence of only seven well-defined and consistent biogeographical regions in sub-Saharan Africa, proposing that the best approach might be to recognize, as White [57]

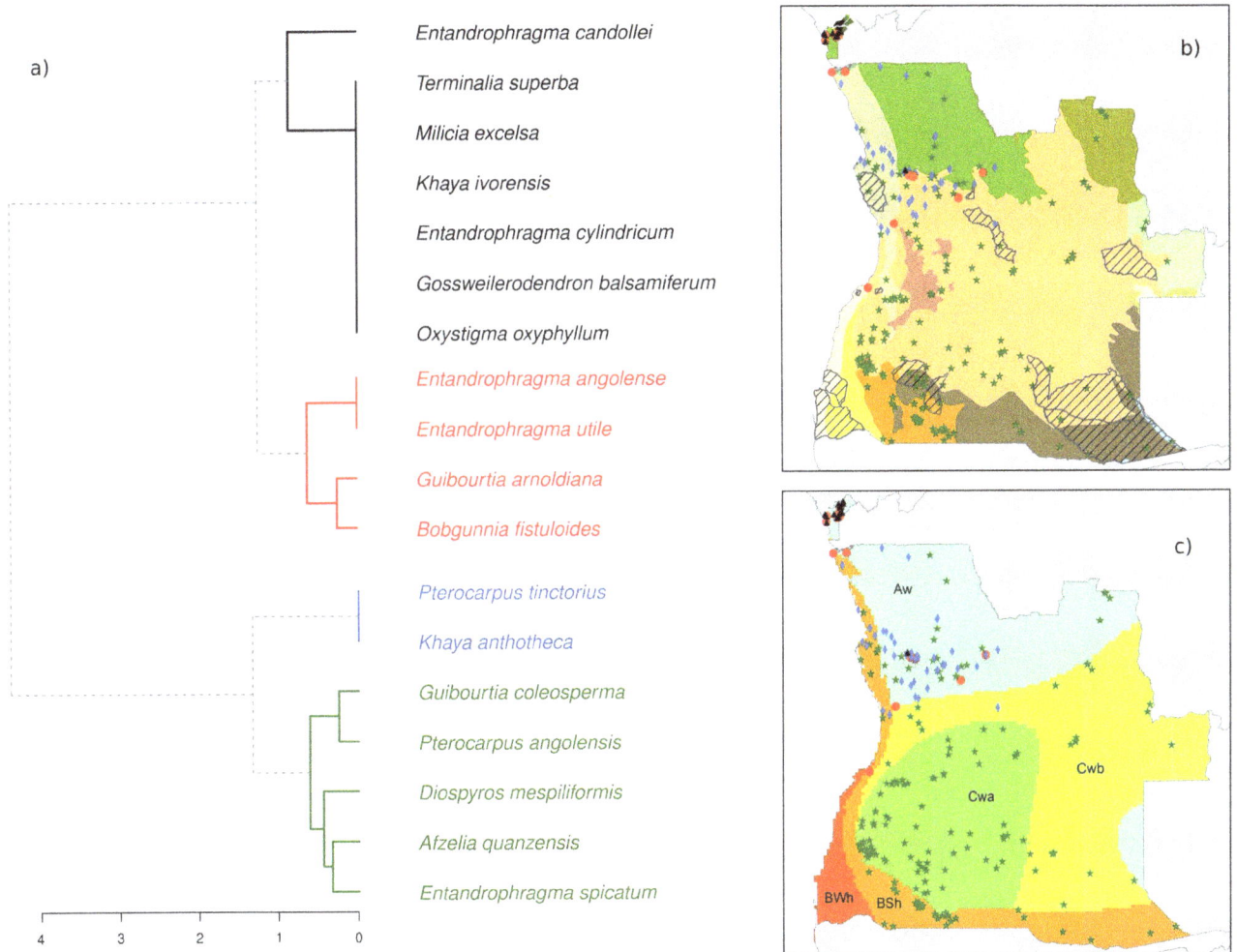

Figure 2. Biogeographical patterns of Angolan forest species selected for this study. a) Dendrogram of a cluster analysis based on the distribution (presence/absence) of timber tree species in each of the ecoregions. Distribution of collection localities of the four species groups identified in this study, in relation to: **b)** protected areas and the 15 WWF ecoregions represented in Angola; and **c)** Köppen–Geiger climate classification. Clustering was based on the Jaccard index of similarity and on the Ward agglomeration algorithm.

did, a small number of very broad biogeographical regions in Africa that can reflect the patterns found in both vertebrates and plants [56].

Species conservation priorities

Our results suggest that at least 11 of the species evaluated should be given high conservation priority in Angola. These include: (1) globally threatened or near-threatened species with small ranges in Angola and largely restricted to Cabinda or, to a much lesser extent, the north-western regions (*E. angolense, E. candollei, E. cylindricum, E. utile, G. balsamiferum, K. anthotheca, K. ivorensis* and *M. excelsa*); and (2) species from dry forests with a large proportion of their global range concentrated in Angola, and occurring in areas that are affected at present by high deforestation rates (*E. spicatum, G. coleosperma* and *P. angolensis*).

From the first group, all but two species (*E. angolense* and *K. anthoteca*) are concentrated in Cabinda's Maiombe forest, with a very small EOO in Angola, though they are widely distributed elsewhere in Guineo-Congolian rainforests. The herbarium collections studied of all these species were made in locations still retaining a relatively extensive forest cover (46.7 to 84.4%), and

where deforestation rates between 2000 and 2012 were lower than in other areas of Angola. The overlap between the EOO and protected areas for these species is negligible (<0.5%), except in the case of *G. balsamiferum* (10.5%). These species may thus remain largely unprotected in Angola despite the recent creation of the Maiombe National Park, which was specifically designed to protect the Cabinda's forest. This is worrying in view of ongoing logging activities in Cabinda, where these species may be under increasing pressure [25,58].

The second group of conservation priority species includes timber trees that are mainly found in dry forests, and that are particularly important from a conservation perspective because their Angolan range represents a large proportion (17.7–55.4%) of their global range. Although these species have a relatively large overlap between their EOO and protected areas in Angola (12.9–16.4%), they occur in areas where tree cover is among the lowest for timber trees in Angola (9.3–20.4%). Further, the collection localities of these species have suffered high recent deforestation rates, amounting to >10% in 12 years for *G. coleosperma* and *P. angolensis*. These species should therefore merit national conservation attention, though only *P. angolensis* has been considered

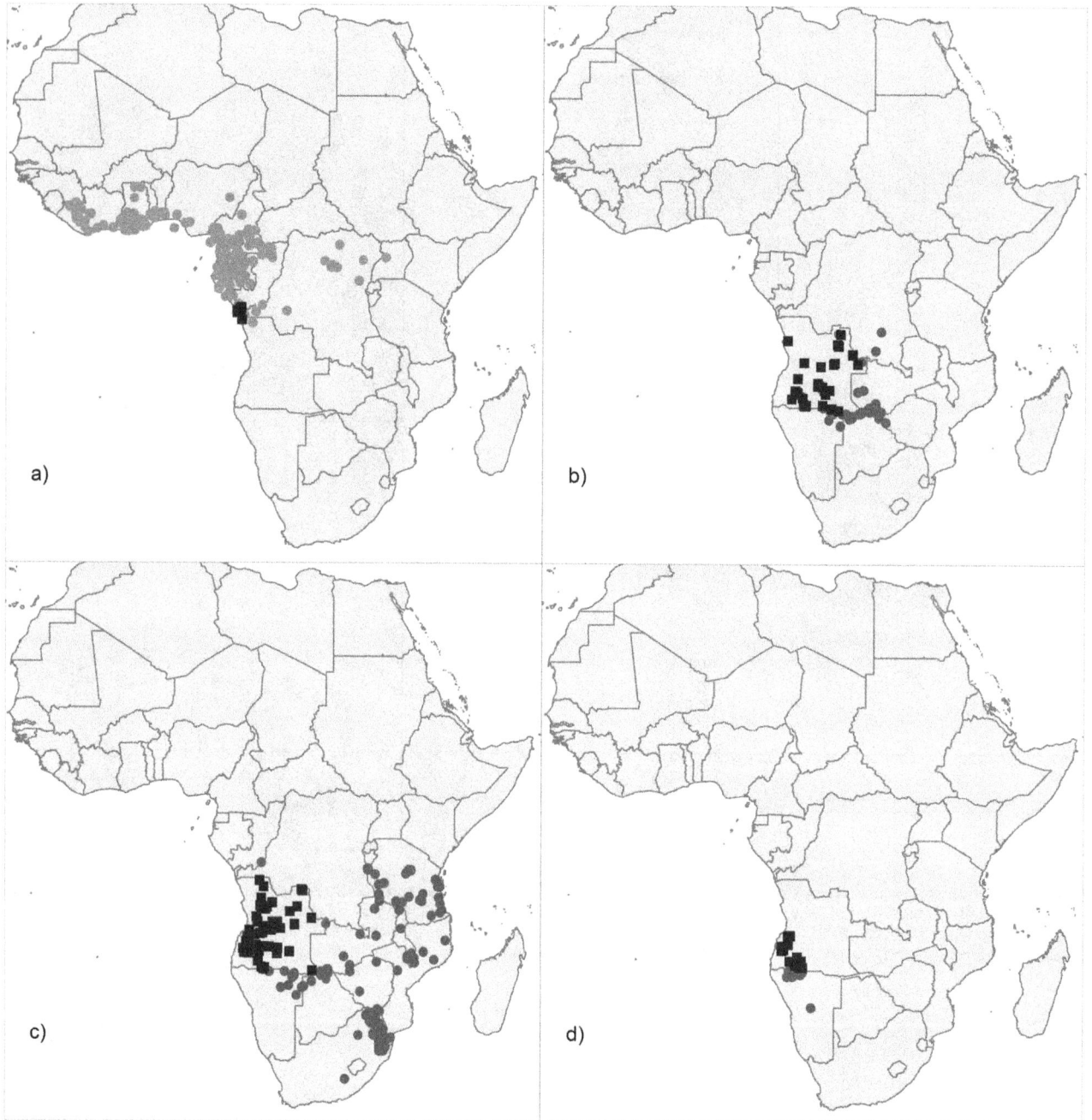

Figure 3. Geographic distribution in Africa of representative timber trees from Angola. (**a**) Species typical of Guineo-Congolian rainforests that reach their southern limit in Cabinda (*E. cylindricum; E. utile; G. arnoldiana; G. balsamiferum; K. ivorensis; O. oxyphyllum; T. superba*); and species with >15% of their global range concentrated in Angola, including (**b**) *G. coleosperma*, (**c**) *P. angolensis*, (**d**) *E. spicatum*. Black squares: studied specimens from Angola, housed in LISC, LISU, COI, BM and K; grey circles: data obtained via the GBIF portal.

near-threatened in IUCN global assessment [45], while the evaluation of *E. spicatum* and *G. coleosperma* is lacking.

The conservation priorities identified in this study are limited because herbarium data may not reflect current distribution, the ecological information for most species is scarce and there are virtually no data on present logging pressure. Notwithstanding, we believe that our approach provides a first approximation for timber tree species prioritization in Angola, which may be a useful guide to conservation decisions until more detailed information becomes available.

Conservation implications

Our study clearly underlines the need to take urgent action to protect the Cabinda's Maiombe forest, where there is a significant concentration of threatened timber species of high conservation priority. At present, these forests may be better conserved than similar Guineo-Congolian forests in neighbouring countries (see Fig. S2 in Supporting Information) where deforestation rates are high and concessions for industrial logging are expanding [27,62]. However, the Cabinda's forest may be under increasing legal and illegal logging pressure, though data to quantify this problem are

Table 2. Number of specimen collection localities and estimates of the global and national (Angola) extent of occurrence (EOO) for the selected timber tree species.

Species	Number of localities			Extent of Occurrence			%EOO	%EOO protected
	Global	Angola	Protected areas	Global ($\times 10^6$ km^2)	Angola ($\times 10^3$ km^2)	Protected areas ($\times 10^3$ km^2)		
Afzelia quanzensis	107	34	3	5.2	429.6	56.7	8.3	13.2
Bobgunnia fistuloides	47	5	1	0.8	0.2	0.0	0.0	0.0
Diospyros mespiliformis	272	37	3	12.4	464.5	96.5	3.8	20.8
Entandrophragma angolense	67	8	1	2.7	169.6	0.5	6.2	0.3
Entandrophragma candollei	42	1	1	1.6	a	a	a	a
Entandrophragma cylindricum	50	3	1	1.9	0.01	0.0	0.0	0.0
Entandrophragma spicatum	13	21	1	0.1	75.1	12.3	52.5	16.4
Entandrophragma utile	46	2	1	1.6	a	a	a	a
Gossweilerodendron balsamiferum	30	9	3	1.0	1.4	0.1	0.1	10.5
Guibourtia arnoldiana	10	5	0	0.1	10.1	0.0	11.2	0.0
Guibourtia coleosperma	36	32	5	1.3	697.0	89.6	55.4	12.9
Khaya anthotheca	76	8	0	6.0	58.4	0.0	1.0	0.0
Khaya ivorensis	49	5	0	1.0	0.6	0.0	0.1	0.0
Milicia excelsa	156	3	1	7.1	0.1	0.0	0.0	0.0
Oxystigma oxyphyllum	34	6	2	1.3	1.1	0.3	0.1	28.1
Pterocarpus angolensis	135	61	2	4.8	856.2	113.3	17.7	13.2
Pterocarpus tinctorius	45	38	1	2.1	158.6	10.3	7.5	6.5
Terminalia superba	98	6	2	2.0	0.7	0.1	0.0	14.3

% EOO is the percentage contribution of Angola to the global extent of occurrence; % EOO protected is the percentage of the EOO within Angola which is included in protected areas network.
[a]EOO not estimated due to insufficient data (≤2 locations).

Table 3. Forest cover (in 2000) and forest cover changes (2000–2012) estimated in 5-km buffers around the herbarium collection localities for each Angolan timber tree species.

Species	N	Tree Cover (%)	Forest Gain (%)	Forest Loss (%)	Deforested (%)
Afzelia quanzensis	6	23.8	0.1	2.3	9.6
Bobgunnia fistuloides	8	60.6	0.8	4.5	7.4
Diospyros mespiliformis	18	13.6	0.1	0.6	4.6
Entandrophragma angolense	9	62.1	0.9	5.5	8.9
Entandrophragma candollei	13	77.2	1.6	6.6	8.6
Entandrophragma cylindricum	10	68.6	1.4	3.8	5.6
Entandrophragma spicatum	11	9.3	0.0	0.8	8.2
Entandrophragma utile	12	46.7	1.1	3.5	7.6
Gossweilerodendron balsamiferum	1	69.0	0.6	2.5	3.6
Guibourtia arnoldiana	7	55.9	0.8	1.5	2.8
Guibourtia coleosperma	3	18.9	0.0	5.1	27.1
Khaya anthoteca	14	52.8	0.4	4.9	9.2
Khaya ivorensis	15	48.8	0.5	1.7	3.6
Milicia excelsa	16	84.4	0.9	4.2	5.0
Oxystigma oxyphyllum	2	76.8	0.7	2.8	3.6
Pterocarpus angolensis	4	20.4	0.0	3.0	14.7
Pterocarpus tinctorius	5	42.1	0.2	3.3	7.8
Terminalia superba	17	72.8	0.3	1.6	2.2

Tree cover, forest gain and forest loss are percentages expressed in relation to total buffer area. Deforestation rate is computed as the percentage of forest loss in 2012, in relation to total tree cover in 2000. Estimates were based on data extracted from Hansen et al. [27].

scarce [25]. According to Buza et al. [58], Cabinda is the largest producer of timber from Angola, being responsible for 33.9% of total timber exportations between 1990 and 1995; from 1996 to 2000 the external market consumed 85% of the logged timber. An important step towards the conservation of these forests has been the recent creation of the Maiombe National Park in a trans-frontier conservation area, as a result of an international cooperation between Angola, Congo and the Democratic Republic of the Congo [63]. Despite its value, however, this new protected area may provide an incomplete representation of priority timber species, as most of the historically known populations are located outside the Park boundaries. Eventual refinements to the limits of the Park may thus be desirable, which, together with prevention of illegal logging, could greatly assist in the protection of threatened timber tree species in Cabinda's Maiombe forest.

Urgent consideration should also be given to dry forests of Angola, where there are at least three timber species of high conservation priority, and where deforestation rates are increasing rapidly [27]. This is in line with the growing perception that tropical dry forests should be given high conservation priority, as they have a high biodiversity value in Sub-Saharan Africa [64,65]. The current exploitation of dry forest trees of high conservation priority is unknown in Angola, but in the past they were all valued timber species [43], and they are exploited elsewhere in Africa [24]. Although some of these species may be represented in protected areas in the south of Angola, the degree of on-the-ground protection that these areas presently afford is very uncertain. It is thus recommended that particular conservation attention should be given to timber trees from dry forests, in the context of ongoing efforts to strengthen the network of protected areas in Angola [25,33].

Lack of recent information is one of the key problems affecting biodiversity conservation in Angola. Shortage of data is probably more serious in Angola than in most places elsewhere in Africa, because of the prolonged war of independence (1961–1974) and post-independence civil war (1975–2002) which left the country largely inaccessible to most researchers. In these circumstances, studies such as the present one, based on accumulated historical information, may provide initial guidance on the identification of conservation priorities and problems, providing useful insights that would otherwise be very difficult to obtain [1,2,5,6,66]. However, it is now more than one decade after the end of the conflicts in Angola and, given the growing prosperity and development in the country, it is essential that new field surveys are undertaken to document contemporary species distributions and conservation challenges. In the particular case of tree species, it is also essential to collect quantitative information on species identity and places of origin of timber exports, which can then be used to guide new surveys and conservation assessments. Collecting this information would be essential to provide a solid basis for the conservation and sustainable use of forest resources in Angola.

Supporting Information

Figure S1 Temporal profile of herbarium records of the selected timber species of Angola. For each species, the temporal range of herbarium specimens housed in the selected herbaria is indicated (grey horizontal line).

Figure S2 Raster data of Maiombe forest cover in the lower Congo basin, showing Cabinda (Angola) and adjacent areas of Congo (upper) and Democratic Republic of the Congo (lower). The pin bullet indicates a transition where a change in the forest cover density across the border is identified, with the higher density being on the Cabinda side. Maps were produced using data available on-line from: http://earthenginepartners.appspot.com/science-2013-global-forest ([27] Hansen et al. 2013. High-Resolution Global Maps of 21st-Century Forest Cover Change. Science 342. 850–853).

Table S1 Georeferenced vouchers or bibliographic records for selected timber species in Angola. Data from the LISC Herbarium is available through GBIF at http://www.gbif.org/dataset/231c5bcf-1b56-4905-a398-6d0e18f6de1a.

Table S2 Sixty-two datasets from GBIF providers queried for the 18 timber species, producing 2253 records. Accessed 29 April 2013.

Table S3 Data provider per species accessed through GBIF data portal for each of the 18 timber species considered in this study. Accessed 29 April 2013.

Table S4 Characteristics of the 18 timber species studied, including information about their family, synonymy, common names, habit and ecology, and timber characteristics and uses.

Acknowledgments

We thank to the Academic Editor, Prof. Giovanni Vendramin, and to the reviewers (Prof. Berthold Heinze and one anonymous reviewer) for the valuable comments and suggestions, that improved the manuscript. We are also grateful to Patrícia Rodrigues for helping with GIS editing and Filipa Monteiro for technical assistance. We dedicate this paper to Prof. Maria Salomé Pais (Full Professor - BioFIG/FCUL) who made significant contributions to our understanding on Plant Biology and always encouraged the first author to work on the fascinating plant groups of tropical regions.

Author Contributions

Conceived and designed the experiments: MR. Analyzed the data: RF. Wrote the paper: MR RF MCD PB ID.

References

1. Krishtalka L, Humphrey PS (2000) Can natural history museums capture the future? BioScience 50: 611–617.
2. Rivers MC, Taylor L, Brummitt NA, Meagher TR, Roberts DL, et al. (2011) How many herbarium specimens are needed to detect threatened species? Biol Conserv 144: 2541–2547.
3. Brummitt N, Bachman S (2010) Plants under pressure a global assessment. The first report of the IUCN sampled red list index for plants. London: Natural History Museum. Available: http://www.kew.org/ucm/groups/public/documents/document/kppcont_027709.pdf. Accessed 2013 Dec 10.
4. Bebber DP, Carine MA, Wood JRI, Wortley AH, Harris DJ, et al. (2010) Herbaria are a major frontier for species discovery. Proc Natl Acad Sci USA 107: 22169–22171.
5. Costion CM, Liston J, Kitalong AH, Iida A, Lowe AJ (2012) Using the ancient past for establishing current threat in poorly inventoried regions. Biol Conserv 147: 153–162.
6. Lavoie C (2013) Biological collections in an ever changing world: herbaria as tools for biogeographical and environmental studies. Perspect Plant Ecol Evol Syst 15: 68–76.
7. Rands MR, Adams WM, Bennun L, Butchart SH, Clements A, et al. (2010) Biodiversity conservation: challenges beyond 2010. Science 329: 1298–1303.
8. Putz FE, Zuidema PA, Synnott T, Peña-Claros M, Pinard MA, et al. (2012) Sustaining conservation values in selectively logged tropical forests: the attained and the attainable. Conserv Lett 5: 296–303. http://dx.doi.org/10.1111/j.1755-263X.2012.00242.x.
9. FAO (2005) State of the World's forests 2005. Food and Agriculture Organization of the United Nations, Rome.
10. Balmford A, Bond W (2005) Trends in the state of nature and their implications for human well-being. Ecol Lett 8: 1218–1234.
11. Wallace KJ (2007) Classification of ecosystem services: problems and solutions. Biol Conserv 139: 235–246.
12. Giurca A, Jonsson R, Rinaldi F, Priyadi H (2013) Ambiguity in timber trade regarding efforts to combat illegal timber trade: potential impacts on trade between South-East Asia and Europe. Forests 4: 730–750.
13. Barrett MA, Brown JL, Morikawa MK, Labat JN, Yoder AD (2010) CITES designation for endangered rosewood in Madagascar. Science 328: 1109–1110.
14. Grogan J, Blundell AG, Landis RM, Youatt A, Gullison RE, et al. (2010) Over-harvesting driven by consumer demand leads to population decline: big-leaf mahogany in South America. Conserv Lett 3: 12–20.
15. Cerrillo RMN, Agote N, Pizarro F, Ceacero CJ, Palacios G (2013) Elements for a non-detriment finding of *Cedrela* spp. in Bolivia - A CITES implementation case study. J Nat Conserv 21: 241–252.
16. Pitman NC, Terborgh JW, Silman MR, Nunez PV, Neill DA, et al. (2001) Dominance and distribution of tree species in upper Amazonian terra firme forests. Ecology 82: 2101–2117.
17. Toledo M, Peña-Claros M, Bongers F, Alarcón A, Balcázar J, et al. (2012) Distribution patterns of tropical woody species in response to climatic and edaphic gradients. J Ecol 100: 253–263.
18. Lai J, Mi X, Ren H, Ma K (2009) Species-habitat associations change in a subtropical forest of China. J Veg Sci 20: 415–423.
19. Li L, Huang Z, Ye W, Cao H, Wei S, et al. (2009) Spatial distributions of tree species in a subtropical forest of China. Oikos 118: 495–50.
20. Schmitt CB, Denich M, Demissew S, Friis I, Boehmer HJ (2010) Floristic diversity in fragmented Afromontane rainforests: altitudinal variation and conservation importance. Appl Veg Sci 13: 291–304. doi: 10.1111/j.1654-109X.2009.01067.x.
21. Blom B, Cummins I, Ashton MS (2012) Large and intact forests: drivers and inhibitors of deforestation and forest degradation. In: Ashton MS, Tyrrell ML, Spalding D, Gentry B, editors. Managing forest carbon in a changing climate. Springer Press. pp. 285–304.
22. Dauby G, Hard OJ, Leal M, Breteler F, Stévart T (2014) Drivers of tree diversity in tropical rain forests: new insights from a comparison between littoral and hilly landscapes of Central Africa. J Biogeogr 41: 574–586. doi: 10.1111/jbi.12233.
23. Bodart C, Brink AB, Donnay F, Lupi A, Mayaux P, et al. (2013) Continental estimates of forest cover and forest cover changes in the dry ecosystems of Africa between 1990 and 2000. J Biogeogr 40: 1036–1047.
24. Chidumayo EN, Gumbo DJ (2010) The dry forests and woodlands of Africa: managing for products and services. London: Earthscan.
25. USAID (2008) 118/119 Biodiversity and tropical forest assessment for Angola. Biodiversity Analysis and Technical Support (BATS) Program. Washington, DC.
26. Blaser J, Sarre A, Poore D, Johnson S (2011) Status of tropical forest management 2011, ITTO Technical Series 38. International Tropical Timber Organization, Yokohama, Japan.
27. Hansen MC, Potapov PV, Moore R, Hancher M, Turubanova SA, et al. (2013) "High-Resolution Global Maps of 21st-Century Forest Cover Change." Science 342: 850–853.
28. Ferreirinha MP (1954) Notas sobre as madeiras do Ultramar - 1ª Série. Estudos e Informação 21. Lisboa: Direcção Geral dos Serviços Florestais e Aquícolas. 12 p.
29. Ferreirinha MP (1959) Madeiras do Ultramar português. Garcia de Orta 7: 363–365.
30. Ferreirinha MP (1962) Madeiras de Angola - 2ª Série. Garcia de Orta 10: 111–123.
31. Ferreirinha MP, Reis JEB (1969) Madeiras de Angola - 3ª Série. Garcia de Orta 17: 289–298.
32. Freitas MC (1961) Madeiras de Angola - 1ª Série. Garcia de Orta 9: 699–712.
33. Huntley BJ, Matos EM (1994) Botanical biodiversity and its conservation in Angola. Strelitzia 1: 53–74.
34. Olson DM, Dinerstein E, Wikramanayake ED, Burgess ND, Powell GV, et al. (2001) Terrestrial ecoregions of the world: a new map of life on earth. BioScience 51: 933–938.
35. Peel MC, Finlayson BL, McMahon TA (2007) Updated world map of the Köppen-Geiger climate classification, Hydrol. Earth Syst Sci 11: 1633–1644, doi:10.5194/hess-11-1633-2007.
36. Grandvaux-Barbosa LA (1970) Carta fitogeográfica de Angola. Luanda: Instituto de Investigação Científica de Angola. 323 p.
37. Exell AW, Mendonça FA (1951) Meliaceae. In: Exell AW, Mendonça FA, editors. Conspectus Florae Angolensis 1. Lisboa: Junta de Investigações do Ultramar. pp. 305–320.
38. Gossweiler J (1953) Nomes indígenas de plantas de Angola. Agronomia Angolana 7: 1–587.
39. Torre AR, Hillcoat D (1956) Caesalpinioideae. In: Exell AW, Mendonça FA, editors. Conspectus Florae Angolensis 2. Lisboa: Junta de Investigações do Ultramar. pp. 162–253.

40. Sousa EP (1966) Papilionoideae Tribo VIII-Dalbergieae. In: Exell AW, Fernandes A, editors. Conspectus Florae Angolensis 3. Lisboa: Junta de Investigações do Ultramar. pp. 344–372.

41. Exell AW, Garcia JG (1970) Combretaceae. In: Exell AW, Fernandes A, Mendes EJ, editors. Conspectus Florae Angolensis 4. Lisboa: Junta de Investigações do Ultramar. pp. 44–93.

42. Mambo A (1990) Taxonomia florestal de Cabinda. Técnicas de herbariologia. Lisboa: Instituto de Investigação Científica Tropical. 98 p.

43. Diniz AC (1991) Angola. O meio físico e potencialidades agrárias. Lisboa: Instituto para a Cooperação Económica. 189 p.

44. Barreto LS (1963) Madeiras ultramarinas. Lourenço Marques: Instituto de Investigação Científica de Moçambique. 52 p.

45. IUCN (2012) IUCN Red List of threatened species. Version 2012.2. Available: www.iucnredlist.org. Accessed 2013 Oct 20.

46. Straw HT (1956) Gazetteer n° 20 Angola. United States Board on Geographic Names. Washington. 234 p.

47. Environmental Systems Research Incorporated (2011) ArcGIS 10.0. Environmental Systems Research Incorporated, Redlands, CA. USA.

48. Ladle RJ, Whittaker RJ (editors) (2011) Conservation biogeography. Oxford: Wiley-Blackwell. 320 p.

49. Legendre P, Legendre L (1998) Numerical ecology, 2nd English edition. Amsterdam: Elsevier.

50. Salvador S, Chan P (2004) Determining the number of clusters/segments in hierarchical clustering/segmentation algorithms. Proceedings of the 16th IEEE – International Conference on Tools with Artificial Inteligence, pp. 576–584. Institute of Electrical and Electronics Engineers, Piscataway, NJ.

51. R Development Core Team (2013) R: A language and environment for statistical computing. R Foundation for Statistical Computing. Vienna, Austria. Available: http://www.R-project.org/

52. IUCN (2001) IUCN Red List Categories and Criteria: Version 3.1. IUCN Species Survival Commission. IUCN, Gland, Switzerland and Cambridge, UK. 30 p.

53. Bachman S, Moat J, Hill AW, de la Torre J, Scott B (2011) Supporting Red List threat assessments with GeoCAT: geospatial conservation assessment tool. In: Smith V, Penev L, editors. e-Infrastructures for data publishing in biodiversity science. ZooKeys 150: 117–126.

54. IUCN and UNEP-WCMC (2013) The world database on protected areas (WDPA) [On-line]. Cambridge, UK: UNEP- WCMC. Available: www. protectedplanet.net. Accessed 2013 Dec 1.

55. The Plant List (2010) Version 1. Published on the Internet; http://www. theplantlist.org/. Accessed 10 Dec 2013.

56. Linder HP, de Klerk HM, Born J, Burgess ND, Fjeldså J, et al. (2012) The partitioning of Africa: statistically defined biogeographical regions in sub-Saharan Africa. J Biogeogr 39: 1189–1205. doi: 10.1111/j.1365-2699.2012. 02728.x.

57. White F (1983) The vegetation of Africa. Paris: UNESCO. 356 p.

58. Buza AG, Tourinho MM, Silva JN (2006) Caracterização da colheita florestal em Cabinda, Angola. Rev Ciênc Agrár Belém 45: 59–78.

59. Ijang TP, Cleto N, Ewane NW, Chicaia A, Tamar R (2012) Transboundary dialogue and cooperation: first lessons from igniting negotiations on joint management of the Maiombe forest in the Congo Basin. Int J Agric For 2: 121–131.

60. Sexton JP, McIntyre PJ, Angert AL, Rice KJ (2009) Evolution and ecology of species range limits. Annu. Rev Ecol Evol Syst 40:415–36.

61. Hulme M, Doherty R, Ngara T, New M, Lister D (2001) African climate change: 1900–2100. Climate research 17: 145–168.

62. Laporte NT, Stabach JA, Grosch R, Lin TS, Goetz SJ (2007) Expansion of industrial logging in Central Africa. Science 316: 1451–1451.

63. Kuedikuenda S, Xavier MNG (2009) Framework report on Angola's biodiversity. Luanda: Republic of Angola, Ministry of Environment. 60 p.

64. Miles L, Newton AC, DeFries RS, Ravilious C, May I, et al. (2006) A global overview of the conservation status of tropical dry forests. J Biogeogr 33: 491–505.

65. Rudel TK (2013) The national determinants of deforestation in sub-Saharan Africa. Phil Trans R Soc B: 368(1625), 20120405.

66. Pyke GH, Ehrlich PR (2010) Biological collections and ecological/environmental research: a review, some observations and a look to the future. Biol Rev Camb Philos Soc 85: 247–266. doi: 10.1111/j.1469-185X.2009.00098.x.

14

Dating the End of the Greek Bronze Age: A Robust Radiocarbon-Based Chronology from Assiros Toumba

Kenneth Wardle[1]*, Thomas Higham[2], Bernd Kromer[3]

1 Department of Classics, Ancient History and Archaeology, University of Birmingham, Birmingham, United Kingdom, 2 Oxford Radiocarbon Accelerator Unit, Research Laboratory for Archaeology and the History of Art, University of Oxford, Oxford, United Kingdom, 3 Akademie der Wissenschaften Heidelberg, Heidelberg, Germany

Abstract

Over 60 recent analyses of animal bones, plant remains, and building timbers from Assiros in northern Greece form an unique series from the 14th to the 10th century BC. With the exception of Thera, the number of ^{14}C determinations from other Late Bronze Age sites in Greece has been small and their contribution to chronologies minimal. The absolute dates determined for Assiros through Bayesian modelling are both consistent and unexpected, since they are systematically earlier than the conventional chronologies of southern Greece by between 70 and 100 years. They have not been skewed by reference to assumed historical dates used as priors. They support high rather than low Iron Age chronologies from Spain to Israel where the merits of each are fiercely debated but remain unresolved.

Editor: John P. Hart, New York State Museum, United States of America

Funding: This work was supported by Natural and Environmental Research Council for a NERC Radiocarbon Fund Grant(NRCF 2010/1/2) to KAW (http://www.c14.org.uk/embed.php?File=index.html) and Institute for Aegean Prehistory to KAW (http://www.aegeanprehistory.net/). The funders had no role in study design, data collection and analysis, decision to publish, or preparation of the manuscript.

Competing Interests: The authors have declared that no competing interests exist.

* Email: k.a.wardle@bham.ac.uk

Introduction

Until very recently the chronology of the later part of the Aegean Bronze Age was entirely based on 'historical' dates derived from Egypt with the aid of exported or imported objects such as Minoan or Mycenaean pottery or Egyptian scarabs. Dates based on ^{14}C dating methods have had wide error margins and the complexities of the calibration curve for the final centuries of the second millennium BC preclude the precise dating of a single sample using ^{14}C techniques alone. Even where the samples and dating techniques are more varied, as in the case of the array of absolute dates determined for the Thera eruption, these have been viewed by some archaeologists with suspicion, particularly since they are offset from the conventional chronology by around 100 years and remain the subject of lively debate [1]. Recent analyses of material from Egypt have, however, confirmed that the Egyptian ^{14}C and historical chronologies are compatible and strengthen our conviction that the Thera ^{14}C dates are correct [2–5] Studies of material from Argos [6] and Aegina [7] in Greece and more widely in the Eastern Mediterranean [8] all lead to similar conclusions. In Greece substantial pieces of wood charcoal suitable for dendrochronological determination are exceptionally rare and it is not yet possible to link those available with the near-absolutely placed Anatolian conifer (core) sequence. Similarly, few sites have provided more than a handful of charcoal samples, usually far from ideal for dating purposes. Although the precision of ^{14}C measurements has improved steadily and Bayesian modelling has provided a powerful tool for the analysis of the results, these results can be no better than the quality of the samples available.

At Assiros in northern Greece [9] (Fig. 1), however, a combination of meticulous excavation, careful sample selection and good fortune has provided the first long, robust sequence of determinations from Greece for the later part of the Bronze Age and the start of the Iron Age. A large number of high precision ^{14}C determinations have been obtained from samples of three different types: charred building timbers, charred seeds and the collagen extracted from a stratified sequence of domestic animal bones (Section A in File S1). An uninterrupted stratigraphic sequence of building levels covers more than 400 years, while the preservation of substantial charred structural timbers from four phases has enabled precise dates to be established for the cutting of these timbers, using the technique of dendrochronological wiggle-matching (DWM). Quantities of crop seeds from a series of granaries have also been closely dated. Determinations of well-stratified animal bone samples representing every phase allowed us to test, through the application of Bayesian modelling techniques, whether the 'old wood effect', often cited as the reason for preferring historical dates to scientific ones based on wood charcoals, could be ruled out for the timbers at Assiros.

This series of dates from well-stratified long- and short-life samples from a single site, is unique in the Eastern Mediterranean and has radically improved our picture of ^{14}C-based chronology for the Greek Bronze Age. It is shown in Fig. 2 in diagrammatic form as summed probability distributions for each of the phases. It provides, for the first time, a sequence of absolute dates which are in no way mediated by reference to historical context or predicted duration of any phase. The robust nature of this sequence, offset from the existing conventional chronologies by between +100 and

Figure 1. Assiros Toumba: the site and its location. The 14 m high tell is formed from the debris of an unbroken, thousand year long sequence of building levels dating between c 2000 and 1000 BC.

+70 years, requires us to reconsider dates based on tenuous links with distant historical chronologies, especially for the Mycenaean and Proto-Geometric sequences (Section C in File S1). As at Thera, they call into question traditional assumptions about historical chronologies. They are especially important at the end of the Greek Bronze Age since they impact upon the vigorous debates surrounding the absolute dates of developments in Israel, in the circum-Alpine region and the Iberian peninsula [10,11] (Section E in File S1). In the same way a recently published Bayesian analysis of short life samples from Southern Italy helps to establish an absolute chronology for the Central Mediterranean Bronze Age independently of Aegean ceramic-based chronologies [12].

Materials

The site and the archaeological materials (Section B in File S1)

Assiros Toumba is a settlement mound situated at the NE end of the Langadas Basin, some 25 km inland from modern Thessaloniki. Its form and history is typical of the many such mounds to be found at intervals of 5–10 km in lowland Central Macedonia. Measuring 100×70 m at its base, it is of average size. The steep-sided profile, rising to 14 m above the surrounding area, is the result of the repeated re-construction of the substantial terrace banks around its perimeter which served both for defence and to support the buildings on the summit. First established c 2000 BC, the settlement appears to have been continuously occupied until early in the Iron Age (Phase 1.5) – perhaps until c 1000 BC on the basis of the ^{14}C determinations presented in this paper.

The successive buildings were of mud-brick framed with oak and suffered (fortunately, from the perspective of the archaeological chronologist) regular destruction as the result of fire, earthquake or natural decay, and the immediate or almost immediate reconstruction on the debris of the preceding phases steadily increased the height of the mound and required, in consequence, the raising of the perimeter banks. Within the settlement, it was possible to distinguish between interiors which were kept more or less clean, unroofed yards where rubbish was allowed to accumulate and streets where gravel and broken pottery was regularly strewn in order to maintain a firm surface in all weathers.

Excavation was directed from 1975–1989 by K.A. Wardle on behalf of the University of Birmingham and the British School at Athens. Ten separate phases of occupation (Phases 9–2, 1.5 and 1, see Figs. S3–S5 in File S1) have been explored in the centre of the mound to reach a depth of c 4.0 m from the surface. Earlier levels have been tested at the edge of the mound but could not be continued further in for practical reasons. A series of preliminary reports and studies of different aspects of the discoveries have

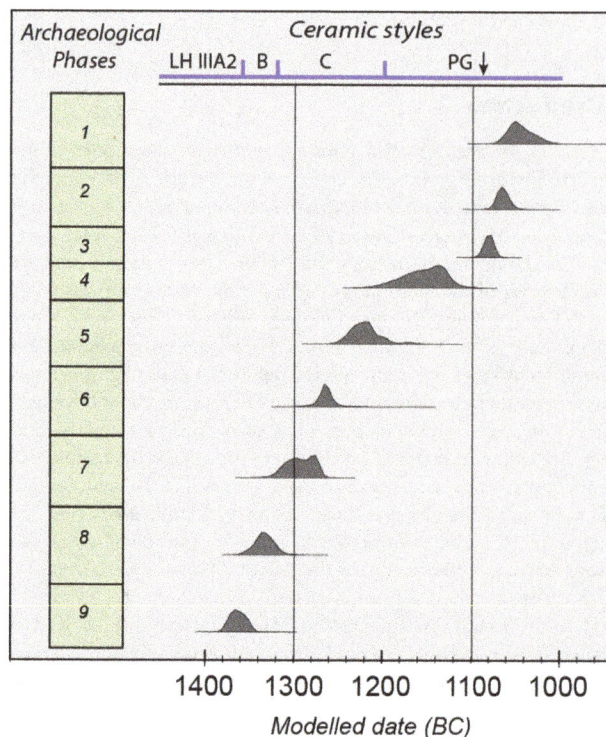

Figure 2. Modelled dates for each phase of the Assiros sequence. The date ranges are the summed probability distributions for each of the dated phases. These are shown for illustrative purposes only. See Figure S2 in File S1 for the full Bayesian model and Table S11 in File S1 for the model code.

already been published, while a comprehensive illustrated overview was published in 2007.

The regular fires had left charred timbers *in situ*, as well as charred crop seeds in several granaries. Animal bones of the principal domesticated species were present in every level of occupation debris. Mycenaean pottery occurred in sufficient quantity to permit absolute dates obtained at Assiros to inform southern Greek Late Bronze Age chronologies, whilst a distinctive Proto-Geometric amphora enables precise dating of the beginning of the Greek Early Iron Age for the first time.

Permits, sample identifiers and sample location

All necessary permits for excavation and permission to export samples for 14C determinations were obtained from the Greek Archaeological Service for the described study, which complied with all relevant regulations.

The index numbers of the archaeological samples which provide the basis for the determinations discussed in this paper are tabulated in Section A in File S1: the Ox-A numbers, analytical data, identification and context information for the animal bones in Table S1 in File S1; the sample numbers and context information for Dendrochronology and Dendrochronological wiggle-matching in Tables S2 and S3 in File S1; for crop seeds and their contexts in Table S4 in File S1.

All archaeological materials from the excavation at Assiros are housed in the Archaeological Museum of Thessaloniki, Greece (GR 54013). Dendrochronological samples are kept for reference in the Malcolm and Carolyn Wiener Laboratory for Aegean and Near Eastern Dendrochronology at Cornell University, Ithaca (NY 14853). ^{14}C samples for dendrochronological wiggle-matching and of crop seeds sent to Heidelberg for high precision determinations have been consumed in the analytical process (other crop seeds from the same contexts are stored in Thessaloniki). Animal bones used partially for sampling are currently held at ORAU Oxford, UK (OX1 3QY) pending eventual return to the Archaeological Museum of Thessaloniki.

The charred building timbers: Tables S2 and S3 in File S1

Charred timbers were preserved in the earliest level explored (Phase 9) and two ^{14}C determinations taken at a 40 year interval on a single timber with waney edge preserved gave a cutting date of 1385–1347 BC (2σ, 95.4% probability) (Fig. S6 in File S1). Three timbers found in the destruction level of Phase 6 were shown, using dendrochronology, to have been growing at the same time and one retained traces of bark. A sequence of five ^{14}C determinations covering a span of 50 years gave a cutting date of 1300–1260 BC once DWM was applied (Fig. 3). Four timbers growing at the same time and used in Phases 3 and 2 had been felled as two pairs ten years apart. Two retained their original circular cross section and are most unlikely to have been trimmed beyond the bark layer. A sequence of seven ^{14}C determinations covering a span of 90 years gave a cutting date of 1083–1062 BC for those used in Phase 2 [13] (Fig. 4). With all these determinations, we can be confident that the ring sequence from which the individual samples were taken reflects the full life span of the tree concerned and that the dates established are true cutting dates.

In addition, dates obtained for the 104 year ring sequence of the samples from Phases 3 and 2, using dendrochronology cross-matching against the Anatolian master sequence, provided close agreement with the ^{14}C dates but increased confidence depends on establishing Aegean oak-based sequences rather than Anatolian juniper-based ones [14].

All of these dates are between 100 and 70 years earlier than anticipated in terms of the accepted chronology. The archaeological evidence, especially in the case of the Phase 3 and Phase 2 samples, makes it unlikely that these are all reused timbers. The complete dating sequence, obtained by Bayesian modelling and including the animal bone determinations undertaken in 2011, enables us to rule out this possibility.

The charred crop seeds: Table S4 in File S1

The granaries of Phase 9 yielded a rich harvest of different crop seeds – einkorn, emmer and spelt wheat, barley, vetch and millet. Given the likelihood that these had recently been harvested and probably originated from a single year's harvest, seven discrete samples of four species were analysed to clarify the cutting dates of the associated timbers. The tight series of determinations provided an earliest date for the destruction, in which they were charred, of 1378–1343 and a latest of 1370–1334 BC. We measured the span of time represented by the samples using OxCal's interval command and determined that it corresponded to a period of 0–28 years (at 2σ, 95.4% probability) (Table S4 in File S1). The same procedure has been used wherever an interval has been calculated.

The animal bones and the Oxford Laboratory methods

42 animal bones (*Sus, Bos* and *Ovis*) were prepared for AMS dating at the ORAU, and 36 were successfully dated. All phases (apart from Phase 9), were represented by three or more samples. They were selected from occupation levels in every area of the site on the basis of their large size and the lack of post-discard gnawing. They had therefore been deposited while 'fresh' and are likely to be from animals raised for meat and under 5 years old.

Treatment was undertaken using the methods outlined by Brock *et al* [15]. Bone samples were dated with an additional ultrafiltration treatment using a pre-cleaned 30kD MWCO ultra-filter manufactured by Vivaspin^{TM}. The recovered filtered gelatin was freeze-dried ready for combustion in a CHN analyser. The sample CO_2 was reduced over an iron catalyst in an excess H_2 atmosphere at 560°C prior to AMS radiocarbon measurement using the ORAU 2.5MV HVEE accelerator. Radiocarbon dates of bone and their context information are reported in Table S1 in File S1. All bones were very well preserved in terms of collagen, with only one <than 1% wt. collagen (the effective threshold in the ORAU). All other analytical parameters measured, including the carbon to nitrogen atomic ratio, were acceptable.

δ^{15}N and δ^{13}C isotope values obtained during the preparation for ^{14}C determinations showed that the pigs and sheep and some of the cattle were consuming a normal C3 plant-based diet. The cattle from later levels (Phase 5 onwards), however, were almost all consuming plants derived via a C4 rather than the normal C3 pathway (Fig. 5) but the δ^{15}N values for these rule out grazing in a salt-marsh environment which was suggested for the animals sampled from the contemporary site of Kastanas 35 km away, and used *inter alia* to explain the unexpectedly 'old' dates at that site [16]. The most likely explanation seems to us to be that cattle at this period were being stall-fed with a crop such as millet, which is a C4 plant, on a very regular basis, since the natural grazing in the region could not have produced this effect. There is, moreover, no systematic discrepancy between the dates for the cattle which were eating a C4 plant diet and for the pigs in the same strata which were not, nor any evidence for reservoir offsets. Assiros is some 30 km from the nearest coast and salt-marsh grazing is, in any case, extremely unlikely.

OxCal v4.2.2 Bronk Ramsey (2013); r:5 Atmospheric data from Reimer et al (2009);

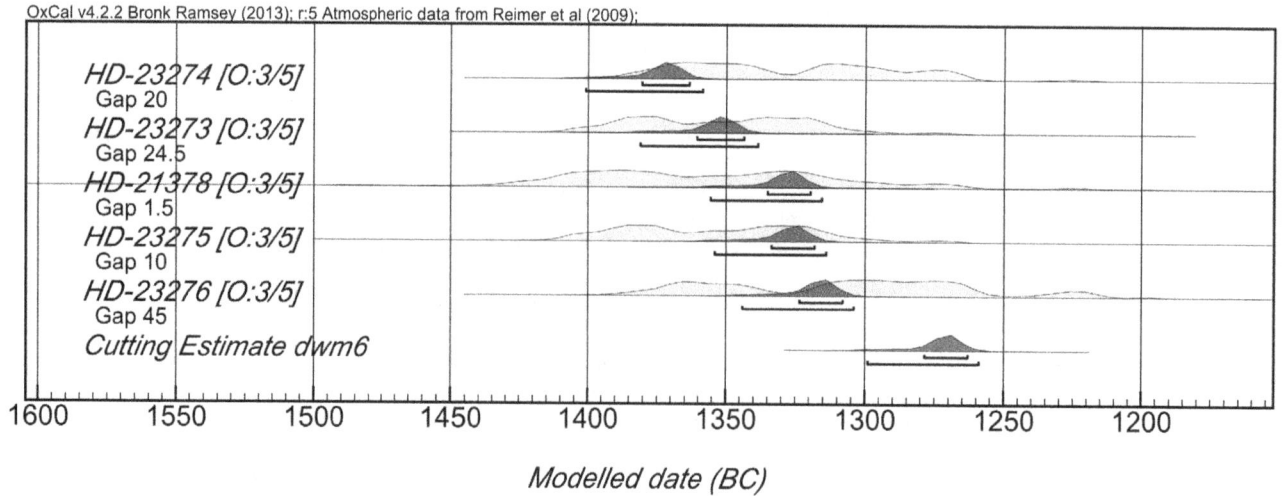

Figure 3. Dendro wiggle match diagram for Phase 6 timbers.

Method

The stratigraphic sequence, the Bayesian model and absolute dates

The stratigraphic sequence at Assiros was continuous, at least from Phases 10–1.5. Each phase was defined as a closed sequence of construction, use and destruction and, as far as can be determined, reconstruction took place shortly after destruction in almost every instance. The only clear hiatus occurred between the end of Phase 1.5 and the start of Phase 1. Phase 9 (Fig. S3 in File S1), the earliest from which samples were available, is dated absolutely by construction timbers, by animal bones from its use and by crop seeds from its destruction fire. Phases 8 and 7 are dated by the animal bones from the periods of use. Phase 6 (Fig. S4 in File S1) is dated by construction timbers, by animal bones from its use and by a single sample of crop seeds from its destruction

fire. Phases 5 and 4 are dated by animal bones from the period of use, whilst Phases 3 and 2 (Figs. S5 & S8 in File S1) are dated by both construction timbers and by animal bones from their use. No samples have been analysed from Phase 1.5, a short period of reoccupation after a fire, and the only samples from Phase 1 are bones. It is not possible to determine whether the animal bones date to the initial stage of use of any structure or indeed to its final phase of use. Given that the building phases at Assiros in most cases are of short duration this uncertainty normally falls within the range of the radiocarbon determinations.

The local Macedonian pottery, which forms more than 90% of any assemblage, is hand-made and changes only slowly in terms of shape and decoration. The end of the *Macedonian* Bronze Age and start of the Iron Age at Assiros is marked by the introduction in Phase 4 of thinner, harder-fired pottery with rather more angular shapes and a characteristic incised or stamped decoration,

OxCal v4.2.2 Bronk Ramsey (2013); r:5 Atmospheric data from Reimer et al (2009);

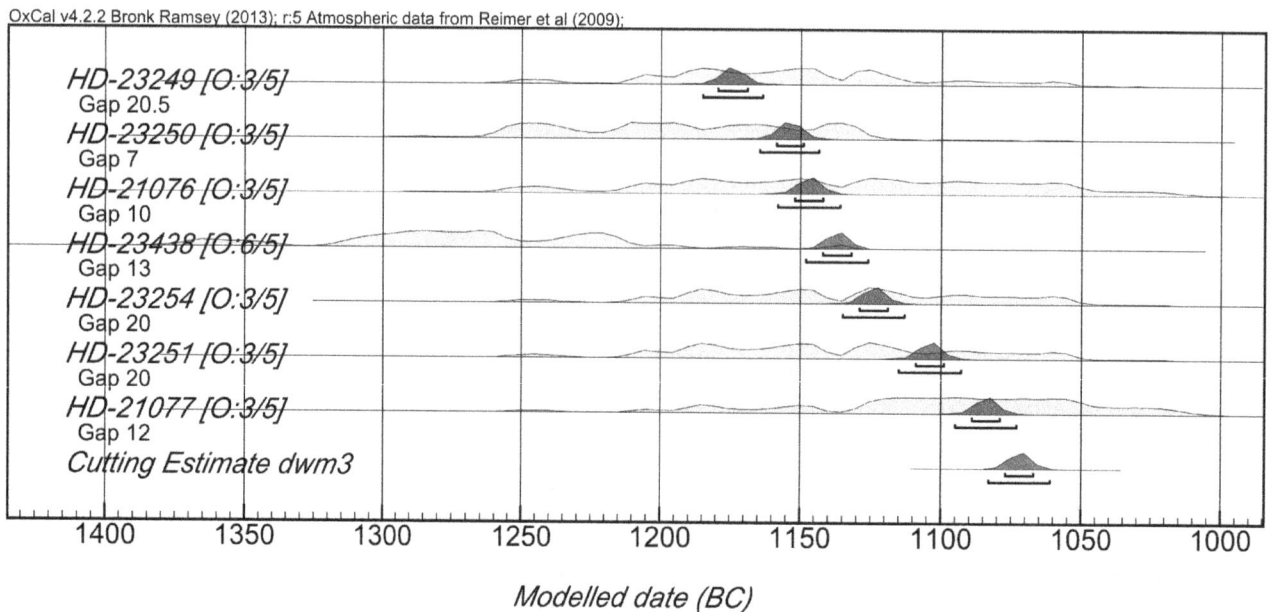

Figure 4. Dendro wiggle match diagram for Phase 3/2 timbers.

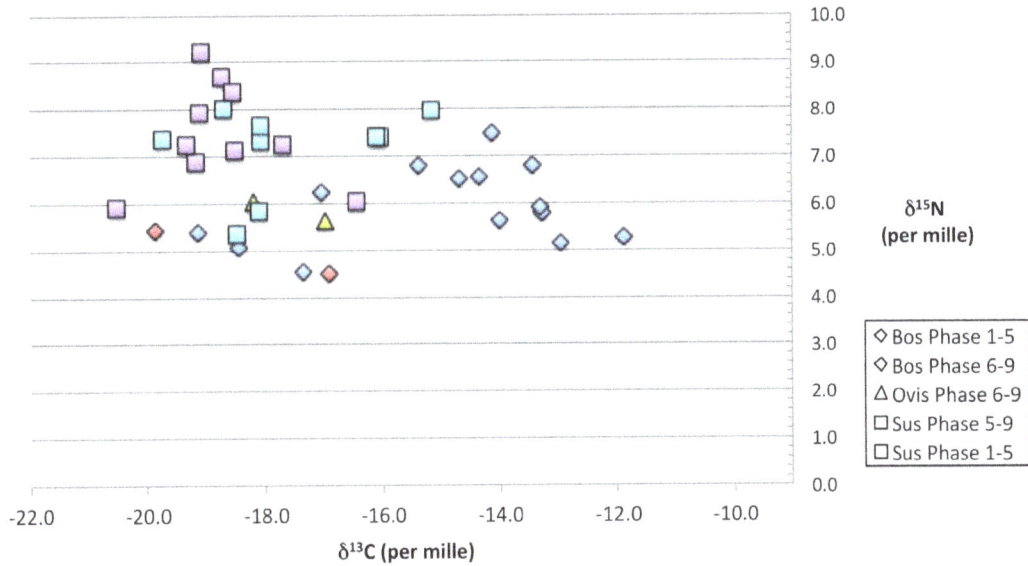

Figure 5. Stable isotope (C and N) values from Assiros. The values are divided between Phases 9–6 and 5–1 (see Table S1 in File S1 for values). Most cattle (*Bos*) from Phase 5 and later ate an atypical diet with high levels of C-4 plants. Errors for the isotope measurements are ±0.2 for Carbon and ±0.3 for Nitrogen. These are not shown on the figure.

but long-established local traditions clearly continue. Mycenaean pottery, which can be related to southern Greek sequences, is found in increasing quantities from Phase 10 onwards, with a maximum presence in Phase 6. At first the pottery was entirely imported to Macedonia but later 'locally-made' vessels became quite frequent. Phase 5 has far fewer pieces and no complete vessels, suggesting the possibility that much, if not all, the Mycenaean pottery from that phase is residual. Apart from a few worn Mycenaean sherds, there is no independently datable pottery in Phase 4. A single but highly significant Early Proto-Geometric amphora (Fig. 6) was broken in the Phase 3 destruction fire, whilst the pottery from Phases 2, 1.5 and 1 is local in character and not as yet precisely datable.

The end of the Bronze Age and start of the Iron Age in southern Greece is marked, among other characteristics, by the appearance of Proto-Geometric pottery, but cannot be directly related to the equivalent transition in Macedonia. At each of the sites where extensive excavation has taken place (Assiros, Kastanas and Toumba Thessalonikis) the local repertoire of fabrics and shapes is slightly different and direct comparisons between the assemblages at each of the sites are difficult. At Toumba Thessalonikis the distinctive features that make up the local Iron Age repertoire are

introduced gradually from Level 3 onwards and found with Late Helladic IIIC late pottery, whilst the concentric circle decoration on wheel-made pottery, which is the hallmark of the Proto-Geometric style, does not appear until Level 2A [17]. The relative chronologies of key levels at Assiros and Kastanas are set out in Table S10 in File S1 together with those from Toumba Thessalonikis. At Kastanas, where the first local Iron Age pottery is found in a single level (Schicht 12) together with Mycenaean and Proto-Geometric pottery, it cannot be determined whether local EIA pottery came into use during the same period as Mycenaean or at the same time as Proto-Geometric.

At Assiros, as already noted, imported pottery is absent from the relatively long Phase 4 and we cannot, therefore, determine on the basis of the pottery whether this phase at Assiros should be considered contemporary with the final Bronze Age (Late Helladic IIIC late) in southern Greece or with the initial stage of the Iron Age (Early Proto-Geometric). The local EIA pottery style was already well established by the time the EPG amphora, which was shattered in the Phase 3 destruction fire, reached Assiros. The position of this phase in relation to southern Greek sequences is important in considering the impact of the absolute dates obtained

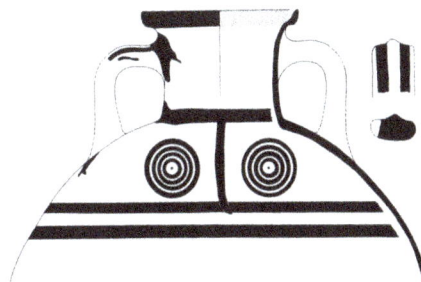

Figure 6. Assiros Proto-Geometric amphora from Phase 3. For find spots see Fig. S8 in File S1.

on any estimation of the absolute start date of Proto-Geometric (see discussion of results below).

All dated samples, from timbers, crop seeds and animal bones have been included in a Bayesian chronometric model (Section A in File S1, Fig. S2 in File S1, Tables S5 & S11 (the model code) in File S1) based on the stratigraphic sequence information described above.

We used OxCal 4.2.2 [18] and the INTCAL13 calibration curve [19] to achieve this. Bayesian modelling allows the archaeological stratigraphic information to be incorporated in the chronometric modelling, along with the radiocarbon likelihoods [20,21]. The model framework reflects the series of phases, destruction layers and archaeological strata excavated through the sequence of the Assiros tell. For each model a start and end boundary is included to bracket each archaeological phase throughout the sequence. Two are included where we need to account for the hiatus between Phases 1.5 and 1, otherwise they simply demarcate the separation of one phase from another. The boundary posterior distributions allow us to determine probability distribution functions (PDF) for the beginning and ending of these phases and also to interrogate information about the span of time that has elapsed between them. Within the sequence of phases we incorporated the dating evidence derived from high precision wiggle-matched radiocarbon sequences, dendro-dated calendar estimates, high precision radiocarbon dates of cereal grains and the AMS dates from identified domestic animal bones. The Bayesian model permits us to estimate the duration of each phase (Fig. S1 in File S1). This suggests that each building phase is marked by the passage of one or perhaps two human generations. The full model plot is shown in Fig. S2 in File S1.

We ran the model multiple times at several million iterations to assess reproducibility. Initial runs of the model were either slow or the MCMC stopped. To solve this problem we included several uniform boundaries with specific ranges, reducing the runtime for the algorithm to find a proper fit over a wider temporal range. The posterior data were very reproducible. Low convergence occurs when MCMC algorithms experience problems in calculating the solution, and incompatible solutions mean the algorithm is slow to calculate, or converge. The convergence values for Assiros models were consistently high and averaged 99.5%.

The Bayesian modelling shows that the animal bone determinations are consistent in terms of matching the cutting dates for the timbers from the same building phase (within the standard error values). The timbers must therefore have been freshly cut, not reused. The assertion by Weninger and Jung [22] 'that the beams do not originally stem from the architectural phases in which they were found stratified' is therefore entirely unfounded and can be definitively rejected. The modelled sequence permits a series of key transitions to be established.

In consequence, a long sequence of absolute dates has been obtained without any reference to 'historical' dates or any attempt to predict the length of any phase to constrain the modelling. In particular, the 36 domestic animal bone samples, formed a consistent sequence with only two outliers [23] (Table S6 in File S1). This pattern is statistically robust and is not the product of bones which had been moved around at random during successive phases of construction.

Results and Discussion of the Chronological Significance for Southern Greece and the Wider Mediterranean Region

Phase 9, with extensive granaries and imported pottery of the Mycenaean late LH IIIA 2 period, starts between 1395–1346 BC

and ends with destruction by fire 1378–1343 BC, a period of 0–19 years (at 2σ, 95.4% probability). Phase 7, with LH IIIC style pottery which should correspond to the period immediately after the destruction of palatial Mycenaean society in southern Greece, starts 1341–1282 BC and ends 1312–1264 BC with a span of 0–46 years (at 2σ, 95.4%). Phase 3, the second phase of the Iron Age at Assiros including a Proto-Geometric amphora (Fig. 6) starts 1096–1074 BC and ends 1087–1064 BC with an overall span of time of 0–15 years (at 2σ, 95.4%). Each of these dates is considerably older than expected on conventional grounds as shown in Table 1. There is no argument for rejecting this sequence in terms of old wood, old bones or special diets.

Given the clear sequence of absolute dates from Assiros and their importance, our next step was to compare them with the determinations obtained from Kastanas which have been used to support the conventional chronology (Section D in File S1, Figs. S9 & S10 in File S1). We reassessed both the stratigraphic evidence and the pattern of dates with a similar Bayesian model without any input from historical dates or hypothetical phase durations (Figs. S11 & S12 in File S1, Tables S9 & S12 (the model code) in File S1). The sequences resulting from modelling the ^{14}C data sets at both Assiros and Kastanas are reasonably consistent with each other, allowing for the difficulties in matching, through ceramic parallels, building levels at two sites excavated with different methods. Previous attempts to reject the self-evident offset at Kastanas from the expected historical dates rely on special pleading about the character of the Kastanas sample base rather than firm evidence [16] and, moreover, ignore some of the pottery evidence from each stratum (Section D in File S1, Tables S7 & S8 in File S1).

The ceramic parallels from Assiros provide relative chronological links between the ceramic phases in northern and southern Greece. In consequence, the absolute dates from the Assiros contexts can be transferred to equivalent contexts in southern Greece, and used to re-date the successive pottery phases there. Their significance is most readily seen in respect of the start date for the Proto-Geometric period, for which there is no evidence-based historical chronology but rather a 'best guess' of 1050 or 1025 BC. This has been determined on the hypothetical duration of the pottery styles of the final Mycenaean period and on the occurrence of Proto-Geometric pottery in the coastal regions of Syria, Lebanon and Israel (Fig. S7 in File S1). There is, however, no agreement about the dating of the levels where this pottery has been found but rather a vigorous debate about the dates in relation to Biblical history. The discussion of a high or low chronology in this region has recently had particular prominence [10,11]. Indeed the presence of Proto-Geometric pottery has often been used to support dates in both schemes without regard for circularity of argument.

The Assiros Proto-Geometric amphora (Fig. 6) belongs to a category discovered at other Macedonian sites, at Troy, in central Greece and at Lefkandi on Euboea (Fig. S7 in File S1). This category is generally assumed to have derived from Attic prototypes after the passage of some years [24]. There is surely no doubt that the Proto-Geometric style had evolved before the amphora at Assiros was manufactured, unless the style originated with this vessel, which would be surprising. Our model provides an estimate for the age of the amphora in Phase 3, and consequently a minimum age for the start of the Proto-Geometric style, of 1095–1070 BC (at 2σ, 95.4% probability). Given this estimate for the Assiros example, it should follow that Attic Proto-Geometric, as the archetype of the whole style, originated before then, perhaps by several decades. This might indicate a date nearer 1100 BC for the origin of the style if, as is normally accepted, the style was first

Table 1. Modelled dates from Assiros: offsets from conventional dates.

Phase	Pottery Period	Expected/conventional date	Absolute/^{14}C date BC (at 2σ, 95.4%)	
			Start	End
1		?? 750–650	1072–1024	1067–1004
Hiatus				
2		950–900	1081–1056	1072–1024
3	Early Proto-Geometric	1000–950	1096–1074	1087–1065
4	**	1050–1000	1232–1145	1140–1078
5	Late Helladic IIIC	1100–1050	1265–1203	1232–1145
6	Late Helladic IIIC	1150–1100	1300–1253	1265–1203
7	Late Helladic IIIC	1200–1150	1341–1282	1312–1264
8	Late Helladic IIIB	1300–1200	1370–1334	1341–1282
9	Late Helladic IIIA2	1350–1300	1395–1346	1378–1343

**See discussion above for the position of Assiros Phase 4 relative to southern Greek sequences.

developed in Athens and was then spread by exports and imitation to the northern Aegean and to Cyprus and the Near East to reach that region at approximately the same time as it reached Macedonia.

Although Weninger and Jung question the attribution of the Proto-Geometric vessel to Assiros Phase 3 and accordingly the date deduced for its manufacture, the majority of the fragments came from Phase 3 floors and a single piece was incorporated into a wall when the reconstruction defined as Phase 2 took place (Fig. S8 in File S1). The assertion by Weninger and Jung that these circumstances result from complex reverse taphonomic processes is therefore untenable [22]. Since the amphora must have been broken in the Phase 3 destruction, before the Phase 2 structures were erected, the *terminus ante quem* of 1083–1062 BC for its manufacture, given by the DMW date for the Phase 2 construction timbers, remains valid.

The absolute date for the start of the preceding Phase 4 (Fig. S13 in File S1), which has some of the deepest deposits of occupation debris at the site, cannot be directly related to the southern Greek sequences since the level contains neither imports nor imitations of southern Greek wheel-made pottery (Mycenaean or Proto-Geometric) as noted above. There is, as yet, no equivalent absolute date from a southern Greek site to which it (or, indeed, any of the other absolute dates from Assiros) can be correlated. If, as seems most probable, Assiros Phase 4 is contemporary with the final Bronze Age of southern Greece, then the absolute date for the start of Early Proto-Geometric should be based on the date obtained for Phase 3, ie earlier than c 1080 BC. If, however, Phase 4 is contemporary with the early stages of Proto-Geometric, an earlier, perhaps much earlier, start date would be indicated, at least c 1120 BC.

The date for the introduction of Proto-Geometric derived from the finds at Assiros fits well with the high Levantine chronology whilst the conventional duration of Attic Proto-Geometric between 1025 and 900 BC, is regularly used as support for the low chronology. Although numerous ^{14}C samples from Israel have now been processed with the goal of establishing an absolute chronology for that region, their significance is hotly debated [10,11].

We learnt, in the final stages of preparation of this paper, of 15 new radiocarbon determinations from Lefkandi, Kalapodi and Corinth [25] which have been modelled using Bayesian analysis

and purport to show that they are compatible with the conventional 'low' chronology for LH IIIC - Proto-Geometric. There are, however, several grounds for not accepting this conclusion which we will explore in detail in a later paper. In summary: a) only a small number of samples have been measured from each of the three sites; b) these samples are related stylistically not stratigraphically and it would therefore have been a better approach to model the data from each site separately before combining the results; c) the error margins of the determinations are around 50 yrs +/-, and the 95.4% probability spans range from 117 yrs to 352 yrs with a mean of 251 yrs; d) some of the samples show such poor agreement that the authors have 'moved' two samples to where they fit better; e) they reject the dates provided by 'only two' samples for the early stages of Geometric as too early but accept the single 'late' date for the Early – Middle stage of LH IIIC without comment. Close examination of their data suggests that their assertion that the results support the conventional chronology for the LH IIIC to Proto-Geometric periods is over-optimistic.

Conclusions

Although the rationale of the conventional dates currently used for the later phases of the Greek Bronze Age has been set out in detail by Warren and Hankey [26], Weninger and Jung [16] and others, the fact remains that the dates from Assiros and Kastanas are systematically offset from these to approximately the same value as those from Thera at the beginning of the Late Bronze Age. It may reasonably be asked how these discrepancies arise. Can the Assiros dates be reconciled with historical dates by careful re-examination of the links between different areas at different periods? If not, which chronology should be preferred and why? To give priority to the historical dates is to challenge 60 years of research into, and improvement of, the ^{14}C methodology and the development of a series of accurate calibration curves. To give priority to the ^{14}C dates calls into question much of the conventional historical chronology for this period in the Eastern Mediterranean.

Recent studies of the dating of the Egyptian Old to New Kingdoms have demonstrated that historical dates and ^{14}C-derived dates are compatible where reliable samples are selected and the correct methodologies applied [2–4]. The consistency of the results relating to the Thera eruption demonstrates the importance of a range of different sample types, although ideally they would have had chronological depth as well as geographical breadth. The exceptionally robust character of the Assiros sequence is based on 1) the length of the stratified sequence, 2) the number and variety of samples, 3) the accuracy of DWM as applied to building timbers and 4) on the confirmation from the animal bones that the timbers are not reused. It thus provides a series of anchor points for future ^{14}C and dendrochronological studies in the Aegean area and challenges long-established assumptions about historical chronology in the region.

Supporting Information

File S1 Supporting figures and tables. Figure S1. Intervals calculated for each Assiros phase. **Figure S2.** Bayesian model for Assiros; *left* Phases 9–6, *right* Phases 5–1. **Figure S3.** Assiros Phase 9 with granaries. **Figure S4.** Assiros Phase 6 plan. **Figure S5.** Plan of village in Phase 2. **Figure S6.** Dendro wiggle match diagram for Phase 9 timber. **Figure S7.** *Left:* Distribution of Group 1 Proto-Geometric amphorae *after* Catling 1998, 156. *Right:* Distribution of Proto-Geometric pottery in the Levant *after* Lemos 2002, 229, Map 8. **Figure S8.** Find spots diagram for pieces of the PG amphora (green) and the 4 timbers used for dendrochronology and the DWM (red). **Figure S9.** In the Kastanas plot preferred by Weninger and Jung, the determinations are given their expected historical ages – resulting in a systematic offset from the calibration curve of around 100 years. **Figure S10.** In the Kastanas plot rejected by Weninger and Jung, the dates fit with reasonable conformity to the calibration curve. **Figure S11.** Bayesian model of Kastanas ^{14}C data, without prior assumptions about period length or assumed 'historical' date. **Figure S12.** Kastanas: The start and end dates for stratum/Schicht 14b as derived from the Bayesian model. **Figure S13.** The start dates of Assiros Phase 4 and Kastanas Schicht 12 compared. **Table S1.** Radiocarbon AMS dates from Assiros obtained from bones with associated analytical and context data. **Table S2.** Oak timbers from Assiros used for dendrochronology. **Table S3.** Oak timbers from Assiros used for dendrochronological wiggle-matching (DWM). **Table S4.** Crop seeds from Assiros Phase 9 Granary Room 9, and pithos in Phase 6 Room 20. **Table S5.** Assiros: Results of the Bayesian modelling. **Table S6.** Assiros: Results of the Bayesian outlier modelling. **Table S7.** Kastanas: summary of date on basis of Jung's rejection of the LH IIIC Late

parallels for catalogue number 91. **Table S8.** Kastanas: summary of date if LH IIIC Late parallels for catalogue number 91 are accepted. **Table S9.** Kastanas: The ^{14}C derived date ranges for the samples from each stratum/Schicht plotted against the destruction dates for each level. **Table S10.** The relative chronology of key building levels at Assiros, Kastanas and Toumba Thessalonikis. **Table S11.** Appendix 1: Model code for the Bayesian age model of Assiros. **Table S12.** Appendix 2: Model code for the Bayesian age model of Kastanas.

Acknowledgments

Without the many members of the Assiros excavation team and the efforts of the local workmen, none of the evidence presented here, based on careful collection and recording of samples throughout the excavation, would have come to light. We would like to thank Richard Hubbard, Jill Carington Smith, Diana Wardle and Nicola Wardle for help with selecting the samples, Maryanne Newton and Peter Kuniholm (*Cornell*) for the dendrochronology determinations. We acknowledge the careful and reliable work undertaken by all of the members of the Oxford Radiocarbon Accelerator Unit (ORAU) and by Sahra Talamo and other members of the Radiocarbon Laboratory at the University of Heidelberg. We are grateful to the late David Smyth and to Diana Wardle for the preparation of illustrative material, to Sturt Manning for comments on the first draft text and to our anonymous reviewer who made many helpful suggestions for improvement, including pointing us to the recent publication by Alberti on Italian dating independent of Aegean chronologies and asking us to clarify, if possible, the position of Assiros Phase 4 in relation to the Proto-Geometric period. Stelios Andreou of the Aristotelian University of Thessaloniki kindly provided unpublished information about the sequence of levels at Toumba Thessalonikis and the place in them of both Proto-Geometric and local Iron Age pottery types.

We are very grateful to the United Kingdom Natural and Environmental Research Council for a NERC Radiocarbon Fund (NRCF 2010/1/2) grant to enable the analysis of 46 animal bone samples at ORAU and to The Institute for Aegean Prehistory, The British School at Athens and the Greek Archaeological Service for supporting this phase of the research at Assiros.

Author Contributions

Conceived and designed the experiments: KW TH BK. Performed the experiments: KW TH BK. Analyzed the data: KW TH BK. Contributed reagents/materials/analysis tools: KW TH BK. Contributed to the writing of the manuscript: KW TH BK.

References

1. Cherubini P, Humbel T, Beeckman H, Gärtner H, Mannes D, et al. (2014) Bronze Age catastrophe and modern controversy: dating the Santorini eruption, Antiquity 88, 267–291.
2. Friedrich WL, Kromer B, Friedrich M, Heinemeier J, Pfeiffer T, et al. (2006) Santorini Eruption Radiocarbon Dated to 1627–1600 BC. Science 312, 548.
3. Manning SW, Bronk Ramsey C, Kutschera W, Higham T, Kromer B, et al. (2006) Chronology for the Aegean Late Bronze Age 1700–1400 BC. Science 312, 565–569.
4. Bronk Ramsey C, Dee MW, Rowland J, Higham T, Harris SA, et al. (2010) Radiocarbon-Based Chronology for Dynastic Egypt, Science 328, 1554–1557.
5. Dee MW, Brock F, Harris SA, Bronk Ramsey C, Shortland AJ, et al. (2010) Investigating the likelihood of a reservoir offset in the radiocarbon record for ancient Egypt, Journal of Archaeological Science 37, 687–693.
6. Voutsaki S, Nijboer AJ, Philippa-Touchais A, Touchais G, Triantaphyllou S (2006) Analyses of Middle Helladic Skeletal Material from Aspis, Argos. Bulletin de Correspondance Hellénique 130, 613–625.
7. Wild EM, Gauß W, Forstenpointner G, Lindblom M, Smetana R, et al. (2010) ^{14}C dating of the Early to Late Bronze Age stratigraphic sequence of Aegina Kolonna, Greece. Nuclear Instruments and Methods in Physics Research B 268, 1013–1021.
8. Manning SW, Kromer B (2011) Radiocarbon dating archaeological samples in the Eastern Mediterranean, 1730 to 1480 BC: further exploring the atmospheric radiocarbon calibration record and the archaeological implications. Archaeometry 2, 413–439.
9. Wardle KA, Wardle D (2007) Assiros Toumba: A brief history of the settlement, in The Struma/Strymon River Valley in Prehistory. Proceedings of the International Symposium Strymon Praehistoricus, Kjustendil–Blagoevgrad (Bulgaria), Serres–Amphipolis (Greece) 27.09–01.10.2004. (In The Steps of James Harvey Gaul Volume 2) eds. Todorova H, Stefanovich M, Ivanov G, Sofia, 451–479.
10. Bruins HJ, Nijboer A, Van der Plicht J (2011) Iron Age Mediterranean Chronology: A Reply, Radiocarbon 53(1), 199–220.
11. Fantalkin A, Finkelstein I, Piasetzky E (2011) Iron Age Mediterranean Chronology: A Rejoinder, Radiocarbon 53(1), 179–198.
12. Alberti G (2013) Issues in the absolute chronology of the Early-Middle Bronze Age transition in Sicily and southern Italy: a Bayesian radiocarbon view, Journal of Quaternary Science 28(6) 630–640.
13. Newton MW, Wardle KA, Kuniholm PI (2005) A Dendrochronological [14]C Wiggle-Match for the Early Iron Age of north Greece: A contribution to the debate about this period in the Southern Levant, in Levy and Higham eds. The Bible and Radiocarbon Dating: Archaeology, Text and Science, 104–113.
14. Newton MW, Wardle KA, Kuniholm PI (2005) Dendrochronology and Radiocarbon determinations from Assiros and the beginning of the Greek Iron Age, Archaiologikon Ergon Makedonias kai Thrakis 17 (2003), 173–190.
15. Brock F, Higham T, Ditchfield P, Bronk Ramsey C (2010) Current Pretreatment Methods for AMS Radiocarbon Dating at the Oxford Radiocarbon Accelerator Unit (ORAU). Radiocarbon 52(1), 103–112.
16. Weninger B, Jung R (2009) Absolute chronology of the end of the Aegean Bronze Age, LH IIIC Chronology and Synchronisms III: LH IIIC Late and the

Transition to the Early Iron Age. Proceedings of the International Workshop at the Austrian Academy of Sciences at Vienna, February 23rd and 24th, 2007, Veröffentlichungen der Mykenischen Kommission, Band 30, eds. Deger-Jalkotzy S. and Baechle AE, Vienna, 373–416.

17. Andreou S (2014) Personal communication.

18. Bronk Ramsey C (2001) Development of the radiocarbon calibration program OxCal, Radiocarbon 43, 355–363.

19. Reimer PJ, Bard E, Bayliss A, Beck JW, Blackwell PG, et al. (2013) IntCal13 and Marine13 Radiocarbon Age Calibration Curves 0–50,000 Years cal BP. Radiocarbon 55(4), 1869–1887.

20. Bronk Ramsey C (2009) Bayesian analysis of radiocarbon dates, Radiocarbon 51(1), 337–360.

21. Buck CE, Cavanagh WG, Litton CD (1996) Bayesian approach to interpreting archaeological data. John Wiley and Sons, Chichester.

22. Weninger B, Jung R (2009) Absolute chronology of the end of the Aegean Bronze Age, LH IIIC Chronology and Synchronisms III: LH IIIC Late and the Transition to the Early Iron Age. Proceedings of the International Workshop at the Austrian Academy of Sciences at Vienna, February 23rd and 24th, 2007, Veröffentlichungen der Mykenischen Kommission, Band 30, eds. Deger-Jalkotzy S. and Baechle AE, Vienna, 388.

23. Bronk Ramsey C (2009) Dealing with outliers and offsets in radiocarbon dating. Radiocarbon, 51(3), 1023–1045.

24. Catling RWV (1998) The Typology of the Protogeometric and Subgeometric pottery from Troia and its Aegean context, Studia Troica 8, 153–164.

25. Toffolo MB, Fantalkin A, Lemos IS, Felsch RCS, Niemeier W-D, et al. (2013) Towards an Absolute Chronology for the Aegean Iron Age: New Radiocarbon Dates from Lefkandi, Kalapodi and Corinth, www.plosone.org/article/authors/info%3Adoi%2F10.1371%2Fjournal.pone.0083117;

26. Warren PM, Hankey V (1989) Aegean Bronze Age Chronology, Bristol.

15

Evaluation of Methods to Estimate Understory Fruit Biomass

Marcus A. Lashley[1][*][¤], Jeffrey R. Thompson[2], M. Colter Chitwood[1], Christopher S. DePerno[1], Christopher E. Moorman[1]

1 Department of Forestry and Environmental Resources, North Carolina State University, Raleigh, North Carolina, United States of America, 2 SAS Institute Inc., Cary, North Carolina, United States of America

Abstract

Fleshy fruit is consumed by many wildlife species and is a critical component of forest ecosystems. Because fruit production may change quickly during forest succession, frequent monitoring of fruit biomass may be needed to better understand shifts in wildlife habitat quality. Yet, designing a fruit sampling protocol that is executable on a frequent basis may be difficult, and knowledge of accuracy within monitoring protocols is lacking. We evaluated the accuracy and efficiency of 3 methods to estimate understory fruit biomass (Fruit Count, Stem Density, and Plant Coverage). The Fruit Count method requires visual counts of fruit to estimate fruit biomass. The Stem Density method uses counts of all stems of fruit producing species to estimate fruit biomass. The Plant Coverage method uses land coverage of fruit producing species to estimate fruit biomass. Using linear regression models under a censored-normal distribution, we determined the Fruit Count and Stem Density methods could accurately estimate fruit biomass; however, when comparing AIC values between models, the Fruit Count method was the superior method for estimating fruit biomass. After determining that Fruit Count was the superior method to accurately estimate fruit biomass, we conducted additional analyses to determine the sampling intensity (i.e., percentage of area) necessary to accurately estimate fruit biomass. The Fruit Count method accurately estimated fruit biomass at a 0.8% sampling intensity. In some cases, sampling 0.8% of an area may not be feasible. In these cases, we suggest sampling understory fruit production with the Fruit Count method at the greatest feasible sampling intensity, which could be valuable to assess annual fluctuations in fruit production.

Editor: Nina Farwig, University of Marburg, Germany

Funding: We thank the United States Department of Defense and Fort Bragg Military Installation for financial contributions to this research. The funders had no role in study design, data collection and analysis, decision to publish, or preparation of the manuscript.

* E-mail: marcus_lashley@ncsu.edu

¤ Current address: 110 Brooks Avenue, Raleigh, North Carolina, United States of America

Introduction

Because plant regeneration is strongly dependent upon seed dissemination, many plants have adapted fruits with edible and nutritive fleshy pulps to encourage consumption by animals for seed dispersal [1]. Seeds of some fruit-bearing flora have adapted mechanisms to avoid damage from digestive enzymes and may even require scarification for germination [2]. Concomitantly, mammals, birds, and reptiles have adapted dietary niches to consume fleshy fruits (also referred to as soft mast; hereafter fruit) as a primary or supplementary energy source [3], leading to a dynamic relationship between seed dispersers and respective fruit-bearing flora [1].

Measuring fruit biomass may be important for evaluating a variety of research questions. For example, estimating fruit biomass may be useful to evaluate mechanisms regulating animal populations, particularly frugivores [4]. Also, fruit biomass may explain variations in animal behavior and could be used to evaluate the potential effects of forests silvicultural practices on fruit consumers, plant dissemination, plant regeneration, and conservation of plant communities [1]. When used to address research hypotheses, accurate estimates of fruit biomass will be needed. Furthermore, an accurate and practical method for estimating fruit production may be necessary for land managers to monitor fruiting responses to management regimes because fruit abundance may change rapidly with disturbance and forest succession [4–13].

A variety of protocols to monitor fruit production have been developed, with the majority focusing on estimating fruit biomass in the forest canopy [14–16]. Visual estimations and fruit traps commonly are used to evaluate fruit biomass in the overstory [14], [16–21]. However, Parrado-Rosselli et al. [15] reported fruit traps were ineffective at monitoring fruit production in tropical forest canopies and suggested observing fruit in the overstory along systematically located transects was the only time-efficient way to monitor overstory fruit production [22]. Plant coverage and visual estimations have been used to monitor understory fruit production [9], [10], [13], [23]. Some studies estimated understory fruit availability in terms of percent land cover of fruit-producing plants [24], [25], but plant cover may not be a reliable measure of fruit production [8]. Though visual estimation or collection of fruits has been used to measure understory fruit availability in many studies, the reported protocols were inconsistent [9], [10], [13], [23].

The problems with current methods are two-fold: 1) protocols are not standardized; and 2) little is known about the relative accuracy and time efficiency of each. Accuracy and feasibility contribute equally to the adequacy of the methods used to quantify fruit biomass. Therefore, a comparative evaluation of both the accuracy and practicality among protocols is needed. We compared 3 methods of estimating understory fruit availability. Our study was designed to test each method in 3 measures of utility listed in order of importance:

1. Estimation ability
2. Sampling intensity necessary for estimation ability
3. Time commitment to conduct each method

Materials and Methods

Ethics Statement

This research was performed in accordance with the United States Department of Defense and Fort Bragg Military Installation. No animals were handled in this study. No permits were required for the described study, which complied with all relevant regulations.

Site Description

We conducted our study at Fort Bragg Military Installation (~65,000 ha, hereafter Fort Bragg), located in the Sandhills physiographic region in southeastern North Carolina, USA (358170N, 828470W), during July 2011. Roughly 65% of Fort Bragg was forested, which was second and third growth longleaf pine (*Pinus palustris*; [26]). Dominant understory fruit producing plant species included blueberries (*Vaccinium* spp.), huckleberries (*Gaylussacia* spp.), gallberries (*Ilex* spp.), grapes (*Vitis* spp.), Atlantic poison oak (*Toxicodendron pubescens*), greenbriers (*Smilax* spp.), and blackberries (*Rubus* spp.). Fort Bragg was managed by the United States Department of Defense under a 3-yr fire-return interval, with prescribed burning primarily during the growing season. Timber was managed on a 120-year rotation for longleaf pine and a 100-year rotation for other pines [26], [27].

Study Design

In 2011, we randomly established 30 25-m transects in each of 4 cover types (upland hardwood, hereafter UH; bottomland hardwood, hereafter BH; upland pine following dormant-season fire, hereafter DSUP; and upland pine following growing-season fire, hereafter GSUP) (n = 120). We systematically centered 7 different sampling plots on each transect (Figure 1). Plot A comprised the entire area along the transect, 1 m wide and 25 m long. Within Plot A, 3 Plots B (3-m^2 each) were placed at 5, 12.5, and 20 m along the transect, and 3 Plots C (0.1-m^2 each) were placed at 7.5, 15, and 22.5 meters along the transect.

We used 3 methods to estimate fruit biomass: Fruit Count, Stem Count, and Plant Coverage. We counted fruits, stems, and

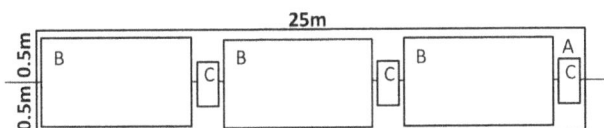

Figure 1. Diagram of sampling scheme for evaluating 3 methods of estimating fruit production. Plot A representing entire 25-m^2 sampling area and Plots B (3 m^2) and C (0.1 m^2) nested within. Note: Plots A, B, and C in the diagram are not drawn to scale for ease of display

coverage of blackberries, huckleberries, blueberries, and gallberries. We chose these genera based on plant fruiting phenology, wide occurrence across the temperate zone, high wildlife food value, and frequent occurrence at Fort Bragg [28].

Fruit Count

We used methods outlined in Jackson et al. [23] to guide the Fruit Count protocol. Jackson et al. [23] collected fruits in 2-m-wide plots along 50-m transects. We modified the protocol by shortening the transects from 50 m to 25 m, decreasing width from 2 m to 1 m, and counted fruits rather than collecting fruits, which decreased our time commitment per plot. We modified the methods used by Jackson et al. [23] because the additional time commitment associated with collecting, drying, and weighing fruits is not practical in most cases. Furthermore, the purpose of our proposed Fruit Count method was to evaluate whether a more time efficient visual count of fruit could accurately estimate the actual fruit biomass present. We counted fruits within 0.5 m along each side of the 25-m transect (Plot A, Figure 1), noting the number of fruits occurring in Plots B and C to test required sampling intensity to accurately estimate fruit production (see *Sampling Intensity* section). Observers counted fruits at a walking pace with minimal pause for counting to reduce time per transect. Therefore, fruit counts are not absolute counts but the best estimate of the number of fruits present.

Stem Count

We used methods outlined in previous literature to guide the stem count protocol [21], [29], [30]. These include 3 systematically located 3-m^2 plots along a 50-m transect. Originally, the Stem Count method was designed to measure diet selection in herbivores, so we modified the method by counting only fruit producing plants. We counted stems of fruit producers only within Plot B (Figure 1). The plot size of plot B was chosen because larger plots may inflate observer error and result in an unreasonable time commitment per plot.

Plant Coverage

We used methods outlined in Daubenmire [31] to guide the plant coverage protocol. Daubenmire [31] used a 0.1-m^2 frame to estimate plant coverage. The original method was designed to sample ground cover, but we modified the method to measure cover of fruit bearing species. We estimated plant coverage of fruit producers within Plot C (Figure 1). We only assessed plant coverage in Plot C because plot sizes greater than 0.1 m^2 may increase observer error [31]. The same 2 observers were used for all plots to reduce bias.

Absolute Biomass

After counts from all three methods were recorded, we collected all fruits within Plot A (i.e., total area surveyed). We kept fruits from each respective plot type separate, which yielded 7 bags (e.g., 1 bag for plot A excluding plots B and C, 3 bags for plot B, and 3 bags for plot C). The absolute fruit biomass (total fruits in plot A) was the sum of all bags collected. We dried fruits in a convection oven at 50 °C for 48 hours and weighed them with seeds (by species) to the nearest 0.01 g keeping biomass from each bag separate. The fruits were not collected as part of the protocol for the three methods; rather, they only were collected so we could compare the accuracy of each method in estimating the actual biomass present.

Testing Accuracy

We used each method to estimate absolute fruit biomass for the respective sampled area (e.g., plant coverage estimated absolute fruit biomass for Plot C). We used fruit biomass as the response variable rather than fruit count, but the same results should be expected with either metric (see Greenberg et al. [13]; Figure 1). Likewise, fruit biomass and fruit count would have yielded identical results in our study because fruit weights were all 0.05 g. We chose a continuous distribution of responses and analyzed absolute fruit biomass under the assumption of normality. Initially, we ran a simple correlation to determine if fruit counts, stem density, and Daubenmire scores were correlated to fruit biomass. Then, to determine if a method could predict fruit biomass, we estimated each model using the QLIM (qualitative and limited dependent variable model) procedure, which is the appropriate SAS procedure to perform with censored distributions for a continuous response variable [32]. This procedure is part of the SAS/ETS (econometric, time series) procedures. The linear regression model is as follows: $Y = \beta_1 X + \beta_2 I_{BH} + \beta_3 I_{DSUP} + \beta_4 I_{GSUP} + \beta_5 I_{UH} + \varepsilon$, where Y = biomass of berries, X = measurement method, I_{BH} is an indicator of specific cover type (so it equals 1 if cover type is BH and equals 0 otherwise), the other indicator variables are defined in a similar way for their respective cover types, and ε is a normally distributed error with mean 0. We used a censored normal distribution for the response Y (fruit biomass). This means fruit biomass in an area of forest follows a normal distribution but is censored at 0, so fruit biomass cannot be less than 0. The assumption of normality was verified by looking at histograms for each measurement method. We set alpha = 0.05. A non-significant result (large P-value associated with the model parameter which leads to not rejecting the null hypothesis of $\beta_1 = 0$) indicated the method was not useful in predicting fruit biomass. A significant result (small P-value rejecting the null hypothesis of $\beta_1 = 0$) indicated the measurement method was useful for predicting fruit biomass.

Then, we used the Akaike information criterion (AIC) as a measure of the relative "goodness of fit" of each significant statistical model [33]. The AIC is based on the concept of information entropy, in effect offering a relative measure of the information lost when a given model is used to describe reality. Furthermore, we used AIC which was corrected for finite sample sizes (AICc) as a secondary measure of model fit to corroborate the AIC model ranking [34]. Burnham and Anderson [34] recommended using AICc, rather than AIC, if n is small or k (the number of parameters in the model) is large because it gives more penalty to small sample sizes and additional parameters. However, AICc ranked the competing models identically to that of AIC and therefore was not reported.

Sampling Intensity

We estimated the fruit biomass of Plot A with Fruit Counts from various combinations of Plots B and C (i.e., all C plots equal 1.5-m^2 area or 1.2% sampling intensity) to test the effectiveness of the Fruit Count method at 0.4%, 0.8%, 1.2%, 4%, 8%, and 12% sampling intensities. At each sampling intensity, we used QLIM procedure as before to test the accuracy of the Fruit Count method. We used a Bonferroni correction to control the overall Type 1 error rate. We did not test sampling intensity of Stem Count or Plant Coverage methods because the Fruit Count method was the superior measurement method based on results of the accuracy test described above.

Table 1. Effectiveness of Fruit Count, Plant Coverage, and Stem Count methods at predicting understory fruit biomass at varying sampling intensities on Fort Bragg Military Installation, NC (July 2011).

Method	100% Sample		12% Sample		8% Sample		4% Sample		1.2% Sample	0.8% Sample	0.4% Sample
	P-Value	AIC	P-Value	AIC	P-Value	AIC	P-Value	AIC	P-Value	P-Value	P-Value
Fruit Count[a]	<0.0001[b]	51.99	<0.0001	289.89	<0.0001	304.92	0.0008	313.04	<0.0001	0.0024	0.2437
Stem Count[a]	<0.0001	197.28	<0.0001	308.57	0.0015	313.98	0.0066	316.8	N/A	N/A	N/A
Plant Coverage	0.2090	N/A									

[a] Fruit Count and Stem Count were extended to smaller sampling intensities for comparison

[b] The P-value is testing whether the method accurately estimated the fruit biomass for the sampled area. A Bonferroni correction was used and P-values are significant if P≤0.008.

Results

Fruit Count ($R^2 = 0.90$, $P<0.001$) and Stem Density ($R^2 = 0.18$, $P<0.001$) were correlated with fruit biomass, indicating both methods were potentially useful even though Fruit Count was substantially more correlated than Stem Density. However, Plant Coverage ($R^2<0.001$, $P = 0.82$) was not correlated to fruit biomass. Fruit Count and Stem Density could be used to accurately estimate fruit biomass at the 100% sampling intensity (Table 1). Plant Coverage did not have significant predictive value at a 100% sampling intensity ($P = 0.21$).

Fruit Count and Stem Count were viable predictors of fruit biomass when sampling as little as 4% of the area and Fruit Count was useful even as small as 0.8% of the area (Table 1). However, Fruit Count was superior to the Stem Count at all sampling levels in terms of P-value magnitude and AIC scores (Table 1). Also, the Fruit Count method (9 ± 0.5 seconds/m^2) was more time efficient than the Stem Count method (18 ± 1 seconds/m^2), requiring half the time commitment for sampling.

Discussion

We showed Plant Coverage was not a viable method of estimating understory fruit biomass, likely because of variations in fruit production annually and among individual plant or plant species [13]. Although the Stem Density method could be flawed for the same reason, Stem Density may have shown more promise because it takes into account the number of fruit producing plant stems rather than the coverage of fruit producing plants. Conversely, the Fruit Count method worked better than the other 2 methods because fruiting phenology, genetics, and environmental influences on an individual plant's fruit production do not inflate error when estimating fruit availability. Although the Fruit Count method is affected by spatio-temporal variability in fruit biomass, increasing sampling intensity can improve the precision of estimates.

Because sampling at least 0.8% of the land area may be required to accurately estimate fruit biomass, quantification of fruit production on larger landholdings may not be feasible. For instance, Fort Bragg includes about 42,000 forested ha. To accurately estimate fruit production for the entire site, sampling would have to cover 336 ha, which is not practical. However, where fruit availability is more homogeneous, smaller portions of the landscape may be sampled to accurately estimate fruit biomass, which may be more practical particularly if a set amount of time is allotted per transect [35]. Additionally, the use of permanent transects to sample yearly may allow managers to monitor relative rather than absolute fruit biomass; this strategy would allow managers to track the direction of change in fruit production following management prescriptions using lower sampling intensity, though sampling would still be required yearly

unless habitat characteristics remained similar. Finally, sampling efforts could be framed on a smaller scale (e.g., vegetation type) to focus inferences where a plant species and associated fruit production are more abundant or consistent.

Researchers and practitioners must consider their objectives to determine whether estimating fruit biomass is necessary. For example, if an experiment was designed to evaluate the effects of silvicultural treatments on fruit production, estimating fruit abundance may be useful to address the hypotheses. However, in many cases researchers and practitioners may be able to effectively address questions or monitor management practices by measuring relative changes in fruit production rather than estimating actual fruit biomass. Furthermore, only presence or absence data over multiple years may be sufficient to guide timber harvest recommendations in some cases [36]. For example, Lashley et al. [36] demonstrated monitoring the presence or absence of mast on trees in the overstory could effectively identify target trees to retain while harvesting timber. In any case, we recommend the Fruit Count method over other methods to monitor understory fruit biomass because the estimates from this method are more robust to variations in fruit densities and the method requires relatively less time to conduct surveys.

Conclusions

Fruits are an important component in the diets of many wildlife species. Monitoring shifts in fruit biomass is important to ensure fruit production is maintained when managing plant communities for wildlife. The Fruit Count method provides an easy and repeatable protocol that allows researchers and land managers to measure fruit biomass with a small relative time commitment, while maintaining accurate estimates that are comparable across studies and years within a study. We recommend using the Fruit Count method at the greatest feasible sampling intensity to estimate fruit production. When target sampling intensities are not feasible, use of the Fruit Count method with permanent plots to evaluate trends in year-to-year fruit biomass may suffice.

Acknowledgments

We thank the Fort Bragg Wildlife Branch for technical and logistical support. Thanks to J. Chvosta for statistical consultation. Special thanks to A. Prince, M. Elfelt, and E. Kilburg for assistance in data collection and entry.

Author Contributions

Conceived and designed the experiments: MAL. Performed the experiments: MAL MCC CSD CEM. Analyzed the data: MAL JRT. Contributed reagents/materials/analysis tools: MAL. Wrote the paper: MAL JRT MCC CSD CEM.

References

1. Jordano P (2000) Fruits and frugivory. In: Fenner, M, editor. Seeds: the ecology of regeneration in plant communities, 2nd edition. Wallingford, UK CABI Publ
2. Samuels IA, Levey DJ (2005) Effects of gut passage on seed germination: do experiments answer the questions they ask? Func Eco 19: 365–368.
3. Howe HF (1986) Seed dispersal by fruit-eating birds and mammals. In: Murray DR, editors. Seed dispersal. Academic Press, New York, New York, USA. pp. 123–190.
4. McCarty JP, Levey DJ, Greenberg CH, Sargent S (2002) Spatial and temporal variation in fruit use by wildlife in a forested landscape. For Eco Mgt 164: 277–291.
5. Herrera CM, Jordano P, Guitián J, Traveset A (1998) Annual variability in seed production by woody plants and the masting concept: reassessment of principles and relationship to pollination and seed dispersal. Am Nat 152: 576–594.
6. Johnson AS, Landers JL (1978) Fruit production in slash pine plantations in Georgia. J Wildl Mgt 42: 594–606.
7. Campo JJ, Hurst GA (1980) Soft mast production in young loblolly plantations. SEAFWA 34: 470–475.
8. Perry RW, Thill RE, Peitz DG, Tappe PA (1999) Effects of different silvicultural systems on initial soft mast production. Wildl Soc Bull 27: 915–923.
9. Reynolds-Hogland MJ, Mitchell MS, Powell RA (2006) Spatio-temporal availability of soft mast in clearcuts in the southern Appalachians. For Eco Mgt 237: 103–114.
10. Greenberg CH, Levey DJ, Loftis DL (2007) Fruit production in mature and recently regenerated upland and cove hardwood forests of the southern Appalachians. J Wildl Mgt 71: 321–329.
11. Greenberg CH, Perry RW, Harper CA, Loftis DJ (2011) The role of young, recently disturbed upland hardwood forest as high quality food patches. In

Greenberg CH, Collins BS, and Thompson FR, editors. Sustaining young forest communities: ecology and management of early successional habitats in the central hardwood region, New York: Springer USA.

12. McCord JM, Harper CA, and Greenberg CH (2014) Brood cover and food resources for wild turkeys following silvicultural treatments in mature upland hardwoods. Wildl Soc Bull: DOI: 10.1002/wsb.403.

13. Greenberg CH, Levey DJ, Kwit C, McCarty JP, Pearson SF, et al. (2012) Long-term patterns of fruit production in five forest types of the South Carolina coastal plain. J Wildl Mgt 76: 1036–1046.

14. Chapman CA, Chapman LJ, Wrangham RW, Hunt K, Gebo D, et al. (1992) Estimators of fruit abundance of tropical trees. Biotropica 24: 527–531.

15. Parrado-Roselli A, Machado J, Prieto-Lopez T (2006) Comparison between two methods for measuring fruit production in a tropical forest. Biotropica 38: 267–271.

16. Greenberg CH, Warburton GS (2007) A rapid hard-mast index from acorn presence-absence tallies. J Wildl Mgt 73: 1654–1661.

17. Schupp EW (1990) Annual variation in seedfall, postdispersal predation, and recruitment of a neotropical tree. Eco 71: 504–515.

18. Borchert R (1998) Responses of tropical trees to rainfall seasonality and its long-term changes. Clim Change 39: 381–393.

19. Galetti M, Aleixo A (1998) Effects of palm heart harvesting on avian frugivores in the Atlantic rain forest of Brazil. J Appl Eco 35: 286–293.

20. Wright SJ, Carrasco C, Calder'on O, Paton S (1999) The El Nino southern oscilation, variable fruit production, and famine in a tropical forest. Eco 80: 1632–1647.

21. Lashley MA, Harper CA, Bates GE, Keyser P (2011) Deer forage available following silvicultural treatments in upland hardwood forests and warm-season plantings. J Wildl Mgt 75: 1467–1476.

22. Chapman CA, Wrangham R, Chapman LJ (1994) Indices of habitat-wide fruit abundance in tropical forest. Biotropica 26: 160–171.

23. Jackson SW, Basinger RG, Gordon DS, Harper CA, Buckley DS, et al. (2007) Influence of silvicultural treatments on eastern wild turkey habitat characteristics in eastern Tennessee. Proc Nation Wild Turkey Symp 9: 199–207.

24. Mitchell MS, Zimmerman JW, Powell RA (2002) Test of a habitat suitability index for black bears in the Southern Appalachians. Wildl Soc Bull 30: 794–808.

25. Mitchell MS, Powell RA (2003) Response of black bears to forest management in the Southern Appalachian Mountains. J Wild Mgt 67: 692–705.

26. Cantrell MA, Brithcher JJ, Hoffman EL (1995) Red-cockaded woodpecker management initiatives at Fort Bragg Military Installation. In: Kulhavy DL, Hooper RG, Costa R, editors. Red-cockaded Woodpecker: Recovery, Ecology and Management. Austin State University, Nacodoches, Texas, USACenter for Applied Studies in Forestry, College of Forestry, Stephen F89–97.

27. Carter JH III, Walters JR, Doerr PD (1995) Red-cockaded woodpeckers in the North Carolina Sandhills: a 12-year population study in D.L. Kulhavy, R.G. Hooper, and R. Costa, editors. Red-cockaded Woodpecker: Recovery, Ecology and Management. Center for Applied Studies in Forestry, College of Forestry, Stephen F. Austin State University, Nacodoches, Texas, USA. pp. 248–258.

28. Miller JH, Miller KV (1999) Forest plants of the southeast and their wildlife uses.Southern Weed Science Society.

29. Shaw CE, Harper CA, Black MW, Houston AE (2010) Initial effects of prescribed burning and understory fertilization on browse production in closed-canopy hardwood stands. J Fish Wildl Mgt 1: 64–71.

30. Lashley MA, Harper CA (2012) The effects of extreme drought on native forage nutritional quality and white-tailed deer diet selection. SE Nat 11: 699–710.

31. Daubenmire R (1959) A canopy-coverage method of vegetational analysis. NW Sci 33: 43–64.

32. SAS Institute Inc (2012). SAS/ETS 12.1 User's Guide. SAS Institute Inc., Cary, NC, USA.

33. Akaike H (1977) On entropy maximization principle. In: Krishnaiah PR, editor. Applications of Statistics, North-Holland, Amsterdam. pp. 27–41.

34. Burnham KP, Anderson DR (2002) Model selection and multi-model inference: a practical information-theoretic approach. New York: Springer.

35. Lashley MA, Chitwood MC, Prince A, Elfelt MB, Kilburg EL, et al. (2014) Subtle effects of a managed fire regime: a case study in the longleaf pine ecosystem. Ecol Ind 38: 212–217.

36. Lashley MA, McCord JM, Greenberg CH, Harper CA (2009) Masting characteristics of white oaks: implications for management. SEAFWA 63: 21–26.

16

A National and International Analysis of Changing Forest Density

Aapo Rautiainen[1], Iddo Wernick[2]*, Paul E. Waggoner[3], Jesse H. Ausubel[2], Pekka E. Kauppi[1]

1 Faculty of Biosciences, University of Helsinki, Helsinki, Finland, 2 Program for the Human Environment, The Rockefeller University, New York, New York, United States of America, 3 Department of Forestry and Horticulture, Connecticut Experimental Agricultural Station, New Haven, Connecticut, United States of America

Abstract

Like cities, forests grow by spreading out or by growing denser. Both inventories taken steadily by a single nation and other inventories gathered recently from many nations by the United Nations confirm the asynchronous effects of changing area and of density or volume per hectare. United States forests spread little after 1953, while growing density per hectare increased national volume and thus sequestered carbon. The 2010 United Nations appraisal of global forests during the briefer span of two decades after 1990 reveals a similar pattern: A slowing decline of area with growing volume means growing density in 68 nations encompassing 72% of reported global forest land and 68% of reported global carbon mass. To summarize, the nations were placed in 5 regions named for continents. During 1990–2010 national density grew unevenly, but nevertheless grew in all regions. Growing density was responsible for substantially increasing sequestered carbon in the European and North American regions, despite smaller changes in area. Density nudged upward in the African and South American regions as area loss outstripped the loss of carbon. For the Asian region, density grew in the first decade and fell slightly in the second as forest area expanded. The different courses of area and density disqualify area as a proxy for volume and carbon. Applying forestry methods traditionally used to measure timber volumes still offers a necessary route to measuring carbon stocks. With little expansion of forest area, managing for timber growth and density offered a way to increase carbon stocks.

Editor: William Bauerle, Colorado State University, United States of America

Funding: Funding for this work was provided by Rockefeller University, the Academy of Finland (IFEE 1117822), and the University of Helsinki (HENVI 470149031). The funders had no role in study design, data collection and analysis, decision to publish, or preparation of the manuscript.

Competing Interests: The authors have declared that no competing interests exist.

* E-mail: iwernick@rockefeller.edu

Introduction

Measuring forests tells their spatial extent and the density of the trees that occupy that extent. Traditionally foresters measured the attribute of merchantable timber, also called growing stock, on a given stand to assess its commercial value. As concerns expand from timber volume to include the attributes of biomass as well as the carbon sequestered in that mass, the attributes must be connected and new coefficients measured.

The Forest Identity [1] connecting those attributes shows timber volume equals area times density, and biomass equals volume times the ratio of growing stock volume to the biomass of crown, foliage, and roots. Finally, carbon mass equals the biomass times its carbon concentration.

Both nature and humanity affect forest area. While climate and geography determine potential forest area, humans determine the hectares they spare from farms, logging, settlement, and transportation.

As for density in a natural forest, climate and geography also affect forest productivity measured as cubic meters of timber growth per hectare. Humans may degrade forests and deplete their timber, biomass, and carbon, or they can manage them by planting faster growing trees, improving sites, and sparing mature and dead trees [2,3,4]. Because trees grow for decades, the resulting rise in density becomes apparent decades after degradation. Managing intensively, humans can take advantage of techniques developed to speed growth and increase density used in tree plantations [5].

Here we examine the effectiveness of changing forest area and density to change timber volume and carbon in a nation with a continuing forest inventory and in a global inventory of nations encompassing a little over two thirds of global forests. We begin with the inventory of a single nation that avoids some discrepancies caused by national differences in method, continuity, and capacity [6,7]. We then proceed to a less certain global inventory.

Results

Forests in a single nation with continuing inventory

The United States represents a single nation with a continuing inventory. We examined measurements from 1953 to 2007 by the United States Forest Service (USFS) [8]. The USFS has published estimates of forest area, timberland area, and growing stock on timberland using a standard, continuing system. Timberland, which comprises about two-thirds of U. S. forest area, is not legally reserved from timber harvest and is capable of annually increasing density per hectare by a minimum of 1.4 cubic meters of industrial wood.

In US regions since 1953, timberland area generally changed little, with excursions up and down always less than 10%, Figure 1. Overall timberland area grew 1% between 1953 and 2007 in the

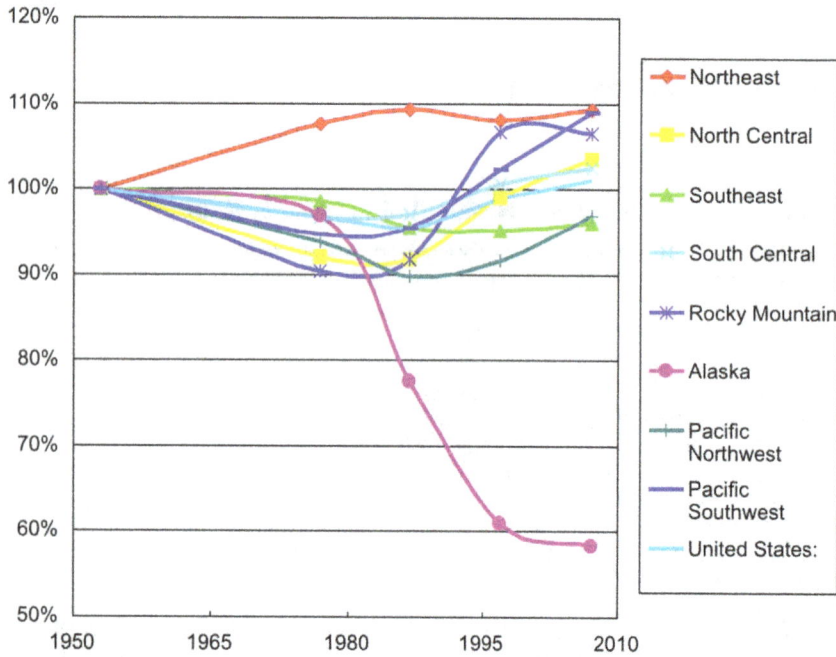

Figure 1. United States timberland area by region, 1953–2007. The conservation measure of reclassifying large areas as non-forest or wilderness caused the exceptional case of declining timberland in Alaska. Index 1953 = 100%.

United States. Total forest land in the United States rose by 0.5% over the same period.

During the same years, the volume of growing stock on the same land generally rose. Demonstrating the greater change in volume than area, the vertical axis for charting volume in Figure 2 extends several times that for charting area in Figure 1. In the two North and the South Central regions, volume rose sharply during a decades-long restoration [1]. In the Southeast, the restoration of volume that began in the 19th century continued but slowed. In the Pacific regions, volume recovered from a dip in the 1980s. All in all, the combined national volume swelled 51%. Clearly, growing timber volume on a stable area indicates growing density.

The Forest Identity shows volume, V equals the area, A times density, D or $V = D \times A$. We denote the annual % changes in

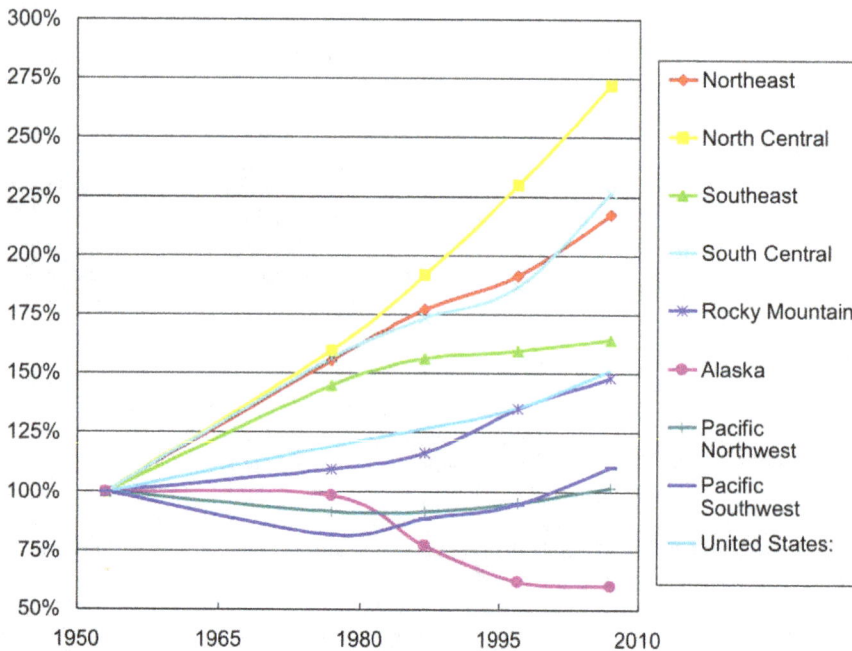

Figure 2. United States volume of growing stock on timberland by region 1953–2007. The conservation measure of reclassifying large areas as non-forest or wilderness caused the exceptional case of declining volume in Alaska. Index 1953 = 100%.

volume, area, and density using small case letters v, a, and d. Because the annual percentage changes are small, it can be shown that the sum of changing area a and density d nearly equals changing volume $v = a+d$ (see Materials and Methods section). Figures 3a–b show the changing area, a and density, d and the consequent changing volume, v during 1953–1987 and during 1987–2007. These two periods were selected based on the available data for the year 1953 and decadal data for 1977–2007. In the first period in the two Eastern, two Central, and the Rocky Mountain regions, growing density increased volume by overwhelming small area changes. In the Western regions and Alaska, losses in area combined with changes in density to shrink volume. For the United States as a whole during the first period, the average annual 0.83% more density less 0.13%/yr less area increased average volume 0.70%/yr.

During 1987–2007, density continued to rise in the Eastern, Central, and Rocky Mountain regions, though at a slower pace. Volume fell in Alaska and the western regions despite small density

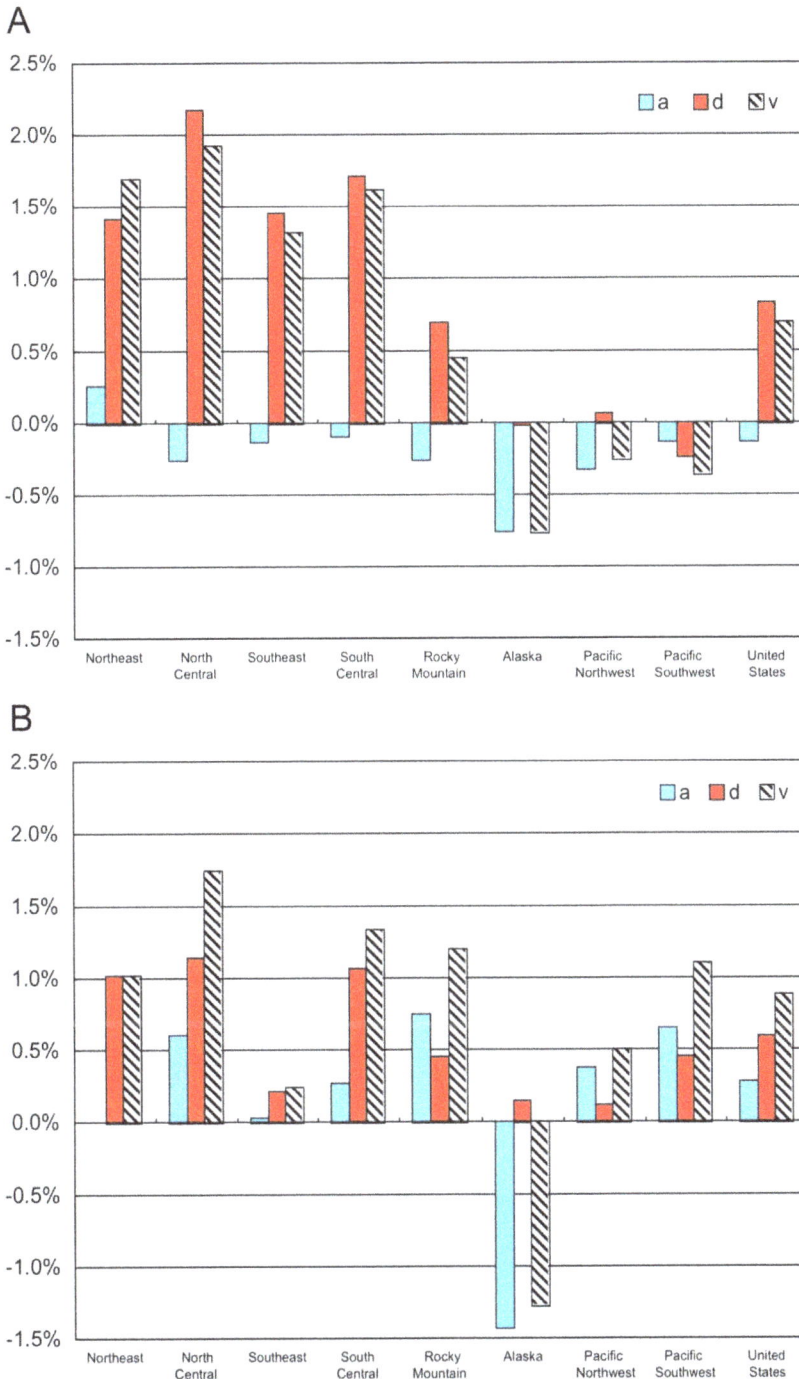

Figure 3. Annual change in timberland area a, density d, and timberland volume v in the United States (a) 1953–1987 (b) 1987–2007.

gains in Alaska and the Pacific Northwest and a large area addition in the Pacific Southwest. For the United States as a whole during the second period, the average annual 0.60% more density plus an additional 0.28%/yr area increased average volume by 0.88%/yr.

International Forests

For a broader understanding of changing area and density, we analyzed international data compiled by the United Nations Food and Agriculture Organization (UNFAO) in the 2010 Global Forest Resources Assessment [9]. Difficulties creating reliable time series from UNFAO reports stem from 1) inconsistent reporting criteria and data quality from member countries, 2) frequent retroactive revisions by the UNFAO, and 3) changing definitions of forest attributes [6]. To address these problems our analysis relies on the latest 2010 publication, which provides a consistent data series for the years 1990–2010.

The UNFAO reports forest area rather than timberland. Because the 2010 report published sequestered carbon but not growing stock volume for 1990–2010, we analyzed carbon rather than volume. Sequestered carbon is the product of volume and two variables: the ratio of biomass to volume and of the concentration of carbon in the biomass. If the ratio and concentration are nearly constant during 1990–2010, the annual percentage change of carbon that we present nearly equals the change of growing stock volume.

Countries meeting data quality criteria were included in the analysis as described in the Materials and Methods section. Table S1 provides a list of the 68 countries included in this analysis by region. These countries provided a global sample that accounted for 72% of the reported global forest area and 68% of the reported global carbon. The following summarizes global results by placing the 68 countries in the 5 regions named for continents as listed in Table S1.

Only 10 countries in the continent of Africa met the quality criteria. 80% of the forest area and carbon mass in the Russian

Federation were included in the Asia region and 20% in the Europe region roughly corresponding to the share of each [10]. Australia was included in the Asian region, where it accounted for 10% of total forest carbon in that region.

Countries in the South American and African regions lost close to 10% of their forest area during the two decades, Figure 4. Asian and European forest area expanded several percent while the area in North America changed little.

The changing carbon mass graphed in Figure 5 combined with the area reflects changing density. During the second of the two decades, carbon mass in the Asian region changed little, while area expanded, indicating falling carbon density. For analyzed countries in the Africa and South America regions, carbon declined slightly less than area, reflecting small density increases. North America and Europe gained carbon well in excess of any area additions.

The Forest Identity shows the mass of sequestered carbon, Q equals the area, A times the density, D' of carbon per area or $Q=D' \times A$. Because the annual percentage changes denoted by lower case letters are small, the changing area, a plus changing carbon density, d' nearly equals changing mass $q=a+d'$ (see Materials and Methods section). Figures 6a–b show the changing area a and density d' and the consequent changing mass q for two decades, 1990–2000 and 2000–2010. This analysis parallels that of U.S. timber volume, replacing changes in timber density and volume with changes in carbon density, d' and carbon mass, q'.

During 1990–2000 carbon density grew in all regions. While area changed little in the Asian and North American regions, rising density increased sequestered carbon. In the European region, increasing area plus carbon density together grew the mass of carbon. The data for the African and South American regions indicate shrinking carbon volume but slight gains in carbon density during the 1990s.

During the second period, 2000–2010, the Asian region displayed the greatly altered pattern. The great loss of density and sequestered carbon in Indonesia obscured the rising density

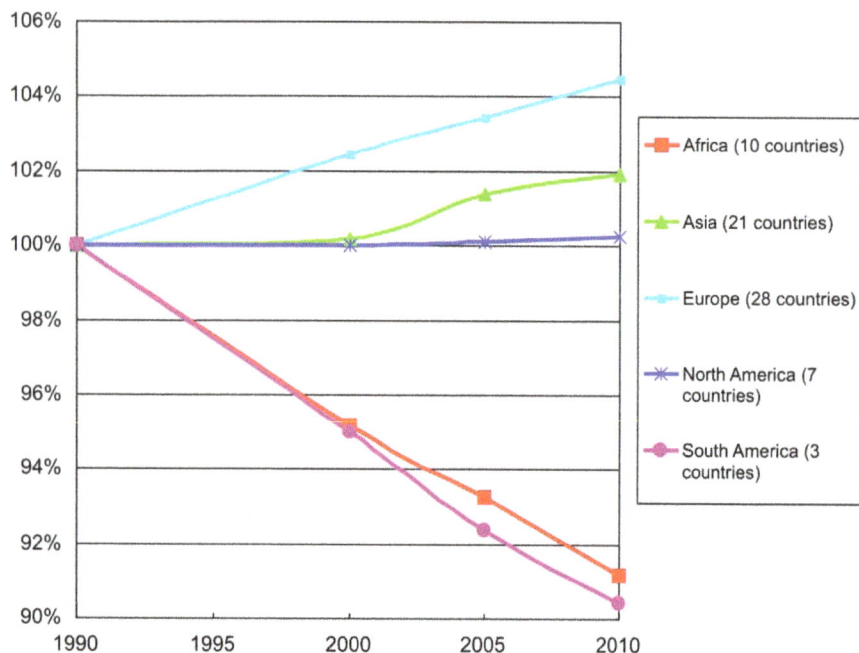

Figure 4. Forest area by region 1990–2010.

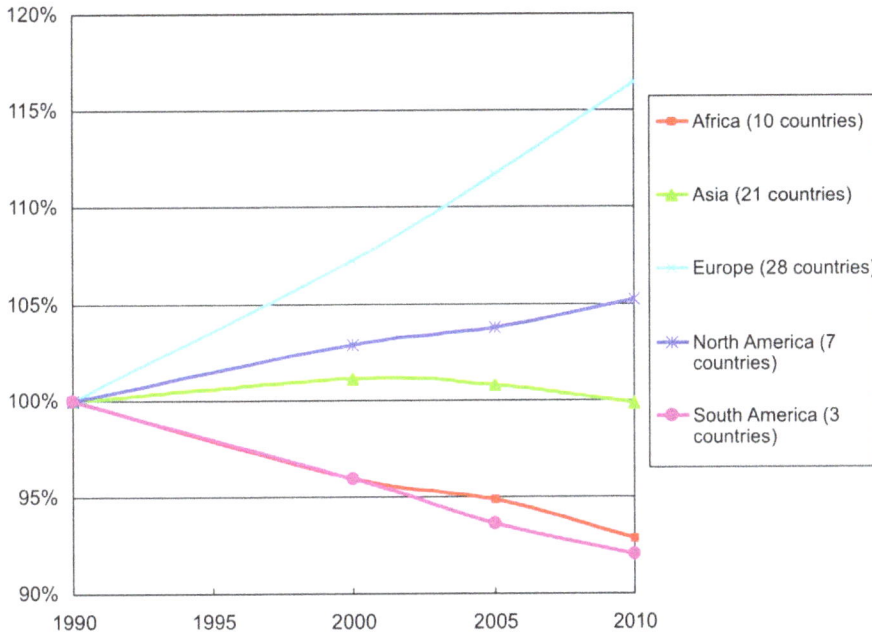

Figure 5. Carbon mass by region 1990–2010.

in ten of the twenty-one nations included in the region. As during the preceding period, rising density contributed most to increased sequestered carbon in the European and North American regions. African and South American losses of sequestered carbon mass were again tempered by slightly rising carbon density.

In all regions apart from Asia, the sign of change in forest area and the stock of sequestered carbon was the same. However, their magnitude differed significantly, especially in Europe and North America where most of the change was attributable to increasing carbon density (Figs. 6a and 6b). As $q = a+d'$, area is a suitable proxy for forest carbon only if carbon density remains constant ($d' = 0$). The discrepancy between the rates of area and density change becomes even more apparent, when the combinations of a and d' are examined for individual countries instead of regional aggregates (Figure 7). Forest area and carbon changed in tandem only in those countries whose points in Figure 7 fall on the diagonal axis.

Materials and Methods

Data for the national analysis for forests in the United States come from the 2009 edition of Forest Resources of the United States, 2007 [8] published by the US Forest Service. International data come from the 2010 Global Forest Resources Assessment [9] published by the UNFAO.

We relied on the Forest Identity [1], which equates growing stock density $D = V/A$ to the volume V and area A, and equates carbon density $D' = Q/A$ to the mass of carbon Q per area A.

To estimate annual changes denoted by small case letters, we use the convention $x = \ln(X_2/X_1)/(f-i)$ where X_f and X_i denote the value of X for final and initial year of the period being analyzed and $(f-i)$ is the number of years in the interval. This operation is justified for small changes i.e., $\Delta X \ll X$. Doing so, we get the equation $v = a+d$ for changing timber volume and $q = a+d'$ for changing carbon mass.

The 68 nations for analysis were selected from the 2010 Global Forest Resources Assessment by the following criteria:

- Data on carbon mass Q must be available for the years 1990, 2000, 2010.
- Carbon density Q/A must have changed sufficiently (≥ 0.1 tons/ha) during both periods (1990–2000, 2000–2010) to avoid nations where Q was likely extrapolated using a constant Q/A ratio.
- Carbon mass Q must have changed sufficiently ($\geq 0.1\%$) during both periods to avoid nations that reported constant Q for all years.
- The rate of change of Q must have changed sufficiently during both periods to avoid countries that linearly extrapolated or/interpolated Q for all years.

For borderline cases, we examined national reports and made some exceptions based on the description of their methods. Table S1 shows the nations analyzed.

Discussion

Whether measured by area or carbon content, the countries included in this study encompass slightly over two thirds of the global forests, leaving one third of the planet's forests unaccounted for. Nonetheless, despite uncertainties, especially among international compilations of national forest inventories, a large principle emerges. Forest area and density change independently with consequences for timber volumes and carbon sequestration.

For measuring two-dimensional forest area, remote sensing (i.e., satellite monitoring) offers the best solution for comprehensive standardized data collection. Reduced grid sizes will offer even greater precision [11,12]. Tests of the ground truth of satellite results for forest area yield results like those found in India [13] establishing accuracies as high as 96%.

For measuring three-dimensional volume, however, accurate estimates rely on ground level measures of the size distribution of trees. Without field measurement or another indication of tree size, implying forest health, or carbon storage, from area alone will continue to be highly uncertain and even misleading [14].

A

B

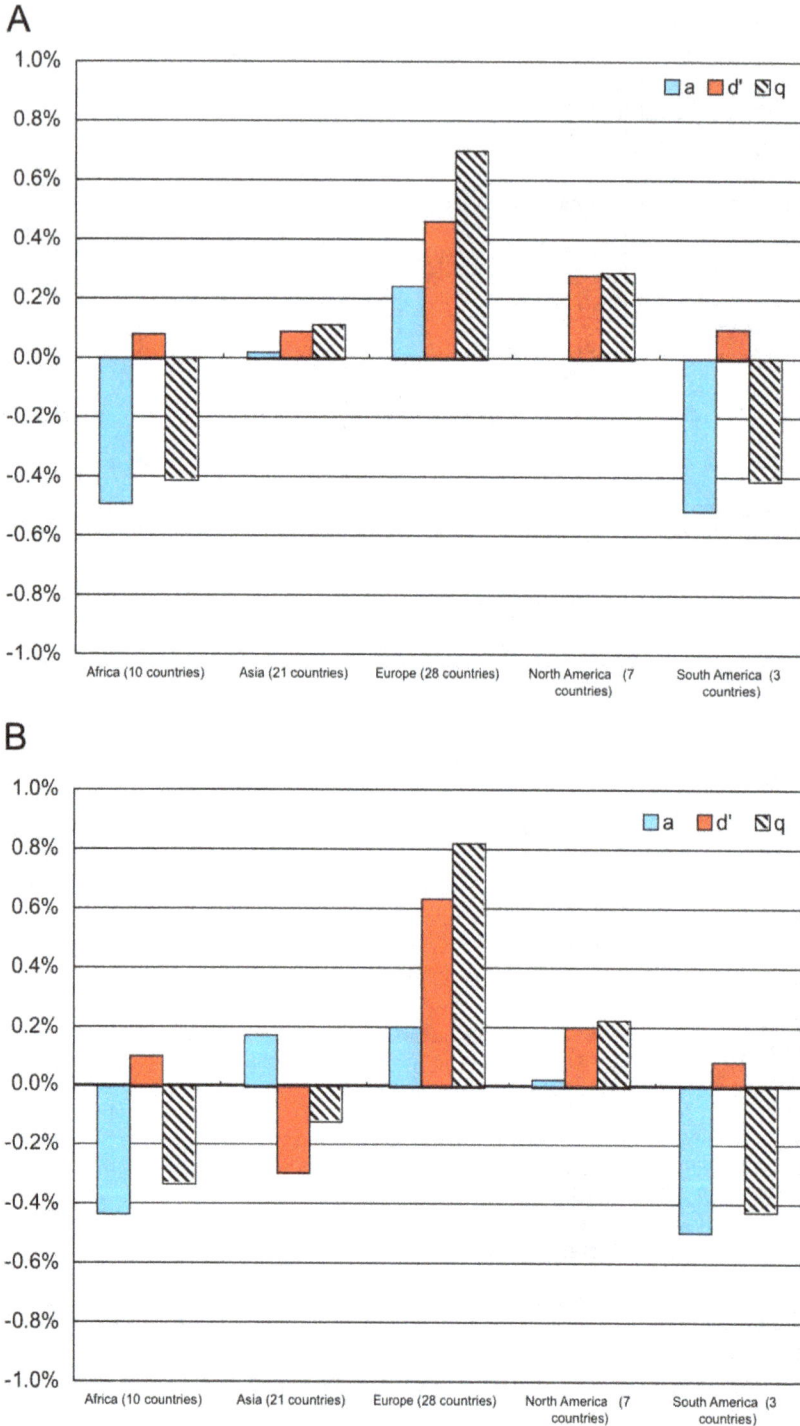

Figure 6. Annual change in forest area _a_, carbon density _d'_, and carbon mass _q_ by region (a) 1990–2000 (b) 2000–2010.

Remote sensing that incorporates estimation of forest height (i.e., LIDAR) as well as area offers promise for reducing the difficulties involved with collecting data on three dimensional as opposed to two dimensional attributes [15,16]. Currently, forest inventories based on field measurements provide the most accurate appraisals of the development over time of timber volume and forest carbon, as well as the most sound basis for anticipating future inventories.

For forests in a world with a growing population, with growing needs, but a fixed expanse of land, faster increases of volume or carbon than decreases of area during recent decades are encouraging. Technological improvements have been shown to improve greatly the efficiency of producing commercial timber products using less land [17,18]. The major stresses on forest land are thus attributable to humanity's appetite for forest land, not forest products.

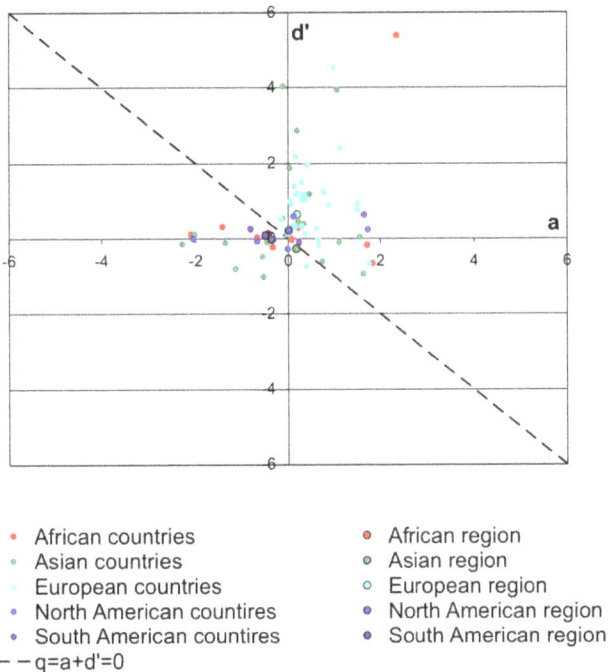

- African countries
- Asian countries
- European countries
- North American countires
- South American countires
- − − q=a+d'=0
- African region
- Asian region
- European region
- North American region
- South American region

Figure 7. Annual change in forest area _a_ and carbon density _d′_ 2000–2010. The forest in countries above the diagonal line (v = a+d′ = 0) gained carbon mass, while countries below it lost carbon mass.

To stop the loss of forest land in countries still experiencing deforestation, lifting crop yields as well as stabilizing population and switching fuels can reduce pressure on forest land [19,20]. Forest management to increase carbon density by encouraging growth of young forests and improving degraded forests offer effective levers for higher global forest carbon sequestration.

Inventories taken steadily by a single nation, the United States, and other inventories gathered recently from many nations by the United Nations confirm the asynchronous effects of changing area and density. While assessing the environmental impacts from forest management has become integral to national environmental policies and the global debate over climate change, current global forest statistics continue to suffer from deficiencies in the availability, accuracy, and precision of measurements of key attributes. The need to address these issues with scientific rigor calls for increased attention to forest density measurement to produce a consistent global data set that acknowledges the importance of forest density, in addition to area, as a decisive factor.

Author Contributions

Conceived and designed the experiments: AR IW PEW JHA PEK. Performed the experiments: AR IW PEW JHA PEK. Analyzed the data: AR IW PEW JHA PEK. Contributed reagents/materials/analysis tools: AR IW PEW JHA PEK. Wrote the paper: AR IW PEW JHA PEK.

References

1. Kauppi PE, Ausubel JH, Fang J, Mather AS, Sedjo RA, et al. (2006) Returning forests analyzed with the forest identity. Proc Natl Acad Sci USA 103: 17574–17579.
2. Sedjo RA (1987) Forest resources of the world: forests in transition. In: Kallio M, Dykstra DP, Binkley CS, eds. The global forest sector: an analytical perspective Wiley. pp 7–31.
3. Sedjo RA (1984) An economic assessment of industrial forest plantations. Forest Ecology and Management 9: 245–257.
4. Sedjo RA (1986) Forest plantations in the tropics and southern hemisphere and their implications for the economics of temperate climate forestry. In: Kallio M, Andersson AE, Seppälä R, Morgan A, eds. Systems analysis in forestry and forest industries Elsevier. pp 55–72.
5. Kirilenko AP, Sedjo RA (2007) Climate change impacts on forestry. Proc Natl Acad Sci USA 104: 19697–19702.
6. Grainger A (2008) Difficulties in tracking the long-term global trend in tropical forest area. Proc Natl Acad Sci USA 105: 818–823.
7. Waggoner PE (2009) Forest inventories. Discrepancies and uncertainties. Discussion Paper RFF DP 09-29. Resources for the Future, Washington DC. Online at http://www.rff.org/RFF/Documents/RFF-DP-09-29.pdf.
8. Smith WB, Miles PD, Perry CH, Pugh SA (2009) Forest Resources of the United States, 2007. Gen. Tech. Rep. WO-78. Washington, DC: U.S. Department of Agriculture, Forest Service.
9. United Nations Food and Agriculture Organization (2010) Global Forest Resources Assessment 2010. Online at http://www.fao.org/forestry/fra/en/. Country Reports can be found at http://www.fao.org/forestry/fra/67090/en/.
10. UNECE/UNFAO (2003) Geneva Timber and Forest Discussion Paper 27, Russian Federation Forest Sector Outlook Study. Online at http://www.unece.org/timber/docs/dp/dp-27.pdf.
11. Sánchez-Azofeifa GA, Castro-Esau KL, Kurz WA, Joyce A (2009) Monitoring carbon stocks in the tropics and the remote sensing operational limitations: From local to regional projects. Ecological Applications 19: 480–94.
12. Kalacskaa M, Sanchez-Azofeifaa GA, Rivarda B, Calvo-Alvaradob JC, Quesadac M (2007) Baseline assessment for environmental services payments from satellite imagery: A case study from Costa Rica and Mexico. Journal of Environmental Management 88: 348–59.
13. Forest Survey of India, Ministry of Environment and Forests (2008) State of Forest Report 2003. Chapter 7. Dehradun.
14. Wernick IK, Waggoner PE, Kauppi PE, Sedjo RA, Ausubel JH (2010) Quantifying forest change. Proc Natl Acad Sci USA;10.1073/pnas1008669107.
15. Næsset E, Gobakken T (2008) Estimates of above- and below-ground biomass across regions of the boreal forest zone using airborne laser. Remote Sensing of the Environment 112: 3079–90.
16. Fagan M, DeFries R (2009) Measurement and Monitoring of the World's Forests: A Review and Summary of Remote Sensing Technical Capability, 2009–2015. Resources for the Future, Washington DC. Online at http://www.rff.org/Publications/Pages/PublicationDetails.aspx?PublicationID=20971.
17. Wernick IK, Waggoner PE, Ausubel JH (2000) The forester's lever: industrial ecology and wood products. Journal of Forestry 98(10): 8–14.
18. Ausubel JH, Waggoner PE (2008) Dematerialization: Variety, caution and persistence. Proc Natl Acad Sci USA 105: 12774–12779. Online at http://www.pnas.org/content/early/2008/08/25/0806099105.full.pdf.
19. Waggoner PE, Ausubel JH (2001) How Much Will Feeding More and Wealthier People Encroach on Forests? Pop Dev Rev 27: 239–257.
20. Victor D, Ausubel JH (2000) Restoring the forests. Foreign Affairs 79(6): 127–144.

17

Causes and Correlates of Calf Mortality in Captive Asian Elephants (*Elephas maximus*)

Khyne U. Mar[1], Mirkka Lahdenperä[2], Virpi Lummaa[1,3]*

1 Department of Animal and Plant Sciences, University of Sheffield, Sheffield, United Kingdom, 2 Section of Ecology, Department of Biology, University of Turku, Turku, Finland, 3 Wissenschaftskolleg zu Berlin, Institute for Advanced Study, Berlin, Germany

Abstract

Juvenile mortality is a key factor influencing population growth rate in density-independent, predation-free, well-managed captive populations. Currently at least a quarter of all Asian elephants live in captivity, but both the wild and captive populations are unsustainable with the present fertility and calf mortality rates. Despite the need for detailed data on calf mortality to manage effectively populations and to minimize the need for capture from the wild, very little is known of the causes and correlates of calf mortality in Asian elephants. Here we use the world's largest multigenerational demographic dataset on a semi-captive population of Asian elephants compiled from timber camps in Myanmar to investigate the survival of calves ($n = 1020$) to age five born to captive-born mothers ($n = 391$) between 1960 and 1999. Mortality risk varied significantly across different ages and was higher for males at any age. Maternal reproductive history was associated with large differences in both stillbirth and liveborn mortality risk: first-time mothers had a higher risk of calf loss as did mothers producing another calf soon (<3.7 years) after a previous birth, and when giving birth at older age. Stillbirth (4%) and pre-weaning mortality (25.6%) were considerably lower than those reported for zoo elephants and used in published population viability analyses. A large proportion of deaths were caused by accidents and lack of maternal milk/calf weakness which both might be partly preventable by supplementary feeding of mothers and calves and work reduction of high-risk mothers. Our results on Myanmar timber elephants with an extensive keeping system provide an important comparison to compromised survivorship reported in zoo elephants. They have implications for improving captive working elephant management systems in range countries and for refining population viability analyses with realistic parameter values in order to predict future population size of the Asian elephant.

Editor: Jane M. Waterman, University of Manitoba, Canada

Funding: This work was financially supported by the National Environmental Research Council, UK (VL), Royal Society of London (VL), Nando Peretti Foundation (VL), Rufford Small Grant for Nature (KUM), International Foundation for Science (KUM), Finnish Cultural Foundation (ML), and Kone Foundation (ML). The funders had no role in study design, data collection and analysis, decision to publish, or preparation of the manuscript.

Competing Interests: The authors have declared that no competing interests exist.

* E-mail: v.lummaa@sheffield.ac.uk

Introduction

Life-history analyses of captive animal populations are of both scientific and practical importance for species such as the Asian elephant (*Elephas maximus*) with a significant proportion of the world's population now living in captivity. Studies on the mortality and fertility rates of such populations contribute to our knowledge of how management practices affect population sustainability. In many species conditions in captivity such as stable food availability, medical interventions, improved hygiene and lack of predation are associated with early age at first reproduction, short inter-birth intervals and increased adult longevity in comparison with wild counterparts. However, efforts to achieve self-sustaining captive populations of many species have been hampered by high juvenile mortality rates caused by factors such as obesity, poor adaptation to climate, inbreeding, social stress and incompatible group structure and size [1].

The Asian elephant is classified as Endangered on the International Union for Conservation of Nature (IUCN) Red List of threatened species. In parallel with wild populations, many captive populations are also facing rapid decline and extinction because of current low fertility and high calf mortality

rates [2]. A large proportion of remaining Asian elephants live in captivity in range countries (22–30% or 14500–16000 individuals [3–4]). Despite the need for detailed data on calf mortality in these remaining captive populations to sustainably manage populations and to minimize the need for capture from the wild, little is known of general trends in Asian elephant calf mortality or its causes and correlates in captive populations. Accurate data on calf mortality is also vital for population viability modeling (e.g. [5]) and is critical to the conservation of remaining often fragmented wild populations as well as captive management. The limited published data on the age-specific levels of mortality from different captive populations are highly variable: while the first year mortality of intensively kept zoo elephants in North America [2] and Europe [6–8] is about 30%, first year mortality of timber camp elephants in South India was estimated at 24% for female and 16% for male calves [9]. Such estimates are much higher than the 10–15% first-year mortality reported for wild African elephants (*Loxodonta africana*) of Amboseli, Kenya [10–11], but no detailed data has been published of the underlying causes of such differences, or the correlates of varying neonatal or juvenile mortality rates in Asian elephants from the range states.

The paucity of information on the determinants of calf mortality in Asian elephants stems from the rarity of long-term demographic datasets. Studies on captive elephant populations in zoological collections use studbook data but both the sample size and data available on causes of calf mortality in such datasets are often limited. Demographic studies on wild Asian elephants mainly use distance sampling techniques to estimate animal population densities, for example via dung counts or ratios of calves to adults, but no longitudinal, individual-based data exists. Limited information is available for captive Asian elephants [3,12] kept in South-East Asia under two systems, "extensive" and "intensive" (details in [13]). However, in many of these countries (e.g. Thailand and India), most captive elephants are owned privately as small-holdings and their population size or demography is impossible to monitor due to a lack of systematic registration or governmental influence over movement and changes of ownership of elephants in captivity [3].

The one exception is Myanmar (Burma), home to the largest captive elephant population ($n>5000$) in the world [14]. Half of the elephant population is owned by the government that has recorded and monitored life-history biodata from individuals for over a hundred years. The government-owned captive population has remained stable at 2700 elephants at least from 1999 until 2009 [15,16]. In Myanmar, elephant draught power has traditionally been extensively used in timber harvesting [17]. Approximately half of the timber elephants used in Myanmar are captive-born, and half are caught from the wild. The elephants live in forest camps, where they are used during the day as riding, transport and draft animals. At night the elephants forage in forests in their family groups unsupervised where they find food and encounter tame and wild conspecifics. Most calves are thought to be sired by wild bulls, and calves born in captivity are cared for by their biological and allomothers, and suckled until lactation no longer supports their demands [18]. Wild-capture of timber elephants was banned in the 1994/95 fiscal year [19] although smaller scale capture of wild elephants (focused on those involved in human-elephant conflict) was resumed again in 2003/04. Given the slow life-history of elephants, calf survival is thus now the major mean of recruitment and determinant of the captive population size. Nevertheless, no studies currently document the causes of calf mortality, compare age-specific mortality rate of males and females, or investigate the effects of maternal age, parity, inter-birth intervals or other maternal attributes on calf success.

Here we use the largest multigenerational demographic dataset on a captive population of Asian elephants, compiled from 260 timber camps in Myanmar over several decades and covering up to four generations of timber elephants. We investigate survival of calves ($n = 1020$) born to captive mothers ($n = 391$), and mortality causes and correlates. Specifically, given the high risk of stillbirths in many captive elephant populations [6,20], we first examine the overall prevalence of stillbirths, as well as the maternal risk factors associated with the probability of a stillbirth. Second, we document age-specific survival of calves from birth until independence (at age 5) for all liveborn calves produced by captive-born females over their lifespan. Whereas calf mortality is difficult to determine and is likely to be underestimated in field studies on elephants because calves may die and disappear from population before being recorded, records of the Myanmar timber elephants cover all livebirths and most if not all stillbirths. The longitudinal nature of the dataset also allows us to investigate both general between-mother differences in calf mortality, as well as within-mother effects of maternal age, parity and inter-birth intervals, and any calf sex-specific differences therein. We control for geographic variation due to different logging camps, and investigate time trends in calf mortality across our 40-year study period. Finally, detailed cause of death records for nearly all dead calves in our dataset ($n = 272$) allows for documenting for the first time the range and prevalence of causes for death for captive Asian elephant calves in their range country.

Materials and Methods

Study population

We used a unique, extensive demographic dataset available on a semi-captive Asian elephant population from Myanmar, covering the full life-history of generations of captive timber elephants. This dataset has been collated from the elephant log-books and annual extraction reports archived and maintained by the Myanmar Timber Enterprise. The traditional elephant log-books are equivalent to the 'studbooks' kept in Western zoos. State ownership of thousands of elephants enables recording data of all registered individual elephants from the log-books on: registration number and name; origin (wild-caught or captive-born); date and place of birth; mother's registration number and name; method, year and place of capture (if wild-captured); year or age of taming; dates and identities of all calves born; date of death or last known date alive; and cause of death. The individual elephant log-books are maintained and updated by local veterinarians and regional extraction managers at least bi-monthly in order to check the health condition and ability of each elephant to work. The multiple sources of data recorded by the Myanmar Timber Enterprise (individual elephant log-books coupled with annual extraction reports and end of the year reports from each region) allows effective cross-checking of any apparent errors. Between-individual variation in workload or rest periods is limited by law: all state-owned elephants are subject to the same regulations set by central government for hours of work per week, working days per year, and tonnage to extract per elephant. For example, in 2010 all mature elephants ($>17–55$) worked 3–5 days a week (depending on weather and forage availability) 5–6 hrs a day (maximum 8 hrs) with a break at noon. All elephants finish their work season by mid-February each year, and work resumes around mid-June depending on the arrival of monsoon. The maximum tonnage of logs allowed to be dragged in a year per elephant was 400 in 2010. The ages of captive-born elephants are exact because precise dates of birth are recorded, and this study concentrates only on the records of captive-born mothers and their offspring in order to have accurate data on maternal age and previous reproductive history, which are incomplete for most wild-captured mothers.

In the wild, elephant calves are raised by their mothers with the help of allomothers and other members of their family unit [21–23]. Female calves remain with small numbers of close kin, presumably for their entire lives, while male calves disperse during adolescence (c. 10 years) [24]. In the study population, working females are given rest from mid-pregnancy (11 months into gestation) until the calves reach their first birthday [17]. Mothers are then used for light duties but allowed to nurse the calves on demand. Mahouts or human caretakers do not intervene nor assist in the calving/nursing processes. The calves are generally weaned at the age of four or earlier if they are capable of independent foraging. They are separated from the maternal herd and tamed between ages four and five and then given a mahout, name, a log-book and registration number and trained and used for light work as baggage elephants until age 17. By age 17, they are put into work-force as full working elephants, and retired at age 55 [17,18].

The entire studbook includes 8006 elephants born and/or captured between 1925 and 2000; data from 2000 onwards is not available to us at the time of this study. Because of the lack of data from 2000 onwards, all calves born in 2000 were removed ($n = 66$) from the analyses, given those born in 2000 would have been censored in all analyses under age one. The remaining sample includes records of a total of 1138 calves born to captive-born mothers, of which 45 calves were recorded as stillborn and included in the analyses of factors associated with the risk of stillbirth. Excluding the stillbirths and animals with missing information on sex or dates of birth/death/lost/escapes, 975 liveborn animals born from 1960 to 1999 ($F = 484$, $M = 491$) remained for further analyses from 391 captive-born mothers, themselves born during 1941–1990. These elephants come from 32 timber extraction areas within eight regions (out of a total of 14) in Myanmar: Ayeyarwaddy, Bago, Chin, Kachin, Magway, Mandalay, Sagaing and Shan. The percentage of elephants from the initial sample that were included in the analyses (after excluding individuals with missing or erroneous data in different areas, cohorts or maternal backgrounds) is presented in Table 1. The youngest first-time mother in the sample was 5.3 years (mean age at first reproduction: 18.3 ± 4.7 years; median: 17.8 years) and the oldest reproducing female was 53 years. The average inter-birth interval was 5.4 ± 2.7 years (range = 1.8–18.1 years, median = 4.9) and the maximum lifetime number of calves was 10. These life-history patterns mirror those reported for wild Asian elephants with the earliest age at first reproduction of 6–9 years, a mean age of first reproduction of 17–18 years, the mean inter-birth interval of 2.5–4 years, and maximum number of calves as 12 [4].

Statistical analyses

We used a generalized linear mixed model (GLMM) to investigate the stillbirth probability and a discrete time survival analysis (also known as event history analysis) to investigate the calf mortality from birth to age of five. The discrete time survival analysis allows a detailed analysis of the effects of time-dependent variables on the calf's probability of dying over discrete time intervals, while controlling for repeated terms of the mother identity and area [25]. This approach also benefits from including data for some ages also for those individuals that have not been followed until the end of the study period or have not yet reached all the ages investigated in this study (censored), thus avoiding biasing the sample towards those dying young or those with complete records only.

We investigated the effects of calf age (discrete time model), sex, birth-order (1st borns vs. later borns), previous inter-birth interval (short, medium, long as defined below, or firstborn with no interval), birth cohort (decade of birth), mother's age at the birth of the calf as a continuous variable and the repeated measures of the mother and living area (mother's area, $n = 32$). We also included quadratic functions of maternal age in the models. The inter-birth interval preceding the birth of the focal individual was grouped into four classes based on the distribution of interbirth intervals in the population: 'short' if the previous birth-interval length was <3.7 years (25% quartile), 'medium' if it was 3.7–6.6 years, 'long' if it was >6.6 years (75% quartile) and 'firstborn or missing' if the calf was first to its mother (most calves in this category) or a laterborn with missing information on birth interval length ($n = 23$). We tested interactions between the fixed variables especially with calf age (discrete time model) to investigate whether any effects change with calf age. The data were restricted to only single-born calves, because of the poor survival prospects and low number of twin births ($n = 8$). Descriptive statistics and follow-up percentages (illustrating data quality) with regard to the variables included in the analyses are given in Table 1 ($n = 1138$, i.e.

Table 1. Descriptive statistics of the data used in the analyses (stillbirth and liveborn survival models) of Asian elephants illustrating the sample size and data completeness.

Category	Subcategory	n	Follow-up (%)
Sex	Male	587	88.1
	Female	550	91.5
Cohort	1960	94	72.3
	1970	260	82.7
	1980	420	93.1
	1990	364	95.1
Interbirth Interval	Short	167	90.4
	Medium	371	93.8
	Long	173	95.4
	First born/Missing	427	83.4
Birth order	First born	404	84.2
	Later born	734	92.6
Maternal age	<15	61	73.8
	15–24	499	87.2
	25–34	400	93.0
	35–44	155	93.5
	>44	23	100.0
Region	Ayeyarwa	7	85.7
	Bago	75	92.0
	Chin	4	75.0
	Kachin	59	86.4
	Magway	123	95.1
	Mandalay	292	88.4
	Sagaing	473	89.2
	Shan	104	89.4

Initial sample includes all calves born to captive-born mothers before year 2000 $n = 1138$. Follow-up percentage refers to the percentage of elephants in the initial cohort included in the final analyses due to the exclusion of individuals with missing or erroneous data. 'Short' interbirth interval refers to cases where the previous birth-interval length was <3.7 years (25% quartile), 'medium' if it was 3.7–6.6 years, 'long' if it was >6.6 years (75% quartile) and 'firstborn' if the calf was first to its mother. Note that logging area ($n = 32$) instead of region ($n = 8$) was used in the statistical models, but values for regions are shown here for simplicity. Firstborn/Missing-category for interbirth interval length includes all firstborns ($n = 404$) and 23 laterborns with missing birth-interval information.

including 45 stillbirths and animals with missing information on sex or dates of birth/death/lost/escapes that were subsequently excluded from the analysis of livebirths). All analyses were conducted using SAS (SAS Institute, release 9.2, 2002–2008). Significance levels were set at $\alpha = 0.05$. All biologically interesting interactions were tested, but omitted if not statistically significant. In all analyses, selection of significant terms retained in the best final models was determined using a backward elimination approach.

(i) Stillbirth probability. Stillbirth was defined as the birth of a lifeless fetus. The mother's probability of having a stillbirth was investigated using a generalized linear mixed effects model (GLMM) fitted to a binomial error structure with logit link function. The effects of linear and quadratic functions of maternal age, calf sex, birth-order and birth cohort were investigated on stillbirth mortality risk, fitted as fixed terms in the model. Because of the low number of stillbirths, it was not possible to simultaneously estimate the effects of birth intervals and birth-order. Inclusion of only birth-order in our final model was thus chosen, because analyzing both variables separately suggested that birth-order may be more important of the two. Non-independence of stillbirth probabilities of calves from the same mother were accounted by including the mother-id as a repeated term. Mother's area was included as a repeated term to control for similar living conditions of calves from the same area. The analysis was conducted on 1020 calves, of which 45 were stillbirths and 975 live-births.

(ii) Calf mortality. We first documented the overall survival pattern (survival distribution function) of calves from birth to age of 18 (average age at first reproduction for females and enrollment into work-force as full working elephant). We then investigated calf mortality from birth (all live-births) to five years using discrete time survival analysis. This analysis allowed us to estimate the calf's risk of dying in each year from birth while investigating the effects of fixed and repeated terms. For each year from birth to 5 years (5 time intervals for each calf; 0–1, 1–2, 2–3, 3–4, 4–5 years), the survival of each calf was coded as survived versus died during the observation year (1/0) or missing (when already dead or disappeared). Following restrictions (see above), this analysis was carried out on 4121 data points ($n = 975$ calves, $n = 378$ mothers).

The analysis investigated the effects of linear and quadratic terms (non-linear effects) of calf age and mother age, previous birth-interval length, calf sex, birth cohort and birth-order and the repeated terms of mother and area. The assessment of interactions between a fixed variable and calf age provides an indication of whether calves are more likely to die at a given age in relation to the variable and thus, whether the effect changes with calf age. Therefore, all interactions with calf age were first included in the model but removed if they did not reach statistical significance level ($P = 0.05$) using the backward elimination approach. We used GLMM with binomial error structure and logit link function in GENMOD procedure of SAS.

(iii) Death causes for calves under age 5. The cause of death was recorded for all but 3 liveborn calves out of the total of 227 individuals that died before five years. We classified the death causes into six groups: accidents (including falls, attacks by wild-elephant, tiger or older sibling, drowning, snake bite, strangulation on chains, choking, jammed between trees, head injury, food poisoning); diseases (including unspecified infectious diseases, anthrax, hemorrhagic septicemia, pneumonia, various parasitic infestations and gastrointestinal complications such as constipation, diarrhea, enteritis, colic and bloat); mother agalactia (lack of milk after parturition) and related general weakness; heat stroke; taming-related injury; and others including unknown. We then investigated the distribution and generality of the causes of death for calves below age five.

Results

(i) Stillbirth probability

Overall, 4% (45/1020) of all calves were stillborn. The probability of stillbirth increased significantly with maternal age (estimate ± S.E. 0.073 ± 0.024, $\chi^2_1 = 5.55$, $P = 0.019$; maternal age^2 $\chi^2_1 = 0.15$, $P = 0.70$) Fig. 1A) reaching 9.5% at age 50. Older mothers (≥ 35 years old) had 3.43 (CIs: 1.08, 10.85) times higher risk of having a stillborn calf than younger mothers in their twenties. Stillbirth probability dropped from being 11.3% (CIs: 7.4, 16.9) for firstborn calves to only 1.8% (CIs: 1.0, 3.4) for laterborn calves, and firstborn calves thus were 6.83 (CIs: 2.86, 16.33) times more likely to be stillborn than subsequent calves of the same mother ($\chi^2_1 = 14,61$, $P = 0.0001$). There was no

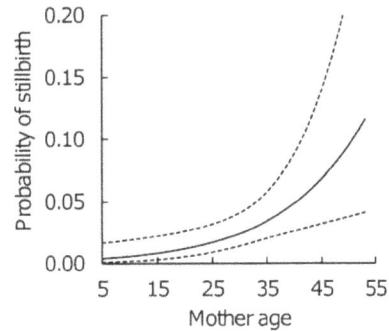

Figure 1. The probability of stillbirth according to maternal age at the birth of calf in Asian elephants. The probability of stillbirth increased significantly with maternal age especially after age 35 (n = 1020 births). Figure shows predicted values and 95% confidence intervals from the model.

significant difference in stillbirths between males and females (5.1% (CIs: 3.5, 7.5) vs. 4.1% (CIs: 2.6, 6.5), respectively; n male $= 517$, n female $= 503$ including 26 stillborn males and 19 females; $\chi^2_1 = 0.55$, $P = 0.46$). Finally, stillbirth probability appeared slightly higher among calves born in 1990s as compared to the other birth cohorts (1960: 3.8% (CIs: 1.1, 12.2), 1970: 3.2% (CIs: 1.3, 7.2), 1980: 3.4% (CIs: 1.9, 6.2), 1990: 7.1% (CIs: 4.6, 10.9)) but the differences between birth cohorts were not statistically significant ($\chi^2_3 = 4.08$, $P = 0.25$).

(ii) Calf mortality

Generally, the mortality risk of liveborn calves from birth until age 18 was highest during the first two years of life, followed by another rise in mortality risk between ages four and five when the calves are weaned and their training begins (Fig. 2A). Given the mortality risk quickly leveled off after age five (Fig. 2B), in the following we concentrate on examining factors associated with the high risk of mortality during the first five years of life.

Of the 975 liveborn calves, 25.6% died before reaching age five. The mortality risk varied significantly across different ages in a curvilinear pattern, rising to 6–7% per year during the first two years of life as well as between ages four and five (Fig. 2B, Tables 2 and 3).

Calves born to older mothers tended to have an increased mortality risk at all ages from birth to five years (Table 3; Fig. 2C). Calf mortality risk increased with maternal age so that for calves born to mothers aged 50 or above, the mortality risk was already 3.5-fold (CIs: 0.82, 15.14) compared to those born to mothers in their twenties. For example, calves born to mothers in their twenties had a 81.1% probability of survival to age 5 whereas calves born to mothers ≥ 50 years of age had on average only a 67.7% probability of survival to age 5.

Producing calves at short inter-birth intervals resulted in increased calf mortality risk, but this effect varied significantly according to the age of the calf (Table 3), so that the differences were largest during the first year of life. Calves born to mothers who had produced their previous calf less than 3.7 years earlier (25% quartile of the birth interval distribution) had a 1.5-fold (CIs: 0.87, 2.74) higher mortality risk during their first year (13.1% (CIs: 8.0, 20.6)) as compared to calves born with long interbirth intervals (6.9% (CIs: 3.9, 11.8)) (Fig. 2D). Such differences largely disappeared with the age of the calf.

Firstborn calves had 4.4 (CI: 0.62, 31.3) times higher mortality risk than did later born calves from the same mothers, and such

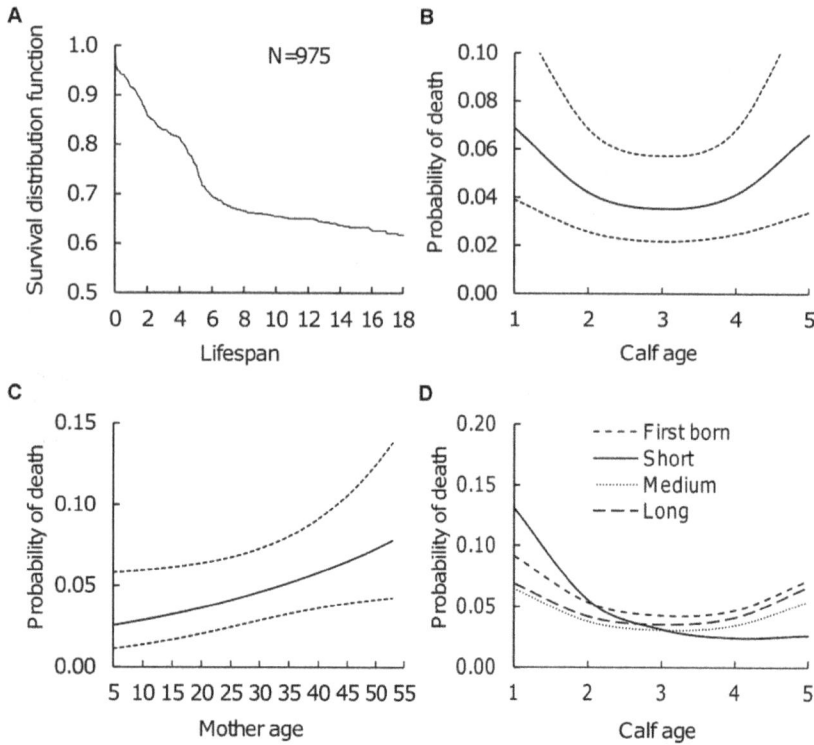

Figure 2. Overall survival distribution function of Asian elephants from birth to age 18 (A) and correlates of mortality of liveborn calves from birth to age of 5 (B–D) (Table 3). The mortality risk was highest during the first two years of calf life, and rose again between ages four and five (B). Calves born to older mothers had an increased probability of death irrespective of their own age (C) and short inter-birth interval put calves to higher risk of dying during their first two years of life (previous birth interval and calf age interaction, D). The sample size to which the discrete time survival analysis (B–D) was based was 975 calves with total of 4121 observations and correlated terms also included calf sex, birth order, birth cohort and repeated terms of the same mother and living area (mother id and mother area, see Methods and Table 3). Each year of calf age (1–5) illustrates the probability of dying within the previous full year (B,D, see Methods). The classification of inter-birth interval to first born, short, medium and long was based on the distribution of the birth interval length in the population and to 25% and 75% quartiles (D, see Methods). Figures show predicted values and 95% confidence intervals (B,C) from the model.

differences remained from birth until age five (no significant age x birth-order interaction, Table 3). This translated to firstborn calves having only a 51.9% survival probability to age 5, whereas laterborn calves had a 77.7% probability to survive to age 5. There was some tendency for particularly high mortality risk of firstborn calves produced by relatively old first-time mothers, but the interaction between maternal age and parity did not reach statistical significance (Table 3).

Male calves had 1.3 (CI: 1.00, 1.68) times higher mortality risk than female calves across all ages since birth until age five (Table 3).

Table 2. Mortality of Asian elephant calves by each age group from birth to age of 5 years.

Calf age	n died	n survived	% died
0–1	80	895	7.06
1–2	56	824	6.36
2–3	28	775	3.49
3–4	14	735	1.87
4–5	49	665	6.86
total(0–5)	227	665	25.6

n = 975, includes all calves used in the time-discrete survival model.

As a consequence, male calves had a 69.2% probability to survive to age 5, whereas female calves had a 75.0% probability to survive to age 5. The calf mortality risk also varied significantly across the study period with the best first year survival in 1980s (93.2% (CI: 88.3, 96.2)) and worst first year survival in 1960s (84.2% (CI: 73.5, 91.8)), but no consistent time pattern emerged across all age-groups (Table 3).

(iii) Death causes for calves under age 5

The main cause of death for calves under five was accidents (42.4%) which were distributed across all ages (Table 4). One in four (26.3%) died, usually during the first year of life, from mother agalactia (lack of milk) and/or general weakness of the newborn. Various diseases representing mainly gastrointestinal complications accounted for 22.8% of all deaths. Other causes such as taming-related injuries were relatively rare (4.5%). There were no clear overall sex-specific differences in any of the causes of death (Table 4).

Discussion

The worldwide populations of both wild and many captive Asian elephants will continue to be in danger of disappearing unless their current fertility and calf mortality rates improve [2]. A quarter of all remaining Asian elephants live in captivity in range countries in Asia and a further one thousand in zoos around the

Table 3. Discrete time survival model of the Asian elephant calf mortality risk from birth to 5 years ($n = 975$ calves with 4121 observations (227 deaths)).

Term	Category	Estimate±SE	Statistic (χ^2_{df})	P value	OR(CIs)
Calf age		-0.69 ± 0.19	29.24_1	<0.0001	
Calf age^2		0.17 ± 0.046	12.25_1	0.0005	
Sex			3.85_1	0.050	
	female	0.00 ± 0.00			1.00
	male	0.26 ± 0.13			1.30(1.00,1.68)
Birth-order			5.34_1	0.021	
	1st born	1.48 ± 1.00			4.40(0.62,31.3)
	later born	0.00 ± 0.00			1.00
Mother's age		0.024 ± 0.012	3.74_1	0.053	
Birth cohort			9.71_3	0.021	
	1960	0.69 ± 0.40			1.99(0.91,4.33)
	1970	0.16 ± 0.29			1.18(0.66,2.09)
	1980	-0.46 ± 0.23			0.63(0.40,0.99)
	1990	0.00 ± 0.00			1.00
Prev. birth interval			10.16_3	0.017	
	short	0.71 ± 0.35			2.03(1.03,4.02)
	medium	-0.07 ± 0.30			0.93(0.51,1.68)
	1st born	-1.17 ± 1.04			0.31(0.041,2.36)
	long	0.00 ± 0.00			1.00
Age*Prev.birth interval		Fig. 2d	9.06_3	0.029	
Age*Birth cohort			13.61_3	0.0035	
Constant		-3.23 ± 0.51			
Mother's age^2		-0.0013 ± 0.0011	1.59_1	0.21	
Sex*Birth-order			1.13_1	0.29	
Age*Mother's age		-0.011 ± 0.0070	2.49_1	0.11	
Age*Sex		-0.044 ± 0.091	0.23_1	0.63	
Age*Birth-order		20.092 ± 0.14	2.13_1	0.14	
Mum age*Birth order		0.036 ± 0.024	2.24_1	0.13	

Estimates with standard errors (S.E.) (positive estimate reflects increasing mortality risk) are provided for all variables (except interactions with several categories) and odds ratios (OR) with 95% confidence intervals (CIs) for categorical variables. Terms retained in the final model are shown above the constant, whereas examples of those that were rejected from the final model are shown below it. Mother's identity and area were fitted as repeated terms.

world [13]. The poor sustainability of many captive elephant populations is particularly constrained by a high calf mortality rate. We used the largest multigenerational demographic dataset in the world on a captive population of Asian elephants compiled from timber camps in Myanmar to investigate calf mortality patterns in detail for the first time. This is to our knowledge the only existing longitudinal dataset on captive Asian elephant from their range countries, and allowed us to explore not only general

Table 4. Death causes of Asian elephant calves dying before age 5 by sex and age group ($n = 224$).

Death cause	Females by age group (n)					Males by age group (n)				
	0–1	1–4	4–5	total	%	0–1	1–4	4–5	total	%
Accidents	11	26	5	42	43.3	19	27	7	53	41.7
General weakness and mother agalactia	20	7	1	28	28.9	20	4	7	31	24.4
Diseases	2	9	8	19	19.6	4	20	8	32	25.2
Taming-related injury	0	0	5	5	5.1	0	1	4	5	3.9
Heat stroke	0	0	1	1	1.0	1	2	0	3	2.4
Others	1	1	0	2	2.1	1	1	1	3	2.4

calf-specific differences in mortality, but also both between-mother and within-mother risk factors and temporal and geographic variation therein, as well as specific causes of death. Our results on extensively managed Myanmar timber elephants provide an important comparison to compromised survivorship reported in zoo elephants. They have implications for improving the management of captive working elephants in Asia as well as for refining population viability analyses with realistic parameter values.

Our finding that the general levels of stillbirth and pre-weaning mortality among Myanmar working elephants were considerably lower than those reported for zoo elephants is of interest. The recent progress in our understanding of physiology [26–27] and assisted reproductive technology in elephants [28–29] have lead to enhanced birth rates in zoo elephants [20]. However, such success is mitigated by the high calf mortality rate in zoo-born elephants (details in [20,30–31]. While zoo elephants experience high mortality due to sub-optimal climate, social group and feeding regime combined with prevalent reproductive problems [6,8,32–33], the wellbeing of captive elephants in timber camps of Southeast Asia living in an environment with more appropriate climate, social group and foraging/feeding regime could be compromised by husbandry and management [3,34] and work-related stress [18]. Unfortunately little data exist to compare our mortality figures with those of wild Asian elephants [4]. Our results on the working timber elephants thus add a significant comparison point for the zoo data, because although data exist on the reproductive patterns of other intensively maintained captive Asian elephants of South India and Sri Lanka (e.g. [9,30]), limited information has previously been available on the proximate causes of neonatal mortality in captive Asian elephants and virtually no previous study exists that follows the reproductive events of the same females throughout their lives.

One striking difference between zoo populations and working elephants in Myanmar is the stillbirth rate. During the past two decades, the reports on stillbirth rate in zoo-born Asian elephants have ranged between 16% [6–7] and 35% [20] of total births. In contrast, stillbirths have been thought to be rare in extensively kept and wild–living elephants [6,30] but few previous data have been published on this. Among the timber elephants in Myanmar, only 4% of both all male and female births resulted in stillbirths across the 40-year study period. Such a large difference as compared to zoo elephants is unlikely to be caused by undetected pregnancies in Myanmar because mahouts check their elephant health on a daily basis [17]. Mortality of liveborn calves was also lower among the working elephants than in zoo populations: 7% of calves died under one year of age and altogether a quarter before the age of five. Our data suggests a U-shaped pattern of survival during early years, with a high risk of mortality during the first two years of age and another peak between ages four and five years when calves are weaned and taming begins; similar double pulse in mortality of young elephants is also evident for a number of wild African elephant populations [35]. By comparison, the most recently estimated first year mortality is 30% for Asian elephants of both North American and European zoos (e.g. [2,8], which is about 6 times higher than our figure. No detailed data exists on wild Asian elephants, but in the Amboseli African elephants, 11% of females and 15% of males died in their first two years of life and altogether 19% of known calves died by the age of five [11]. It appears that the high calf mortality rates reported for zoo populations may be specific to populations kept under such conditions, and elephants kept under extensive keeping system in range countries as well as possibly also wild populations experience far lower mortality rates [see also 8,30]. Although the results on Myanmar timber elephants

are not directly relevant for example to zoo populations, a comparison between the different keeping systems might help to determine improvements that are required for the welfare of captive Asian elephants across the world [30].

Our results offer possibilities to improve the management and sustainability of the large captive working elephant populations in Asia. Given that in the wild the Asian elephant is listed as endangered and the captive populations form a large proportion of the worldwide remaining Asian elephants [3–4], understanding the factors which influence the ability of females to reproduce successfully both in natural conditions and in captivity, and finding ways in which to mitigate calf mortality risk, will help to maintain the large demands for captive populations of elephants in many countries without a need to capture elephants from the wild. This is of importance because for example, although many Asian elephant range countries have banned logging, captive working elephants remain important in the forestry industry in Myanmar. Our study identified several mother-related risk factors for increased stillbirth rate and calf mortality. Primiparous mothers (in particular those over the median age of first birth) had over 4-times higher risk of losing their liveborn calf, as compared to their later births, and they also had nearly 7-times higher risk of a stillbirth. Furthermore, calves born to older mothers had an increased risk of stillbirth and postnatal mortality, and short inter-birth interval put calves to higher risk of dying during their first two years of life. The effects reported here of maternal status, age as well as offspring sex on calf survival in Asian elephants generally support those reported previously for shorter-lived mammals, in particularly in ungulates [36–37]. That the results from Asian elephants confirm those for other large herbivores is of interest because mammals with extended longevity are under-represented in the current studies on evolutionary ecology and most theories are not tested with species living more than ~15 years. It is possible that repeated pregnancies without systematic provisioning in female timber elephants, coupled with lactational demands from previous calves could cause malnutrition in older Myanmar elephants that rely solely on natural foraging. These under-nourished older females may fail to maintain full-term pregnancy or to bear a healthy foetus to term, and may also struggle to meet the nutritional demands of live born calves.

It thus appears that maternal health and survival of calves could be improved by systematic supplementation and or along with reduced work-load of females in the risk groups. In order to enhance survival of calves, it might be necessary to ensure that older females and those with short birth-intervals take a long enough rest to replenish their body reserves after producing and suckling an offspring. It may also be advisable to supplement primiparous pregnant females and/or reduce their workload to avoid excess calf mortality. Furthermore, it is likely that social networks with older herd members are particularly important for the breeding success of first-time elephant mothers, given that in the wild, calves are typically raised by their mothers with the help of other related females [21–23]. In timber camps, elephants are grouped into units comprised of 6–7 animals. Calves are allowed to stay with and be nursed by their mothers around the clock until weaning. Although not all adult females live with their siblings or mothers in the same area/camp, the support of kin could be crucial at least in early reproductive attempts. Further research is needed on the effects of relationships between related and unrelated females on raising calves [11] to understand how physical and social attributes impact reproduction and offspring survival in Asian elephants in captivity.

The largest fractions of deaths were caused by accidents and by maternal agalactia or innate calf weakness. These latter causes might be partly preventable by supplementary feeding and/or work reduction of the high risk first time or ageing mothers. Over 40% of all liveborn calves died in accidents. They were often linked with a calf whose mother did not produce enough milk straying away from the mother in search of additional food, and might thus be avoidable by improving the nutrition of such calves and their mothers. The main causes for calf-weakness related deaths are inability of mothers to produce adequate milk following pregnancy and nutritional stress, as well as parasite infections of calves [11,38]. Males are known to be more susceptibility to parasitism than females [39], and they may also die more often from various diseases between ages one and four (Table 4). Overall, male calves had higher mortality rate at all ages as compared to females, likely caused by their higher infection rate as well as the fact that they were more likely to encounter accidents when searching for supplementary food as male calves are larger, require more energy to sustain adequate growth and are more expensive for mothers to produce [40]. Studies on Amboseli African elephants show that mother-calf distance increases with calf age and this occurs more rapidly if the mothers do not produce enough milk, and male calves tend to break the mother-offspring relationship earlier than female calves [11,40]. A third of all accidents in our study occurred during the summer months, when water and fodder are not as abundant and as nutritious as in other seasons and mothers are struggling to nurse their calves, despite not working and being able to forage ad-libitum along with their calves.

These results have implications for refining population viability analyses with realistic parameter values in order to understand the sustainability of captive elephant populations. Several attempts have been made recently to estimate the future population of the Asian elephant. For example, Leimgruber et al. [5] applied a Population Viability Analysis (PVA) using estimated demographic rates and population sizes to investigate the demographic link between captive and wild elephant populations in Myanmar, and the past, present and future consequences of continued live capture for remaining wild populations. They concluded that elephants will likely disappear from the wild in Myanmar in 31 years. A major weakness in such analyses has been the lack of detailed life-history data, in addition to disagreements over the wild population size and off-take from the wild. Coulson et al. [41] concluded that PVAs are good at predicting population dynamics and extinction probabilities only when extensive and reliable demographic data are available. The assumption of Myanmar elephant mortality schedules in Leimgruber et al. [5] was based on scattered information on wild elephant demography from other countries [42–43]. Our data show that assumptions such as equal mortality rates for females and males or correspondence of survivorship patterns in wild and captive population, as well as a lack of systematic assessment on the age groups most vulnerable to mortality may be inaccurate and may lead to biased population viability estimates. For example, the calf mortality rate for the captive population of Myanmar by Leimgruber et al. [5] corresponds to 34% of calves of both sexes dying by age 5, whereas in our study 25.6% of total liveborn calves died before age five with 1.3 times higher mortality risk among males. Future investigations are needed to project population growth rate for both captive and wild populations, taking into account the effects of female population age structure and parity on their calf mortality rate and sex differences therein. Another factor to consider is that the birth sex-ratios are close to 50:50, and while males do not contribute to population parameters, they do contribute to the work done in the timber industry. Therefore the differential mortality of males might contribute to a need to capture more males for work, affecting the breeding success of the wild populations even further.

Acknowledgments

We thank J. Ballou, A. Courtiol, P.C. Lee, H. Mumby, M. Rowcliffe, the editor and three anonymous referees for helpful comments and/or advice with statistics.

Author Contributions

Conceived and designed the experiments: KUM VL. Performed the experiments: KUM. Analyzed the data: ML KUM. Wrote the paper: KUM ML VL.

References

1. Mason GJ (2010) Species differences in responses to captivity: stress, welfare and the comparative method. Trends Ecol Evol 25: 713–721.
2. Weise RJ (2000) Asian elephants are not self-sustaining in North America. Zoo Biology 19: 299–309.
3. Lair RC (1997) Gone astray: The care and management of the Asian elephant in domesticity. Bangkok: FAO Regional Office for Asia and Pacific.
4. Sukumar R (2003) The living elephants: evolutionary ecology, behaviour and conservation. Oxford: Oxford University Press.
5. Leimgruber P, Senior B, Uga M, Aung M, Songer A, et al. (2008) Modeling population viability of captive elephants in Myanmar (Burma): implications for wild populations. Anim Conserv 11: 198–205.
6. Kurt F, Mar KU (1996) Neonate mortality in captive Asian elephants (Elephas maximus). Z Saugetierk 61: 155–164.
7. Schmid J (1998) Status and reproductive capacity of the Asian elephant in zoos and circuses in Europe. Int Zoo News 45: 341–351.
8. Clubb R, Rowcliffe M, Lee P, Mar KU, Moss C, et al. (2008) Compromised survivorship in zoo elephants. Science 322: 1649–1649.
9. Sukumar R, Krishnamurthy V, Wemmer C, Rodden M (1997) Demography of captive Asian elephants (Elephas maximus) in southern India. Zoo Biology 16: 263–272.
10. Moss CJ (2001) The demography of an African elephant (Loxodonta africana) population in Amboseli, Kenya. J Zool 255: 145–156.
11. Moss CJ, Croze H, Lee PC (2011) The Amboseli elephants: A long-term perspectives on a long-lived mammal. Chicago: The University of Chicago Press.
12. Sukumar R (1998) The Asian elephant: Priority populations and projects for conservation. Report to WWF-US.
13. Kurt F, Mar KU, Garai M (2008) Giants in chains: History, biology and preservation of Asian elephants in Asia. In: Wemmer C, Christen CA, eds.

Elephants and ethics: Toward a morality of coexistence. Maryland: The John Hopkins University Press. pp 327–345.
14. Sukumar R (2006) A brief review of the status, distribution and biology of wild Asian elephants Elephas maximus. Int Zoo Yearbook 40: 1–8.
15. Aung TU, Nyunt TU (2001) A country study on Asian elephant in Myanmar. In: Baker I, Kashio M, eds. Giants on our hands: proceedings of the international workshop of the domesticated Asian Elephant. Bangkok, Thailand: Dharmasarn Co. Ltd. pp 89–101.
16. Myint MU (2009) The analysis on the status, care and management of elephants in Myanmar. Yangon, Myanmar: Departmental circulation, Myanma Timber Enterprise.
17. Toke Gale U (1971) Burmese timber elephants. Yangon: Trade Corporation.
18. Mar KU (2007) The demography and life-history strategies of timber elephants of Myanmar. PhD thesis, University College of London, London.
19. Schmidt MJ, Mar KU (1996) Reproductive performance of captive Asian elephants in Myanmar. Gajah 16: 23–42.
20. Saragusty J, Hermes R, Göritz F, Schmitt DL, Hildebrandt TB (2009) Skewed birth sex ratio and premature mortality in elephants. Anim Rep Science 115: 247–254.
21. Rapaport L, Haight J (1987) Some observations regarding allomaternal caretaking among captive Asian elephants (Elephas maximus). J Mammal 68: 438–442.
22. Lee PC (1987) Allomothering among African elephants. Anim Behav 35: 278–291.
23. McComb K, Shannon G, Durant SM, Sayialel K, Slotow R, et al. (2011) Leadership in elephants: the adaptive value of age. Proc R Soc Lond B. In press.
24. Mellen J, Keele M (1994) Social structure and behaviour. In: Mikota S, Sargent EL, Ranglack GS, eds. Medical management of the elephant. Michigan: Indira Publishing House. pp 19–26.

25. Singer JB, Willett JB (2003) Applied longitudinal data analysis: Modeling change and event occurrence. New York: Oxford University Press.

26. Brown JL, Hildebrandt TB (2003) The science behind elephant artificial insemination. Biol Reprod 68: MS17.

27. Brown JL, Kersey DC, Freeman EW, Wagener T (2010) Assessment of diurnal urinary cortisol excretion in Asian and African elephants using different endocrine methods. Zoo Biol 29: 274–283.

28. Hermes R, Saragusty J, Schaftenaar W, Goritz F, Schmitt DL, et al. (2008) Obstetrics in elephants. Theriogenology 70: 131–144.

29. Hildebrandt TB, Goritz F, Hermes R, Reid C, Denhard M, et al. (2006) Aspects of the reproductive biology and breeding management of Asian and African elephants *Elephas maximus* and *Loxodonta africana*. Int Zoo Yearbook 40: 20–40.

30. Taylor VJ, Poole TB (1998) Captive breeding and infant mortality in Asian elephants: A comparison between twenty western zoos and three eastern elephant centers. Zoo Biology 17: 311–332.

31. Dale R (2010) Birth statistics for African (*Loxodonta africana*) and Asian (*Elephas maximus*) elephants in human care: history and implications for elephant welfare. Zoo Biology 29: 87–103.

32. Clubb R, Rowcliffe M, Lee P, Mar KU, Moss C, et al. (2009) Fecundity and population viability in female zoo elephants: problems and possible solutions. Anim Welfare 18: 237–247.

33. Mason GJ, Veasey JS (2010) How should the psychological well-being of zoo elephants be objectively investigated? Zoo Biology 29: 237–255.

34. Ramanathan A, Mallapur A (2008) A visual health assessment of captive Asian elephants (*Elephas maximus*) housed in India. J Zoo Wildlife Med 39: 148–154.

35. Shrader AM, Pimm SL, van Aarde RJ (2010) Elephant survival, rainfall and the confounding affects of water provision and fences. Biodiv & Conserv 19: 2235–2245.

36. Clutton-Brock TH, Albon SD, Guinness FE (1985) Parental investment and sex differences in juvenile mortality in birds and mammals. Nature 313: 131–133.

37. Hamel S, Côté SD, Festa-Bianchet M (2010) Maternal characteristics, environment, and costs of reproduction in female mountain goat. Ecology 91: 2034–2043.

38. Fowler ME, Mikota SK (2006) Biology, medicine and surgery of elephants. Iowa: Blackwell Publishing Professional.

39. Marriott I, Huet-Hudson YM (2006) Sexual dimorphism in innate immune responses to infectious organisms. Immunol Res 34: 177–192.

40. Lee PC, Moss CJ (1986) Early maternal investment in male and female African elephant calves. Behav Ecol Sociobiol 18: 353–361.

41. Coulson T, Mace GM, Hudson E, Possingham H (2001) The use and abuse of population viability analysis. Trends Ecol Evol 16: 219–221.

42. Sukumar R, Santiapillai C (1993) Asian elephant in Sumatra; population and habitat viability analysis. Gajah 11: 59–63.

43. Tilson R, Soemarna K, Ramono W, Sukumar R, Seal U, et al. (1994) Asian elephant in Sumatra: population and habitat viability analysis report. Bandar Lampung, South Sumatra: CBSG.

18

Influence of Forest Management Regimes on Forest Dynamics in the Upstream Region of the Hun River in Northeastern China

Jing Yao[1,2]*, Xingyuan He[1]*, Anzhi Wang[1], Wei Chen[1], Xiaoyu Li[1], Bernard J. Lewis[1], Xiaotao Lv[1]

1 State Key Laboratory of Forest and Soil Ecology, Institute of Applied Ecology, Chinese Academy of Sciences, Shenyang, People's Republic of China, 2 Graduate University of Chinese Academy of Sciences, Beijing, People's Republic of China

Abstract

Balancing forest harvesting and restoration is critical for forest ecosystem management. In this study, we used LANDIS, a spatially explicit forest landscape model, to evaluate the effects of 21 alternative forest management initiatives which were drafted for forests in the upstream region of the Hun River in northeastern China. These management initiatives included a wide range of planting and harvest intensities for *Pinus koraiensis*, the historically dominant tree species in the region. Multivariate analysis of variance, Shannon's Diversity Index, and planting efficiency (which indicates how many cells of the target species at the final year benefit from per-cell of the planting trees) estimates were used as indicators to analyze the effects of planting and harvesting regimes on forests in the region. The results showed that the following: (1) Increased planting intensity, although augmenting the coverage of *P. koraiensis*, was accompanied by decreases in planting efficiency and forest diversity. (2) While selective harvesting could increase forest diversity, the abrupt increase of early succession species accompanying this method merits attention. (3) Stimulating rapid forest succession may not be a good management strategy, since the climax species would crowd out other species which are likely more adapted to future climatic conditions in the long run. In light of the above, we suggest a combination of 30% planting intensity with selective harvesting of 50% and 70% of primary and secondary timber species, respectively, as the most effective management regime in this area. In the long run this would accelerate the ultimate dominance of *P. koraiensis* in the forest via a more effective rate of planting, while maintaining a higher degree of forest diversity. These results are particularly useful for forest managers constrained by limited financial and labor resources who must deal with conflicts between forest harvesting and restoration.

Editor: Ben Bond-Lamberty, DOE Pacific Northwest National Laboratory, United States of America

Funding: This study was financed by the National Forestry Public Benefit Research Foundation of China (No. 200804001) and the National Natural Science Foundation of China (No. 40971272). The funders had no role in study design, data collection and analysis, decision to publish, or preparation of the manuscript.

Competing Interests: The authors have declared that no competing interests exist.

* E-mail: avril_y@foxmail.com (JY); hexy@iae.ac.cn (XH)

Introduction

Conservation and responsible utilization are two essential facets of the harmonious relationship between humans and nature, as well as two interrelated strategies of sustainable forest development. Forests around the world are shrinking due to over-exploitation [1–3]. Because forests play a critical role in water conservation [4,5], prevention of soil erosion [6,7] and climate regulation [8], ecological problems such as massive soil erosion, catastrophic flooding, and severe dust storms often follow forest degradation. As a reaction to this, a host of theories and methods on forest restoration have been explored by researchers [9–11]. Still, some research has questioned whether the strategy to restore forest ecosystems following the pathway of historical forest succession is a proper one under conditions of uncertain variation of climatic patterns induced by climate change [12,13]. Ravenscroft et al. [14] have also pointed out that high diversity of conditions and species within forest landscapes is the most effective means of ensuring the future resistance of ecosystems to climate-induced declines in productivity.

Meanwhile, some research on the effects of utilization especially timber harvesting, on forest ecosystems has found that while there is an initial setback of forest succession after the forest is harvested, harvesting can increase ecosystem diversity [15–17]. Yet relatively little research has explored how to ensure a balance between restoration and harvesting [18], which is important in most regions where forest industry is dominant.

Due to rapid population growth, coupled with agricultural development, urban construction, and unsound forest management, the degradation of forest resources in China, particularly in the North, has been accelerating, with the resultant deterioration of the environment [3]. To prevent this, the Three-North (which includes northwestern, north and northeastern of China) Shelterbelt project (http://www.forestry.gov.cn) was launched in 1978, aiming to prevent soil erosion and desertification by increasing forest coverage through afforestation and protection of farmland, as well as enhance urban eco-security in North China.

In the Northeast, forests have been degraded due to unsound timber harvest and farming [3,19]. The forest in the upstream area of the Hun River (UHR) in Qingyuan county of Liaoning province is an important focus of the Three-North Shelterbelt Project, since

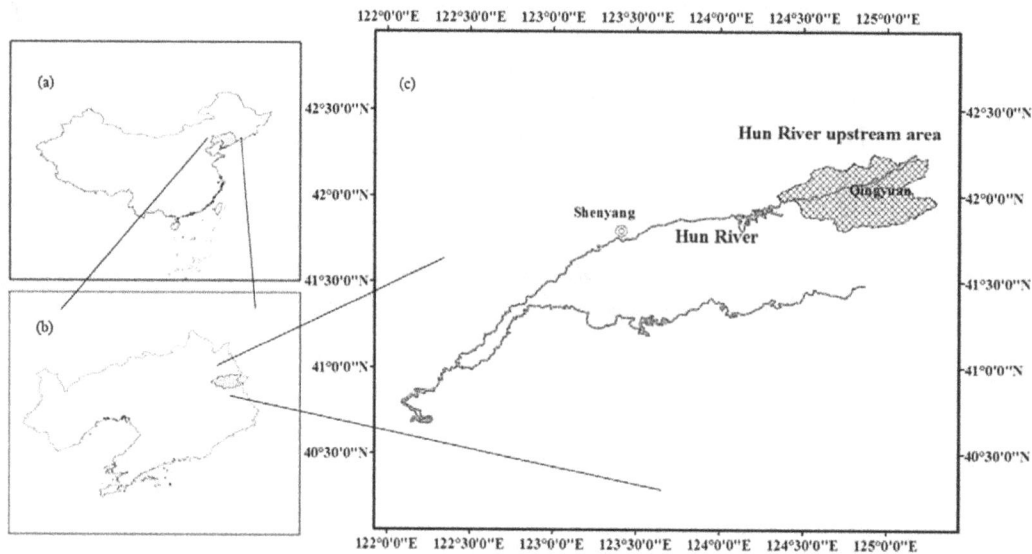

Figure 1. Location of the study area. a– Liaoning province in Northeast China; b– Qingyuan County in Liaoning province; and c– Hun River upstream area.

it benefits the eco-environment of the downstream area of Hun River (DHR), a region of heavy industry and the population center of Northeast China. Unfortunately, the forests in the UHR were over-exploited in the last century and degraded from mixed broad-leaved *P. koraiensis* forest [20] to secondary forests dominated by *Quercus. mongolica*, *Betula spp.*, *Populus spp.* and other early or mid-successional species. Thus forests in the UHR are in a transitional stage from early- to mid-succession and accelerating the successional process in this region is urgent for ecological security of the urban centers of the DHR. At the same time, forest industry is still one of the basic industries in Qingyuan county, just as in other forested areas of northeast China. In accordance with the Three-North Shelterbelt project and the needs of local forest industry, half of the forest in Qingyuan county is reserved as public benefit forest where timber harvesting is forbidden; while the remaining half serves as a timber resource.

Given the above context, there are two key points of concern in this article: (1) the choice of planting intensity for accelerating the successional path toward the climax forest; and (2) at the landscapes scale, whether the harvest in forests designated for timber harvesting would contribute to the degradation of the overall forest ecosystem.

LANDIS is a powerful tool for evaluating alternative forest management strategies at a landscape scale [21–25] due to its ability to simulate forest variation at large spatial and temporal scales, a capability beyond the limits of traditional studies [22,26,27]. In the past decade, LANDIS has been used by researchers across North America, Europe and China [28] as a tool for decision making in forest management. For example, Wang et al. [27] examined effects of different planting densities on forest restoration, providing a guide for forest management decisions which has proven to be one of the most cost-effective and least labor-intensive in North China. Cairns et al. [23] considered alternative restoration strategies for insect-affected landscapes by LANDIS and suggested that it is necessary to consider the patterns of hosts on the landscape as well as the landscape composition.

In this research, we utilized LANDIS to examine the effects of different planting regimes and harvest regimes on forest succession

to explore the balance between restoration and harvesting. There are 3 specific questions we address in this paper. (1) Which planting intensity is the most effective for restoring the forest ecosystem while taking diversity and limited financial and labor resources into account? (2) How does harvesting affect forest dynamics – i.e., does harvesting increase biodiversity of the landscape and which harvest intensity should be selected? (3) How do individual species respond to forest management regimes which involve different planting and harvest intensities?

Methods

1 Study Area

The upper Hun River area extends from 41°47′52″ ∼ 42°28′25″N,124°20′06″ ∼ 125°28′58″E (Fig. 1), and is characterized by a temperate continental monsoon climate. The mean annual temperature is 6.6°C and mean annual precipitation is 788 mm (data provided by the Qingyuan County Forestry Bureau). The average altitude of this region is 470 m, ranging from 150 m to 1086 m. The total study area encompasses 2.5×10^5 ha. The forest in this area was at one time the climax community of the *P. koraiensis* and deciduous broad-leaved mixed forest. However, the natural forest was severely affected by human disturbance in the early 20th century. It has now been replaced by a secondary mixed forest which includes *P. koraiensis*, *Q. mongolica*, *Larix olgensis*, *Pinus tabulaeformis*, *Pinus densiflora*, *Pinus sylvestris var. mongolica*, *Fraxinus rhynchophylla*, *Fraxinus chinensis*, *Juglans mandshurica*, *Betula platyphylla*, *Populus davidiana*, *Acer pictum subsp. mono*, *Ulmus pumila*, *Tilia amuresis*, *Abies nephrolepis*, *Picea asperata* and other lesser species.

2 Description of LANDIS

LANDIS is a spatially explicit, stochastic, raster-based landscape model which facilitates the study of the effects of natural and anthropogenic disturbances, vegetational succession, management strategies and their interactive effects on forest landscapes [22,26,29]. It simulates species-level forest dynamics by tracking the presence or absence of species age cohorts at 10-year time steps under natural and anthropogenic disturbances, including fire, wind throw, insects and disease, harvesting, and fuel management. It simulates forest landscape change over large spatial (10^3–10^7 ha) and temporal (10^1–10^3 years) scales with flexible resolutions (10–500 m pixel size). The model is described in detail elsewhere [25,26,30]. The version used in this research is LANDIS 6.0.

2.1 Parameterization of LANDIS. There are two types of inputs in LANDIS: necessary inputs and optimal inputs. The necessary inputs include species life history attributes, species composition maps with associated presence/absence and age information for species, land type maps and species establishment coefficients for each land type. The optimal inputs include disturbance and management parameters including harvesting, planting and fire.

2.2 Species attributes and species composition maps. Sixteen common tree species in the study area are included in our LANDIS simulation (Table 1). Species life history attributes were derived from the literature on species characteristics in this region [31–33], the parameterization of other research on northeastern China, and consultations with local experts [22,34,35]. The species composition map was derived from an extant stand map of 2006 and a stand attribute database, the latter of which was a component of 2006 forest inventory data provided by the Qingyuan County Forestry Bureau. The forest stand map recorded the boundaries of stands. The stand attribute database provided information on the relative percentage of canopy species, the average age of dominant canopy species, timber production, and crown density. The forest composition map was processed at a resolution of 60 m×60 m, which yielded 1320 rows ×836 columns. Each cell contained the presence/absence and age cohorts of all of the 16 tree species. For each cell in a stand, a stand-based assignation (SBA) approach [36] was used to stochastically assign species age cohorts to that cell based on forest inventory data.

2.3 Land type map. In LANDIS the heterogeneous landscape is stratified into relatively homogeneous units (land types or eco-regions) in LANDIS. Within each land type, environments for species establishment are assumed to be similar [26]. In this study, we first extracted the water body and city out of the land type map. Then we derived seven land types (Fig. 2), primarily based on terrain attributes of the 2006 forest inventory data and the 1992 Digital Elevation Model (DEM) of Qingyuan County at a resolution of 30 m×30 m which was downloaded from http://www1.csdb.cn/. Seven kinds of terrain were delineated: North Ridge (NR), South Ridge (SR), North Slope (NL), South Slope (SL), North Slope of valley (NV), South Slope of valley (SV) and terrace (T). There were no non-active land types in our land type map. The seven active land types accounted for 0.04%, 0.04%, 50.38%, 44.19%, 0.13%, 0.37% and 4.85% of the total area, respectively.

The species establishment coefficients of a land type are critical. They estimate the probability of a species successfully establishing on that land type, given the environmental conditions encapsulated by that type. We estimated the species establishment coefficients of land types from the literature on species characteristics in this region [31–33] and from parameterizations in other research on northeastern China [22,34,35].

3 Simulation Scenarios

The initial forest composition and land type maps including species/age classes realistically represented the status of the forests in the study area in 2006. We simulated 21 scenarios, including the natural succession process without planting and harvest, five levels of planting intensity and fifteen different combinations of the five planting intensity levels and three selective harvest intensity levels (Table 2). *P. koraiensis* was planted under broadleaved trees that were >9 years old and whose canopies were broad enough to provide a shaded environment for the seedlings of *P. koraiensis* [37]. Harvest age of species in this study followed National Forest Resources Continuous Inventory Technique Formula (Table 3). Three replicates of each scenario were simulated. All scenarios were simulated up to 300 years to examine the long-term effects of planting intensity and harvest intensity on forest succession. The

Table 1. Species' life attributes for forests in the upstream area of the Hun River in northeastern China.

Species	LONG	MTR	ST	FT	ED	MD	VP	MVP
Pinus koraiensis	400	40	5	1	50	200	0	0
Pinus tabulaeformis	200	30	2	1	100	500	0	0
Pinus densiflora	200	30	2	1	100	500	0	0
Pinus sylvestris var. mongolica	250	40	2	2	30	100	0	0
Larix olgensis	300	30	1	5	100	400	0	0
Picea asperata	300	30	5	3	80	150	0	0
Abies nephrolepis	250	40	5	3	80	150	0	0
Populus davidiana	100	8	1	2	−1	−1	1	10
Betula platyphylla	150	15	1	1	200	4000	0.8	50
Ulmus pumila	250	10	2	4	300	1000	0.3	60
Fraxinus chinensis	250	30	3	3	50	150	0.3	80
Fraxinus rhynchophylla	250	30	3	3	50	150	0.3	80
Juglans mandshurica	250	15	3	4	50	150	0.9	60
Quercus mongolica	350	40	3	5	20	200	0.9	60
Acer pictum subsp. mono	250	10	4	2	120	350	0.3	50
Tilia amuresis	300	30	4	4	50	100	0.9	30

Long– longevity (years); MTR–age of maturity (years); ST–shade tolerance class; FT–fire tolerance class; ED–effective seeding distance (m); MD–maximum seeding distance (m); VP–vegetative reproduction probability; MVP–minimum age of vegetative reproduction (years).

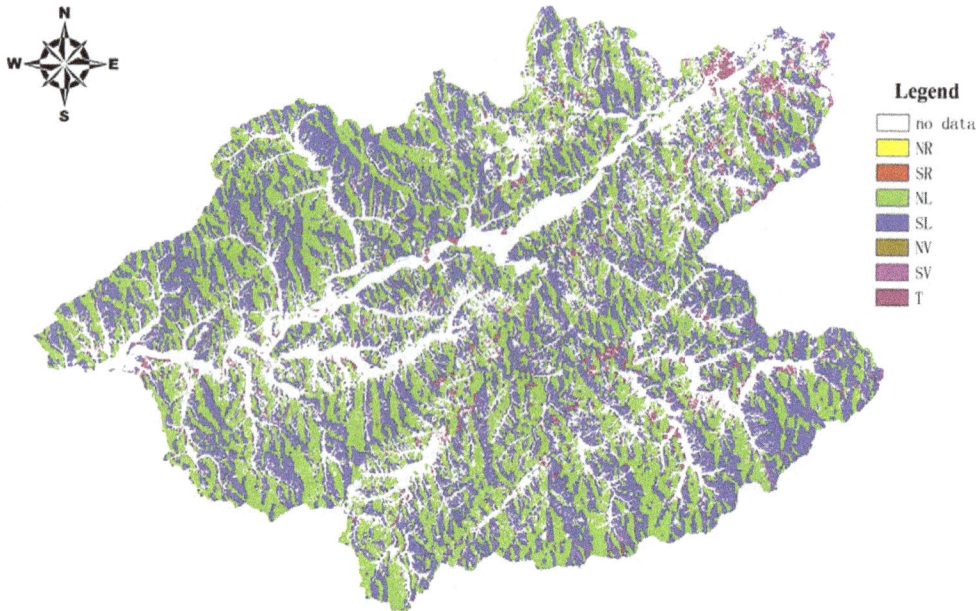

Figure 2. Landtype map of the Hun River upstream area. NR–North Ridge; SR–South Ridge; NL–North Slope; SL–South Slope; NV–North Slope of valley; SV–South Slope of valley; and T–terrace.

study area was divided into 10 management areas (MA) identified via planting and harvest options practiced there: (1) short-rotation timber of broadleaved trees which are >9 years old (MA 1); (2) short-rotation timber of all trees other than those included in MA 1 (MA 2); (3) fast-growing timber of broadleaved trees which are >9 years old (MA 3); (4) fast-growing timber of all trees except those included in MA 3 (MA 4); (5) public forest of broadleaved trees that are >9 years old (MA 5); (6) public forest of all trees except those included in MA 5 (MA 6); (7) general natural timber of broadleaved trees that are >9 years old (MA 7); (8) general natural timber of all trees except those included in MA 7 (MA 8); (9) general plantation timber of broadleaved trees that are >9 years old (MA9); and (10) general plantation timber of all trees except those included in MA 9 (MA 10). The harvest regimes were implemented on all management areas except MA5 and MA6, where harvesting of public forests is forbidden according to the Three-North Shelterbelt project.

4 Analysis Methods

Utilizing SPSS 18.0, the area percentage (AP) of each species was calculated from the output map for each 10-year step in the LANDIS output statistical program to depict the trend for each species over the 300 simulated years.

We analyzed the AP of each species utilizing multivariate analysis of variance (MANOVA) with planting intensity and harvest intensity in SPSS 18.0. Pillai's Trace statistic was used to test the hypotheses that planting intensity and harvest intensity affect the area of species in the study area, because it is the least sensitive of the four multivariate tests provided by SPSS with respect to the heterogeneity of variance assumption of MANOVA [27]. Shannon's Diversity Index of each scenario in the 300th year was calculated in FRAGSTATS 3.0. Then we analyzed differences in Shannon's Diversity values induced by different harvest intensities using one-way ANOVA in SPSS 18.0. The LSD was used to test the hypotheses that different harvest intensities induce differences in Shannon's Diversity values.

We also developed and calculated planting efficiency to test the response of *P. koraiensis* coverage to different planting intensities via the following formula: , where PE is planting efficiency; A_i is the area (cell) of *P. koraiensis* coverage at year 300 under different planting intensity scenarios; A_N is the area (cell) of *P. koraiensis* coverage at year 300 in the natural succession scenario without any planting and harvesting; and A_j is the overall planting area (cell) in different planting scenarios. Because planting is a way of restoration to increase the seed source [38], the PE indicates how many cells of the target species at year 300 benefit from per-cell of planting *P. koraiensis* under the different planting intensities.

Results

In the N scenario (Table 2), in which there was no planting or harvesting, mid- and late-succession species (*P. koraiensis, P. asperata, A. nephrolepi, U. pumila, A. pictum subsp. mono and T. amuresis*) showed increasing trends in percentage of total area (Fig.3 F), whereas early succession species (*P. sylvestris var. mongolica, L. olgensis and B. platyphylla*) showed decreasing trends (Fig.3 D). *Q. mongolica*, which is a mid-succession species, also demonstrated decreasing trends in area percentage (Fig.3 A). Trends for *P. tabulaeformis, P. densiflora, F. chinensis, F. rhynchophylla and J. mandshurica*, which are mid-tolerant species, first increased and then decreased (Fig.3 E). The area percentage of *P. koraiensis* was 4.16% in the first year and rose to 19.53% by year 300. The area percentages of *Q. mongolica* and *L. olgensis* were 44.69% and 37.18%, respectively, in the initial year, and fell to 30.94% and 0%, respectively, by year 300. Although *Q. mongolica* was still the most abundant species in the study area at year 300, the composition of the Hun River upstream forest changed from one dominated by *Q. mongolica* and *L. olgensis* to one dominated by *Q. mongolica* and *P. koraiensis*. Thus the successional trajectory of the forest was slowly heading toward the climax forest in this region.

In the five P scenarios (Table 2) in which there were planting, most species displayed the same area percentage trends as they did in the N scenarios (Fig. 3). Nevertheless, the planting of *P. koraiensis*

Table 2. The scenarios simulated by LANDIS 6.0.

Scenario	Planting intensity of *P. koraiensis*	Selective Harvest (general timber forest)	Selective Harvest (short-rotation forest and fast-growing forest)
N	–	–	–
P1	5%	–	–
P2	10%	–	–
P3	30%	–	–
P4	50%	–	–
P5	70%	–	–
P1H1	5%	10%	30%
P1H2	5%	30%	50%
P1H3	5%	50%	70%
P2H1	10%	10%	30%
P2H2	10%	30%	50%
P2H3	10%	50%	70%
P3H1	30%	10%	30%
P3H2	30%	30%	50%
P3H3	30%	50%	70%
P4H1	50%	10%	30%
P4H2	50%	30%	50%
P4H3	50%	50%	70%
P5H1	70%	10%	30%
P5H2	70%	30%	50%
P5H3	70%	50%	70%

Note: *P. koraiensis* was planted under broadleaved trees which were >9 years old, whose canopies were broad enough to provide a shaded environment for the seedlings of *P. koraiensis* [37].

Table 3. Harvest age (years) of species in this study according to National Forest Resources Continuous Inventory Technique Formula (China).

Species	SRT	FGT	GPT	GNT
Pinus koraiensis	>40	–	>80	>120
Pinus tabulaeformis	–	–	>40	>60
Pinus densiflora	–	–	>40	>100
Pinus sylvestris var. mongolica	>20	–	>40	>100
Larix olgensis	>20	>20	>40	>100
Picea asperata	–	–	>80	>120
Abies nephrolepis	–	–	>40	>100
Populus davidiana	>10	>20	>20	>20
Betula platyphylla	>10	>20	>40	>60
Ulmus pumila	–	–	>40	>60
Fraxinus chinensis	>20	–	>50	>80
Fraxinus rhynchophylla	>20	–	>50	>80
Juglans mandshurica	>20	–	>50	>80
Quercus mongolica	–	–	>50	>80
Acer pictum subsp. mono	>30	–	>50	>80
Tilia amuresis	–	–	>50	>80

SRT– short-rotation timber; FGT– fast-growing timber; GPT– general plantation timber; GNT– general natural timber.

did suppress the increasing trends of other species (Fig. 3) and accelerate the successional process toward a climax forest dominated by *P. koraiensis* (Fig. 3 B). Increasing trends for *A. pictum subsp. mono*, *U. pumila*, *T. amuresis* and *A. nephrolepis* were either diminished or reversed (Fig. 3 F). At year 300, the area percentage of *P. koraiensis* was nearly same as that of *Q. mongolica* under the 5% planting intensity scenario and reached 56% under the 30% planting intensity scenario (Fig. 4). Although the area percentage of *P. koraiensis* at year 300 increased with increasing planting intensity, the sensitivity of response of *P. koraiensis* to planting intensity decreased with increasing intensity (Table 4).

In the PH scenarios (Table 2), in which there was both planting and harvesting, trends in area percentage of species (not shown in the paper) were similar to those in the P scenarios, except for *P. davidiana*, as exemplified by comparing values for 5% planting intensity with those for combinations of 5% planting intensity and different harvest regimes (Fig. 5). The dynamic of area percentage for *P. davidiana* was stable until there was an abrupt increase at year 250 under 5% planting intensity scenario without harvesting. However, a similar increase occurred around year 50 under the PH scenarios.

The test of MANOVA on PH scenarios showed that both P regimes and H regimes had significant effects on the dynamics of the forest in the study area (Table 5). For individual species, P regimes had significant effects on most species, with the exception of *L. olgensis* and *P. sylvestris var. mongolica;* while H regimes only had significant effects on *Q. mongolica*, *L. olgensis*, *P. davidiana*, *B. platyphylla* and *U. pumila* (Table 5). For more detail, the effects on species between different harvest levels

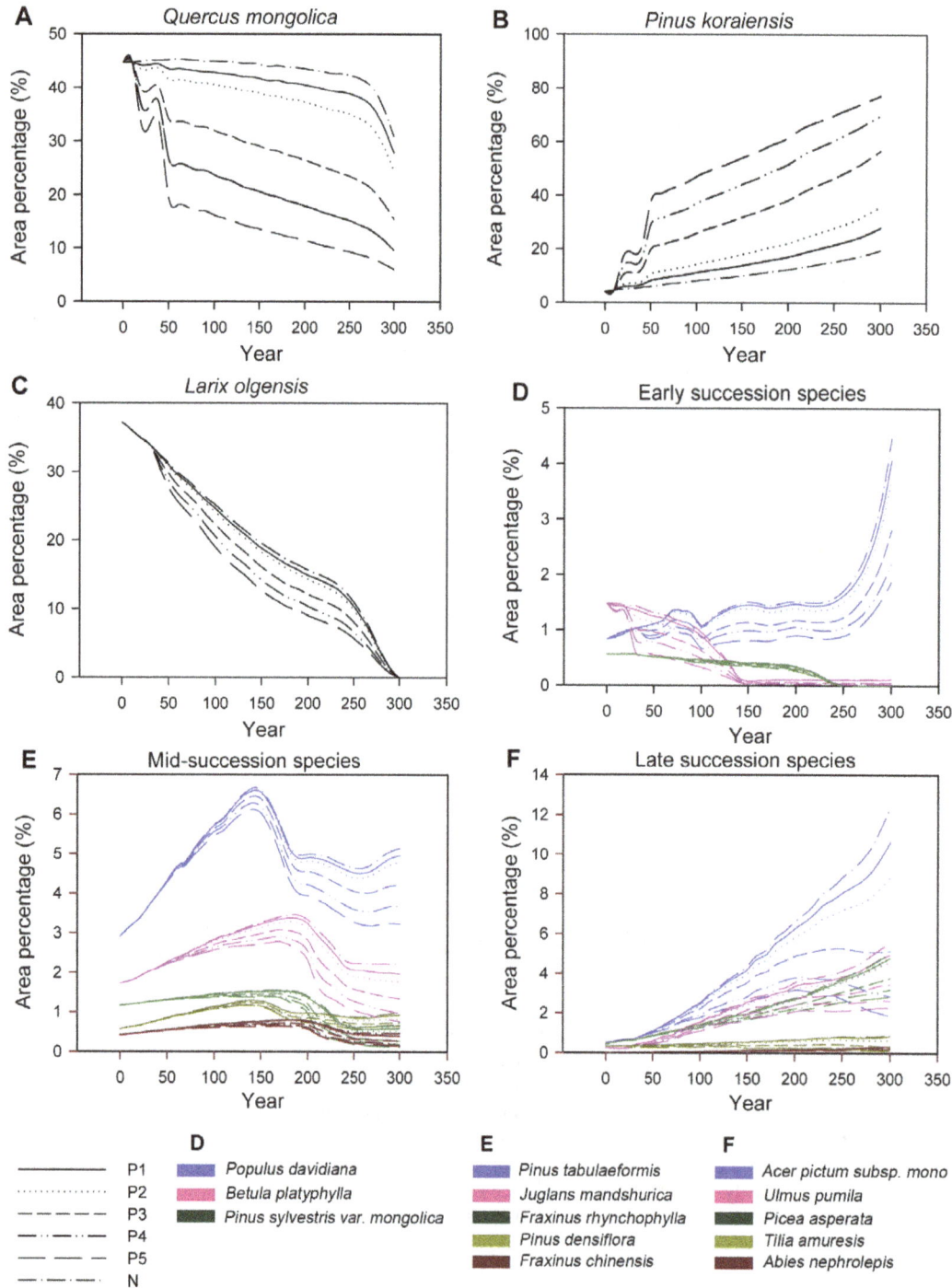

Figure 3. Response of area proportion of different species to different planting intensities. P1–P5 Definitions are given in Table 2.

were tested by the custom hypothesis test (simple contrast, the H1 which is referenced to Table 2 was chosen as the reference) and showed that increasing harvest intensity had a significant positive effect on *P. davidiana*, *B. platyphylla*, *P. densiflora*, *U. pumila* and *T. amuresis* and a negative effect on *L. olgensis* and *Q. mongolica* (Table 6).

Shannon's Diversity Index value showed that increased planting intensity of *P. koraiensis* was followed by decreasing diversity of

forest composition, while harvest regimes could slightly increase forest diversity (Fig. 6).

Discussion

1 The Effects of Planting

Planting is an effective method of accelerating forest succession [11,17,27], but the most effective and proper planting intensity depends on the forest management strategy and investment

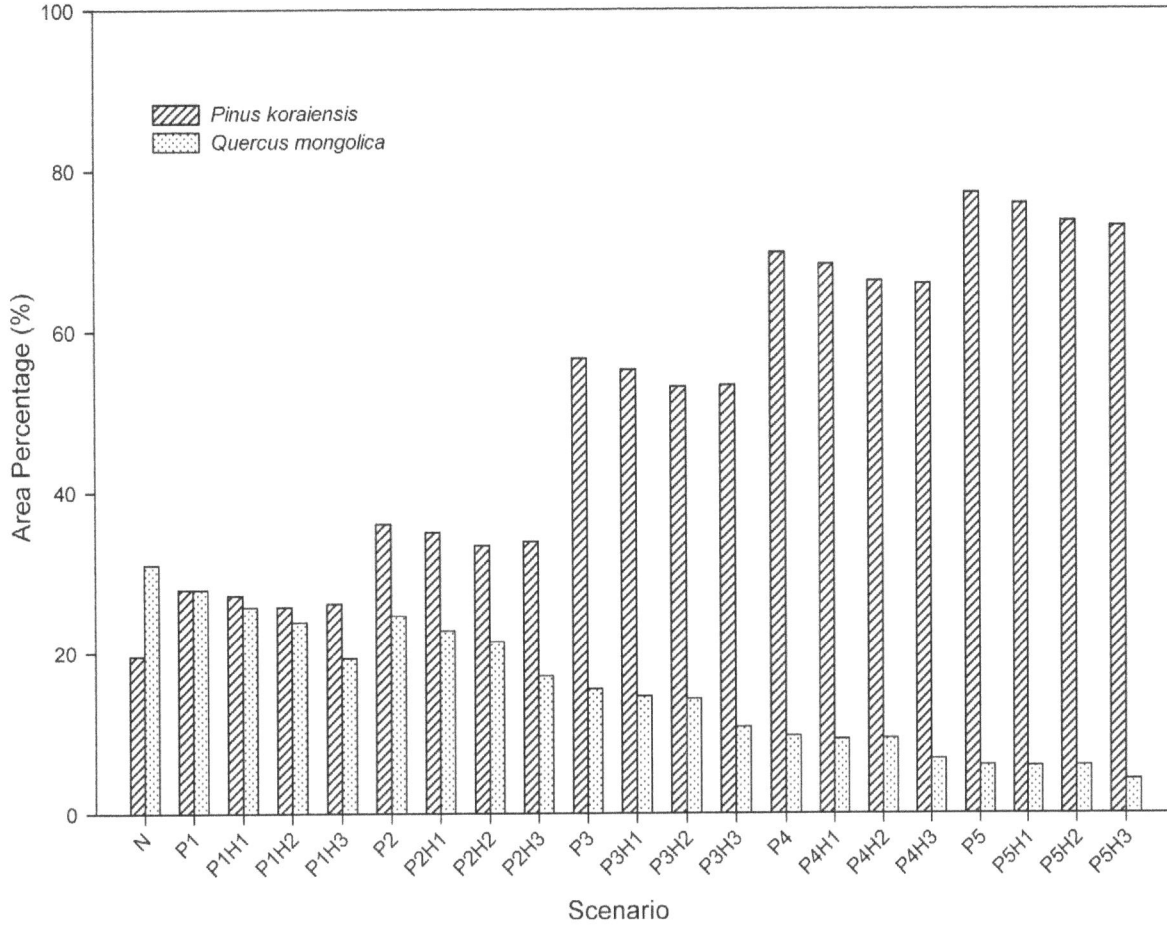

Figure 4. Area percentage of *Pinus koraiensis* **and** *Quercus mongolica* **in simulated scenarios at year 300.** The definitions of the scenarios may be found in Table 2.

budget. We suggest that a 30% planting intensity is the proper planting regime. On one hand, in the long run, it can accelerate the forest to be absolutely dominated by *P. koraiensis* via a more effective rate of planting. On the other hand, it could maintain a higher degree of forest diversity than would be achieved with higher planting intensities such as 50% and 70% (Fig. 6). Our research showed that although the area percentage of *P. koraiensis* at year 300 increased with increasing planting intensity, the sensitivity of response of *P. koraiensis* to planting intensity decreased with increasing intensity (Table 4). Wang et al. [27]

also found that species abundance is more sensitive to low intensity planting. The reason for this is because the planted trees occupy the spaces which are probably the living spaces for natural regeneration of trees, then, it turns out that the planting efficiency decreases with the increasing planting intensity. In other words, in a low planting intensity, more space is saved for natural regeneration of trees, so the planting efficiency is higher, but more time is required to increase the coverage of the target species. With a high planting intensity, less space is saved for the seeds of trees, so the planting efficiency is lower, but the time to

Table 4. Planting yield of different planting intensity in the 300th year of our simulation.

Scenario	Planting Intensity (%)	Area Percentage at Year 300 (%)	Area Percentage Increase (%)	Planting Efficiency
P1	5	27.87	8.34	3.60
P2	10	35.99	16.46	3.54
P3	30	56.63	37.10	2.67
P4	50	69.76	50.23	2.17
P5	70	77.14	57.61	1.77

The definition of the scenarios is in Table 2 of this paper.
The definition of Planting Efficiency is in the section 2.4 of this paper.

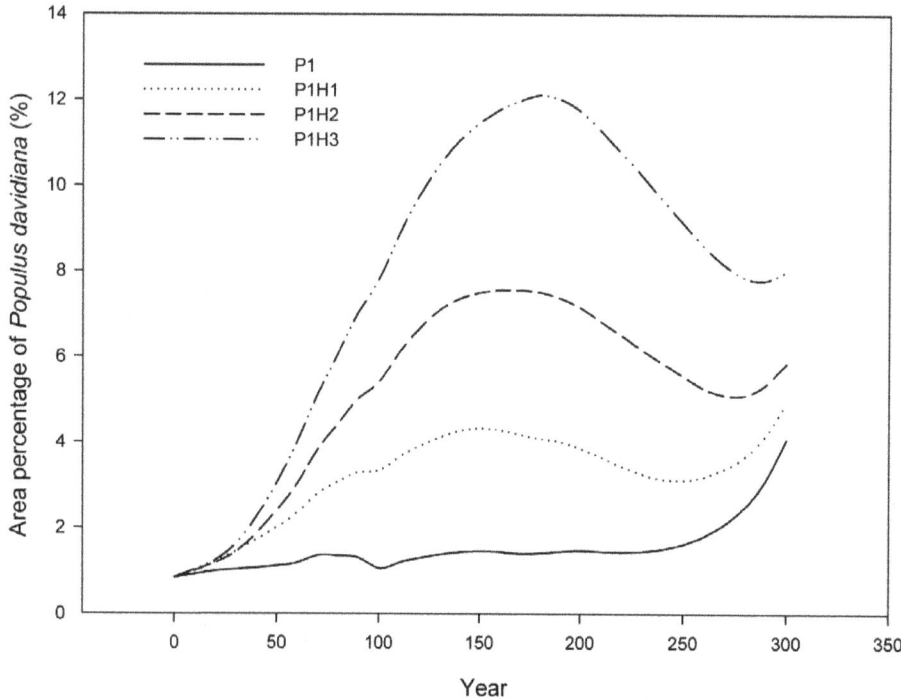

Figure 5. Response of area percentage of *Populus spp.* to different harvest intensities. P1, P1H1, P1H2, P1H3: Definitions may be found in Table 2.

increase the coverage of the target species is less, because planting trees substitutes for the process from natural regeneration to establishment.

Table 5. MANOVA and ANOVA results for area proportion of species of our study area as a function of planting intensity and harvest intensity.

Effect	Planting				Harvest			
	Pillai's trace /type III SS	df	F	P	Pillai's trace /type III SS	df	F	P
MANOVA test	1.977	68	26.648	<0.001	1.122	34	32.355	<0.001
ANOVA test								
Pinus koraiensis	85378.164	4	111.771	<0.001	42.035	2	0.110	0.896
Pinus tabulaeformis	45.402	4	12.872	<0.001	1.699	2	0.963	0.382
Pinus densiflora	1.817	4	13.381	<0.001	0.194	2	2.851	0.059
Pinus sylvestris var. mongolica	0.038	4	0.239	0.916	0.038	2	0.476	0.621
Larix olgensis	939.107	4	1.965	0.099	1401.289	2	5.865	0.003
Picea asperata	20.663	4	4.146	<0.001	0.196	2	0.078	0.925
Abies nephrolepis	0.093	4	5.361	<0.001	0.009	2	1.018	0.362
Populus davidiana	289.569	4	17.732	<0.001	1161.638	2	142.270	<0.001
Betula platyphylla	8.837	4	14.670	<0.001	1.659	2	5.508	0.004
Ulmus pumila	177.853	4	13.394	<0.001	71.537	2	10.775	<0.001
Fraxinus chinensis	1.651	4	23.867	<0.001	0.016	2	0.459	0.632
Fraxinus rhynchophylla	5.990	4	9.878	<0.001	0.010	2	0.034	0.966
Juglans mandshurica	41.342	4	33.690	<0.001	0.618	2	1.006	0.367
Quercus mongolica	26094.143	4	91.088	<0.001	877.550	2	6.127	0.002
Acer pictum subsp. mono	554.685	4	29.920	<0.001	0.366	2	0.039	0.961
Tilia amuresis	7.453	4	161.711	<0.001	0.070	2	3.051	0.048

Table 6. Contrast estimate of effects of different harvest intensity on area percentage of species for individual species.

Dependent variables	Level 2 vs. level 1		Level 3 vs. level 1	
	Contrast Estimate	P	Contrast Estimate	P
Pinus koraiensis	−0.627	0.690	−0.648	0.680
Pinus tabulaeformis	0.065	0.545	0.148	0.167
Pinus densiflora	0.150	0.467	0.049	0.020
Pinus sylvestris var. mongolica	−0.020	0.366	−0.017	0.442
Larix olgensis	−2.127	0.087	−4.252	0.001
Picea asperata	0.008	0.949	0.047	0.711
Abies nephrolepis	6.452E−5	0.993	−0.009	0.219
Populus davidiana	1.548	<0.001	3.847	<0.001
Betula platyphylla	0.054	0.220	0.145	0.001
Ulmus pumila	0.493	0.018	0.961	<0.001
Fraxinus chinensis	0.003	0.819	0.014	0.358
Fraxinus rhynchophylla	−0.003	0.942	0.008	0.857
Juglans mandshurica	0.022	0.723	0.086	0.172
Quercus mongolica	−0.567	0.555	−3.156	0.001
Acer pictum subsp. mono	0.053	0.830	0.065	0.792
Tilia amuresis	0.004	0.771	0.028	0.024

Level1: selectively harvesting 10% of general timber forest and 30% of other timber forest
Level2: selectively harvesting 30% of general timber forest and 50% of other timber forest.
Level3: selectively harvesting 50% of general timber forest and 70% of other timber forest.

2 The Effects of Harvesting

Selective harvest regimes increased the forest diversity, while the planting regimes promoted the homogenization of forest composition (Fig. 6). Moderate disturbance has proven to be a way of increasing ecosystem diversity [39]. Elliott and Knoepp [40] found that there is greater species diversity after harvesting. Hall et al. [41] also found that harvesting could provide a sustainable management strategy for biodiversity conservation. There are two sources of forest diversity, the evenness of abundance of different types species and the number of types of species [42], acording to Shannon's Diversity Index, but the source of the diversity of our research is the evenness of abundance of different types species. Two reasons for this are: (1) harvesting creates gaps for the establishment of early-successional species [43] and leads to the evenness of abundance of different types of species; (2) the types of species in the model are set according to the vegetation map required for the running of LANDIS, thus there will not be new species migrating into the study area.

The forest diverstity increased with the increasing harvesting intensity (Fig. 6). We propose that the combination of selective harvesting 50% of general timber forest and selectively harvesting 70% of other timber forest could be the choice of harvest intensity, because its negative effect on forest composition is small. For example, P. davidiana is the species most sensitive to harvest regimes (Table 6), but even in the scenario P1H3 (Table 2), which is the combination of the lowest planting intensity and the highest harvest intensity, the variability of area percentage of P. davidiana is at a low level of around 10% during the simulated 300 years (Fig. 5). This is because harvesting is limited in timber forests (57%

of total forest area) and selective harvesting is limited exclusively to the trees at the harvestable ages. Thus while harvest intensity may be high at the patch scale, it is not high at the landscape scale [44].

3 The Dynamics of Area Percentage of Some Species

The area percentage of Q. mongolica decreased due to the planting of P. koraiensis, which is more shade-tolerant than the former, while the dynamics of area percentage of Q. mongolica differed from that of other mid-successional species such as P. tabulaeformis, P. densiflora, F.chinensis, F. rhynchophylla and J. mandshurica (Fig. 3). We assume this is because: (1) the existing area percentage of Q. mongolica exceeds that level typical of a climax forest and as a result there are likely no shade-intolerant species around them; and (2) the planted P. koraiensis occupies some of the area that would have been taken up by Q. mongolica. Although this Q. mongolica forest is not the climax forest for this region and would eventually be replaced by P. koraiensis forest according to history [20], a rapid rate of forest succession toward the P. koraiensis forest would not be ideal because climate change appears to be creating a different environment [45,46].

The early successional species P. davidiana showed an abrupt increase in area percentage. In the N and P scenarios, this increase, which occurred around the year 250, would benefit from the death of some species whose longevity is 250 years and establishment efficiency is low, such as P. sylvestris var. mongolica and L. olgensis. In the PH scenarios, the increase would benefit from the harvesting. Because the seed dispersal ability of P. davidiana is strong and the seed can adapt to various environments [47], the gaps which are created by either death of other species or harvesting would have a high probability of being occupied by P. davidiana. It is reported that suitable habitat and seed dispersal are key to the distribution and abundance of a species [48,49]. Due to their high seed dispersal and establishment ability, the early successional species can help restore the ecosystem in case of serious disturbance [49–51]. Therefore adequate seed sources for these species should be maintained in the forest.

The area percentage of L. olgensis showed a decreasing trend and this species eventually disappeared in this region, because it is shade-intolerant and its establishment ability is low. Although L. olgensis is currently distributed widely in the study area, it is all in the form of plantations. The investigation of the vegetation in the mountains of eastern Liaoning province and the predictions of different models in the Changbai Mountain area show that the L. olgensis forest is a declining population [52–54]. Zhu et al. [32] pointed out that L. olgensis has difficulty in natural regeneration. In light of the above, we assume that from a management perspective L. olgensis is not a proper species for restoration in mid- to climax successional stages.

4 Caveats

The dynamics of Q. mongolica and P. davidiana stimulated in our research merit attention because a trend of higher temperature and less precipitation has been occurring in this region over the last 44 years [55], although the prediction of climate variation is uncertain [56]. (1) Q. mongolica is crowed out by P. koraiensis in our simulation, but previous research has found that Q. mongolica is more resistant to climate warming than P. koraiensis [57]. These lead us to recommend the adoption of conservative measures from a management perspective: (a) Promote forest succession via adopting a low planting intensity; (b) Understand and follow the responses of species to the emerging pattern of climate variation, so that management regimes can be altered in time to adapt to these climatic changes. (2) Although the presence of P. davidiana would insure restoration after disturbances, it will be a challenge

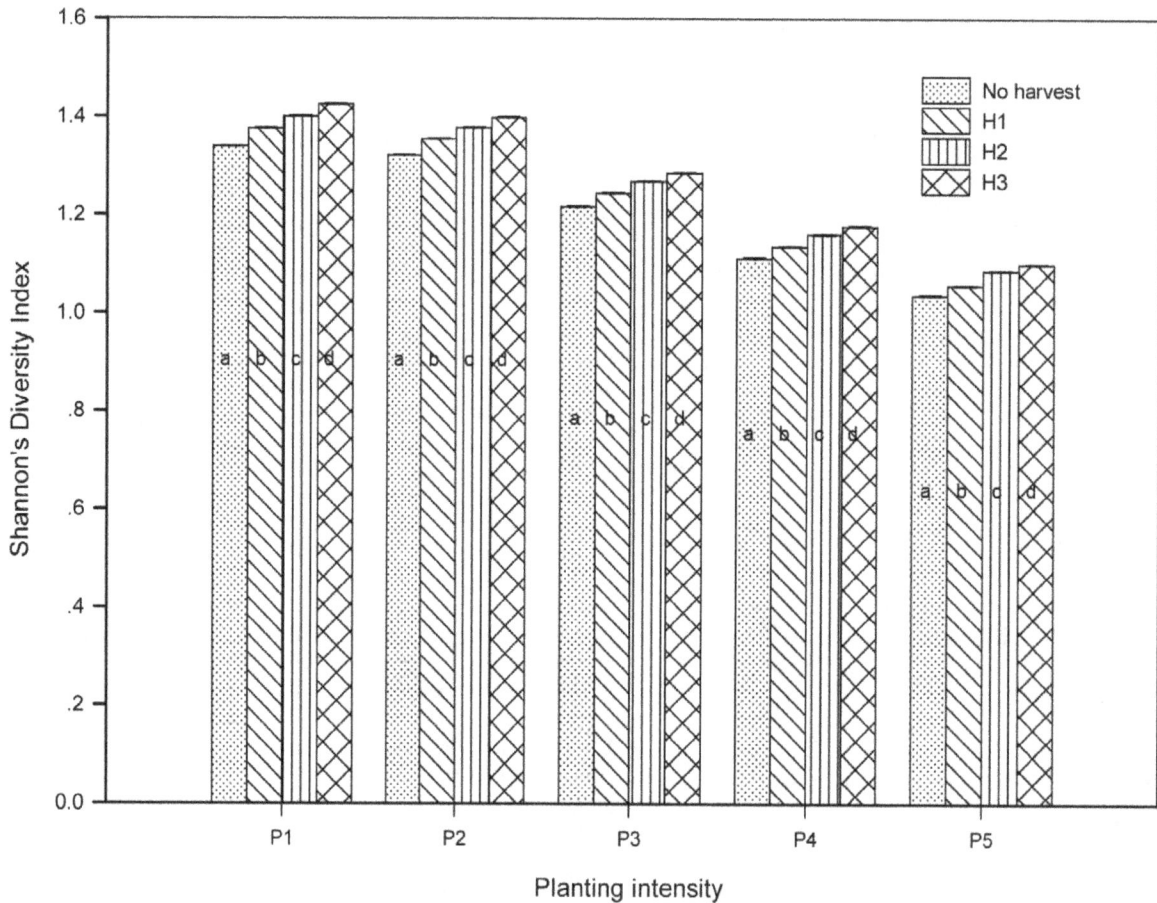

Figure 6. Shannon's Diversity Index of Hun River upstream forest in simulated scenarios. P1–P5 and H1, H2 and H3: Definitions are given in Table 2; a–d: significant difference (LSD, P<0.01) of Shannon's Diversity Index within same planting intensity but different harvest intensity.

for maintaining ecosystem water balance, especially under climate-induced drought conditions, because consumption of water by early successional species such as *Populus spp.* is high [58]. The pros and cons of harvesting should be carefully evaluated for the abrupt increase of *P. davidiana*.

Conclusions

With increasing planting intensity the coverage of *P. koraiensis* increases and planting efficiency decreases. This is important information for forest management in the context of limited financial and labor resources. In addition, diversity will decrease with increasing planting intensity. Taking both forest diversity and labor and financial constraints into account, a low planting intensity, such as 30%, in which the forest is restored to the climax forest over the long run, is a better management strategy for restoration.

When timber forests occupy about 50% of the total forest area, an intensity level of selectively harvesting 50% of the general timber forest and selectively harvesting 70% of the remaining timber forest could be appropriate, because the negative effects of

this harvest intensity are small at the landscape scale and landscape diversity increases with the increased harvest intensity.

From the dynamic of species, we note two important caveats. (1) Encouraging a rapid pace of forest succession may not be a good management strategy, because the climax species would crowd out other species, some of which would likely be more adapted to future climate conditions in the long run. (2) Careful evaluation of the pros and cons of harvesting is needed. That is because although harvesting can increase forest diversity, we should pay attention to the abrupt increase of early successional species such as *P. davidiana* after harvest, due to their characteristics of high water consumption which will be a challenge for maintaining ecosystem water balance, especially under climate-induced drought conditions.

Author Contributions

Conceived and designed the experiments: JY XH AW. Performed the experiments: JY. Analyzed the data: JY. Contributed reagents/materials/analysis tools: JY WC XL. Wrote the paper: JY BJL XTL.

References

1. Weiss G (2004) The political practice of mountain forest restoration - comparing restoration concepts in four European countries. For Ecol Manage 195: 1–13.
2. Shin JH, Lee DK (2004) Strategies for restoration of forest ecosystems degraded by forest fire in Kangwon Ecoregion of Korea. For Ecol Manage 201: 43–56.
3. Li WH (2004) Degradation and restoration of forest ecosystems in China. For Ecol Manage 201: 33–41.
4. Alexander HD, Arthur MA (2010) Implications of a predicted shift from upland oaks to red maple on forest hydrology and nutrient availability. Can J Forest Res 40: 716–726.
5. Bent GC (2001) Effects of forest-management activities on runoff components and ground-water recharge to Quabbin Reservoir, central Massachusetts. For Ecol Manage 143: 115–129.

6. Bhatti JS, Fleming RL, Foster NW, Meng FR, Bourque CPA, et al. (2000) Simulations of pre- and post-harvest soil temperature, soil moisture, and snowpack for jack pine: comparison with field observations. For Ecol Manage 138: 413–426.

7. Bonell M, Purandara BK, Venkatesh B, Krishnaswamy J, Acharya HAK, et al. (2010) The impact of forest use and reforestation on soil hydraulic conductivity in the Western Ghats of India: Implications for surface and sub-surface hydrology. Journal of Hydrology 391: 49–64.

8. Carlson M, Chen J, Elgie S, Henschel C, Montenegro A, et al. (2010) Maintaining the role of Canada's forests and peatlands in climate regulation. Forest Chron 86: 434–443.

9. Youngblood A, Metlen KL, Coe K (2006) Changes in stand structure and composition after restoration treatments in low elevation dry forests of northeastern Oregon. For Ecol Manage 234: 143–163.

10. Zerbe S (2002) Restoration of natural broad-leaved woodland in Central Europe on sites with coniferous forest plantations. For Ecol Manage 167: 27–42.

11. Chen XW, Li BL, Lin ZS (2003) The acceleration of succession for the restoration of the mixed-broadleaved Korean pine forests in Northeast China. For Ecol Manage 177: 503–514.

12. Liu G, Fu B (2001) The influence of global climate change on forest ecosystem. Journal of Natural Resources 16: 71–78.

13. Tausch RJ, Wigand PE, Burkhardt JW (1993) Viewpoint - plant community thresholds, multiple steady-states, and multiple successional pathways - legacy of the quaternary. J Range Manage 46: 439–447.

14. Ravenscroft C, Scheller RM, Mladenoff DJ, White MA (2010) Forest restoration in a mixed-ownership landscape under climate change. Ecol Appl 20: 327–346.

15. Shinneman DJ, Cornett MW, Palik BJ (2010) Simulating restoration strategies for a southern boreal forest landscape with complex land ownership patterns. For Ecol Manage 259: 446–458.

16. Pietsch SA, Hasenauer H (2002) Using mechanistic modeling within forest ecosystem restoration. For Ecol Manage 159: 111–131.

17. Duncan RS, Chapman CA (2003) Consequences of plantation harvest during tropical forest restoration in Uganda. For Ecol Manage 173: 235–250.

18. Steenberg JWN, Duinker PN, Bush PG (2011) Exploring adaptation to climate change in the forests of central Nova Scotia, Canada. For Ecol Manage 262: 2316–2327.

19. He J, Zhou ZM, Weyerhaeuser H, Xu JC (2009) Participatory technology development for incorporating non-timber forest products into forest restoration in Yunnan, Southwest China. For Ecol Manage 257: 2010–2016.

20. He HS, Hao ZQ, Larsen DR, Dai LM, Hu YM, et al. (2002) A simulation study of landscape scale forest succession in northeastern China. Ecol Modell 156: 153–166.

21. Mehta S, Frelich LE, Jones MT, Manolis J (2004) Examining the effects of alternative management strategies on landscape-scale forest patterns in northeastern Minnesota using LANDIS. Ecol Modell 180: 73–87.

22. Bu R, He HS, Hu YM, Chang Y, Larsen DR (2008) Using the LANDIS model to evaluate forest harvesting and planting strategies under possible warming climates in Northeastern China. For Ecol Manage 254: 407–419.

23. Cairns DM, Lafon CW, Waldron JD, Tchakerian M, Coulson RN, et al. (2008) Simulating the reciprocal interaction of forest landscape structure and southern pine beetle herbivory using LANDIS. Landsc Ecol 23: 403–415.

24. Franklin J, Syphard AD, He HS, Mladenoff DJ (2005) Altered fire regimes affect landscape patterns of plant succession in the foothills and mountains of southern California. Ecosystems 8: 885–898.

25. Gustafson EJ, Shifley SR, Mladenoff DJ, Nimerfro KK, He HS (2000) Spatial simulation of forest succession and timber harvesting using LANDIS. Can J Forest Res 30: 32–43.

26. He HS, Mladenoff DJ (1999) Spatially explicit and stochastic simulation of forest-landscape fire disturbance and succession. Ecology 80: 81–99.

27. Wang XG, He HS, Li XZ, Chang Y, Hu YM, et al. (2006) Simulating the effects of reforestation on a large catastrophic fire burned landscape in Northeastern China. For Ecol Manage 225: 82–93.

28. Mladenoff DJ (2004) LANDIS and forest landscape models. Ecol Modell 180: 7–19.

29. Chang Y, He HS, Hu YM, Bu RC, Lia XZ (2008) Historic and current fire regimes in the Great Xing'an Mountains, northeastern China: Implications for long-term forest management. For Ecol Manage 254: 445–453.

30. He HS, Mladenoff DJ, Boeder J (1999) An object-oriented forest landscape model and its representation of tree species. Ecol Modell 119: 1–19.

31. Li J, Nie S, An B (2005) Stump sprouting of the main broad-leaved tree species of secondary forest in Eastern area of Northest China. Scientia silvae sinicae 41: 72–77.

32. Zhu J-J, Liu Z-g, Wang H-x (2008) Obstacles for natural regeneration of *Larix olgensis* plantations in montane regions of eastern Liaoning province, China. Chin J Appl Ecol 19.

33. Zhang J, Hao Z, Song B, Ye J, Li B, et al. (2007) Spatial distribution patterns and associations of *Pinus koraiensis* and *Tilia amurensis* in broad-leaved Korean pine mixed forest in Changbai Mountains Chin J Appl Ecol 18: 1681–1687.

34. He HS, Hao ZQ, Mladenoff DJ, Shao GF, Hu YM, et al. (2005) Simulating forest ecosystem response to climate warming incorporating spatial effects in north-eastern China. J Biogeogr 32: 2043–2056.

35. Chang Y, He HS, Bishop I, Hu YM, Bu RC, et al. (2007) Long-term forest landscape responses to fire exclusion in the Great Xing'an Mountains, China. Int J Wildland Fire 16: 34–44.

36. Xu CG, He HS, Hu YM, Chang Y, Larsen DR, et al. (2004) Assessing the effect of cell-level uncertainty on a forest landscape model simulation in northeastern China. Ecol Modell 180: 57–72.

37. Li J, Li J (2003) Regeneration and restoration of broad-leaved Korean pine forests in Lesser Xing'an Mountains of Northest China. Acta ecologica sinica 23.

38. Will R, Hennessey T, Lynch T, Holeman R, Heinemann R (2010) Effects of Planting Density and Seed Source on Loblolly Pine Stands in Southeastern Oklahoma. For Sci 56: 437–443.

39. Feyrer IJ, Duffus DA (2011) Predatory disturbance and prey species diversity: the case of gray whale (Eschrichtius robustus) foraging on a multi-species mysid (family Mysidae) community. Hydrobiologia 678: 37–47.

40. Elliott KJ, Knoepp JD (2005) The effects of three regeneration harvest methods on plant diversity and soil characteristics in the southern Appalachians. For Ecol Manage 211: 296–317.

41. Hall SJ, Lindig-Cisneros R, Zedler JB (2008) Does harvesting sustain plant diversity in Central Mexican wetlands? Wetlands 28: 776–792.

42. Schumann ME, White AS, Witham JW (2003) The effects of harvest-created gaps on plant species diversity, composition, and abundance in a Maine oak-pine forest. For Ecol Manage 176: 543–561.

43. Yosi CK, Keenan RJ, Fox JC (2011) Forest dynamics after selective timber harvesting in Papua New Guinea. For Ecol Manage 262: 895–905.

44. Berry NJ, Phillips OL, Ong RC, Hamer KC (2008) Impacts of selective logging on tree diversity across a rainforest landscape: the importance of spatial scale. Landsc Ecol 23: 915–929.

45. Bertrand R, Lenoir J, Piedallu C, Riofrio-Dillon G, de Ruffray P, et al. (2011) Changes in plant community composition lag behind climate warming in lowland forests. Nature 479: 517–520.

46. Yu M, Gao QO (2011) Leaf-traits and growth allometry explain competition and differences in response to climatic change in a temperate forest landscape: a simulation study. Ann Bot-london 108: 885–894.

47. Moss EH (1937) Longevity of seed and establishment of seedlings in species of Populus. Botanical Gazette 99: 529–542.

48. He HS, Mladenoff DJ (1999) The effects of seed dispersal on the simulation of long-term forest landscape change. Ecosystems 2: 308–319.

49. van Loon AH, Soomers H, Schot PP, Bierkens MFP, Griffioen J, et al. (2011) Linking habitat suitability and seed dispersal models in order to analyse the effectiveness of hydrological fen restoration strategies. Biol Conserv 144: 1025–1035.

50. Donath TW, Holzel N, Otte A (2003) The impact of site conditions and seed dispersal on restoration success in alluvial meadows. Appl Veg Sci 6: 13–22.

51. Bakker JP, Poschlod P, Strykstra RJ, Bekker RM, Thompson K (1996) Seed banks and seed dispersal: Important topics in restoration ecology. Acta Botanica Neerlandica 45: 461–490.

52. Yu XM, Qu HJ (2009) Succession trend of Larch plantation communities in forest regions of Northeast China. Journal of Northeast Forestry University 37: 18–19,53.

53. Leng WF, He HS, Bu RC, Dai LM, Hu YM, et al. (2008) Predicting the distributions of suitable habitat for three larch species under climate warming in Northeastern China. For Ecol Manage. Amsterdam: Elsevier Science Bv. 420–428.

54. Hao ZQ, Dai LM, He HS (2001) Potential response of major tree species to climate warming in Changbai Mountain, Northeast China. Chin J Appl Ecol 12: 653–658.

55. Zhao DS, Zheng D, Wu SH, Wu ZF (2007) Climate changes in northeastern China during last four decades. Chin Geogr Sci 17: 317–324.

56. Xu C, Gertner GZ, Scheller RM (2009) Uncertainties in the response of a forest landscape to global climatic change. Glob Change Biol 15: 116–131.

57. Wu JL, Wang M, Lin F, Hao ZQ, Ji LZ, et al. (2009) Effects of precipitation and interspecific competition on Quercus mongolica and Pinus koraiensis seedlings growth. Chin J Appl Ecol 20: 235–240.

58. Wang J (2010) The comparative study on photosynthesis water consumption and drought-resistance of tree species. Forestry science and technology 35: 10–13.

19

Population Dynamics and Range Expansion in Nine-Banded Armadillos

William J. Loughry[1]*, **Carolina Perez-Heydrich**[2], **Colleen M. McDonough**[1], **Madan K. Oli**[3]

1 Department of Biology, Valdosta State University, Valdosta, Georgia, United States of America, 2 Carolina Population Center, University of North Carolina, Chapel Hill, North Carolina, United States of America, 3 Department of Wildlife Ecology and Conservation, University of Florida, Gainesville, Florida, United States of America

Abstract

Understanding why certain species can successfully colonize new areas while others do not is a central question in ecology. The nine-banded armadillo (*Dasypus novemcinctus*) is a conspicuous example of a successful invader, having colonized much of the southern United States in the last 200 years. We used 15 years (1992–2006) of capture-mark-recapture data from a population of armadillos in northern Florida in order to estimate, and examine relationships among, various demographic parameters that may have contributed to this ongoing range expansion. Modeling across a range of values for γ, the probability of juveniles surviving in the population until first capture, we found that population growth rates varied from 0.80 for $\gamma = 0.1$, to 1.03 for $\gamma = 1.0$. Growth rates approached 1.0 only when $\gamma \geq 0.80$, a situation that might not occur commonly because of the high rate of disappearance of juveniles. Net reproductive rate increased linearly with γ, but life expectancy (estimated at 3 years) was independent of γ. We also found that growth rates were lower during a 3-year period of hardwood removal that removed preferred habitat than in the years preceding or following. Life-table response experiment (LTRE) analysis indicated the decrease in growth rate during logging was primarily due to changes in survival rates of adults. Likewise, elasticity analyses of both deterministic and stochastic population growth rates revealed that survival parameters were more influential on population growth than were those related to reproduction. Collectively, our results are consistent with recent theories regarding biological invasions which posit that populations no longer at the leading edge of range expansion do not exhibit strong positive growth rates, and that high reproductive output is less critical in predicting the likelihood of successful invasion than are life-history strategies that emphasize allocation of resources to future, as opposed to current, reproduction.

Editor: Christopher Joseph Salice, Texas Tech University, United States of America

Funding: Field work was supported by Earthwatch, the American Philosophical Society, and faculty research awards from Valdosta State University. Publication supported by the University of Florida Open Access Publishing Fund. The funders had no role in study design, data collection and analysis, decision to publish, or preparation of the manuscript.

Competing Interests: The authors have declared that no competing interests exist.

* E-mail: jloughry@valdosta.edu

Introduction

Understanding why some species are able to successfully colonize new areas while others do not is a key question in ecology and conservation biology [1], [2]. A number of critical features of successful invaders have been proposed; among these are possession of certain life-history characteristics [3], [4], ecological release from former predators and/or pathogens [5], and various anatomical and behavioral features that may increase adaptability to novel environments [6].

In addition to the aforementioned, intrinsic features of animal populations must inevitably play some role in determining the success of any invasion. For example, for a range to expand it is only logical to assume that populations produce sufficient individuals such that some leave current areas to colonize new ones. This could be accomplished by high reproductive output, high survivorship, or some combination of the two. Consequently, models to estimate population growth rates, coupled with prospective and retrospective perturbation analyses to identify parameters that most influence these rates, can provide valuable insights into the factors that might promote range expansion in a particular species.

Among mammals, the nine-banded armadillo (*Dasypus novemcinctus*; hereafter referred to as "armadillo") is a dramatic example of a successful invader. Although widely distributed across much of the Americas [7], armadillos have colonized the United States only recently. First recorded in the Rio Grande valley of Texas in the 1840 s [8], the species has subsequently expanded its range quite rapidly, so that it is now found from eastern New Mexico [9] to South Carolina [10], and as far north as Nebraska [11], southern Illinois and Indiana [12-14], and the Cumberland Plateau of Tennessee [15]. No quantitative assessments have been conducted but speculation about factors promoting this extensive range expansion have focused on the seemingly high tolerance of armadillos to human disturbance, which underscores their flexibility in adapting to a wide range of environmental conditions, and the occurrence of polyembryony, whereby females produce litters of genetically identical quadruplets from a single fertilized egg each year when they reproduce, thus generating an apparently high reproductive rate (at least relative to other species of armadillos; see reviews in [16-18]).

In this paper we use 15 years of capture-mark-recapture (CMR) data from a population of armadillos in northern Florida in an attempt to explore various demographic parameters that might contribute to range expansion. Specifically, we build on a previous study that focused on estimating apparent annual survival rate and transition probabilities between reproductive and non-reproductive states to estimate population growth rates. We then performed prospective perturbation analyses to quantify the relative influence of various demographic parameters on these estimates. A potential concern with these analyses was how our estimates might have been impacted by a three-year program of hardwood removal that eliminated much preferred habitat for armadillos at our study site [19]. Consequently, we used life-table response experiment (LTRE) analyses to decompose decreases in population growth rate due to logging into contributions from various demographic variables. Our findings represent the first rigorous analysis of population dynamics in nine-banded armadillos, and, thus, also provide the first formal attempt at identifying potential demographic mechanisms that might underlay the ongoing range expansion occurring in the United States. More broadly, our analyses provide data relevant to several theoretical issues in the study of biological invasions.

Materials and Methods

Ethics Statement

Permission to conduct fieldwork was provided by the Director of Research, Tall Timbers Research Station. All field procedures followed American Society of Mammalogists guidelines [20] and were approved by the animal care and use committee at Valdosta State University.

Field Methods

Details of the field site and sampling methods can be found in [18], [19]. Briefly, data were collected at the Tall Timbers Research Station, located just north of Tallahassee, Florida during the summers (May-August) of 1992–2003. Within each year, we

attempted to capture and mark, or in the case of previously marked individuals, identify all animals discovered during nightly censuses. Armadillos were captured using long dip nets. Once caught, individuals were weighed, sexed, measured, marked for temporary visual identification with various shapes and colors of reflective tape glued to different areas of the carapace, and marked for permanent identification by injection of a passive induced transponder (PIT) tag under the front carapace at its juncture with the neck.

Body mass was used to assign captured animals to one of three age categories: juveniles (young of the year) were individuals weighing <2 kg, yearlings weighed 2–3 kg, and adults weighed >3 kg [21]. Reproductive status of adult females was determined from inspection of the nipples as (1) definitely lactating, (2) possibly lactating, or (3) definitely not lactating [21]. We treated the first two categories as representing the reproductively active females in the population each year, however, because all adult males are physiologically capable of reproduction [22] we were unable to distinguish between reproductive and non-reproductive individuals (see [19]).

Although our fieldwork ended in 2003, some data were available from 2004–2006 because of the harvesting of armadillos at Tall Timbers in order to remove nest predators of northern bobwhite (*Colinus virginianus*; see [23]). We were granted access to these specimens in order to identify any individuals that had been captured and marked as part of our earlier sampling, and data from those animals are included here.

Matrix Population Model and Deterministic Demographic Analysis

We constructed and analyzed stage-structured matrix population models, focusing on the female segment of the population because, as mentioned above, it was not possible to obtain reliable estimates of reproductive parameters for males.

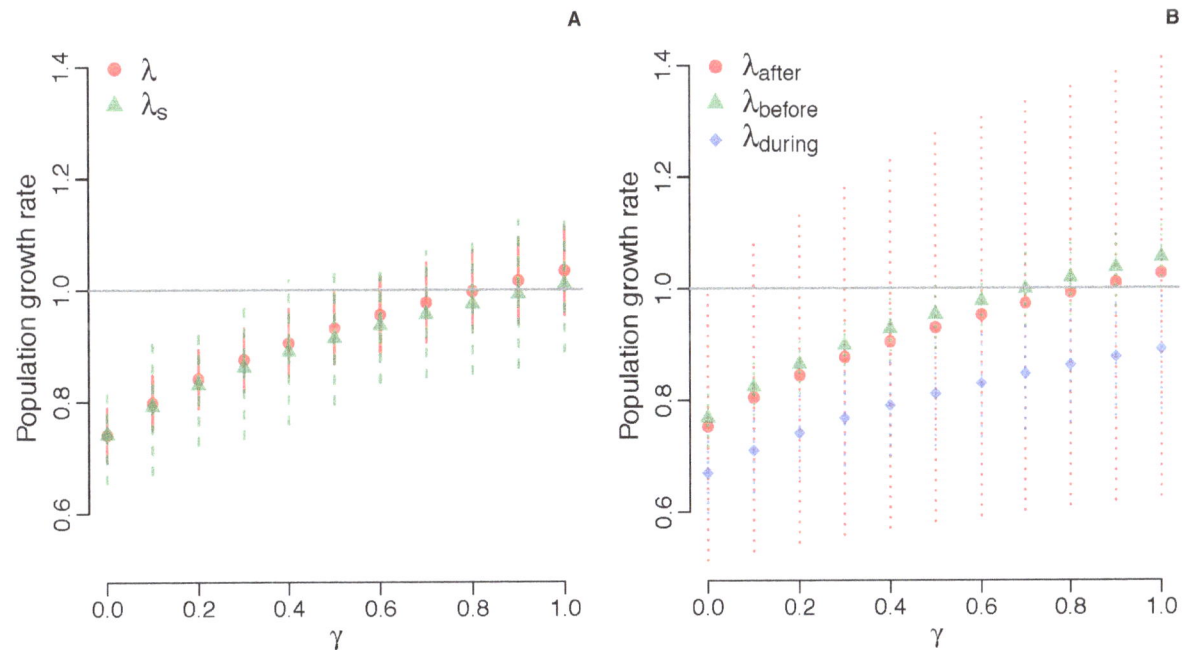

Figure 1. Annual population growth rate estimates as a function of γ. Gamma is the probability of a juvenile surviving to trappable age. (A) Estimates of deterministic (λ) and stochastic (λs) growth rate across all years of the study. (B) Estimates of deterministic growth rate before (λbefore), during (λduring), and after (λafter) hardwood removal. Vertical lines represent ±1 SE. There was considerable overlap in estimates provided by deterministic and stochastic projection models.

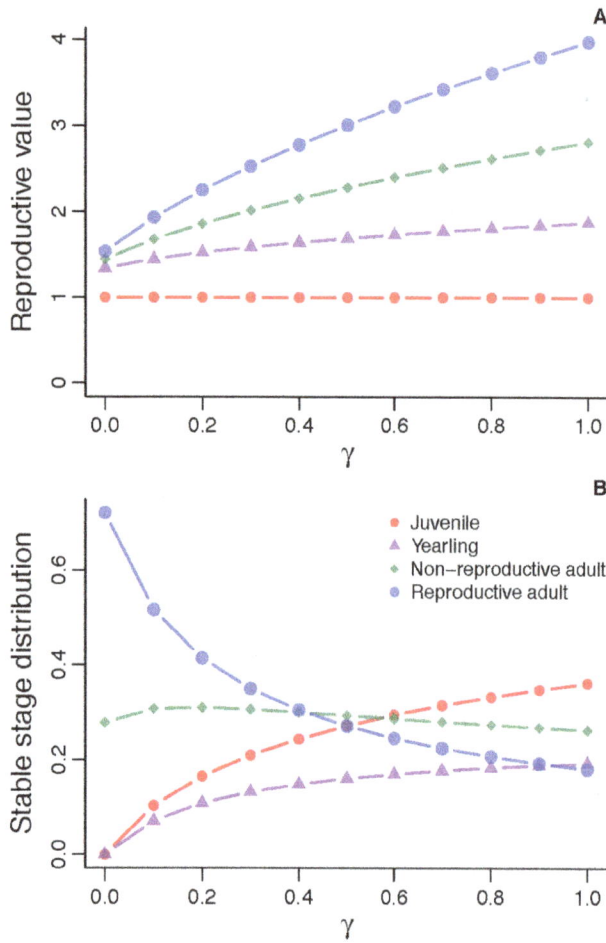

Figure 2. Stage-specific reproductive value (A) and stable stage distribution (B) as a function of γ. Gamma is the probability of a juvenile surviving to trappable age.

We considered 4 stages, based on age and reproductive status: <1 year old = juveniles; ≥1 and <2 year old = yearlings; and ≥2 years old = adults [19]. Juveniles survive with annual survival probability S_j and all survivors become yearlings the following year. Yearlings survive with annual survival rate S_y and all surviving yearlings become non-reproductive adults the following year. Non-reproductive and reproductive adults survive the year with annual survival rate S_n and S_r, respectively. Additionally, non-reproductive adult females that survive the year become reproductive adults the following year with probability ψ_{nr}, and remain non-reproductive with probability $(1 - \psi_{nr})$. Finally, reproductive adult females that survive the year become non-reproductive adults the following year with probability ψ_{rn}, and remain reproductive with probability $(1 - \psi_{rn})$. The stage-structured population projection matrix was of the form:

$$\mathbf{A} = \begin{bmatrix} 0 & 0 & F_n & F_r \\ S_j & 0 & 0 & 0 \\ 0 & S_y & S_n(1-\psi_{nr}) & S_r(1-\psi_{rr}) \\ 0 & 0 & S_n\psi_{nr} & S_r\psi_{rr} \end{bmatrix}$$

where F_n and F_r are fertility rates for non-reproductive and

reproductive adults. Fertility rates were estimated using post-breeding census methods [24] as:

$F_n = 0.5 * LS * \gamma * S_n * \psi_{nr}$ and $F_r = 0.5 * LS * \gamma * S_r * \psi_{rr}$, where LS is litter size and γ is a composite parameter that quantifies the probability of survival until trappable age.

All parameters except LS and γ were estimated using a multistate capture-mark-recapture (CMR) modeling framework [19]. Although the most parsimonious model [19] did not include a sex effect on survival, an equally well supported model ($\Delta AIC = 1.55$) included an additive effect of sex and reproductive states on survival probabilities. Because our population model was limited to females only, and survival estimates obtained from the two models were very similar, we used this latter model to obtain estimates of survival and transition probabilities (and their variances and covariances) for females (see Figure S1).

We did not have reliable, field-based estimates of reproductive parameters. However, all available evidence indicates females give birth just once per year, and invariably produce litters of genetically identical quadruplets from a single fertilized egg (via obligate polyembryony, see [18], [25]). Thus, we assumed that LS was 4. Next, we created a variable, γ, to represent the proportion of quadruplets that survive to trappable age. Trappable age begins at first emergence of juveniles from their natal burrows (at ~ 6–7 weeks old [18]). Note that this is a minimum time interval; time to actual capture can vary considerably beyond the date of first emergence. Multiple lines of evidence indicate that survivorship of all four littermates is low (review in [18]). Thus, it seems unlikely that γ would typically approach 1.0. However, we did not have sufficient data to identify a specific, well-supported point estimate of γ. Consequently, rather than limit our analyses to a single, arbitrarily picked value, we repeated them for a series of values ranging from 0 to 1.0.

Using the population projection matrix thus parameterized, we followed Caswell [24] to estimate deterministic finite population growth rate (λ), stable stage distribution, reproductive values, and elasticity of λ to changes in entries of the population projection matrix, as well as lower-level vital rates. The delta method was used to estimate variance and confidence intervals of λ [24]. For this, we obtained a variance-covariance matrix for stage-specific survival and transition probabilities directly from the CMR analysis [19]. Estimates of variances for LS and γ were not available, and so were assumed to be zero.

During the course of the study an extensive hardwood removal was conducted that eliminated much of the habitat favored by armadillos [26]. Previous work showed that state-specific survival rates of all animals were lower during the logging period than before or after [19]. Thus, in addition to estimating λ across all years of the study as a whole, we also performed demographic analyses separately for the years before (1992–1997), during (1998–2000), and after (2001–2006) hardwood removal.

Life-table Response Experiment (LTRE) Analysis

To further examine the impact of hardwood removal on population dynamics, we used a fixed effect LTRE analysis [24], [27], [28] to decompose any change in λ due to hardwood removal into contributions from various vital rates, primarily, stage-specific survival. We expected lower population growth rate during and after hardwood removal than in the years prior to removal. Consequently, we used vital rates and λ prior to hardwood removal as a reference, and decomposed the difference in λ ($\Delta\lambda$) between the reference and treatments (during or after hardwood removal) as:

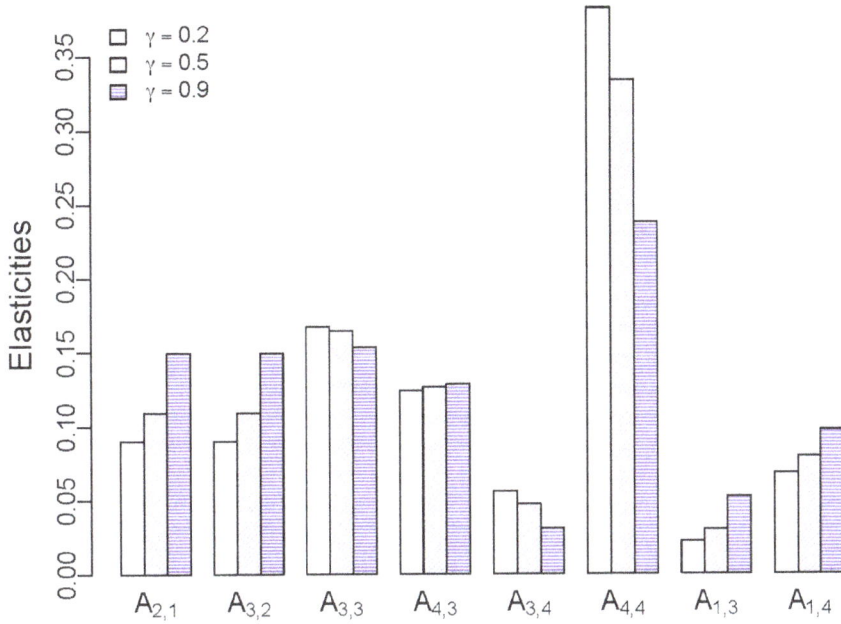

Figure 3. Elasticity of annual deterministic population growth rate (λ). Elasticities are presented as entries of the population projection matrix for three values of γ. X-axis labels (i.e., matrix entries) are: $A_{2,1}$ = survival of juveniles; $A_{3,2}$ = survival of yearlings; $A_{3,3}$ = probability of surviving and remaining in the non-reproductive adult stage; $A_{4,3}$ = probability of surviving and transitioning to the reproductive adult stage; $A_{3,4}$ = probability of surviving and transitioning from the reproductive adult stage to the non-reproductive adult stage; $A_{4,4}$ = probability of surviving and remaining in the reproductive adult stage; $A_{1,3}$ = fertility rate of non-reproductive adults; and $A_{1,4}$ = fertility rate of reproductive adults.

$$\Delta\lambda \approx \sum \left(\pi_i^r - \pi_i^t\right) \frac{\partial\lambda}{\partial\pi_i}\Big|_{\frac{\pi_i^r + \pi_i^t}{2}}$$

[24], [29], [30]; π_i is a lower-level vital rate, and superscripts r and t refer to reference (before hardwood removal) and treatment (during or after hardwood removal). The term $\frac{\partial\lambda}{\partial\pi_i}\big|_{\frac{\pi_i^r + \pi_i^t}{2}}$ indicates that sensitivities were evaluated at the mean values of π_i.

Stochastic Demographic Analysis

Deterministic demographic analyses assume that the environment is constant, and there is no variability in vital demographic parameters. In reality, however, the environment as well as vital rates can vary unpredictably. Stochastic demographic methods allow explicit consideration of variability in vital rates. We used a simulation-based approach (50,000 steps) to estimate stochastic population growth rate and stochastic elasticities [24], [31]. We assumed that demographic parameters estimated using data collected before, during and after hardwood removal represented good, poor and moderate environmental conditions for our study population. We further assumed that these three environmental states were independently distributed with observed probabilities 0.4, 0.2, and 0.4, respectively. The stochastic population growth rate, , was calculated as: $\log \lambda_s = \frac{1}{T}\sum_{t=1}^{T-1} r_t$ where $r_t = \log(n(t+1)/n(t))$ is a one-step population growth rate, $n(t)$ and $n(t+1)$ are projected population sizes at time t and $t+1$, respectively, and $T = 50,000$ steps [24], [31]. Variance of $\log \lambda_s$ was estimated using log-normal approximation [24]. Elasticity of λ_s to matrix entries was calculated as:

where $\mathbf{u}(t)$ and $\mathbf{v}(t)$ are stochastic stage structure and reproductive value vectors at time t, $\lambda(t)$ is 1-time step population growth rate, and the term $\langle\mathbf{v}(t),\mathbf{u}(t)\rangle$ is the scalar product of vectors $\mathbf{v}(t)$

and $\mathbf{u}(t)$. Following [31], [32], we calculated three types of stochastic elasticities: (1) overall stochastic elasticities E_{ij}^S were calculated by setting $C_{ij}(t) = A_{ij}(t)$ for every year t; (2) elasticities of λ_s to the mean of matrix elements $E_{ij}^{S^\mu}$ were obtained by setting $C_{ij}(t) = \mu_{ij}$, and (3) elasticities of λ_s to the variance of the matrix entries $E_{ij}^{S^\sigma}$ were obtained by setting $C_{ij}(t) = \mu_{ij}$, and $C_{ij}(t) = A_{ij}(t) - \mu_{ij}$. Elasticities of λ_s to lower-level vital rates were calculated using methods described by Caswell [33].

All analyses were performed using programs written in MATLAB (Mathworks, Inc., Natick, MA).

Results

Population Dynamics across All Years

Overall estimates of demographic variables for the entire study period are presented in the Supplementary Materials (Figure S1). Across all years of the study, estimates of λ increased from 0.80–1.03, depending on the value of γ (Figure 1). Growth rates ≥ 1.0 were attained only with values of $\gamma \geq 0.80$; the upper limits of 95% confidence intervals for λ were <1.0 for $\gamma \leq 0.50$. Likewise, for $\gamma \leq 0.85$, net reproductive rates were <1 (Figure S2), suggesting that most females did not replace themselves, except in unlikely scenarios where an average of ≥ 3.5 of the quadruplets survived to trappable age. Reproductive values and stable stage distributions also varied with γ, with an increase in reproductive value of adult stages as γ increased (Figure 2), and, as expected, a higher proportion of juveniles in the population with increased values of γ (Figure 2). Estimates of life expectancy indicated that juvenile armadillos were expected to live for 2.98 ± 2.99 (SE) years.

Matrix entry elasticities revealed that λ was proportionately most sensitive to changes in the probability of surviving and remaining in the reproductive adult stage, followed by the probability of surviving and remaining in the non-reproductive

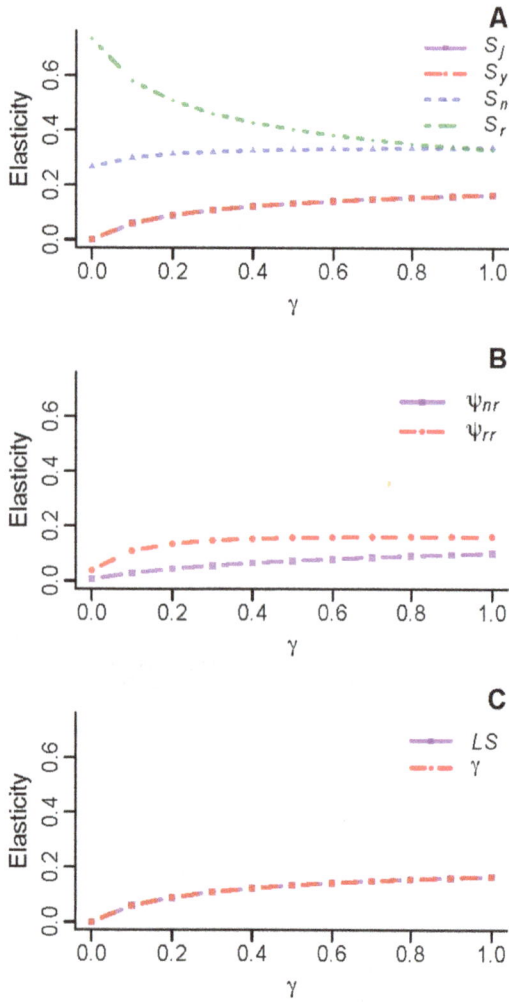

Figure 4. Elasticity of annual deterministic population growth rate (λ) to vital demographic parameters. Elasticities are presented for (A) survival, (B) reproductive transitions, and (C) litter size and gamma. Symbols are: S_j, S_y, S_n, and S_r = survival of juveniles, yearlings, non-reproductive adults and reproductive adults, respectively; ψ_{nr} = probability of transitioning from non-reproductive to reproductive adult stage; ψ_{rr} = probability of reproductive adults remaining reproductive adults; LS = litter size; and γ = probability of surviving to trappable age.

adult stage (Figure 3). As the value of γ increased, elasticity of λ to probability of surviving and remaining in the reproductive adult stage decreased with a corresponding increase in elasticity of λ to other entries of the population projection matrix (Figure 3). The reproductive adult stage was generally the most influential life stage except when γ ≈ 1 (Figure 3).

Elasticity of λ to lower-level vital rates identified S_r, followed by S_n and ψ_{rr}, as the most influential vital rates across all reasonable values of γ (Figure 4). As the value of γ increased, elasticity of λ to S_r decreased, with a corresponding increase in elasticity of λ to other vital rates; elasticity to S_n slightly exceeded that to S_r when γ ≈ 1 (Figure 4). The relative importance of reproductive parameters and survival of younger age classes was generally low, but increased as γ increased (Figure 4).

Effects of Logging

Estimates of population growth rate were highest before, and lowest during, hardwood removal for all values of γ (Figure 1). Before hardwood removal, λ approached 1 for γ ≈ 0.75; λ never approached 1.0 during or after the hardwood removal, even when γ ≈ 1 (Figure 1). However, estimates were less precise for the logging and post-logging time frames (Figure 1), probably because of small sample sizes. Patterns of elasticities were similar to those described previously for the overall population (results not shown).

LTRE analysis revealed that the difference in survival of reproductive adults, followed by that of non-reproductive adults, contributed the most to observed differences in λ. However, the contribution of survival of reproductive adults decreased, and that of non-reproductive adults and juveniles increased, as the value of γ increased; these three vital rates contributed almost equally when γ ≈ 1 (Figure 5). This change in the pattern of vital rate contribution to λ was due primarily to an increase in the sensitivity of λ to the latter two variables (and a corresponding decrease in that to reproductive adults). Results of LTRE analysis comparing demography before and after hardwood removal were generally similar to those described above (results not shown).

Stochastic Analyses

Stochastic population growth rates (λ_s) were slightly lower than deterministic ones, but exhibited a similar relationship with γ (Figure 1). Patterns of elasticity of λ_s to mean vital rates and overall stochastic elasticities were similar to those of deterministic elasticities (Figure S3). However, elasticities of λ_s to all standard deviations of vital rates were negative, indicating that increases in variances of these rates reduced λ_s (Figure S3). Interestingly, λ_s was proportionately most sensitive to both the mean and standard deviation of survival of reproductive adults, followed by that to the mean and standard deviation of survival of non-reproductive adults for most values of γ (Figure S3).

Discussion

The remarkable success of nine-banded armadillos in colonizing much of the southern United States has been puzzling because studies of reproductive success [34] and juvenile mortality [35] seem to indicate low recruitment [36]. The analyses reported here reinforce that view. Indeed, estimates of λ were <1.0 for values of γ ≤0.80 (the probability of a juvenile surviving to trappable age). Field observations suggest that high values of γ are unlikely. For example, data from three sites (including Tall Timbers) each showed that the modal litter size of captured juveniles was one, and that, across all sites, 468 juveniles from 283 litters were captured [18]. Assuming a fixed litter size of four, this means 664 (58.7%) juveniles were not caught. Whether these missing individuals died, dispersed, or remained in the population and somehow evaded capture is unknown, but to the extent these data indicate potentially low values of γ in populations of armadillos, it seems reasonable to conclude that range expansion has been achieved despite low population growth rates.

Such an assertion may be misleading for two reasons. First, irrespective of the species involved, successful biological invasions generally proceed in a more or less predictable sequence [37], [38]. Invasion begins with the establishment phase during which the invasive species colonizes a novel habitat and establishes itself. Once well established, and population density exceeds the Allee threshold, populations exhibit unregulated exponential growth, leading to the expansion phase. During expansion, dispersing propagules spread out from the initial site of invasion, creating an invasion front where the population may continue to grow

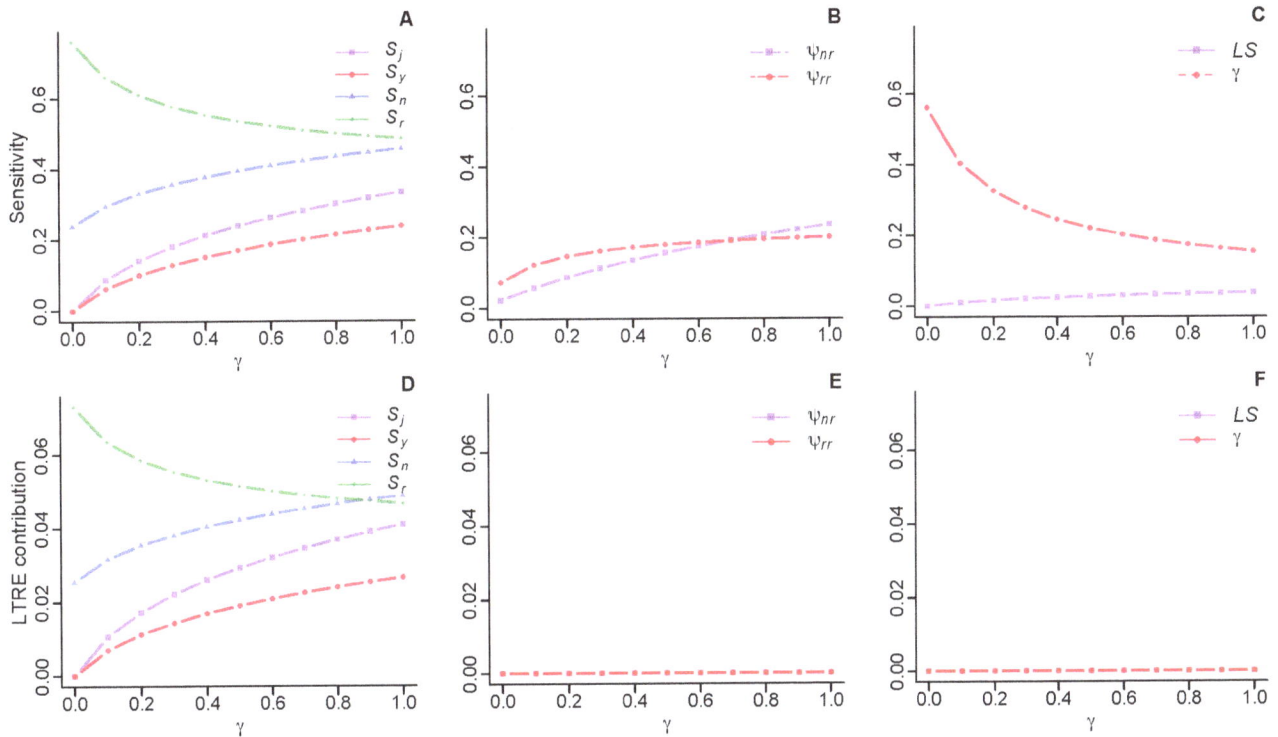

Figure 5. Results of the life table response experiment (LTRE) analysis. Deterministic population growth rates (λ) were compared before versus during hardwood removal. (A-C) Sensitivity of λ to vital rates. (D-F) LTRE contributions of vital rates to observed differences in λ. Note that four vital rates (i.e., transition rates for non-reproductive and reproductive adults, litter size, and γ) did not contribute to observed differences in population growth rate because separate estimates of these vital rates before and during hardwood removal were unavailable.

exponentially. Finally, the spread of invasion may slow down or stop due to environmental constraints or other regulatory mechanisms during the saturation phase [37]. As the invasion front moves forward, population growth may slow down or cease altogether at the interior of the habitat due to density-dependent population regulation [39]. Indeed, theoretical models of invasion dynamics assume density-dependent population dynamics, whereby population growth depends on location relative to the invasion front, and local population density [40-43].

Nine-banded armadillos were first recorded in the Tallahassee area in 1972 [44]. Thus, the population at Tall Timbers was likely in place for about 20 years by the time we began our sampling, and was no longer at the leading edge of range expansion. Consequently, an alternative interpretation of our results might be that features of the Tall Timbers population do not reflect conditions occurring along the invasion front, but are instead more representative of a population in the saturation phase. Although such a proposal is consistent with theoretical expectations regarding biological invasions [37-43], unfortunately, no data are currently available to test this hypothesis because no populations along the northern limit of the species' distribution have been studied. Nonetheless, assuming such a scenario is valid, our data suggest that populations of armadillos may become established quickly, with periods of high population growth being quite brief.

A second consideration is that our estimates of stage-specific survival rates were apparent survival rates. As such, these estimates cannot distinguish between death and dispersal. Given that range expansion is ongoing, dispersal undoubtedly occurs; failure to account for dispersal would lead to underestimation of population growth rate [30]. It is therefore likely that we have underestimated

λ by confounding death and dispersal. Thus, although superficial examination of our results might suggest our population was in decline, and perhaps on the way to local extinction, the population may in fact be stable, as predicted by theoretical models of invasion dynamics [40], [41]. Some support for this position comes from the fact that even though armadillos were systematically culled from Tall Timbers during 2004–2006, the number of armadillos collected each year during this period was remarkably stable [23]. The most likely explanation for this was that large numbers of individuals were available in surrounding areas that immigrated and swiftly replaced removed residents. This would imply that dispersal rates can be quite high, which could in turn lead to maintenance of a relatively stable population, despite apparently low growth rates.

The models of population dynamics developed here were only moderately helpful in explaining the range expansion of nine-banded armadillos in the United States, but they are consistent with recent theoretical expectations regarding biological invasions. For example, as discussed above, because our population was no longer at the leading edge of expansion, we did not find evidence of high population growth rates, just as theory predicts [37], [38], [40], [41]. Likewise, recent theory has deemphasized the importance of reproductive parameters in determining invasive success, instead focusing on life-history strategies that favor investment in future, as opposed to current, reproduction [4]. Our data support this hypothesis. Both matrix entry and lower-level elasticity analyses indicated that survival parameters were generally more influential than reproductive parameters, and that adult stages (both reproductive and non-reproductive) were the most valuable in terms of relative contribution to population growth. Thus, for armadillos, the success of any particular reproductive event may be less critical than the capacity to survive and reproduce again.

Our life expectancy estimate of 3 years might argue against the importance of survival and future reproduction as important components of population growth in armadillos. Loughry and McDonough [18] reported that captive animals may live >20 years, and that ages of some animals at Tall Timbers exceeded ≥10 years. Based on this, they estimated that longevity in the wild might be about 8–12 years. Nonetheless, they also reported that the average (± SD) number of years juveniles recruited into the Tall Timbers population remained there was 3.83±2.34 years, and that the average tenure of animals first caught as adults and retained in the population was 2.89±2.06 years. These data suggest we have not severely underestimated life expectancy. Also, the large variance around our life expectancy estimate suggests some animals may be relatively long-lived, as indicated by the field data. In any case, a challenge for future work will be to determine the details of how survival and future reproduction influence the population dynamics of armadillos.

As argued above, it is debatable whether the Tall Timbers population was in general decline or not. Nevertheless, we did find that hardwood removal was associated with a substantially lower population growth rate (see also [19], [26]). A number of studies have identified bottomland hardwoods as preferred habitat for armadillos (review in [18]), and our analyses reinforce the view that eliminating such areas has serious negative consequences, probably by promoting increased emigration from logged sites. Because hardwood removal was followed by three years of culling armadillos as part of the predator removal experiment, we were unable to fully evaluate the long-term consequences of logging on our population. Even so, the fact that population growth rates were highest prior to logging, lowest during removal, and intermediate after the completion of logging suggests that hardwood removal not only directly impacted the population during the logging period, but that negative effects continued to persist in the population for an extended time thereafter. Thus, hardwood removal may represent one form of human disturbance that nine-banded armadillos do not tolerate well. Not surprisingly, the same may be true for many other species of armadillos also [7].

Fundamentally, for range expansion to occur some individuals must leave the place where they were born to colonize new areas. Populations of armadillos seem to consist of a core of long-term residents that move very little over time, and about an equal number of transients that are caught a few times as they move through an area but are rarely seen again (review in [18]). Presumably, it is transients that contribute the most to range

expansion. Unfortunately, what determines whether an individual becomes a resident or a transient is unknown. Thus, even at the individual level, many of the factors promoting range expansion in nine-banded armadillos remain mysterious. An interesting project for the future would be to integrate population-level analyses of the type reported here with information on behavioral phenotypes (resident versus transient) to investigate whether the proportions of residents and transients are affected by changes in population dynamics. Perhaps such an approach will provide the insights necessary to better understand how armadillos have achieved such impressive success in colonizing the United States.

Supporting Information

Figure S1 Estimates of vital demographic parameters. Symbols are: S_j, S_y, S_n, and S_r = survival of juveniles, yearlings, non-reproductive adults and reproductive adults, respectively; ψ_{nr} = probability of transitioning from non-reproductive to reproductive adult stage; ψ_{rr} = probability of reproductive adults remaining reproductive adults. Bars represent ±1 SE.

Figure S2 Net reproductive rate as a function of γ. Net reproductive rate approaches 1.0 when $\gamma \approx 0.8$.

Figure S3 Elasticity of stochastic population growth rate (λ_s). Elasticities are presented for (A) mean, and (B) standard deviation (SD) of vital demographic parameters for a range of values of γ. Symbols are: S_j, S_y, S_n, and S_r = survival of juveniles, yearlings, non-reproductive adults and reproductive adults, respectively; ψ_{nr} = probability of transitioning from non-reproductive to reproductive adult stage; ψ_{rr} = probability of reproductive adults remaining reproductive adults; LS = litter size; and γ = probability of surviving to trappable age.

Acknowledgments

The staff of Tall Timbers Research Station provided invaluable help and support over the many years of field work associated with this project. C.J. Krebs, C. J. Salice and an anonymous reviewer provided many helpful comments, for which we are grateful.

Author Contributions

Performed the experiments: WJL CMM. Analyzed the data: CPH MKO. Wrote the paper: WJL MKO.

References

1. Kolar CS, Lodge DM (2001) Progress in invasion biology: predicting invaders. Trends Ecol Evol 16: 199–204.
2. Zenni RD, Nuñez MA (2013) The elephant in the room: the role of failed invasions in understanding invasion biology. Oikos 122: 801–15.
3. Pianka ER (1970) On r- and K-selection. Am Nat 104: 592–97.
4. Sol D, Maspons J, Vall-llosera M, Bartomeus I, García-Peña GE, et al. (2012) Unraveling the life history of successful invaders. Science 337: 580–83.
5. Elton CS (1958) The ecology of invasions by animals and plants. London: Methuen. 181 p.
6. Sol D, Bacher S, Reader SM, Lefebvre L (2008) Brain size predicts the success of mammal species introduced into novel environments. Am Nat 172: S63–S71.
7. Abba AM, Superina M (2010) The 2009/2010 armadillo Red List assessment. Edentata 11: 135–84.
8. Audubon JJ, Bachman J (1854) Quadrupeds of North America, 3. New York: V. G. Audubon.
9. Stuart JN, Frey JK, Schwenke ZJ, Sherman JS (2007) Status of the nine-banded armadillo in New Mexico. Prairie Nat 39: 163–69.
10. Platt SG, Snyder WE (1995) Nine-banded armadillo, *Dasypus novemcinctus* (Mammalia: Edentata), in South Carolina: additional records and reevaluation of status. Brimleyana 23: 89–93.
11. Freeman PW, Genoways HH (1998) Recent northern records of the nine-banded armadillo (Dasypodidae) in Nebraska. Southwest Nat 43: 491–504.
12. Van Deelen TR, Parrish JD, Heske EJ (2002) A nine-banded armadillo (*Dasypus novemcinctus*) from central Illinois. Southwest Nat 47: 489–91.
13. Hofmann JE (2009) Records of nine-banded armadillos, *Dasypus novemcinctus*, in Illinois. Trans Ill State Acad Sci 102: 95–106.
14. Whitaker JO Jr. (2010) Mammals of Indiana: a field guide. Bloomington: Indiana University Press. 327 p.
15. Eichler SE, Gaudin TJ (2011) New records of the nine-banded armadillo, *Dasypus novemcinctus*, in southeast Tennessee, and their implications. Edentata 12: 7–13.
16. Humphrey SR (1974) Zoogeography of the nine-banded armadillo (*Dasypus novemcinctus*) in the United States. BioScience 24: 457–62.
17. Taulman JF, Robbins LW (1996) Recent range expansion and distributional limits of the nine-banded armadillo (*Dasypus novemcinctus*) in the United States. J Biogeogr 23: 635–48.
18. Loughry WJ, McDonough CM (2013) The nine-banded armadillo: a natural history. Norman: University of Oklahoma Press. 338 p.
19. Loughry WJ, Perez-Heydrich C, McDonough CM, Oli MK (2013) Population ecology of the nine-banded armadillo in Florida. J Mammal 94: 408–16.

20. Sikes RS, Gannon WL (2011) Guidelines of the American Society of Mammalogists for the use of wild mammals in research. J Mammal 92: 235–53.
21. Loughry WJ, McDonough CM (1996) Are road kills valid indicators of armadillo population structure? Am Midl Nat 135: 53–59.
22. Peppler RD (2008) Reproductive biology of the nine-banded armadillo. In: Vizcaíno SF, Loughry WJ, editors. The biology of the Xenarthra. Gainesville: University Press of Florida. 151–59.
23. McDonough CM, Lockhart JM, Loughry WJ (2007) Population dynamics of nine-banded armadillos: insights from a removal experiment. Southeast Nat 6: 381–92.
24. Caswell H (2001) Matrix population models: construction, analysis, and interpretation. Sunderland: Sinauer Associates. 722 p.
25. Prodöhl PA, Loughry WJ, McDonough CM, Nelson WS, Avise JC (1996) Molecular documentation of polyembryony and the micro-spatial dispersion of clonal sibships in the nine-banded armadillo, *Dasypus novemcinctus*. Proc Roy Soc Lond B 263: 1643–49.
26. McDonough CM, Loughry WJ (2005) Impacts of land management practices on a population of nine-banded armadillos in northern Florida. Wildl Soc Bull 33: 1198–1209.
27. Caswell H (1989) Analysis of life table response experiments. 1. Decomposition of effects on population growth rate. Ecol Modell 46: 221–37.
28. Dobson FS, Oli MK (2001) The demographic basis of population regulation in Columbian ground squirrels. Am Nat 158: 236–47.
29. Rolland V, Hostetler JA, Hines TC, Johnson FA, Percival HF, et al. (2011) Harvest, weather and population dynamics of northern bobwhites in south Florida. Wildl Res 38: 396–407.
30. Hostetler JA, Kneip E, Van Vuren DH, Oli MK (2012) Stochastic population dynamics of a montane ground-dwelling squirrel. PLoS One 7: e34379.
31. Tuljapurkar S, Horvitz CC, Pascarella JB (2003) The many growth rates and elasticities of populations in random environments. Am Nat 162: 489–502.
32. Haridas CV, Tuljapurkar S (2005) Elasticities in variable environments: properties and implications. Am Nat 166: 481–95.
33. Caswell H (2005) Sensitivity analysis of the stochastic growth rate: three extensions. Aust N Z J Stat 47: 75–85.
34. Loughry WJ, Prodöhl PA, McDonough CM, Nelson WS, Avise JC (1998) Correlates of reproductive success in a population of nine-banded armadillos. Can J Zool 76: 1815–21.
35. McDonough CM, Loughry WJ (1997) Patterns of mortality in a population of nine-banded armadillos, *Dasypus novemcinctus*. Am Midl Nat 138: 299–305.
36. Loughry WJ, McDonough CM (2001) Natal recruitment and adult retention in a population of nine-banded armadillos. Acta Theriol 46: 393–406.
37. Arim M, Abades SR, Neill PE, Lima M, Marquet PA (2006) Spread dynamics of invasive species. Proc Nat Acad Sci USA 103: 374–78.
38. Shigesada N, Kawasaki K (1997) Biological invasions: theory and practice. Oxford: Oxford University Press. 205 pp.
39. Royama T (1992) Analytical population dynamics. London: Chapman and Hall. 371 pp.
40. Kot M, Lewis MA, van den Driessche P (1996) Dispersal data and the spread of invading organisms. Ecology 77: 2027–42.
41. Neubert MG, Caswell H (2000) Demography and dispersal: calculation and sensitivity analysis of invasion speed for structured populations. Ecology 81: 1613–28.
42. Ellner EP, Schreiber SJ (2012) Temporally variable dispersal and demography can accelerate the spread of invading species. Theor Popul Biol 82: 283–98.
43. Altwegg R, Collingham YC, Erni B, Huntley B (2013) Density-dependent dispersal and the speed of range expansions. Divers Distrib 19: 60–68.
44. Stevenson HM, Crawford RL (1974) Spread of the armadillo into the Tallahassee-Thomasville area. Fla Fld Nat 2: 8–10.

20

Carnivore Translocations and Conservation: Insights from Population Models and Field Data for Fishers (*Martes pennanti*)

Jeffrey C. Lewis[1]*, Roger A. Powell[2], William J. Zielinski[3]

1 Washington Department of Fish and Wildlife, Olympia, Washington, United States of America, **2** Department of Biology, North Carolina State University, Raleigh, North Carolina, United States of America, **3** Unite States Department of Agriculture Forest Service, Pacific Southwest Research Station, Arcata, California, United States of America

Abstract

Translocations are frequently used to restore extirpated carnivore populations. Understanding the factors that influence translocation success is important because carnivore translocations can be time consuming, expensive, and controversial. Using population viability software, we modeled reintroductions of the fisher, a candidate for endangered or threatened status in the Pacific states of the US. Our model predicts that the most important factor influencing successful re-establishment of a fisher population is the number of adult females reintroduced (provided some males are also released). Data from 38 translocations of fishers in North America, including 30 reintroductions, 5 augmentations and 3 introductions, show that the number of females released was, indeed, a good predictor of success but that the number of males released, geographic region and proximity of the source population to the release site were also important predictors. The contradiction between model and data regarding males may relate to the assumption in the model that all males are equally good breeders. We hypothesize that many males may need to be released to insure a sufficient number of good breeders are included, probably large males. Seventy-seven percent of reintroductions with known outcomes (success or failure) succeeded; all 5 augmentations succeeded; but none of the 3 introductions succeeded. Reintroductions were instrumental in reestablishing fisher populations within their historical range and expanding the range from its most-contracted state (43% of the historical range) to its current state (68% of the historical range). To increase the likelihood of translocation success, we recommend that managers: 1) release as many fishers as possible, 2) release more females than males (55–60% females) when possible, 3) release as many adults as possible, especially large males, 4) release fishers from a nearby source population, 5) conduct a formal feasibility assessment, and 6) develop a comprehensive implementation plan that includes an active monitoring program.

Editor: Mark S. Boyce, University of Alberta, Canada

Funding: The authors have received funding from two different offices of the U.S. Fish and Wildlife Service (Yreka, CA office, and Lacey, WA office) to support this research. The funders had no role in study design, data collection and analyis, decision to publich or preparation of the manuscript.

Competing Interests: The authors have declared that no competing interests exist.

* E-mail: Jeffrey.Lewis@dfw.wa.gov

Introduction

Since the settlement of Europeans in North America, the ranges of many of the continent's carnivores have contracted significantly (e.g., black-footed ferrets [*Mustela nigripes*], wolves [*Canis lupus*], Canada lynxes [*Lynx canadensis*], wolverines [*Gulo gulo*], fishers [*Martes pennanti*], grizzly bears [*Ursus arctos*]; [1–6]). Translocations – the intentional transport and release of animals to reestablish, augment or introduce a population – have been used in attempts to recover extirpated or depleted populations. Translocations, however, are not always successful [7–9]. Carnivore translocations can be expensive, time-consuming and controversial; and their success may depend upon adequate planning, expertise, organization, and cooperation [7,10]. Translocations that fail, even those that are well-planned and well-executed, can erode the support necessary to continue restoration efforts for imperiled species [11]. The International Union for Conservation of Nature (IUCN) has provided guidelines for translocations [10,12] and several sources provide specific recommendations and cautions for carnivore translocations [7–9]. However, specific recommendations are lacking for many species, and wildlife managers may have little to guide them when developing a translocation program. Understanding factors that are associated with translocation success is critically important for developing adequate feasibility studies and effective implementation plans, yet such factors are often unknown.

Having recommendations would be especially valuable for the fisher, which is a candidate for federal endangered status in the Pacific states (California, Oregon, Washington) [13], is listed as an endangered species in Washington [14], and is a target species for recovery efforts in the Pacific states [15,16]. While fishers have been translocated successfully to a number of locations in eastern and central North America, many translocations in western North America failed to re-establish populations [1,3,14,15,17].

The fisher is a mid-sized carnivore in the weasel family (Mustelidae) that occurs only in the temperate and boreal forests of North America [1]. Since the mid-1800s, the fisher's geographic distribution contracted substantially [18], due probably to historical over-trapping, non-compensatory mortality from predator-control campaigns and incidental trapping [19], habitat loss and fragmentation [1], and climatic changes in eastern North America associated with the Little Ice Age [20]. The extremely high prices paid for prime fisher pelts (up to $350/pelt in the early 1900s) [21–24], their vulnerability to trapping [1], and a lack of harvest regulations resulted in unsustainable exploitation of many fisher populations. By the mid-1900s, despite protections established for fisher populations throughout much of their historical range (i.e., range prior to European settlement), many populations did not recover. Moreover, the loss and fragmentation of structurally complex forests due to timber harvest, human development, changes in fire regimes, and climate change likely exacerbated population declines and impeded or prevented the recovery of many populations.

When the fisher's range was most contracted, large portions of its historical range in the US and southern Canada were unoccupied [25]. By the early 1900s, small populations of fishers persisted in only 6 locations in the US: northwestern California and southwestern Oregon; the southern Sierra Nevada; the Bitterroot Mountains in north-central Idaho and west-central Montana; the Big Bog area of northern Minnesota; Adirondack Park in northern New York; and the White Mountains and Moosehead Plateau in northern New Hampshire and northwestern Maine [1,3,25–33].

During the mid-1900s, many resource management agencies and timber companies suffered significant tree losses from unusually large porcupine (*Erithizon dorsatum*) populations, which they attributed to the absence of fishers. This prompted many wildlife and forest management agencies to reintroduce fishers to restore an effective predator of porcupines and a valuable furbearer [29,34–38]. Although many of these reintroduction efforts and their outcomes have been reported in the literature [1,3,17,36,39–45], they have not been evaluated to identify factors that influence the success of fisher translocations.

In this paper we: (1) present a population model for fisher reintroductions and use the model to predict factors that influence translocation success; we also use the model to evaluate the population-level effects of removing fishers from a source area for translocation; (2) summarize data from actual fisher translocations and use those data, combined with demographic data from the literature, to evaluate factors that may influence translocation success; (3) use data from actual translocations to test the predictions of the population model; (4) evaluate the contributions of translocations to fisher conservation; and (5) provide managers with recommendations to increase translocation success.

Methods

We define translocation as the intentional transport and release of animals to reestablish, augment or introduce a population in the wild. We define a reintroduction as an attempt to reestablish a population where one no longer exists within a species' historical range, an augmentation as adding individuals to an existing population, and an introduction as an attempt to establish a population outside the species' historical range [12,46]. We considered translocations successful if the target population was reestablished (i.e., for reintroductions), established (introductions) or growing (augmentations), as determined by the resource agency responsible or as documented in the literature. We concluded that reintroductions or introductions had failed when active monitoring or incidental observations (e.g., road kills, sightings, trapped fishers) indicated a consistent lack of detections of fishers (or lack of population growth, for augmentations) in the vicinity of the release area, as documented in the literature or in unpublished reports.

Population Model for Fishers

We used the population simulation program *VORTEX* [47] to model fisher populations because it allowed us to develop models for both reintroduced and source populations. We used *VORTEX* to explore characteristics of reintroduction programs that might lead to success or failure, assuming that release sites had adequate habitat. We incorporated the values for demographic parameters shown in Table 1, taking values from Powell's review [1] and data for survival reported by Raine [48]. We used these values to develop both a baseline population model and alternative models that represented a number of reintroduction and source population scenarios.

Next, we simulated environmental conditions for a source population occupying a landscape partitioned into a grid of 50 contiguous subsites (the maximum number allowed by *VORTEX*) that we envisioned as hexagons covering the landscape. Each hexagon contained habitat capable of supporting 20, 40 or more fishers, depending on the model conditions (hexagon capacity = $K/50$). Across their range, fishers experience distinctly different habitat configurations at the landscape scale. In the West, fishers occupy forested landscapes that are predominantly a patchwork of private and federal ownership (16). Therefore, we designated 25 hexagons as private land managed for timber with the probability of 2 harvests that removed critical habitat each 100 years, 18 as land managed by USDA Forest Service or USDI Bureau of Land Management for multiple uses with 1 harvest that removed critical habitat each 100 years, and 7 as protected land with no history of harvest. During the year after harvest, we reduced reproduction by 50% and survival by 15% in all hexagons. Because *VORTEX* allows changes for only 1 year after significant events, we exaggerated the reduction of reproduction and survival in the year following harvest to gain a long-term effect. We assigned a 5% probability of a dispersing juvenile moving to each adjacent hexagon, and a 3% probability of moving to other hexagons.

The fisher population in northwestern California is currently being used as the source population for a reintroduction in the northern Sierra Nevada of California. This source population is more restricted in area than the populations in British Columbia, Minnesota and New York, which have commonly been used as source populations for reintroductions. If removing fishers has an effect on a source population, the effects should be most pronounced on smaller source populations. Therefore, we set the carrying capacity (K) for our simulated source population at 2000, a number that is roughly modal for unpublished estimates (no published estimates exist) for the population in northwestern California. We modeled a population that was not subject to trapping or hunting mortality.

We calculated elasticity values for demographic parameters in Table 1 by varying the baseline values ±10%. We did not vary values for sex ratio or age of first reproduction because they were considered the least variable demographic characteristics [1,49]. We ran each set of values for each variable 100 times and used 1 minus the mean probability of extinction as an *index* of population viability (i.e., high index values indicate a high probability of viability), not as a direct estimate.

To evaluate the potential population-level effects of removing fishers from the source population, we removed 20 fishers from the source population for each of 2, 3, 5 or 8 years, removing either 5 fishers from each of 4 different hexagons each year or 1 fisher from each of 20 hexagons. To increase the effects of losing reproductive females from the source population, we removed adults only and removed 3 females for every 2 males.

To compare the effectiveness of different reintroduction approaches, we simulated the release of fishers onto an empty

Table 1. Values for demographic parameters and characteristics of the model source fisher population.[1]

Demographic Variable	Value (± SD)	Elasticity −10%	+10%
Starting population size, N_0	1000	−4	2
Carrying capacity, K	2000±250	−7	8
Mean litter size	2.0±1.0	−23	3
Age (yr) first reproduction	2	—[2]	—[2]
Exponent for density dependence, B	16	0	0
Exponent for Allee Effect, A	0.5	0	0
Survival rates			
Juveniles (age 0–1)[3]	35±25%	−26	6
Yearlings (age 1–2)	75±20%	−9	7
Adults (age ≥2)	88±20%	−15	10
Reproduction after logging	50%	−4	8
Survival after logging	75%	—[2]	—[2]
Local subpopulations (N = 50; and timber harvest/100 yrs)			
On private lands	25 (2 harvests/100 yrs)		
On USFS and BLM lands managed for timber	18 (1 harvest/100 yrs)		
On USFS and BLM protected lands	7 (no harvest)		

[1]Simulations were run 100 times for 100 years using *VORTEX* with stochastic variation as indicated by standard deviations (SD). Elasticity indexes the change in the viability index (1 minus the probability of extinction) when input variables are changed by ±10%.
[2]Elasticity values not calculated.
[3]Juveniles constituted ~45% of the population.

landscape of 50 hexagons in several ways, including: 1) the release of 20 fishers (5 in each of 4 hexagons) for each of 2, 3, 5 or 8 years, releasing fishers into new hexagons each year; 2) releasing 20 to 160 fishers, all released in 1 year (5 per hexagon); and 3) releasing 80 fishers all in 1 year into 1, 2, 3 … 20 hexagons. We assumed that the simulated reintroductions would occur on a landscape with 80% federal land (40 hexagons) and 20% private land (10 hexagons), which is consistent with ownership patterns where fishers were recently reintroduced in Washington and California [15,50]. Consistent with the simulated source population, the simulated reintroduction area was subject to timber harvests. To explore the effects of sex and age ratios on potential reintroduction success, we varied the sex ratio from 4:1 (M:F) to 1:4, and varied the number of juveniles released from 0 to 3 in each group of 5 fishers released in a hexagon. We assumed that the reintroduced population of fishers would initially occupy a smaller area than the source population, and therefore set K at 1,000. Values for other variables were the same as those for the source population. For this modeling exercise, we considered a reintroduction successful if the population persisted for 100 years.

Actual Fisher Translocations

We compiled information on fisher translocations from the scientific and popular literature; theses and dissertations; agency databases, files, and archives; and interviews with individuals that participated in or had knowledge of the translocations, substantially expanding the information reported previously [1,17,51]. We documented the following characteristics of each translocation: (1) translocation type (reintroduction, augmentation or introduction), (2) location (state or province and specific release sites), (3) outcome (success or failure), (4) source population (state or province), (5) purpose of translocation, and (6) years of initiation and completion. Additional factors that could influence translocation success are presented in Table 2.

To evaluate the extent to which successful translocations may have contributed to recent range expansions, we overlaid the locations of release sites on 3 different depictions of the fisher's geographic distribution: the historical range, the range at its most geographically contracted state (hereafter, most-contracted range), and the current range. We developed these 3 range maps based on previously reported ranges [1,18,28,52–54]; data from museum collections, published literature, and unpublished reports; and interviews with agency personnel and local experts.

We considered translocations to be independent if they were separated by ≥200 km (163 km is the largest post-release movement documented; [51]). Translocations were also considered independent if they occurred within 200 km of each other but the earlier translocation failed.

Test of Population Model Predictions

We used data from actual reintroductions to test the predictions of our *VORTEX* model and to determine if we obtained similar results for reintroduction success from these 2 independent approaches (*VORTEX* model vs. data from actual reintroductions). Before testing hypotheses we derived from *VORTEX* model results, we tested for independence of the variables from actual reintroductions using Pearson's correlation coefficient ($\alpha = 0.05$). Using data from actual reintroductions, we calculated Akaike's Information Criterion adjusted for small sample sizes (AIC$_C$; [55]) and evaluated the hypotheses we derived from the *VORTEX* model results. We calculated likelihood values for AIC$_C$ using Proc GENMOD in SAS (SAS 9.2; www.sas.com), and retained all hypotheses with ΔAIC$_C$≤2.0.

Weights of released fishers are included in some translocation reports, but no dependable age or maturity data exist for any of the reintroductions with a known outcome. Sex-ratio data, however, exist for the majority of reintroductions. Therefore, we first evaluated the strengths of the following hypotheses to explain reintroduction success.

Table 2. General characteristics (variables) of fisher translocations that could influence translocation success.

Variables that could influence translocation success	Variable name	VORTEX	Data
Number of fishers released	Number of fishers	Yes	Yes
Number of release sites	Number of sites	Yes	Yes
Number of years fishers released	Number of years	Yes	Yes
Sex ratio of released fishers	Sex-ratio	Yes	Yes
Feasibility assessment prior to release	Feasibility	No	Yes
Genetic diversity of source population (number sources)	Diversity	No	Yes
Genetic relatedness to source population (proximity)	Relatedness	No	Yes
Monitoring post-release	Monitor	No	Yes
Protection from fur-trapping for fishers specifically	Protect1	No	Yes
Protection from incidental fur-trapping	Protect2	No	Yes
Region (Eastern vs Western North America[1])	Region	No	Yes
Season of release	Season	No	Yes
Type of release (hard versus soft)	Type	No	Yes
Ownership of lands where fishers released	Owner	No	No
Trapping re-established following translocation	Trapping	No	No

Four variables were included in our VORTEX simulations and data for 13 of these variables were available from actual translocations.
[1]106° West Longitude, chosen because it is midway within a gap between translocations that occurred in eastern versus western North America (see Figure 6).

(1) Number of fishers released, as a substitute for number of adult females, allowing us to use data for all reintroductions. Our VORTEX model results suggested that the probability of a successful reintroduction should increase with the number of adult females released.

(2) Number of females released, as a substitute for number of adult females.

(3) Number of males released, because our VORTEX model results suggested, perhaps counter-intuitively, that number of adult males released should have no effect on reintroduction success.

(4) Number of release sites, because our VORTEX model results suggested that the probability of a successful reintroduction should increase with the number of release sites used.

(5) Number of release years (controlled for total numbers of fishers released), because our VORTEX model results suggested, again perhaps counter-intuitively, that the number of years over which a given number of fishers was released should have no effect on reintroduction success.

(6–14) The combinations of variables in (1), (2) or (3) with those in (4) and (5).

For the hypotheses with $\Delta AIC_C \leq 2.0$, we evaluated additional factors (variables) from Table 2, taken one at a time, again using AIC_C to rank the performance of our hypotheses. Genetic diversity of the source population and genetic relatedness of the source population to the original population were unknown for most translocations. We used the number of states and provinces that provided source populations (which was known for all translocations) as an index of genetic diversity, and we used proximity of source population to the release site (same part of continent = near, otherwise far) as an index of genetic relatedness. We started with "Region" and noted that this variable has such a large effect on AIC_C values that we evaluated all other single variables along with the "Region" variable.

Results

Population Models for the Fisher

VORTEX simulations of the baseline population model (i.e., no fishers removed) had a 95% viability index. Baseline simulations

also indicated that juvenile survival and litter size had the greatest elasticity values, suggesting that variation in these variables had the greatest potential to affect the viability index (Table 1). Adult survival had intermediate elasticity values; variation in other variables had little effect on the index.

To account for the possibility that our estimates of the present population size or carrying capacity for fishers in northwestern California were overestimates, we also ran simulations with each of these variables reduced to half of its original value. Halving the initial population size or halving K decreased the viability index by about 1%, which is a smaller effect than caused by 10% changes in juvenile survival or litter size.

Model results for the source population indicated that the removal of fishers from the population had little effect on the viability index (Table 3). The index dropped <5% when 20 fishers (5 each from 4 different hexagons each year) were removed from

Table 3. Predicted effects of removing fishers for 3, 5, or 8 years on the viability of a source population.[1]

Removal Scenario			Decrease in Viability Index
Number fishers removed from each hexagon	Number hexagons	Number years	
1	20	3	0%
1	20	5	3%
5	4	3	2%
5	4	5	0%
5	4	8	1%

[1]The source population had a carrying capacity (K) of 2000; viability declines were predicted by 100 runs of VORTEX for each set of values for variables. Each of 50 hexagons was modeled to determine landscape effects; simulated animals could move among the hexagons.

Table 4. Predicted index of successful reintroduction of fishers from 100 runs of *VORTEX* for differing founder population compositions.

Reintroduction Scenario: number released at each of 5 release sites each year					Index of Success
Adult females	Adult males	Juvenile females	Juvenile males	Number of release years	
3	2	0	0	2	2
3	2	0	0	3	20
3	2	0	0	5	42
3	2	0	0	8	82
1	4	0	0	5	3
1	3	0	0	5	4
1	2	0	0	5	3
2	3	0	0	5	29
2	2	0	0	5	26
4	1	0	0	5	66
2	2	1	0	5	40
2	2	0	1	5	27
2	1	1	1	5	40
2	1	2	0	5	54
2	1	0	2	5	17
1	1	3	0	5	55

the model population for each of 8 years; removals of 1 fisher from each of 20 hexagons for 3 years had no effect on the index.

The index of success for simulated populations of reintroduced fishers varied considerably with the reintroduction scenario (Table 4, Figures 1, 2, 3, 4, 5). The index increased with the number of adult females in the release population as long as 1 adult male was also included (Table 4, Figures 1, 2, 3), with the exception that releasing 2 adult females and 2 juvenile females was equivalent to releasing 1 adult female and 3 juvenile females (Figure 3). Releasing juvenile females and males, or more than 1 adult male, reduced reintroduction success by limiting the number of adult females released (Figure 3). The index of success increased with the number of release sites (Figure 4) but the number of years

used to release a fixed number of fishers had no effect on success (Figure 5).

Actual Fisher Translocations and Subsequent Range Expansions

We documented 38 fisher translocations (30 reintroductions, 5 augmentations, and 3 introductions) in 7 Canadian provinces and 15 US states between 1900 and 2011 (Table 5, Figures 6 and 7). Undocumented translocations also occurred [56]. The first documented fisher translocation in North America was the introduction of 2 fishers to Anticosti Island, Quebec, around 1900 [57]. Fishers were translocated to reestablish a component of the native fauna (63%), to control porcupines (37%), to reestablish

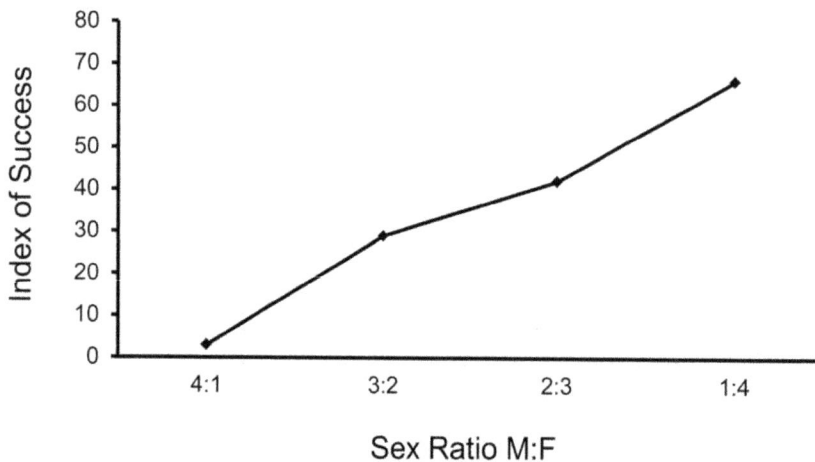

Figure 1. Predicted relationship between sex ratio for fishers that are released and reintroduction success. The *VORTEX* model included stochastic variation and was run for 100 years. Twenty fishers were modeled to be released in each of 5 years in 4 groups of 5, with each group released into a different site.

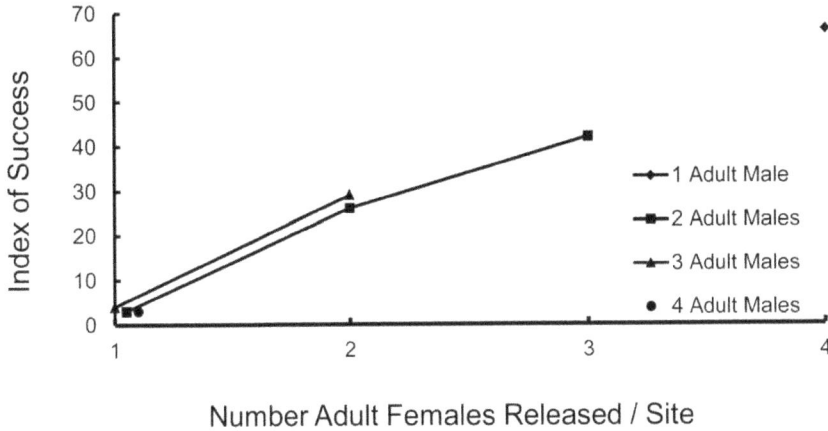

Figure 2. Predicted relationship between numbers of adult female and adult male fishers released and reintroduction success. Fishers were released at each of 4 subsites in each of 5 years. No juvenile fishers were released in these scenarios. The *VORTEX* model included stochastic variation and was run for 100 years. Twenty fishers were modeled to be released each of 5 years in 4 groups of 5, with each group released into a different site.

a valuable furbearer for commercial trapping (16%), to initiate research (5%), or for combinations of these reasons (26%; Table 5).

All 5 augmentations succeeded, whereas 77% (20 of 26) of reintroductions with known outcomes succeeded, and none of the 3 introductions were successful (Table 5). Among reintroductions in eastern North America with known outcomes, 89% (17/19) succeeded, including all that released ≥12 fishers. In contrast, only 43% (3/7) of reintroductions in western North America succeeded; where a release with as few as 17 fishers succeeded but 1 reintroduction of 60 fishers failed. For reintroductions, we obtained data sets for 10 of the 13 variables thought to influence translocation success (Tables 2 and S1). The first 4 variables in Table 2 are, or relate directly to, factors incorporated into our *VORTEX* population model.

Our map of the fisher's geographic distribution (Figure 8) includes the historical range (approximately 5,498,000 km^2), the most-contracted range (~2,343,000 km^2; ~43% of the historical range) and the current range (~3,717,000 km^2; ~68% of the

historical range). The historical range (Figure 8) includes corrections of previous maps by Powell [1] and Gibilisco [18]. Areas where the most-contracted range expanded to form the current range coincide closely with the distribution of successful reintroductions, especially in eastern North America. Several translocation programs used numerous release sites over large areas (e.g., Nova Scotia, Vermont, upper peninsula Michigan, Pennsylvania), and in the case of Nova Scotia, initial reintroductions were followed by augmentations. Fishers were also reintroduced successfully following failed reintroduction attempts in southwestern Oregon and Manitoba.

Range expansion by fishers, however, was not limited to areas where they had been reintroduced. Fishers expanded their range through natural dispersal into areas where they had been extirpated in New York and New England, Ontario and Quebec, Minnesota, and northern California. Fishers also expanded their range naturally into areas where reintroductions failed (e.g., coastal Maine) and into an area outside the historical range in

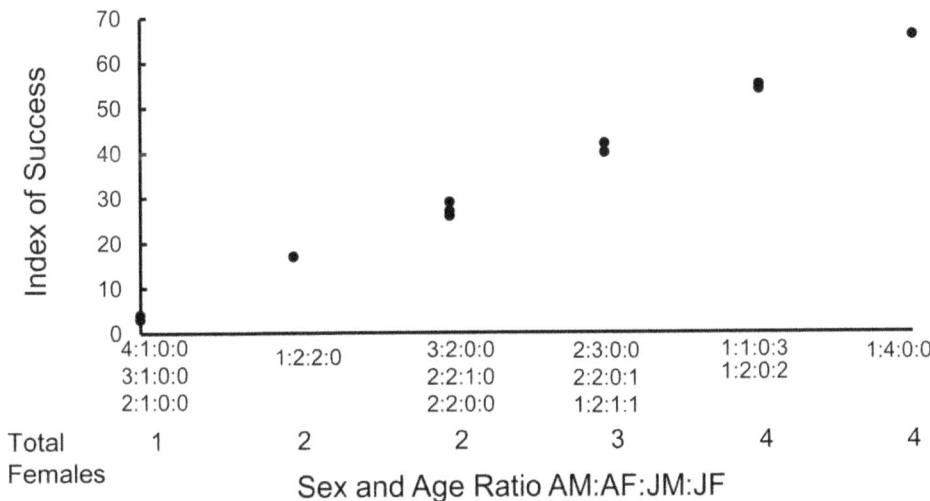

Figure 3. Predicted relationship between numbers of female and male fishers released, including juveniles, and reintroduction success. Fishers were released at each of 4 subsites in each of 5 years. The *VORTEX* model included stochastic variation and was run for 100 years. Twenty fishers were modeled to be released each of 5 years in 4 groups of 5, with each group released into a different site.

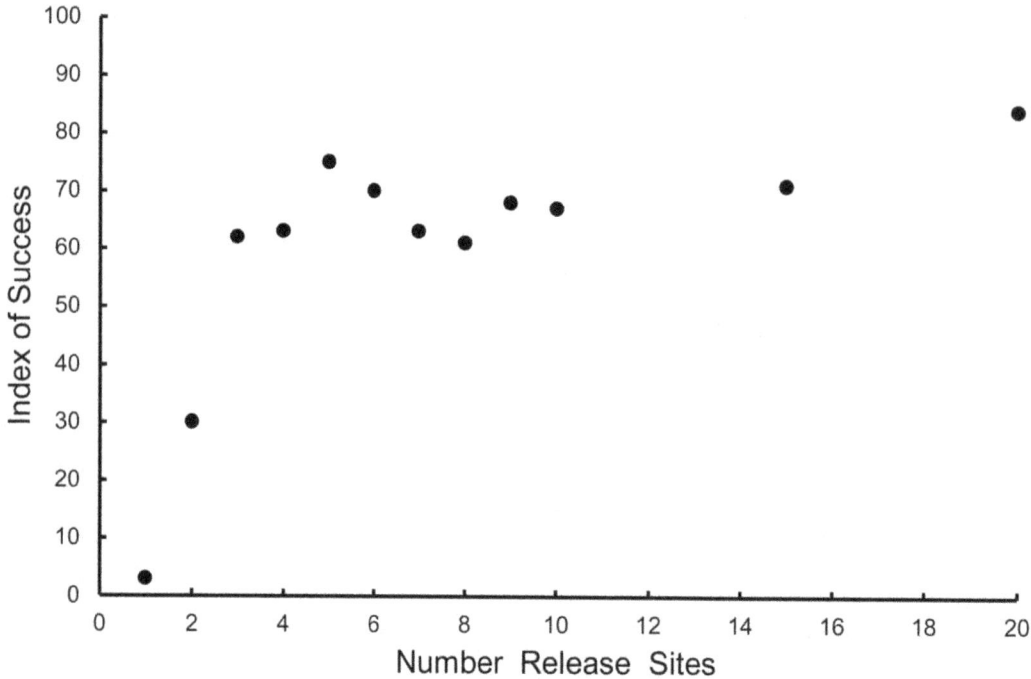

Figure 4. Predicted relationship between the number of release sites for a translocation and reintroduction success. Points represent mean values for 100 simulations.

northwestern British Columbia and Yukon Territory (Figures 6 and 8) [54].

Tests of Predictions from Population Modeling

For actual reintroductions, Number of fishers released, Number of females released and Number of males released were all highly correlated, as expected ($r > 0.97$ for all combinations). Number of

release sites correlated significantly with Number of males released ($r = 0.43$) and with Number of years that fishers were released ($r = 0.47$). Number of years that fishers were released also correlated significantly with Number of fishers released, Number of females released, and Number of males released (correlation coefficients 0.56–0.65). We did not combine variables that were significantly correlated in the same hypothesis without blocking to eliminate the effects of the correlation.

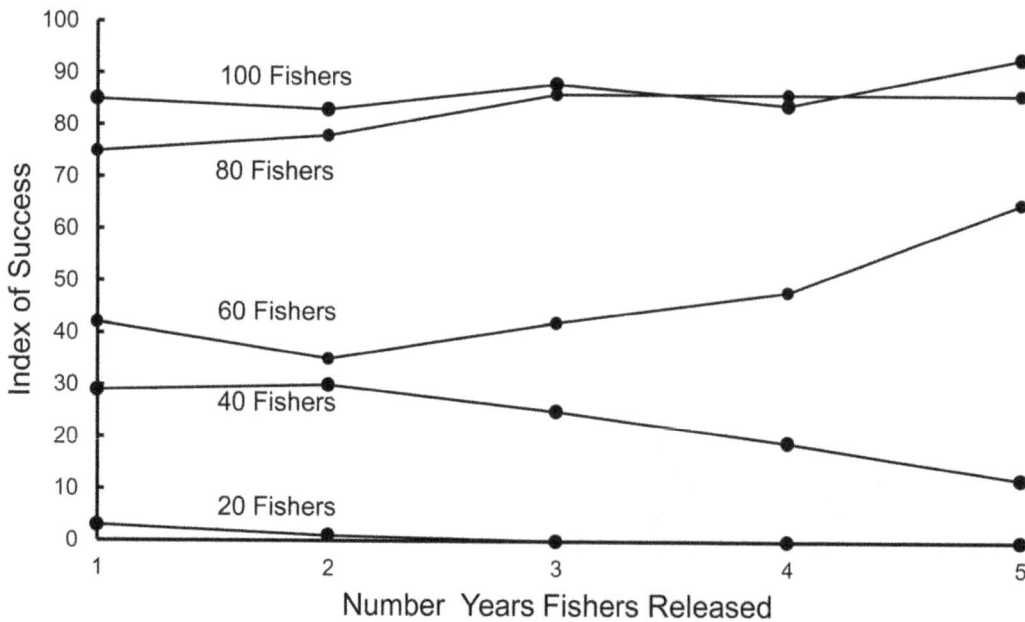

Figure 5. Predicted relationship between the number of years fishers are released and reintroduction success. The index of success was predicted for a fixed number of fishers released all in one year or released over 2, 3, 4 or 5 years (e.g., 60 fishers released in 1 year; 30 in each of 2 years; up to 12 in each of 5 years). Points represent mean values for 100 simulations.

Table 5. Summary of data for 38 fisher translocations, 1896–2010, listed chronologically.

Release location	Source location	Years(s)	Transloc. type[a]	Number released (♀s)	Success Status[b]	Purpose[c]	References
Quebec	Unknown	1896–1914	I	2 (?)	F	IF	[57]
Nova Scotia	Ranch	1947–1948	R	12 (6)	S	Unknown	[40,79]
Wisconsin	New York, Minnesota	1956–1963	R	60 (24)	S	PC	[36,39,80–82]
Ontario	Ontario	1956	R	25 (?)	U	RS	[17,56]
Ontario	Ontario	1956–1963	R	97 (60)	S	RS	[17,56]
Montana	British Columbia	1959–1960	A	36 (20)	S	RS,PC,RF	[32,41,51,83,84]
Vermont	Maine	1959–1967	R	124 (?)	S	PC	[17,85]
Oregon	British Columbia	1961	R	11 (6)	F	PC	[3,86,87]
Oregon	British Columbia	1961	R	13 (8)	F	PC	[3,86,87]
Michigan	Minnesota	1961–1963	R	61 (19)	S	PC	[29,36,88]
Idaho	British Columbia	1962–1963	A	39 (19)	S	RS,PC	[17,89,90,91]
Nova Scotia	Maine	1963–1966	R	80 (51)	S	RS,PC	[79]
Wisconsin	Minnesota	1966–1967	R	60 (30)	S	PC	[80–82]
New Brunswick	New Brunswick	1966–1968	R	25 (15)	S	RS, PC	[43,92,93]
West Virginia	New Hampshire	1969	R	23 (?)	S	RS,RF	[42,94]
Minnesota	Minnesota	1968	R	15 (?)	F	PC	[17,95]
Maine	Maine	1972	R	7 (3)	U	RS	[17,96]
Manitoba	Manitoba	1972	R	4 (?)	F	RS	[17,97]
New York	New York	1976–1979	R	43 (24)	S	RS	[44,98]
Oregon	British Columbia, Minnesota	1977–1981	R	30 (15)	S	PC	[3]
Colorado	Unknown	1978 or 79	I	2 (1)	F	Unknown	[99]
Ontario	Ontario	1979–1981	R	55 (32)	S	RF	[56,100,101]
Ontario	Ontario	1979–1982	R	29 (14)	S	RF	[56,100,101]
Alberta	Alberta	1981–1983	R	32 (16)	F	RS	[45,102,103]
British Columbia	British Columbia	1984–1991	I	15 (4)	F	PC	[104,105]
Montana	Minnesota, Wisconsin	1988–1991	R	110 (63)	S	RS	[51,83]
Michigan	Michigan	1988–1992	R	189 (101)	S	RS,RF	[88]
Connecticut	New Hampshire, Vermont	1989–1990	R	32 (19)	S	RS	[106–109]
Alberta	Ontario, Manitoba	1990	R	17 (11)	S	RS,R	[45,100,110]
British Columbia	British Columbia	1990–1992	A	15 (13)	S	RS,R	[111]
Nova Scotia	Nova Scotia	1993–1995	A	14 (6)	S	RS	[112–115]
Manitoba	Manitoba	1994–1995	R	45 (21)	S	RS	[116]
Pennsylvania	New York, New Hampshire	1994–1998	R	190 (97)	S	RS	[117]
British Columbia	British Columbia	1996–1998	R	60 (36)	F	RS,RF	[66,67]
Nova Scotia	Nova Scotia	2000–2004	A	28 (21)	S	RS	[113,114]
Tennessee	Wisconsin	2001–2003	R	40 (20)	S	RS	[68,118]
Washington	British Columbia	2008–2011	R	90 (50)	O	RS	[50,119]
California	California	2009–2012	R	40 (24)	O	RS	[120]

Additional data for these 38 translocations are included in Table S1.
[a] A = augmentation, I = introduction, R = reintroduction.
[b] F = failure, S = success, U = unknown outcome, O = ongoing.
[c] IF = introduction of furbearer, PC = porcupine control, RS = reestablish species, RF = reestablish furbearer, R = research.

When we combined Number of years with Number of fishers, Number of females or Number of males in a hypothesis, we blocked Number of years by the appropriate variable for Number of fishers when calculating likelihoods. We retained 2 single-variable hypotheses from our evaluation of the hypotheses identified in our *VORTEX* population model as influencing reintroduction success (Table 6). The highest ranked hypothesis included only the Number of males released (42% probability of being the most strongly supported hypothesis), but the second-ranked model, including only the Number of females released, was almost as well supported (30% probability). Neither the Number of release sites nor the Number of years were included in any model that we retained, despite our *VORTEX* model simulations identifying the number of release sites as being important.

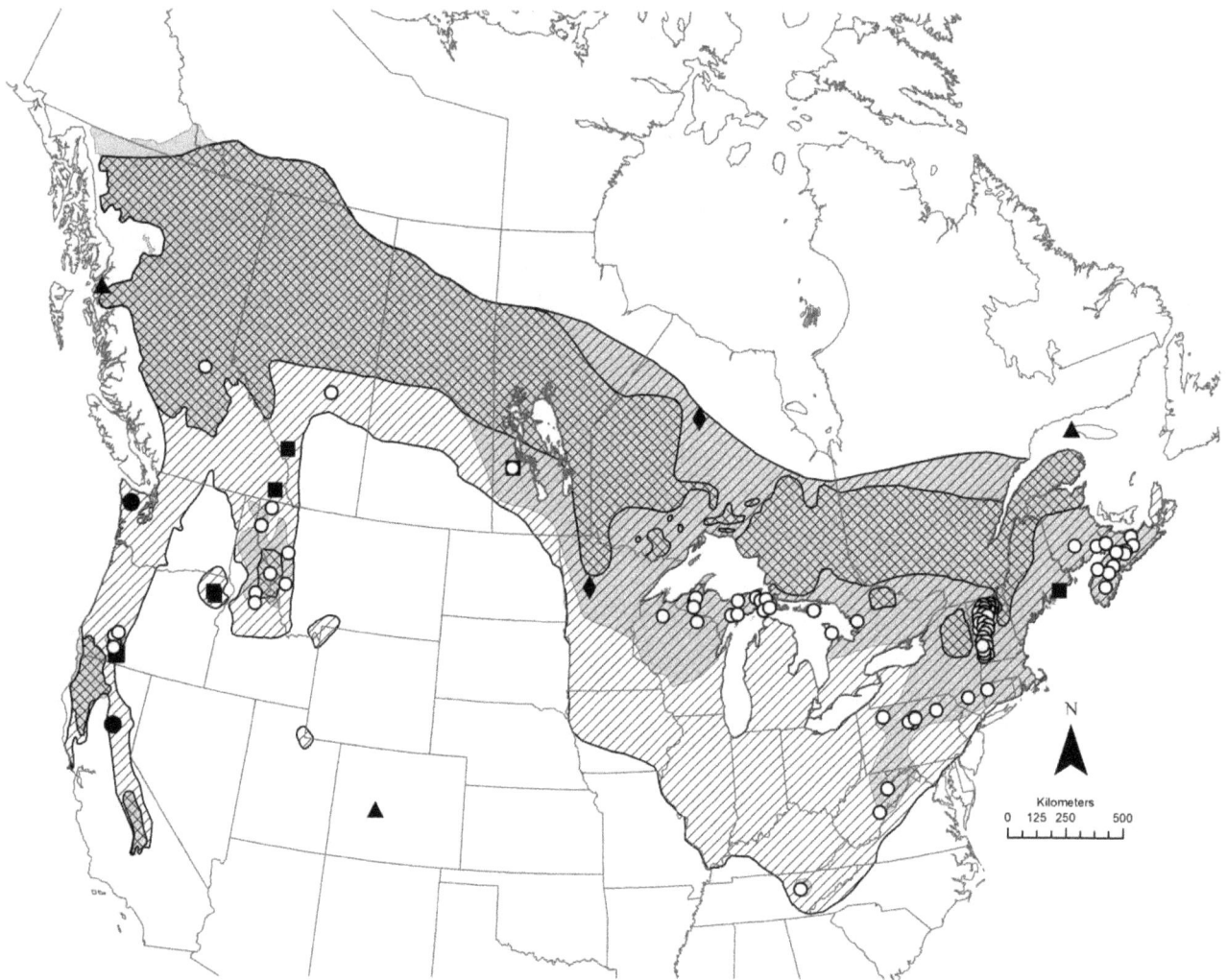

Figure 6. Locations of translocations in relation to the fisher's historical, most-contracted and current range. The historical fisher range occurred prior to European settlement (diagonal hatching), but was reduced to the range at its most contracted state (cross hatching; 43% of the historical range) before expanding to the current range (shaded; 68% of the historical range). White circles represent successful reintroductions or augmentations, black squares represent failed reintroductions, black diamonds represent reintroductions with unknown outcomes, black circles represent ongoing reintroductions and black triangles represent introductions.

We obtained adequate data for 10 additional variables in Table 2 (see full data set in Table S1). We first tested Number of males and Number of females each with Region (eastern vs western North America) and noted that adding Region affected AIC_C values profoundly (Table 7). Therefore, we added the other variables one at a time to hypotheses including Region with either Number of females or Number of males (Table 7). The result was that Relatedness of the source animals (indexed by the proximity of source population to the release site; Tables 5 and S1) had the greatest effect on reintroduction success, once Region was already included in the hypothesis (Table 7). Note, however, that all AIC_C values in Table 7 are less than those in Table 6, implying that all the hypotheses in Table 7 are able to mimic the original data better than the hypotheses of Number of males and Number of females alone (all hypotheses in Tables 6 and 7 had the same structure and are, therefore, comparable using AIC_C). The hypotheses including Numbers of males or females with Region and potential Relatedness of the source animals to the original population both have about 45% probabilities of being the hypothesis best able to mimic the data (i.e., best models; Table 7).

Discussion

The fisher is among the most successfully reintroduced carnivores, yet little information is available to explain this success or to guide managers seeking to reestablish fishers. In many areas of the historical range, protection from historical over-trapping may have been all that was needed to prevent extirpation and to maintain a self-sustaining population of fishers. The great success of fisher reintroductions in eastern and mid-western North America in the mid 1900s suggests that the only thing lacking was the fishers themselves, and reintroductions addressed that problem. Past performance, however, does not guarantee future success, and the costs, risks and uncertainties associated with reintroductions prompted us to investigate how we could help managers tilt the odds in their favor. Our results should be helpful as managers continue to reestablish fishers in the vacant areas within the southern portions of the fisher's historical range.

Increasing Fisher Reintroduction Success

Four factors appear to have a meaningful influence on reintroduction success: the number of females released, number

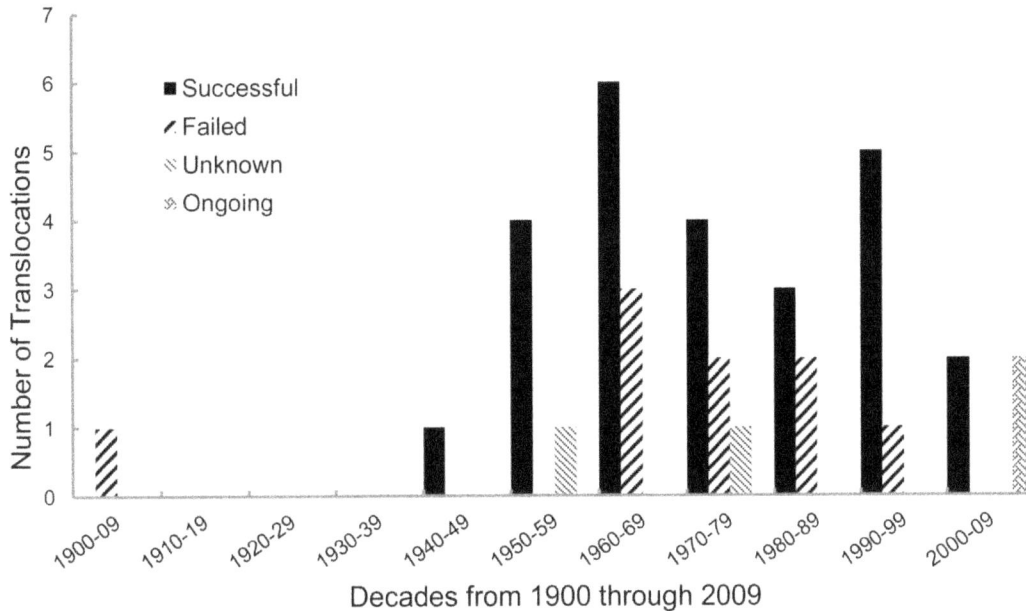

Figure 7. Chronology and success status of 38 fisher translocations.

of males released, region (eastern vs. western North America), and the proximity of the source population to the reintroduction area.

It seems intuitively obvious that releasing more females would result in greater success, because releasing more females is expected to result in greater offspring production and greater population growth. Releasing adult females also makes sense, because adult females are sexually mature, and pregnant females can immediately contribute to population growth. Unlike adult females, juvenile females must survive long enough to become sexually mature, successfully mate and give birth, which may not happen for ≥2 years after they are released. The lack of age data for actual reintroductions prevented us from testing the hypothesis that the proportion of adults in a founder population is a meaningful predictor of reintroduction success.

Why the number of males released would have such a meaningful influence on reintroduction success is less obvious to us. Given the non-monogamous mating system of fishers, where a male can mate with >1 female during a breeding season [1], a founder population dominated by females would be more likely to succeed than a founder population of the same size with an even or male-biased sex-ratio. Nonetheless, our analyses indicated that the number of males was at least as important as the number of females for explaining reintroduction success. We believe this result provides important insights about variation in the reproductive abilities of males.

The age, size, and experience of released males is also likely to affect reintroduction success. In other solitary carnivores with large sexual dimorphism in size, extremely large males can dominate smaller males and secure most of the breeding opportunities [58,59]. Male fishers must survive long enough to reach their maximum body size (at ≥3 years of age), which explains why they represent such a small component of most fisher populations [1]. We hypothesize that the most effective breeding males are large males with pronounced, well-developed musculature on their heads, especially the large temporalis muscles that originate along the sides of the skull and attach to their large sagittal crests [1,60,61]. These muscles enlarge during the breeding season and shrink during summer, suggesting a relationship with breeding

(Powell unpublished data). We hypothesize that such males are the most effective breeders and the strongest competitors for reproductive females, and as such, they would make more significant contributions to translocation success than young males. If a reproductive advantage for large and experienced males exists in translocated fisher populations, then relatively few large males need to be released to achieve reintroduction success. Reintroduction programs that released larger numbers of fishers were likely to include a greater number of these large breeding males, thereby increasing the likelihood of success.

We hypothesize that for other mammals where reproduction of males has a large skew, reintroduction success should depend on the number of males that are effective breeders and not on the total number of males released, as we hypothesize for fishers. Males of many mammals exhibit sexually dimorphic, sexually selected traits that could be the basis for such reproductive skew [e.g. 58, 59]. We encourage biologists who plan reintroductions of such mammals to investigate the logistics of selecting males for release based on their sexually dimorphic traits and determine how those traits affect reproductive success.

We can not explain the difference in success of reintroductions in eastern vs. western North America. Our data, the information available in the literature, and personal communications from agency personnel involved in reintroductions provide no insights into why reintroductions in the East have twice the probability of success as those in the West. Differences in snow cover [62], forest succession, and forest characteristics [63] have been proposed as contributing factors. Forestry practices, predator-prey communities, and genetic characteristics also differ by region and may play a role.

Our analyses of actual reintroductions suggest that releasing fishers from a nearby source population (a population from the same part of the continent) positively affects the probability of reintroduction success (Table 7). Many successful reintroductions used source populations from the same or nearby states and provinces. This result suggests that some local, genetic adaptation may exist within fisher populations.

Releasing animals into suboptimal habitat is a major reason for failure of translocations in general [8,64,65]. We assume that

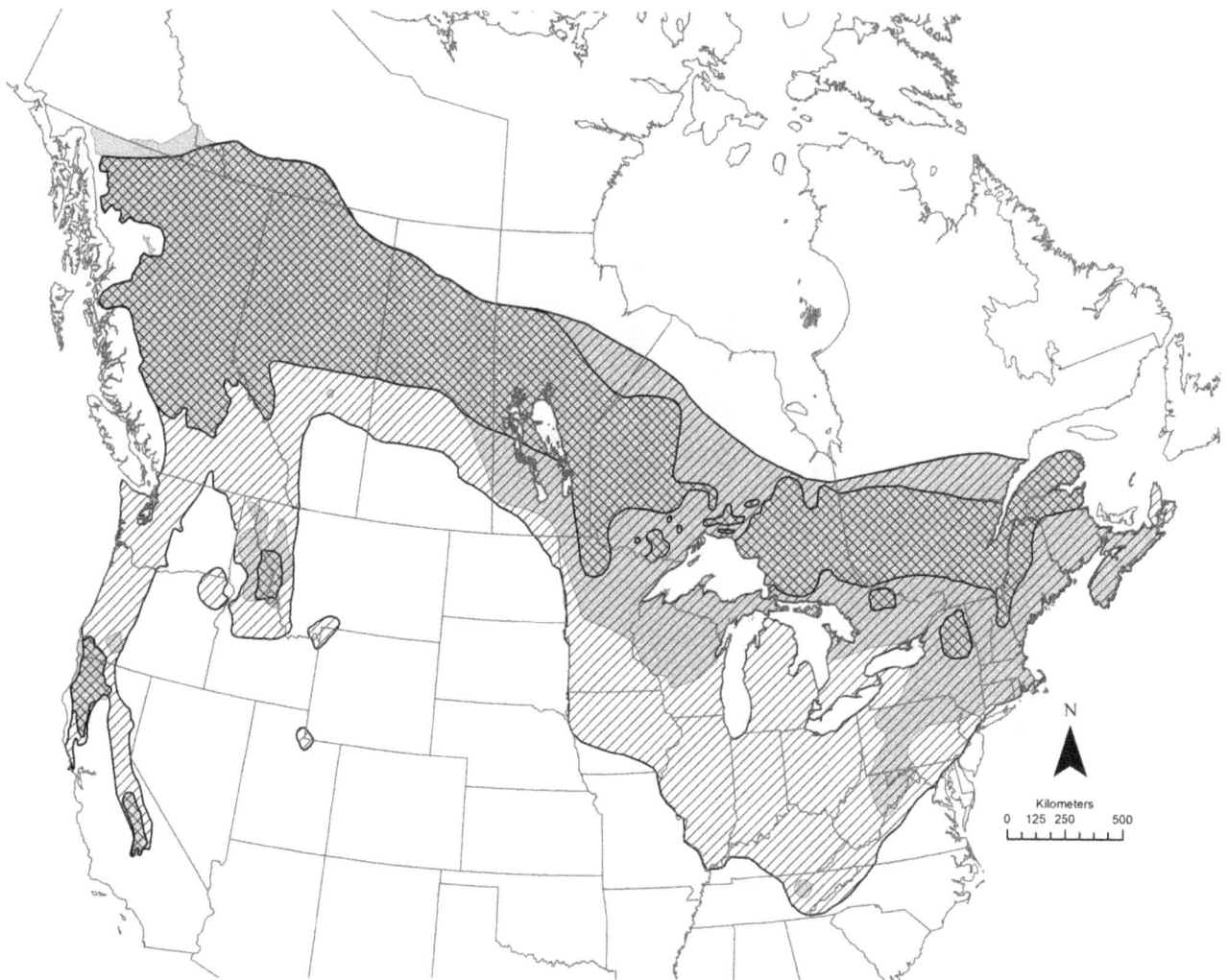

Figure 8. The fisher's historical (diagonal hatching), most-contracted (cross hatching) and current (shaded) ranges. Significant portions of these ranges were obscured by translocation symbols in Figure 6.

project managers released fishers into areas they considered to be suitable habitat, yet only 4 reintroduction projects conducted formal assessments of habitat quality prior to initiating releases (southeastern British Columbia [66,67]; Tennessee [68], Washington [69,70] and California [15]). Consequently, we lack data for one of the most significant factors affecting success (habitat adequacy) for 89% (34 of 38) of documented fisher translocations. Our small sample size of translocations with formal habitat assessments was too small to draw meaningful conclusions from the data, however, the value of a formal habitat assessment seems clear given the uncertainties and the resources at stake.

The 100% success rate for the 5 fisher augmentations suggests that they are more likely to be successful than reintroductions (77% success). While there are several reasons why augmentations could have a greater likelihood of success than reintroductions, the indications of success are also less clear for augmentations. Continued persistence of a small, reestablished population can demonstrate the success of a reintroduction, whereas some indication of population growth, expansion or improvement is necessary to conclude that an augmentation was successful. The presence of a small resident population of fishers could easily increase the likelihood of success for augmentations, as these

resident fishers may serve as a locally adapted foundation for population expansion that does not exist where reintroductions are initiated. Augmentations may also remedy a skewed sex-ratio that limited population growth in a small resident population or add new genotypes to a population with low genetic diversity. The data for augmentations were also complicated by fisher releases in Montana (1959–1960) and Idaho (1962–1963), which were thought to be reintroductions because, at the time, no one knew that a remnant, native population still existed in the Rocky Mountains of eastern Idaho and western Montana [32,33]. We do not know if managers would have hesitated to release new fishers in this area had they known of the remnant native population. Nor do we know if these augmentations prevented the loss of the remnant native population or if they influenced fisher fitness. What is clear, however, is that the fisher range expanded from its most-contracted state in this region following the augmentations.

We documented 3 fisher introductions, and while each failed, each was unique in its circumstances. An introduction of 2 fishers to Anticosti Island, Quebec was conducted around 1900 by the Island's owner, Henry Menier, whose goal was to introduce a number of game species to the Island for future harvest [57]. The second introduction involved the captive-rearing and release of 2

Table 6. Hypotheses generated by the results of our *VORTEX* simulations and their ranking using AIC_C[1].

HYPOTHESES	AIC_C	ΔAIC_C	w_i
Number Males Released	22.91	0.00	0.415
Number Females Released	23.58	0.67	0.297
Number Females Released, Number Release Sites	26.11	3.20	0.084
Number Females Released, Number Release Years	26.14	3.23	0.083
Number Males Released, Number Release Years	26.14	3.23	0.083
Number Fishers Released	28.94	6.02	0.020
Number Fishers Released, Number Release Sites	31.51	8.59	0.000
Number Fishers Released, Number Release Years	31.53	8.61	0.000
Number Release Sites	52.25	9.34	0.000
Number Release Years	32.28	9.36	0.000

See data for all variables in Table S1.

[1]*VORTEX* simulations indicated that the variables in these hypotheses could influence reintroduction success. Only Number of Males and Number of Females were retained for further evaluation (Table 7). In calculating likelihoods, Number of Years was blocked for Number of Fishers, Females or Males released (as appropriate) because these variables were strongly correlated.

fishers (male and female siblings) in west-central Colorado around 1978 by Marty Stouffer as part of his "Wild America" television series (season 3, Fishers in the family, episodes 8 and 9; www. wildamerica.com). Stouffer's purpose for releasing the fishers may have been solely to create an interesting television show while

Table 7. Alternative hypotheses that may affect reintroduction success, and their ranking using AIC_C.

HYPOTHESES	AIC_C	ΔAIC_C	w_i
Number Males Released, Region, Relatedness	−1.76	0.00	0.447
Number Females Released, Region, Relatedness	−1.76	0.01	0.445
Number Males Released, Region, Diversity	3.48	5.25	0.032
Number Females Released, Region, Diversity	3.52	5.29	0.032
Number Males Released, Region, Type	4.23	5.99	0.022
Number Females Released, Region, Type	4.49	6.26	0.020
Number Females Released, Region, Protect2	7.55	9.31	0.000
Number Males Released, Region, Protect2	7.55	9.32	0.000
Number Females Released, Region	12.38	14.14	0.000
Number Males Released, Region	12.47	14.23	0.000
Number Females Released, Region, Feasibility	13.70	15.47	0.000
Number Females Released, Region, Protect1	13.72	15.49	0.000
Number Males Released, Region, Feasibility	13.92	15.69	0.000
Number Males Released, Region, Protect1	14.04	15.80	0.000
Number Females Released, Region, Season	14.70	16.47	0.000
Number Males Released, Region, Season	14.72	16.49	0.000
Number Males Released, Region, Monitor	15.21	16.98	0.000
Number Females Released, Region, Monitor	15.27	17.03	0.000
Number Males Released, Region, Sex ratio	15.33	17.10	0.000
Number Females Released, Region, Sex ratio	15.33	17.10	0.000

The first two hypotheses are considered the best models. See data for all variables in Table S1.

evicting an unruly duo of fishers from his home. The third introduction was conducted from 1984 to 1991 in an area just outside the historical range in northern-coastal British Columbia; 11 males and 4 females (2 of the 4 were badly injured) were released to control porcupines. While we have almost no information to speculate on why these introductions failed or the suitability of habitat at the release sites, the success of an introduction involving the release of only 2 fishers or only 2 uninjured females would be extremely unlikely.

Reintroduction success did not differ for hard *vs* soft releases in our analyses, even though all soft releases were successful. Our test lacked power, however, because the number of soft releases was very small (n = 3). Similarly, our analyses failed to detect any effects of protecting released fishers from trapping or of releasing fishers in different seasons. State and provincial regulations related to protection were diverse and difficult to categorize, and likely prevented us from detecting the true effects of no protection. Moreover, some reintroduced fishers may have benefited from informal protection from trappers that wanted reintroductions to succeed. In addition, our analyses could not detect effects of post-release monitoring on reintroduction success. An evaluation of monitoring effects was complicated by the lack of formal monitoring efforts in the earlier fisher translocations, and agency reports and publications consistently failed to mention management actions that could have been triggered by monitoring results to increase the likelihood of success.

Our simulations suggest that the number of years over which a targeted number of fishers are released should not affect the probability of success. Nonetheless, among actual reintroductions, successful reintroductions involved releasing fishers over multiple years, probably because logistical and financial constraints limit the number of fishers that can be obtained and released in a given year. For example, a reintroduction project could target 100 fishers for release, but limitations on trapping conditions (e.g., suitable weather, adequate road access) and trapping success may require trapping over a number of years to reach the target number of animals to release. Thus, despite evidence to the contrary, releasing fishers over multiple years may be required out of practical necessity.

Our simulations and analyses focused on fisher populations, but the factors that influence fisher reintroduction success are also likely to influence the reintroduction success of other species, including other mustelids, other carnivores, and other wide-ranging mammals that do not live in groups or have other complex social organizations. Consistent with our *VORTEX* simulations, we expect that the number of adult females released, the sex-ratio within a founder population, and the number of release sites to be factors that are likely to influence carnivore reintroduction success. Our analyses of data from actual reintroductions suggest that the number of males released and the proximity of the source population could also be meaningful predictors of carnivore reintroduction success. While some factors that influence reintroduction success are likely to be species- or area-specific, we expect one or more of the factors we have identified to be informative to managers as they plan and implement carnivore reintroductions and that this information will improve their likelihood of success.

The Fisher's Range, Translocations and Conservation

While strict harvest regulations in many states and provinces provided significant protection for fishers from overexploitation across much of their historical range, no one has conducted a range-wide evaluation of how reintroductions and augmentations have contributed to fisher conservation. Our depictions of the historical, most-contracted and current fisher ranges illustrate the vulnerability of this carnivore to a variety of threats [71]; they also provide meaningful baselines for measuring recovery. The expansion of the most-contracted range to the current range was most extensive in the eastern and Great Lake states and provinces. For a number of these states and provinces, range expansions may have resulted from successful reintroductions, natural range expansions, or both (e.g., Maine, New Brunswick, New York, New Hampshire, Ontario, Vermont). Reintroductions, however, were responsible for the reestablishment of fisher populations in Maryland, Michigan, Nova Scotia, Pennsylvania, West Virginia, and Tennessee, long before a natural range expansion could have occurred.

Given the success of individual reintroductions and the range-wide consequences of successful reintroductions, it is clear that reintroductions have made a significant contribution to fisher conservation. Much of the fisher's recovery was the result of uncoordinated reintroduction programs conducted within individual states and provinces that occurred long before the fisher was a candidate for federal listing. However, the recovery of the fisher suggests that if uncoordinated restoration efforts can result in a significant range expansion of a wide-ranging carnivore, coordinated and well-planned restoration efforts could easily result in a similar or greater level of success.

Management Recommendations

Because the number of females and males released is critical to reintroduction success, we recommend that managers release as many fishers as possible. We expect the likelihood of success to be improved by acquiring a founder population that is slightly female-biased (e.g., 55–60% females), and one that is also adult-biased, in an effort to obtain a greater proportion of pregnant females and large males that are effective breeders. Although we are not in a position to recommend a minimum number of fishers to release, we observed that 9 of 10 reintroductions that released ≥60 fishers were successful, and all 4 reintroductions of ≥100 fishers were successful. (We also note that all reintroductions in eastern North America with ≥12 fishers succeeded.)

We recommend that managers select source populations that are close to release sites. Using multiple release sites and diverse source populations may also increase the probability of success. Given the fisher's vulnerability to overharvest and incidental capture [1], protection from direct and incidental harvest should improve the likelihood of reintroduction success, even though our analyses could not verify this effect. While a formal feasibility assessment is recommended for any translocation [10,12], only 5 fisher translocations were preceded by formal assessments; all 5 were among the most recent translocations (1994–2011). Feasibility assessments are essential given the cost, uncertain outcome, and risk to source populations associated with translocations, especially for species at risk. Although a feasibility assessment is useful for evaluating the likelihood that a reintroduction will succeed, an assessment also provides the foundation for a reintroduction implementation plan [15,69,70]. These and other planning efforts (e.g., species status reviews, recovery planning, NEPA analyses) can help managers minimize or avoid controversies associated with reintroductions and, thereby, increase support for reintroduction projects.

Future reintroductions should incorporate active, post-release monitoring with effective adaptive management [10]. Passive monitoring alone (e.g., incidental observations, road-kill mortalities) is insufficient to assess translocation success and to identify avoidable hazards. Reintroductions and monitoring programs are expensive and essential funding may be contingent upon progress demonstrated through monitoring. However, monitoring efforts can be coordinated with research programs, saving time and money.

With growing concern for species persistence in the face of climate change [72–75], wildlife managers and researchers have had to consider new approaches to protecting species at risk. Assisted colonization, (aka, managed relocation) is the translocation of individuals of a species at risk of extinction to a location outside their historical range, where their likelihood of persistence is greater [76–78]. While assisted colonization could be essential for protecting a species whose habitats are disappearing as a result of climate change, it is also controversial in that it could result in a harmful invasion of a species into habitats critical to other species. The status of the fisher may not decline to the point where assisted colonization is required to protect the species from extinction, yet, the factors that we have identified that influence reintroduction success would likely influence the success of assisted colonizations of fishers and other at-risk species.

Supporting Information

Table S1   The complete set of attribute data used in our analyses of 38 fisher translocations, listed chronologically. References in this table refer to sources listed in the reference section of the manuscript.

Program File S1   A program file created for our *VORTEX* simulations is included as supplemental information to interested readers (Program File S1.vpj).

Acknowledgments

We thank B. Anderson, J. Apker, R. Baird, J. Baker, W. Berg, M. Boudreau, T. Dilworth, R. Earle, R. Henry, J. Jorgenson, E. Lofroth, J. Mills, M. Novak, D. Potter, P. Rego, K. Royar, and R. Weir for providing data and insights on fisher translocations. We thank L. Finley, M. Jensen and J. Bush for facilitating support from the U.S. Fish and Wildlife Service for this project. K. Aubry and two anonymous reviewers provided helpful comments on an earlier draft of the manuscript. A *VORTEX* program file (Program File S1 [Program File S1.vpj]) has been provided as supplemental information.

Author Contributions

Conceived and designed the experiments: JCL RAP WJZ. Performed the experiments: JCL RAP WJZ. Analyzed the data: JCL RAP WJZ. Contributed reagents/materials/analysis tools: JCL RAP WJZ. Wrote the paper: JCL RAP WJZ.

References

1. Powell RA (1993) The fisher: Life history, ecology, and behavior. Second edition. University of Minnesota Press, Minneapolis, Minnesota, USA.
2. Miller B, Reading RP, Forrest S, Stanley Price MR (1996) Prairie Night: Black-Footed Ferrets and the Recovery of Endangered Species. Smithsonian, Washington, DC. 254 p.
3. Aubry KB, Lewis JC (2003) Extirpation and reintroduction of fishers (*Martes pennanti*) in Oregon: implications for their conservation in the Pacific states. Biological Conservation 114: 79–90.
4. Boitani L (2003) Wolf Conservation and Recovery. 317– 340 *in* Mech LD, Boitani L, eds. Wolves: behavior, ecology, and conservation University of Chicago Press. 448 p.
5. Laliberte AS, Ripple WJ (2004) Range contraction of North American carnivores and ungulates. Bioscience 54: 123–138.
6. Aubry KB, McKelvey KS, Copeland JP (2007) Distribution and broadscale habitat relations of the wolverine in the contiguous United States. Journal of Wildlife Management 71: 2147–2158.

7. Reading RP, Clark TW (1996) Carnivore reintroductions: an interdisciplinary examination. 296–336 in Gittleman JL, ed. Carnivore behavior, ecology, and evolution. Vol. 2, Cornell University Press, Ithaca, New York, USA.

8. Miller B, Ralls K, Reading RP, Scott JM, Estes J (1999) Biological and technical considerations of carnivore translocation: a review. Animal Conservation 2: 59–68.

9. Breitenmoser U, Breitenmoser-Wursten C, Carbyn LW, Funk SM (2001) Assessment of carnivore reintroductions. 240–281 in Gittleman JL, Funk SM, Macdonald DW, Wayne RK, eds. Carnivore Conservation, Cambridge University Press, New York.

10. IUCN (1995) IUCN/SSC Guidelines for re-introductions. Forty-first meeting of the IUCN Council, Gland, Switzerland. May 1995. 6 p. Available: www.iucnsscrsg.org/download/English.pdf. Accessed 2012 Feb 24.

11. Yalden DW (1993) The problems of reintroducing carnivores. Symposium of the Zoological Society of London 65: 289–306.

12. IUCN (1987) IUCN position statement on translocation of living organisms: introductions, reintroductions, and re-stocking. Available: http://www.iucnsscrsg.org/policy_guidelines.html. Accessed 2012 Feb 24.

13. USFWS (US Fish & Wildlife Service) (2004) Endangered and threatened wildlife and plants: 12-month finding for a petition to list the west coast distinct population segment of the fisher (Martes pennanti). Federal Register 69(68): 18770–18792.

14. Hayes GE, Lewis JC (2006) Washington State recovery plan for the fisher. Washington Department of Fish and Wildlife, Olympia, WA. 62 p.

15. Callas RL, Figura P (2008) Translocation plan for the reintroduction of fishers (Martes pennanti) to lands owned by Sierra Pacific Industries in the northern Sierra Nevada of California California Department of Fish and Game. 80 p.

16. Lofroth EC, Raley CM, Higley JM, Truex RL, Yaeger JS, et al. (2010) Conservation of Fishers (Martes pennanti) in South-Central British Columbia, Western Washington, Western Oregon, and California–Volume I: Conservation Assessment. USDI Bureau of Land Management, Denver, Colorado, USA.

17. Berg WE (1982) Reintroduction of fisher, pine marten and river otter. 159–175 in Sanderson GC, ed. Midwest furbearer management, North Central Section of The Wildlife Society, Bloomington, Illinois.

18. Gibilisco CJ (1994) Distributional dynamics of American martens and fishers in North America. 59–71. in Buskirk S, Harestad A, Raphael M, Powell R, eds. Martens, sables, and fishers: biology and conservation, Cornell University Press, Ithaca, New York.

19. Lewis JC, Zielinski WJ (1996) Historical harvest and incidental capture of fishers in California. Northwest Science 70: 291–297.

20. Krohn WB (In press) Distributional dynamics of Martes in eastern North America: spatiotemporal analyses of historical patterns, 1699–2001. In Aubry KB, Zielinski WJ, Raphael MG, Proulx G, Buskirk SW, eds. Biology and Conservation of Martens, Sables, and Fishers: A New Synthesis, Cornell University Press, Ithaca, New York.

21. Seton ET (1926) Lives of Game Animals. Vol. 2. Doubleday, Doran & Co., New York.

22. Bailey VF (1936) The mammals and life zones of Oregon. North American Fauna 55: 1–416.

23. Grinnell J, Dixon JS, Linsdale JM (1937) Fur-bearing mammals of California: their natural history, systematic status and relations to man. Volume 1. University of California, Berkeley.

24. Dalquest WW (1948) Mammals of Washington. University of Kansas, Lawrence, Kansas.

25. Coulter MW (1966) Ecology and management of fishers in Maine. Dissertation, State University, College of Forestry at Syracuse University, New York.

26. Hall ER (1942) Gestation period in the fisher with recommendations for the animals protection in California. California Fish and Game 28: 143–147.

27. Schorger AW (1942) Extinct and endangered mammals and birds of the Great Lakes Region. Transactions of the Wisconsin Academy of Science, Arts & Letters 34: 24–57.

28. deVos A (1952) Ecology and management of fisher and marten in Ontario. Ontario Department of Lands and Forests. Technical BulletinWildlife Series 1.

29. Brander RB, Books DJ (1973) Return of the Fisher. Natural History 82: 52–57.

30. Ingram R (1973) Wolverine, fisher and marten in central Oregon. Central Region Administration Report No. 73-2. Oregon State Game Commission, Salem, Oregon.

31. Zielinski WJ, Truex RL, Schlexer FV, Campbell LA, Carroll C (2005) Historical and contemporary distributions of carnivores in forests of the Sierra Nevada, California. Journal of Biogeography 32: 1385–1407.

32. Vinkey RS, Schwartz MK, McKelvey KS, Foresman KR, Pilgrim KL, et al. (2006) When reintroductions are augmentations: the genetic legacy of fishers (Martes pennanti) in Montana. Journal of Mammalogy 87: 265–271.

33. Schwartz MK (2007) Ancient DNA confirms native Rocky Mountain fisher (Martes pennanti) avoided early 20th century extinction. Journal of Mammalogy 88: 921–925.

34. Cook DE, Hamilton WJ, Jr. (1957) The forest, the fisher, and the porcupine. Journal of Forestry 55: 719–722.

35. Irvine GW (1960) Progress report on the porcupine problem on the Ottawa National Forest. Unpublished report, Ottawa National Forest, Forest Service, U.S.D.A. Ironwood, Michigan.

36. Irvine GW, Magnus LT, Bradle BJ (1964) The restocking of fishers in lake state forests. Transactions of the North American Wildlife and Natural Resources Conference 29: 307–315.

37. Earle RD (1978) The fisher-porcupine relationship in Upper Michigan. M.S. thesis, Michigan Technological University, Houghton, Michigan.

38. Earle RD, Kramm KR (1982) Correlation between fisher and porcupine abundance in upper Michigan. American Midland Naturalist 107: 244–249.

39. Bradle BJ (1957) The fisher returns to Wisconsin. Wisconsin Conserv. Bull 22: 9–11.

40. Benson DA (1959) The fisher in Nova Scotia. J Mammalogy 40: 451.

41. Weckwerth RP, Wright PL (1968) Results of transplanting fishers in Montana. Journal of Wildlife Management 32: 977–980.

42. Pack JC, Cromer JI (1981) Reintroduction of fisher in West Virginia. 1431–1442 in Chapman JA, Pursley D, eds. Proceeedings of Worldwide Furbearer Conference, Frostburg, Maryland.

43. Dilworth TG (1974) Status and distribution of fisher and marten in New Brunswick. Canadian Field-Naturalist 88: 495–498.

44. Wallace K, Henry R (1985) Return of a Catskill native. The Conservationist 40: 17–19.

45. Proulx GA, Kolenosky J, Badry M, Drescher MK, Seidel K, et al. (1994) Post-release movements of translocated fishers. 197–203 in Buskirk SW, Harestad A, Raphael M, eds. Martens, sables, and fishers: biology and conservation, Cornell University Press, Ithaca, New York.

46. Nielsen L (1988) Definitions, considerations, and guidelines for translocation of wild animals. 12–51 in Nielson L, Brown RD, eds. Translocation of wild animals, Wisconsin Humane Society and Caesar Kleberg Wildlife Research Institute, Milwaukee, Wisconsin. 333 p.

47. Lacy RC, Borbat M, Pollak JP (2005) VORTEX: A Stochastic Simulation of the Extinction Process. Version 9.50. Brookfield, IL: Chicago Zoological Society.

48. Raine RM (1981) Winter food habits, responses to snow cover and movements of fisher (Martes pennanti) and marten (Martes americana) in southeastern Manitoba. MSc thesis. University of Manitoba, Winnipeg. 145 p.

49. Strickland MA, Douglas CW, Novak M, Hunzinger NP (1982) Fisher (Martes pennanti). 586–598 in Chapman JA, Feldhamer GA, eds. Wild Mammals of North America: biology, management, and economics, The Johns Hopkins University Press, Baltimore, Maryland.

50. Lewis JC, Happe PJ, Jenkins KJ, Manson DJ (2011) Olympic fisher reintroduction project: 2010 progress report. Washington Department of Fish and Wildlife, Olympia. Available at: http://wdfw.wa.gov/wlm/diversty/soc/fisher/ (last accessed 24 February 2012).

51. Roy KD (1991) Ecology of reintroduced fishers in the Cabinet Mountains of northwest Montana. Thesis, University of Montana, Missoula, Montana, USA.

52. Hagmeier EM (1956) Distribution of marten and fisher in North America. Canadian Field-Naturalist 70: 149–168.

53. Proulx GA, Aubry KB, Birks J, Buskirk SW, Fortin C, et al. (2004) World distribution and status of the genus Martes in 2004. 21–76 in Harrison DJ, Fuller AK, Proulx G, eds. Martens and fishers (Martes) in human-altered environments: an international perspective, Springer Science and Business Media, New York, New York.

54. Jung TS, Slough BG (2011) The status of fisher (Martes pennanti) at the northwestern edge of their range: are they increasing and expanding in the Yukon? Northwestern Naturalist 92: 57–64.

55. Burnham KP, Anderson DR (2002) Model selection and multimodel inference. Springer-Verlag, New York.

56. Novak M Ontario Ministry of Natural Resources, personal communication.

57. Newsom WM (1937) Mammals on Anticosti Island. Journal of Mammalogy 18: 435–442.

58. Kovach AI, Powell RA (2003) Reproductive success of male black bears. Canadian Journal of Zoology 81: 1257–1268.

59. Erlinge S, Sandell M (1986) Seasonal changes in social organization of male stoats, Mustela erminea: An effect of shifts between two decisive resources. Oikos 47: 57–62.

60. Douglas CW, Strickland MA (1987) Fisher. 511:529 in Novak MJ, Baker A, Obbard ME, Malloch B, eds. Wild furbearer management and conservation in North America, Ontario Ministry of Natural Resources, Toronto, Ontario.

61. Strickland MA, Douglas CW (1981) The status of the fisher in North America and its management in southern Ontario. 1443–1458 in Chapman JA, Pursley D, eds. Proceedings of the worldwide furbearer conference, Worldwide Furbearer Conference Baltimore, Maryland.

62. Krohn WB, Elowe KD, Boone RB (1995) Relations among fishers, snow and martens: Development and evaluation of two hypotheses. The Forestry Chronicle 71: 97–105.

63. Buskirk SW, Powell RA (1994) Habitat ecology of fishers and American martens. 283–296 in Buskirk SW, Harestad AS, Raphael MG, Powell RA, eds. Martens, sables, and fishers: biology and conservation, Cornell University Press, Ithaca, New York, USA.

64. Griffith B, Scott JM, Carpenter JW, Reed C (1989) Translocation as a species conservation tool: status and strategy. Science 245: 477–480.

65. Armstrong DP, Seddon PJ (2008) Directions in Reintroduction Biology. Trends in Ecology & Evolution 23: 20–25.

66. Fontana AJ, Teske IE, Pritchard K, Evans M (1999) East Kootenay fisher reintroduction program, 1996–1999. Ministry of Environment, Lands, and Parks, Cranbrook, British Columbia.

67. Weir RD, Adams IT, Mowat G, Fontana AJ (2003) East Kootenay fisher assessment. British Columbia Ministry of Water, Land, and Air Protection, Cranbrook, British Columbia.

68. Anderson B (2002) Reintroduction of fishers (*Martes pennanti*) to the Catoosa Wildlife Area in Tennessee. Tennessee Wildlife Resources Agency, Crossville, Tennessee.

69. Lewis JC, Hayes GE (2004) Feasibility assessment for reintroducing fishers to Washington. Washington Department of Fish and Wildlife, Olympia. 70 p. Available: http://wdfw.wa.gov/wlm/diversty/soc/fisher/. Accessed 2012 Feb 24.

70. Lewis JC (2006) Implementation plan for reintroducing fishers to Olympic National Park. Washington Department of Fish and Wildlife, Olympia, WA. 32 p. Available: http://wdfw.wa.gov/wlm/diversty/soc/fisher/. Accessed 2012 Feb 24.

71. Naney RH, Finley LL, Lofroth EC, Happe PJ, Krause AL, et al. (2012) Conservation of Fishers (*Martes pennanti*) in South-Central British Columbia, Western Washington, Western Oregon, and California–Volume III: Threat Assessment. USDI Bureau of Land Management, Denver, Colorado, USA.

72. Thomas CD, Cameron A, Green RE, Bakkenes M, Beaumont LJ, et al. (2004) Extinction risk from climate change. Nature 427: 145–148.

73. Schwartz MW, Iverson LR, Prasad AM, Matthews SN, O'Connor RJ (2006) Predicting extinctions as a result of climate change. Ecology 87: 1611–1615.

74. IPCC (Intergovernmental Panel on Climate Change) (2007) Synthesis report of the fourth assessment of the Intergovernmental Panel on climate change. Cambridge University Press, Cambridge.

75. Lawler JJ, Shafer SL, White D, Kareiva P, Maurer EP, et al. (2009) Projected climate induced faunal change in the Western Hemisphere. Ecology 90: 588–597.

76. McLachlan JS, Hellman JJ, Schwartz MW (2007) A framework for debate of assisted migration in an era of climate change. Conservation Biology 21: 297–302.

77. Hoegh-Guldberg O, Hughes L, McIntyre S, Lindenmayer DB, Parmesan C, et al. (2008) Assisted colonization and rapid climate change. Science 321: 345–346.

78. Richardson DM, Hellmann JJ, McLachlan JS, Sax DF, Schwartz MW, et al. (2009) Multidimensional evaluation of managed relocation. Proc Natl Acad Sci 106: 9721–9724.

79. Dodds DG, Martell AM (1971) The recent status of the fisher, *Martes pennanti* (Erxleben), in Nova Scotia. Canadian Field-Naturalist 85: 62–65.

80. Petersen LR, Martin MA, Pils CM (1977) Status of fishers in Wisconsin. Wisconsin Department Natural Resources Report 92: 1–14.

81. Kohn BE, Payne NF, Ashbrenner JE, Creed WE (1993) The fisher in Wisconsin. Wisconsin Department Natural Resources, Technical Bulletin No. 183.

82. Dodge WE (1977) Status of the fisher (*Martes pennanti*) in the conterminous United States. Unpubl. report submitted to US Dept. of the Interior. On file with JCL.

83. Heinemeyer KS (1993) Temporal dynamics in the movements, habitat use, activity, and spacing of reintroduced fishers in northwestern Montana. MS Thesis, University of Montana, Missoula.

84. Vinkey RS (2003) An evaluation of fisher (*Martes pennanti*) introductions in Montana. Master's Thesis. University of Montana, Missoula. 97 p.

85. Royar K Vermont Fish and Wildl Dept, personal communication.

86. Kebbe CE (1961a) Return of the fisher. Oregon State Game Commission Bulletin 16: 3–7.

87. Kebbe CE (1961b) Transplanting fisher. Western Association of State Game and Fish Commissioners 41: 165–167.

88. Earle R (2005) Michigan Dept of Natural Resources, personal communication.

89. Williams RM (1962) Trapping and transplant project, fisher transplant segment. Project W 75-D-9, completion report. Idaho Fish and Game Department, Boise. 5 p.

90. Williams RM (1963) Final segment report, trapping and transplanting. Project W 75-D-10. Idaho Fish and Game Department, Boise. 6 p.

91. Luque M (1984) The fisher: Idaho's forgotten furbearer. Idaho Wildlife 4: 12–15.

92. Drew RE, Hallett JG, Aubry KB, Cullings KW, Koepfs SM, et al. (2003) Conservation genetics of the fisher (*Martes pennanti*) based on mitochondrial DNA sequencing. Molecular Ecology 12: 51–52.

93. Dilworth T (2006) New Brunswick Department of Natural Resources and Energy, personal communication.

94. Wood J (1977) The Fisher is: … National Wildlife 15: 18–21.

95. Berg W (2004) Minnesota Division of Fish and Wildlife, personal communication.

96. Jakubas W (2005) Maine Dept. of Inland Fisheries and Wildlife, personal communication.

97. Baird R (2005) Riding Mountain National Park, Manitoba, personal communication.

98. Henry R (2005) New York State Dept. of Environmental Conservation, personal communication.

99. Apker J (2006) Colorado Division of Wildlife, personal communication.

100. Kyle CJ, Robitaille JF, Strobeck C (2001) Genetic variation and structure of fisher (*Martes pennanti*) populations across North America. Molecular Ecology 10: 2341–2347.

101. Baker J (2005) Ontario Ministry of Natural Resources, personal communication.

102. Davie JW (1984) Fisher introduction. Project Completion Report, Alberta Forest, Lands, and Wildlife, Edmonton. 10 p.

103. Jorgenson J (2005) Alberta Sustainable Resource Develop., personal communication.

104. Weir R (2006) Artemis Wildlife Consultants, British Columbia, personal communication.

105. Lofroth E (2005) British Columbia Ministry of Environment, personal communication.

106. Rego PW (1989) Wildlife investigation: fisher reintroduction, 10/1/88-9/30/89. Project number W-49-R-14 performance report. Connecticut Department of Environmental Protection, Wildlife Division, Burlington. 7 p.

107. Rego PW (1990) Wildlife investigation: fisher reintroduction, 10/1/89-9/30/90. Project number W-49-R-15 performance report. Connecticut Department of Environmental Protection, Wildlife Division, Burlington. 6 p.

108. Rego PW (1991) Wildlife investigation: fisher reintroduction, 10/1/90-9/30/91. Project number W-49-R-16 performance report. Connecticut Department of Environmental Protection, Wildlife Division, Burlington. 3 p.

109. Rego P (2005) Connecticut Dept of Environmental Protection, personal communication.

110. Proulx GA (2005) The fisher in our aspen parklands. Edmonton Nature News 2: 21–22.

111. Weir RD (1995) Diet, spatial organization, and habitat relationships of fishers in south-central British Columbia. Thesis, Simon Fraser University, Burnaby, British Columbia.

112. Potter D (2002) Modelling fisher (*Martes pennanti*) habitat associations in Nova Scotia. M.S. Thesis. Acadia University, Wolfville, Nova Scotia.

113. Potter D (2006) Ontario Ministry of Natural Resources, personal communication.

114. Boudreau M (2006) Nova Scotia Dept. of Natural Resources, personal communication.

115. Mills J (2006) Nova Scotia Dept. of Natural Resources, personal communication.

116. Baird R, Frey S (2000) Riding Mountain National Park fisher reintroduction program 1994–1995. Riding Mountain National Park, Manitoba.

117. Serfass TL, Brooks RP, Tzilkowski WM (2001) Fisher reintroduction in Pennsylvania: Final report of the Pennsylvania fisher reintroduction project. Frostburg State University, Frostburg, Maryland.

118. Anderson B (2005) Tennessee Wildlife Resources Agency, personal communication.

119. Lewis J (2011) Washington Dept of Fish and Wildlife, unpublished data.

120. Powell R (2012) North Carolina State University, unpublished data.

PERMISSIONS

LIST OF CONTRIBUTORS

Howard B. Wilson and Hugh P. Possingham
Australian Research Council Centre of Excellence for Environmental Decisions, School of Biological Sciences, The University of Queensland, Brisbane, Australia

Erik Meijaard
Australian Research Council Centre of Excellence for Environmental Decisions, School of Biological Sciences, The University of Queensland, Brisbane, Australia
People and Nature Consulting International, Jakarta, Indonesia

Oscar Venter
Australian Research Council Centre of Excellence for Environmental Decisions, School of Biological Sciences, The University of Queensland, Brisbane, Australia
Centre for Tropical Environmental and Sustainability Science and the School of Marine and Tropical Biology, James Cook University, Cairns, Australia

Marc Ancrenaz
Kinabatangan Orangutan Conservation Project, Sandakan, Malaysia

Juliann E. Aukema
The National Center for Ecological Analysis and Synthesis, Santa Barbara, California, United States of America

Brian Leung
Department of Biology, McGill University, Montreal, Quebec, Canada
School of Environment, McGill University, Montreal, Quebec, Canada

Kent Kovacs
Department of Applied Economics and Institute on the Environment, University of Minnesota, St. Paul, Minnesota, United States of America

Corey Chivers
Department of Biology, McGill University, Montreal, Quebec, Canada

Kerry O. Britton
U.S. Forest Service, Research and Development, Arlington, Virginia, United States of America

Jeffrey Englin
Morrison School of Agribusiness and Resource Management, Arizona State University, Mesa, Arizona, United States of America

Susan J. Frankel
U.S. Forest Service, Pacific Southwest Research Station, Albany, California, United States of America

Robert G. Haight
U.S. Forest Service, Northern Research Station, St. Paul, Minnesota, United States of America

Thomas P. Holmes
U.S. Forest Service, Southern Research Station, Research Triangle Park, North Carolina, United States of America

Andrew M. Liebhold
U.S. Forest Service, Northern Research Station, Morgantown, West Virginia,United States of America,

Deborah G. McCullough
Department of Entomology and Department of Forestry, Michigan State University, East Lansing, Michigan, United States of America

Betsy Von Holle
Department of Biology, University of Central Florida, Orlando, Florida, United States of America

Claudia Funi
Programa de Pós Graduação em Biodiversidade Tropical, Universidade Federal do Amapá, Macapá, Brazil
Instituto Estadual de Pesquisas do Amapá, Macapá, Brazil

Adriana Paese
Programa de Pós Graduação em Biodiversidade Tropical, Universidade Federal do Amapá, Macapá, Brazil
Conservação Internacional, Belo Horizonte, Brazil

Matti Salo and Samuli Helle
Department of Biology, University of Turku, Turku, Finland

Tuuli Toivonen
Department of Geosciences and Geography, University of Helsinki, Helsinki, Finland

Hania Lada, James R. Thomson, Shaun C. Cunningham and Ralph Mac Nally
School of Biological Sciences, Monash University, Melbourne, Victoria, Australia

David L. A. Gaveau, Mrigesh Kshatriya, Molidena, Elis Arief Wijaya Manuel R. Guariguata and Pablo Pacheco
Center for International Forestry Research, Bogor, Indonesia

Douglas Sheil
Center for International Forestry Research, Bogor, Indonesia
School of Environment, Science and Engineering, Southern Cross University, Lismore, NSW, Australia
Institute of Tropical Forest Conservation (ITFC), Mbarara University of Science and Technology (MUST), Kabale, Uganda

Sean Sloan
Centre for Tropical Environmental and Sustainability Science, School of Marine & Tropical Biology, James Cook University, Cairns, QLD, Australia

Serge Wich
Research Centre in Evolutionary Anthropology and Palaeoecology, School of Natural Sciences and Psychology, Liverpool John Moores University, Liverpool, United Kingdom

Marc Ancrenaz
Sabah Wildlife Department, Kota Kinabalu, Sabah, Malaysia
HUTAN, Kinabatangan Orang-utan Conservation Programme, Kota Kinabalu, Sa,bah, Malaysia
North England Zoological Society, Chester Zoo, Chester, United Kingdom

Matthew Hansen, Peter Potapov and Svetlana Turubanova
Department of Geographical Sciences, University of Maryland, College Park, Maryland, United States of America

Mark Broich
The Climate Change Cluster, University of Technology Sydney, NSW, Australia

Erik Meijaard
Center for International Forestry Research, Bogor, Indonesia
Borneo Futures Project, People and Nature Consulting International, Ciputat, Jakarta, Indonesia
School of Biological Sciences, University of Queensland, Brisbane, Australia

Xiaona Li, Zhiwei Wu and Yu Liang
State Key Laboratory of Forest and Soil Ecology, Institute of Applied Ecology, Chinese Academy of Sciences, Shenyang, People's Republic of China

Hong S. He
State Key Laboratory of Forest and Soil Ecology, Institute of Applied Ecology, Chinese Academy of Sciences, Shenyang, People's Republic of China
School of Natural Resources, University of Missouri-Columbia, Columbia, Missouri, United States of America

Jeffrey E. Schneiderman
School of Natural Resources, University of Missouri-Columbia, Columbia, Missouri, United States of America

Than J. Boves
Department of Forestry, Wildlife, and Fisheries, University of Tennessee, Knoxville, Tennessee, United States of America
Department of Natural Resources and Environmental Science, University of Illinois, Urbana, Illinois, United States of America

David A. Buehler, Patrick D. Keyser and Tiffany A. Beachy
Department of Forestry, Wildlife, and Fisheries, University of Tennessee, Knoxville, Tennessee, United States of America

James Sheehan, Gregory A. George, Molly E. McDermott and Kelly A. Perkins
West Virginia Cooperative Fish and Wildlife Research Unit, Division of Forestry and Natural Resources, West Virginia University, Morgantown, West Virginia, United States of America

Petra Bohall Wood
U.S. Geological Survey, West Virginia Cooperative Fish and Wildlife Research Unit, West Virginia University, Morgantown, West Virginia, United States of America

Amanda D. Rodewald, Felicity L. Newell and Marja H. Bakermans
School of Environment and Natural Resources, Ohio State University, Columbus, Ohio, United States of America

Jeffrey L. Larkin, Andrea Evans and Matthew White
Department of Biology, Indiana University of Pennsylvania, Indiana, Pennsylvania, United States of America

T. Bently Wigley
National Council for Air and Stream Improvement, Inc., Clemson, South Carolina, United States of America

Clint R. V. Otto
Department of Fisheries and Wildlife, Michigan State University, East Lansing, Michigan, United States of America

United States Geological Survey, Northern Prairie Wildlife Research Center, Jamestown, North Dakota, United States of America

Gary J. Roloff and Rachael E. Thames
Department of Fisheries and Wildlife, Michigan State University, East Lansing, Michigan, United States of America

Bill Buffum
Department of Natural Resources Management, University of Rhode Island, Kingston, Rhode Island, United States of America

Christopher Modisette
Ecological Science, United Stated Department of Agriculture, Natural Resources Conservation Service, Warwick, Rhode Island, United States of America

Scott R. McWilliams
Department of Natural Resources Management, University of Rhode Island, Kingston, Rhode Island, United States of America

Andrea F. Currylow, Brian J. MacGowan and Rod N. Williams
Department of Forestry and Natural Resources, Purdue University, West Lafayette, Indiana, United States of America

Huong Nguyen and John Herbohn
School of Agriculture and Food Sciences, The University of Queensland, St Lucia, Australia
Forest Industries Research Centre, The University of the Sunshine Coast, Sippy Downs, Australia

David Lamb
School of Agriculture and Food Sciences, The University of Queensland, St Lucia, Australia
Centre for Mined Land Research, The University of Queensland, St Lucia, Australia

Jennifer Firn
Faculty of Science and Technology, School of Earth, Environmental and Biological Sciences, Queensland University of Technology, Brisbane, Australia

Maria M. Romeiras
Tropical Botanical Garden, Tropical Research Institute (IICT), Lisbon, Portugal
Centre for Biodiversity, Functional and Integrative Genomics (BIOFIG), Faculty of Sciences, University of Lisbon, Lisbon, Portugal

Rui Figueira and Maria Cristina Duarte
Tropical Botanical Garden, Tropical Research Institute (IICT), Lisbon, Portugal
CIBIO - Research Center in Biodiversity and Genetic Resources/InBIO, University of Porto, Vairão, Portugal

Pedro Beja
CIBIO - Research Center in Biodiversity and Genetic Resources/InBIO, University of Porto, Vairão, Portugal

Iain Darbyshire
Royal Botanic Gardens, Kew. Richmond, United Kingdom

Kenneth Wardle
Department of Classics, Ancient History and Archaeology, University of Birmingham, Birmingham, United Kingdom

Thomas Higham
Oxford Radiocarbon Accelerator Unit, Research Laboratory for Archaeology and the History of Art, University of Oxford, Oxford, United Kingdom

Bernd Kromer
Akademie der Wissenschaften Heidelberg, Heidelberg, Germany

Marcus A. Lashley
Department of Forestry and Environmental Resources, North Carolina State University, Raleigh, North Carolina, United States of America
110 Brooks Avenue, Raleigh, North Carolina, United States of America

Jeffrey R. Thompson
SAS Institute Inc., Cary, North Carolina, United States of America

M. Colter Chitwood, Christopher S. DePerno and Christopher E. Moorman
Department of Forestry and Environmental Resources, North Carolina State University, Raleigh, North Carolina, United States of America

Aapo Rautiainen and Pekka E. Kauppi
Faculty of Biosciences, University of Helsinki, Helsinki, Finland

Iddo Wernick and Jesse H. Ausubel
Program for the Human Environment, The Rockefeller University, New York, New York, United States of America

Paul E. Waggoner
Department of Forestry and Horticulture, Connecticut Experimental Agricultural Station, New Haven, Connecticut, United States of America

Khyne U. Mar
Department of Animal and Plant Sciences, University of Sheffield, Sheffield, United Kingdom

Mirkka Lahdenperä
Section of Ecology, Department of Biology, University of Turku, Turku, Finland

Virpi Lummaa
Wissenschaftskolleg zu Berlin, Institute for Advanced Study, Berlin, Germany

Jing Yao
State Key Laboratory of Forest and Soil Ecology, Institute of Applied Ecology, Chinese Academy of Sciences, Shenyang, People's Republic of China
Graduate University of Chinese Academy of Sciences, Beijing, People's Republic of China

Xingyuan He, Anzhi Wang, Wei Chen, Xiaoyu Li, Bernard J. Lewis and Xiaotao Lv
State Key Laboratory of Forest and Soil Ecology, Institute of Applied Ecology, Chinese Academy of Sciences, Shenyang, People's Republic of China

William J. Loughry and Colleen M. McDonough
Department of Biology, Valdosta State University, Valdosta, Georgia, United States of America

Carolina Perez-Heydrich
Carolina Population Center, University of North Carolina, Chapel Hill, North Carolina, United States of America

Madan K. Oli
Department of Wildlife Ecology and Conservation, University of Florida, Gainesville, Florida, United States of America

Jeffrey C. Lewis
Washington Department of Fish and Wildlife, Olympia, Washington, United States of America

Roger A. Powell
Department of Biology, North Carolina State University, Raleigh, North Carolina, United States of America

William J. Zielinski
Unite States Department of Agriculture Forest Service, Pacific Southwest Research Station, Arcata, California, United States of America

Index